Intelligent Engineering Systems
and Computational Cybernetics

J.A. Tenreiro Machado · Béla Pátkai
Imre J. Rudas
Editors

Intelligent Engineering Systems and Computational Cybernetics

Editors

J.A. Tenreiro Machado
Institute of Engineering of the
Polytechnic Institute of Porto
Department of Electrotechnical Engineering
Rua Dr. Antonio Bernardino de Almeida
4200-072 Porto
Portugal
jtm@isep.ipp.pt

Imre J. Rudas
Budepest Tech
Department of Intelligent Engineering Systems
John von Neumann Faculty of Informatics
Bécsi út 96/B
Budapest 1034
Hungary
rudas@bmf.hu

Béla Pátkai
Department of Engineering
Institute for Manufacturing
University of Cambridge
Mill Lane
Cambridge
United Kingdom CB2 2RX
bela@patkai.org

ISBN 978-90-481-7949-7 e-ISBN 978-1-4020-8678-6

Printed on acid-free paper

9 8 7 6 5 4 3 2 1

springer.com

Preface

It is now well accepted that time to market for new products can be reduced and practical value of goods can be increased most effectively by the application of advanced cybernetics and computational intelligence. The above recognition motivated this book addressing, in a single volume, selected contributions, in important areas, from two relevant conferences. The 10th International Conference on Intelligent Engineering Systems (INES 2006) and the 4th International Conference on Computational Cybernetics (ICCC 2006) were held in year 2006 to provide researchers and practitioners from industry and academia with a platform to report on recent developments in cybernetics and computational intelligence. INES and ICCC are organized as annual scientific events and founded by Professor Imre J. Rudas in 1997 and 2003, respectively. In earlier years, classical concept of artificial intelligence proved merely partial solution for problem of intelligent systems. During the recent years, researchers and experts were keen in merging intelligent elements in different engineering and technical systems. Computational Cybernetics is the synergetic integration of Cybernetics and Computational Intelligence techniques. Hence the topics of "Intelligent Engineering Systems and Computational Cybernetics" cover not only mechanical, but biological (living), social and economical systems and for this purpose uses computational intelligence based results of communication theory, signal processing, information technology, control theory, the theory of adaptive systems, the theory of complex systems (game theory, operational research), and computer science. Recently, engineering activities are integrated in virtual spaces. Human scientific knowledge gained importance in tasks such as the creation of complex product descriptions, modeling of processes and working conditions, optimizing parameter sets, and handling problems in large, complex and dynamic systems. In this volume, a thematic selection of significant papers represent of recent emphases among others in topics such as dynamic analysis, fuzzy process model, semantic Web, document clustering, harmonic drive systems, fault diagnosis, Internet traffic, non-linear filters, genetic algorithms, behavioral characteristics, fractional controllers, complex systems, evolutionary optimization, force-impedance control, associative engineering objects, model of natural languages, and fractional representations.

Porto, August 2007 — *J.A. Tenreiro Machado*
Cambridge, August 2007 — *Béla Pátkai*
Budapest, August 2007 — *Imre J. Rudas*

Contents

Part III Computational Intelligence

Part IV Intelligent Mechatronics

Part I

Intelligent Robotics

On-Line Trajectory Time-Scaling to Reduce Tracking Error

Emese Szádeczky-Kardoss[1] and Bálint Kiss[2]

[1] Department of Control Engineering and Information Technology
Budapest University of Technology and Economics
H-1117, Magyar Tudósok krt. 2, Budapest, Hungary
kardoss@seeger.iit.bme.hu

[2] Department of Control Engineering and Information Technology
Budapest University of Technology and Economics
H-1117, Magyar Tudósok krt. 2, Budapest, Hungary
bkiss@iit.bme.hu

Summary. The paper describes an on-line trajectory time-scaling control algorithm for wheeled mobile robots. To reduce tracking errors the controller modifies the velocity profile of the reference trajectory according to the closed loop behavior of the robot. The geometry of the reference trajectory is unchanged, only the time distribution varies during the motion. We give a control algorithm which uses time-scaled reference and a feedback calculated from the linearized error dynamics. The closed loop behavior is also studied together with the controllability of the linearized error dynamics.

1 Introduction

Path planning and tracking control for car-like mobile robots is a popular research field. The goal is to develop vehicles which are able to perform tasks in an autonomous way, or can help humans to make involved manoeuvres. One wishes robots to carry out the prescribed tasks with high precision and high speed which are usually conflicting requirements. Fast motions may cause tracking errors, especially if the path designed off-line is calculated using wrong model parameter values. Slow movements are generally more accurate, since they leave more reaction margin for the controllers. The price to pay is that the time optimal property of the path is lost.

Several algorithms were developed to find the time-optimal trajectories for robots [1, 2]. It is an important question, because robot actuators cannot generate unlimited torques or forces. Thus the minimum time has to be found such that the limits on the actuator torques are not violated. These algorithms are usually applied off-line, before the tracking is started.

The optimal trajectory, which needs the least time to finish the motion, requires at least one actuator to be on its saturation boundary at a given time. The methods for time-optimal motion planning find the switching points where one actuator, which

J.A.T. Machado et al. (eds.), *Intelligent Engineering Systems and Computational Cybernetics,*

has operated in its boundary, decreases the torque or force, and another actuator goes into saturation.

Time-scaling is a commonly used concept to find these optimal trajectories. The first off-line time-scaling methods were used to plan the time optimal trajectory for robot manipulators [3,4]. A similar method was used in [5] for multi-robot trajectories and in [6] to find the optimal path for autonomous mobile vehicles. The problem with these off-line methods is that no sufficient control margins are always assured for the closed loop control during the tracking.

Other algorithms use therefore on-line trajectory time-scaling for robotic manipulators to change the actuator boundaries such that sufficient margin is left for the feedback controller [7,8]. The on-line time-parameterization of the path can also be determined from the prediction of the evolution of the robot [9].

Another concept is to use the tracking error instead of the input bounds in order to modify the time-scaling of the reference path for robot arms [10,11]. These methods change the traveling time of the reference path according to the actual tracking error by decelerating if the movement is not accurate enough and by accelerating if the errors are small or vanish.

Time-scaling of the reference trajectory can also be used to control an underactuated biped robot [12] or a car like wheeled mobile robot with a single steering input, i.e. the driving velocity of the car is not generated by a controller, but by an external source. In such cases the time-scaling function is determined by the longitudinal velocity of the car [13,14].

In this paper we use the tracking error based concept already reported for robot manipulators in [11] but now applied to a different class of problems, namely in the control of wheeled mobile robots (WMR) which are non-holonomic in contrast to the holonomic models of the robot arms. If the time-scaling is done for robot manipulators, then the dynamic equation of the robot is considered with the second time-derivatives of the configuration variables, hence the second time-derivative of the time-scaling function has to be determined. In this work we use the kinematic model of the car for the control and also for the time-scaling, hence we only need to determine the first time-derivative of the time-scaling.

The remaining part of this paper is organized as follows. The next section describes the kinematic model of the car. Then we introduce the notations of the time-scaling. In Section 4 the time-scaling based controller is detailed and its properties are also analyzed. Section 5 shows the results of some simulations. Finally, a short summary concludes the paper.

2 Kinematic Model of the Mobile Robot

Our task is to move a car like wheeled mobile robot (WMR) along a given reference trajectory. The desired movement is defined for the midpoint of the rear axle. This point is denoted by R in Fig. 1.

The configuration of the robot can be described by the position of R in the x-y plane (x, y) and by the orientation of the car (θ). The inputs of the vehicle are the

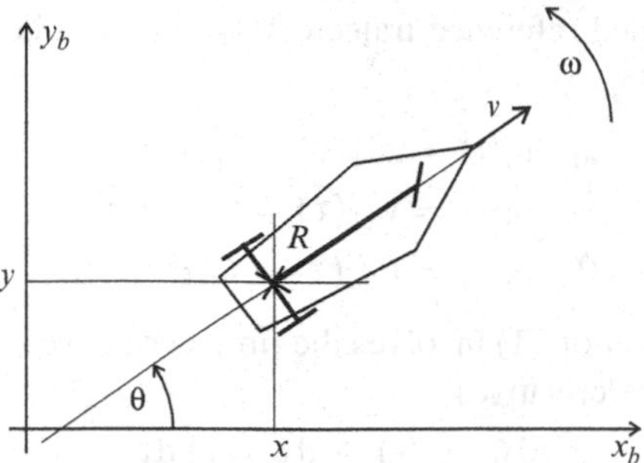

Fig. 1: Notations of the kinematic model

longitudinal velocity (v) and the angular velocity (ω) around its vertical axis. Using these notations the kinematic equation of the robot reads (see for example [15, 16]):

$$\dot{\mathbf{q}}(t) = \begin{bmatrix} \cos\theta(t) & 0 \\ \sin\theta(t) & 0 \\ 0 & 1 \end{bmatrix} \mathbf{u}(t), \tag{1}$$

where

$$\mathbf{q}(t) = \begin{pmatrix} x(t) \\ y(t) \\ \theta(t) \end{pmatrix} \qquad \mathbf{u}(t) = \begin{pmatrix} v(t) \\ \omega(t) \end{pmatrix}. \tag{2}$$

Observe that the linearized system obtained around the equilibrium point ($\mathbf{q} = 0$, $\mathbf{u} = 0$) in not controllable since the rank of the controllability matrix is less than the dimension of the states. Hence other methods are suggested to control the robots described by (1) (see for example [16, 17]). We use a time-scaling based controller based on the linearization of the error dynamics to reduce the tracking errors.

3 Time-Scaling

If time-scaling is used for the reference trajectory, the geometry of the path is left unchanged, only the velocity profile and the time distribution may vary. The time-scaling is defined by a function

$$\tau = f(t), \tag{3}$$

where t denotes the real time and τ is the scaled time. The function f has to satisfy some special properties:

- $f(t)$ is an increasing function of time ($\dot{f} > 0$), since time cannot rewind.
- $f(0) = 0$, because the movement must start from the same initial position.
- $f(t_1) = t_f$ where $t_1 \geq 0$ denotes the real time required to finish the motion along the trajectory, and $t_f \geq 0$ shows the time of the motion according to the original velocity profile.

In this case the scaled reference trajectory ($_{\rm ref,TS}$), can be determined from the original reference ($_{\rm ref}$) as

$$x_{\rm ref,TS}(t) = x_{\rm ref}(\tau) = x_{\rm ref}(f(t)), \tag{4}$$
$$y_{\rm ref,TS}(t) = y_{\rm ref}(\tau) = y_{\rm ref}(f(t)), \tag{5}$$
$$\theta_{\rm ref,TS}(t) = \theta_{\rm ref}(\tau) = \theta_{\rm ref}(f(t)). \tag{6}$$

Since the kinematic equation (1) involves the time derivatives of these variables, one needs to calculate these derivatives

$$\dot{x}_{\rm ref,TS}(t) = \frac{{\rm d}x_{\rm ref,TS}(t)}{{\rm d}t} = \frac{{\rm d}x_{\rm ref}(\tau)}{{\rm d}\tau}\frac{{\rm d}\tau}{{\rm d}t} = x'_{\rm ref}(\tau)\dot{f}, \tag{7}$$
$$\dot{y}_{\rm ref,TS}(t) = \frac{{\rm d}y_{\rm ref,TS}(t)}{{\rm d}t} = \frac{{\rm d}y_{\rm ref}(\tau)}{{\rm d}\tau}\frac{{\rm d}\tau}{dt} = y'_{\rm ref}(\tau)\dot{f}, \tag{8}$$
$$\dot{\theta}_{\rm ref,TS}(t) = \frac{{\rm d}\theta_{\rm ref,TS}(t)}{{\rm d}t} = \frac{{\rm d}\theta_{\rm ref}(\tau)}{{\rm d}\tau}\frac{{\rm d}\tau}{{\rm d}t} = \theta'_{\rm ref}(\tau)\dot{f}, \tag{9}$$

such that according to (1)

$$x'_{\rm ref}(\tau) = \cos\theta_{\rm ref}(\tau)v_{\rm ref}(\tau), \tag{10}$$
$$y'_{\rm ref}(\tau) = \sin\theta_{\rm ref}(\tau)v_{\rm ref}(\tau), \tag{11}$$
$$\theta'_{\rm ref}(\tau) = \omega_{\rm ref}(\tau), \tag{12}$$

where $\dot{x}$ stands for $\frac{dx}{dt}$ and x' denotes $\frac{dx}{\tau}$.

If such a time-scaling is used, the reference input signals, which are required to follow the reference path, also change. Since according to (1)

$$\dot{\theta}_{\rm ref,TS}(t) = \omega_{\rm ref,TS}(t), \tag{13}$$

and using (9) and (12)

$$\omega_{\rm ref,TS}(t) = \omega_{\rm ref}(\tau)\dot{f}. \tag{14}$$

Similarly we get from (1), (6) and (7), (10) that

$$v_{\rm ref,TS}(t) = v_{\rm ref}(\tau)\dot{f}. \tag{15}$$

This shows that the time-scaling of the reference path modify both of the reference inputs (i.e. $v_{\rm ref}$ and $\omega_{\rm ref}$) by the same factor, which is the first time derivative of the time-scaling function ($\dot{f}$). If $\dot{f} < 1$, the reference trajectory is slowed down, while $\dot{f} > 1$ results acceleration of the reference.

The task is now to determine the time-scaling function f. The methods described in the literature use the dynamic equation of the robot and the bounds on the input signals to calculate f. Most often a constant time-scaling function is used thanks to its simplicity:

$$\tau = f(t) = ct, \tag{16}$$

where $c > 0$ is a constant value. If $c < 1$ the movement is slowed down, while values such that $c > 1$ result a faster tracking of the reference. Using the time-scaling function (16) the time derivative $\dot{f}$ is also a simple expression:

$$\dot{\tau} = \dot{f} = c. \tag{17}$$

3.1 Tracking Error Based Time-Scaling

Our proposal is to use the kinematic model and the tracking errors to determine f. We suggest to slow down the motion, if the movement is not accurate enough, and to switch to higher speeds if the errors are small.

We use the following time-scaling function:

$$\tau = f(t,a), \tag{18}$$

where a denotes the absolute value of the tracking error. The derivative of this time-scaling function is more involved:

$$\frac{d\tau}{dt} = \frac{\partial f(t,a)}{\partial t} + \frac{\partial f(t,a)}{\partial a}\frac{\partial a}{\partial t}, \tag{19}$$

shortly

$$\dot{\tau} = \dot{f} = f'_t + f'_a \dot{a}. \tag{20}$$

We know some properties of the terms in (20):

- $f'_t > 0$ the partial time derivative of f is positive, since time cannot rewind.
- $f'_a \leq 0$ since larger errors result slower motion and consequently τ should be decreased.
- $\dot{a}$ describes how the absolute value of the tracking error changes. We do not know the sign of this term, but our goal is to reduce tracking errors, i.e. to get $\dot{a} < 0$.

The time derivative of the tracking error based time-scaling function is involved (19). To get a simpler equation we study a special class of $f(t,a)$ functions, where the time t is multiplied by an error dependent value. We call this function as error-based linear time-scaling function:

$$\tau = f(t,a) = p(a)t. \tag{21}$$

In this case the time derivative of the time-scaling function reads:

$$\dot{\tau} = p + p'_a \dot{a} t. \tag{22}$$

Piecewise Constant Time-Scaling

There are several possibilities to define error-based linear time-scaling functions. In this paper we use the piecewise constant time-scaling. This function is also based on the tracking error, moreover it is almost as simple as (16). In this case the time can be multiplied by three different constants; the error magnitude determines which constant should be used:

$$\tau = p(a)t = \begin{cases} c_1 t, & \text{if } a < a_1, \\ t, & \text{if } a_1 \leq a < a_2, \\ c_2 t, & \text{if } a_2 \leq a, \end{cases} \tag{23}$$

where a_1 denotes the limit on the tracking error, where acceleration is allowed, in this case $c_1 \geq 1$ gives the rate of the time-scaling. If the error is larger than the limit

a_2, then the motion should be slowed down, the rate of deceleration is defined by $0 < c_2 \leq 1$. If the absolute value of the tracking error is between a_1 and a_2, then the velocity profile of the motion is unchanged.

Similarly to the constant time-scaling in (16)–(17) the derivative of the piecewise constant time-scaling (23) can be easily calculated:

$$\dot{t} = \begin{cases} c_1, \text{if} & a < a_1, \\ 1, \text{if} \; a_1 \leq a < a_2, \\ c_2, \text{if} & a_2 \leq a. \end{cases} \tag{24}$$

Constant-Linear Time-Scaling

It can be seen from (14) and (15) that the time derivative of the time-scaling function ($\dot{f}$) determines how the reference input signals change. If $\dot{f}$ is not a continuous function or if it contains jumps in its values, then the velocity and the angular velocity of the car should also change suddenly. In a real system sudden changes are not desirable, since the accelerations cannot be infinity large. If the tracking the reference path requires infinity large accelerations, the real robot will not be able to follow it without errors.

To overcome this problem we introduce a new time-scaling function, which is similar to the piecewise constant one, but it does not require infinity large accelerations. We call this function as constant-linear time-scaling, since $\dot{f}$ contains three parts: the two constant parts (according to the piecewise constant time-scaling) and a linear segment, which connects the two constant sections without any jump (see Fig. 2). The change of the time-scaling constant from an initial value c_A to the end value c_B takes $T_2 - T_1$ time in this case. (If $T_2 = T_1$ the piecewise constant time-scaling is got back.)

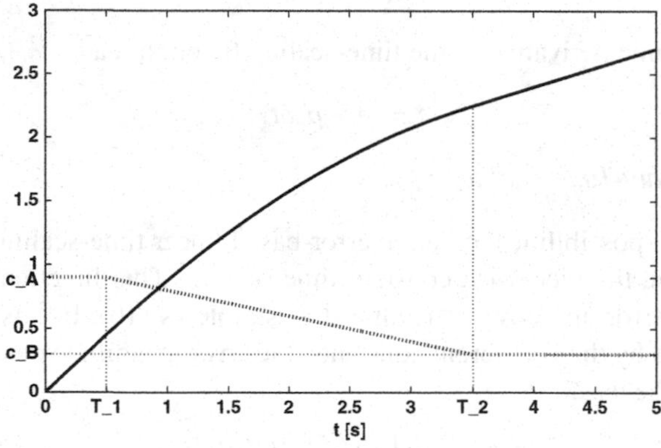

Fig. 2: The constant-linear time-scaling function (*solid line*) and its time derivative (*dotted line*)

To give the equations of the constant-linear time-scaling function we make the calculations in reverse order as we made it before. First the time derivative of the time-scaling function ($\dot{\tau} = \dot{f}$) is defined and from this equation τ is determined. According to the above mentioned consideration $\dot{\tau}$ can be calculated by the following equations for the constant-linear time-scaling (see Fig. 2):

$$\dot{\tau} = \begin{cases} c_A, & \text{if } t < T_1, \\ (c_B - c_A)(t - T_1)/(T_2 - T_1) + c_A, & \text{if } T_1 \leq t < T_2, \\ c_B, & \text{if } T_2 \leq t, \end{cases} \tag{25}$$

where T_1 denotes the time instant, when the time-scaling constant should be modified because of the value of the error. The duration of the change from c_A to c_B is $T_2 - T_1$.

The time derivative of the time-scaling function is given in (25), we integrate it and we take some conditions into consideration to get τ. The initial condition reads $f(0) = 0$, and f has to be continuous:

$$\tau = \begin{cases} c_A t, & \text{if } t < T_1, \\ (c_B - c_A)(t - T_1)^2/(2T_2 - 2T_1) + c_A t, & \text{if } T_1 \leq t < T_2, \\ (c_A - c_B)(T_1 + T_2)/2 + c_B t, & \text{if } T_2 \leq t. \end{cases} \tag{26}$$

4 Time-Scaling Based Control

Our control method is based on a transformation of the tracking error which is expressed in the configuration variables and according to a scaled reference trajectory. The transformation allows to define an error dynamics which is then linearized around the reference path which is the origin of the state space of the error dynamics. The linearized error dynamics is controlled by a state feedback similar to the one reported in [17].

4.1 Transformation of the Tracking Error

Suppose that the desired behavior of the robot is given by $x_{\text{ref}}, y_{\text{ref}}$ and θ_{ref} such that these functions identically satisfy (1) for the corresponding reference input signals v_{ref} and ω_{ref}.

We suggest to scale this reference trajectory to reduce the tracking error. The scaled reference trajectory is then given by (4) – (6), and the velocities along this path are described by (7) – (15).

We define the tracking errors as the difference between the real configuration variables and the scaled reference values given in (4) – (6):

$$e_x(t) = x(t) - x_{\text{ref,TS}}(t), \tag{27}$$
$$e_y(t) = y(t) - y_{\text{ref,TS}}(t), \tag{28}$$
$$e_\theta(t) = \theta(t) - \theta_{\text{ref,TS}}(t). \tag{29}$$

Let us now consider the transformation

$$\begin{pmatrix} e_1(t) \\ e_2(t) \\ e_3(t) \end{pmatrix} = \begin{bmatrix} \cos\theta(t) & \sin\theta(t) & 0 \\ -\sin\theta(t) & \cos\theta(t) & 0 \\ 0 & 0 & 1 \end{bmatrix} \begin{pmatrix} e_x(t) \\ e_y(t) \\ e_\theta(t) \end{pmatrix} = \underline{\underline{T}} \begin{pmatrix} e_x(t) \\ e_y(t) \\ e_\theta(t) \end{pmatrix} \tag{30}$$

of the error vector $(e_x(t), e_y(t), e_\theta(t))$ into a frame fixed to the robot such that the longitudinal axis of the vehicle coincides the transformed x axis.

Differentiating (30) w.r.t time t and using (7) – (9), (27) – (29) we get

$$\begin{pmatrix} \dot{e}_1(t) \\ \dot{e}_2(t) \\ \dot{e}_3(t) \end{pmatrix} = \underline{\underline{\dot{T}T^{-1}}} \begin{pmatrix} e_1(t) \\ e_2(t) \\ e_3(t) \end{pmatrix} + \underline{\underline{T}} \begin{pmatrix} \dot{x}(t) - x'_{\mathrm{ref}}(\tau)\dot{f} \\ \dot{y}(t) - y'_{\mathrm{ref}}(\tau)\dot{f} \\ \dot{\theta}(t) - \theta'_{\mathrm{ref}}(\tau)\dot{f} \end{pmatrix}. \tag{31}$$

Equation (31) can be rewritten in the following form using (1), (10) – (12) and (29)

$$\begin{pmatrix} \dot{e}_1(t) \\ \dot{e}_2(t) \\ \dot{e}_3(t) \end{pmatrix} = \begin{bmatrix} 0 & \omega(t) & 0 \\ -\omega(t) & 0 & 0 \\ 0 & 0 & 0 \end{bmatrix} \begin{pmatrix} e_1(t) \\ e_2(t) \\ e_3(t) \end{pmatrix} + \begin{pmatrix} 0 \\ \sin e_3(t) \\ 0 \end{pmatrix} v_{\mathrm{ref}}(\tau)\dot{f} + \begin{bmatrix} 1 & 0 \\ 0 & 0 \\ 0 & 1 \end{bmatrix} \begin{pmatrix} w_1 \\ w_2 \end{pmatrix}, \tag{32}$$

which describes the evolution of the errors with respect to the reference path. The input w_1 and w_2 are

$$w_1 = v(t) - v_{\mathrm{ref}}(\tau)\dot{f}\cos e_3(t), \tag{33}$$

$$w_2 = \omega(t) - \omega_{\mathrm{ref}}(\tau)\dot{f}. \tag{34}$$

Notice that the inputs w_1 and w_2 allow to calculate the real inputs $v(t)$ and $\omega(t)$ as

$$v(t) = w_1 + v_{\mathrm{ref}}(\tau)\dot{f}\cos e_3(t) = w_1 + v_{\mathrm{ref,TS}}(t)\cos e_3(t), \tag{35}$$

$$\omega(t) = w_2 + \omega_{\mathrm{ref}}(\tau)\dot{f} = w_2 + \omega_{\mathrm{ref,TS}}(t). \tag{36}$$

4.2 Linearization

The nonlinear model of the system (32) can be linearized along the reference trajectory with zero input signals (i.e. for $[e_1(t)\ e_2(t)\ e_3(t)]^T = 0$ and $[w_1\ w_2]^T = 0$).

For the linearization we denote (32) shortly as

$$\dot{\mathbf{e}} = \mathbf{g}(\mathbf{e}, \mathbf{w}), \tag{37}$$

$$\mathbf{e} = \begin{pmatrix} e_1(t) \\ e_2(t) \\ e_3(t) \end{pmatrix}, \tag{38}$$

$$\mathbf{w} = \begin{pmatrix} w_1 \\ w_2 \end{pmatrix}. \tag{39}$$

If we suppose, that the reference inputs $(v_{\mathrm{ref}}, \omega_{\mathrm{ref}})$ are constant, we get the following linearized model for (37):

$$\dot{\mathbf{e}} = \begin{bmatrix} 0 & \omega_{\mathrm{ref}}(\tau)\dot{f} & 0 \\ -\omega_{\mathrm{ref}}(\tau)\dot{f} & 0 & v_{\mathrm{ref}}(\tau)\dot{f} \\ 0 & 0 & 0 \end{bmatrix} \mathbf{e} + \begin{bmatrix} 1 & 0 \\ 0 & 0 \\ 0 & 1 \end{bmatrix} \mathbf{w}. \tag{40}$$

The linearized system (40) is controlled by a state feedback of the form

$$\mathbf{w} = -K\mathbf{e}, \tag{41}$$

such that the gain matrix K puts the eigenvalues of the closed loop system in the left half of the complex plane.

4.3 Analysis

Now we calculate the controllability matrix for the linearized system (40):

$$M_C = \begin{bmatrix} 1 & 0 & 0 & 0 & -\omega_{\mathrm{ref}}^2(\tau)\dot{f}^2 & \omega_{\mathrm{ref}}(\tau)v_{\mathrm{ref}}(\tau)\dot{f}^2 \\ 0 & 0 & -\omega_{\mathrm{ref}}(\tau)\dot{f} & v_{\mathrm{ref}}(\tau)\dot{f} & 0 & 0 \\ 0 & 1 & 0 & 0 & 0 & 0 \end{bmatrix}. \tag{42}$$

The rank of this matrix is 2 or 3, while the dimension of the state $\mathbf{e}$ is 3. The system (40) is controllable by the state feedback (41) if the rank of controllability matrix equals 3 (i.e. $\mathrm{rank}(M_C) = \dim(\mathbf{e}) = 3$). So the requirement is to have at least one nonzero reference input (i.e. $\omega_{\mathrm{ref}} \neq 0$ or $v_{\mathrm{ref}} \neq 0$) and $\dot{f} \neq 0$ (this is always satisfied in (17) and (24)). There is a singularity at zero velocities, which is a common problem of car like robots.

5 Simulation Results

We made some simulations to see the effect of the different time-scaling functions. The closed loop behavior of the robot was studied for a given reference path without time scaling, with piecewise constant and finally with constant-linear time-scaling. The schema of control is depicted in Fig. 3.

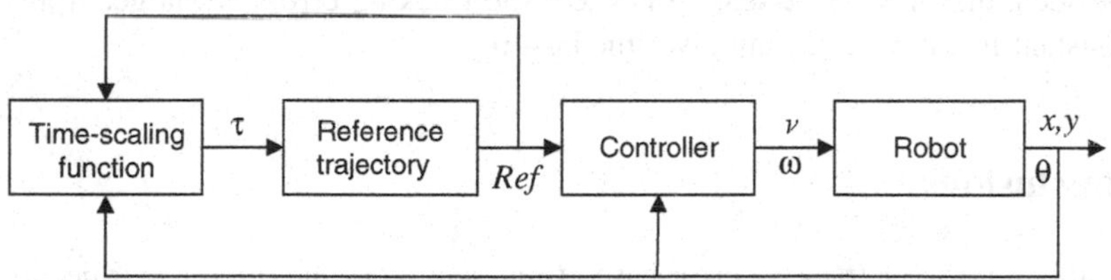

Fig. 3: Scheme of the control using the tracking error based time-scaling

The task was to follow the reference path (see Fig. 4) as accurately as possible despite the small initial error of the orientation (0.1 rad) and the disturbances. (We used a step signal at $t = 1.5$ s with final value 0.075 m to simulate some disturbances in the x direction.) The input signals of the robot were saturated during the simulations ($-1 \leq v \leq 1$ m/s, $-10 \leq \omega \leq 10$ rad/s), moreover, the velocity of their changes was also limited ($-0.8 \leq \dot{v} \leq 0.8$ m/s^2, $-30 \leq \dot{\omega} \leq 30$ rad/s^2).

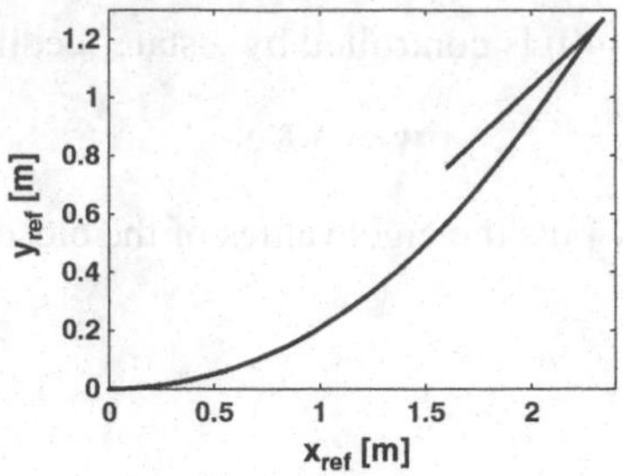

Fig. 4: Reference path in the x-y plane

The control law described in Section 4 was used to track the scaled reference trajectory. If no time-scaling was used, than according to the disturbance and the limitations on the input signals the robot was not able to perform an accurate tracking. To get a better tracking, error based time-scaling was applied to slow down the reference if the tracking error is larger than 0.05 m:

$$\tau = \begin{cases} t, & \text{if } \max(a) < 0.05, \\ 0.5t, & \text{if } \max(a) \geq 0.05. \end{cases} \tag{43}$$

At $t = 1.5$ s due to the effect of the disturbance the error was larger than 0.05 m, hence the movement was slowed down by the time-scaling. Since the reference path changed suddenly, the robot was not able to follow it immediately. It needed some time to perform an accurate tracking. To overcome this problem the constant-linear time-scaling (26) was used with the following parameters: $c_A = 1$, $c_B = 0.5$, $T_1 = \{\min(t)|a(t) \geq 0.05\}$ and $T_2 = T_1 + 0.5$ s. Since the trajectory does not change suddenly in this case, more accurate motion is achieved.

Figure 5 shows the path of the rear axle midpoint in the three different cases. It can be seen, that if no time-scaling is used, the tracking errors are large. Applying the constant-linear time-scaling gives the best result.

6 Conclusions

This paper reported a tracking error based time-scaling of the reference trajectory to control a car like wheeled mobile robot.

The way of determining the time-scaling function from the tracking error was detailed. We also defined some special time-scaling functions. A controller was designed after some mathematical operations (error transformation and linearization), using the scaled reference trajectory. The controllability of the system was also analyzed, and the result showed that the linearized system is controllable except the case of zero velocities.

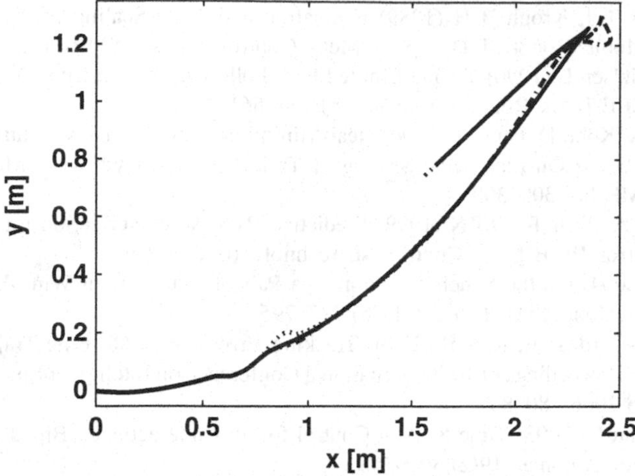

Fig. 5: Results of simulations (without time-scaling (*dashdot line*), with piecewise constant time-scaling (*dotted line*), with constant-linear time-scaling (*solid line*))

Some simulations were also done. Their results are included in this paper. As we expected, if the time-scaling slows down the motion, the tracking errors are smaller at the expense of the traveling time.

Acknowledgments

The research was partially supported by the Advanced Vehicles & Vehicle Control Knowledge Center under grant RET 04/2004 and by the Hungarian Science Research Fund under grant OTKA T068686.

References

1. Zefran M (1994) Review of the Literature on Time-Optimal Control of Robotic Manipulators, GRASP Laboratory, Department of Computer and Information Science, University of Pennsylvania
2. Wu W, Chen H, Woo P-Y (2000) Time Optimal Path Planning for a Wheeled Mobile Robot, J. Robotic Syst. 17(11):585–591
3. Hollerbach J M (1984) Dynamic Scaling of Manipulator Trajectories, J. Dyn. Syst. Meas. Control, Trans. ASME 106(1):102–106
4. Sahar G, Hollerbach J M (1986) Planning of Minimum-Time Trajectories for Robot Arms, Int. J. Robot. Res. 5(3):90–100
5. Moon S B, Ahmad S (1991) Time Scaling of Cooperative Multirobot Trajectories, IEEE Trans. Syst., Man, Cybern. 21(4):900–908

6. Graettinger T J, Krogh B H (1989) Evaluation and Time-Scaling of Trajectories for Wheeled Mobile Robots, J. Dyn. Syst. Meas. Control, Trans. ASME 111:222–231
7. Dahl O, Nielsen L (1990) Torque-Limited Path Following by On-Line Trajectory Time Scaling, IEEE Trans. Robot. Automat. 6(5):554–561
8. Kumagai A, Kohli D, Perez R (1996) Near-Minimum Time Feedback Controller for Manipulators Using On-Line Time Scaling of Trajectories, J. Dyn. Syst. Meas. Control, Trans. ASME 118:300–308
9. Bemporad A, Tarn T-J, Xi N (1999) Predictive Path Parameterization for Constrained Robot Control, IEEE Trans. Contr. Syst. Technol. 7(6):648–656
10. Lévine J (2004) On the Synchronization of a Pair of Independent Windshield Wipers, IEEE Trans. Contr. Syst. Technol. 12(5):787–795
11. Szádeczky-Kardoss E, Kiss B (2006) Tracking Error Based On-Line Trajectory Time Scaling, In: Proceedings of 10th International Conference on Intelligent Engineering Systems (INES 2006), 80–85
12. Chevallereau C (2003) Time-Scaling Control for an Underactuated Biped Robot, IEEE Trans. Robot. Automat. 19(2):362–368
13. Respondek W (1998) Orbital Feedback Linerization of Single-Input Nonlinear Control Systems, In: Proceedings of the IFAC NOLCOS'98, 499–504
14. Kiss B, Szádeczky-Kardoss E (2007) Tracking Control of Wheeled Mobile Robots with a Single Steering Input, In: Proceedings of the 4th International Conference on Informatics in Control, Automation and Robotics, unpublished
15. Rouchon P, Fliess M, Lévine J, Martin Ph (1993) Flatness and Motion Planning: The Car with *n*-Trailers, In: Proceedings of the European Control Conference (ECC'93), 1518–1522
16. Cuesta F, Ollero A (2005) Intelligent Mobile Robot Navigation, In: Springer Tracts in Advanced Robotics, Springer, Berlin, 16
17. Dixon W E, Dawson D M, Zergeroglu E, Behal A (2001) Nonlinear Control of Wheeled Mobile Robots, In: Lecture Notes in Control and Information Sciences, Springer, Berlin/Heidelberg/New York

Intelligent Mobile Robot Control in Unknown Environments

Gyula Mester

Department of Informatics
University of Szeged, Árpád tér 2, H-6720 Szeged, Hungary
gmester@inf.u-szeged.hu

Summary. This paper gives the fuzzy reactive control of a wheeled mobile robot motion in an unknown environment with obstacles. The model of the vehicle has two driving wheels and the angular velocities of the two wheels are independently controlled. When the vehicle is moving towards the target and the sensors detect an obstacle, an avoiding strategy is necessary. We proposed a fuzzy reactive navigation strategy of collision-free motion in an unknown environment with obstacles. First, the vehicle kinematics constraints and kinematics model are analyzed. Then the fuzzy reactive control of a wheeled mobile robot motion in an unknown environment with obstacles is proposed. Output of the fuzzy controller is the angular speed difference between the left and right wheels (wheel angular speed correction) of the vehicle. The simulation results show the effectiveness and the validity of the obstacle avoidance behavior in an unknown environment of the proposed fuzzy control strategy.

1 Introduction

A wheeled mobile robot is a wheeled vehicle which is capable of autonomous motion. Autonomous mobile robots are a very interesting subject both in scientific research and practical applications. This paper gives the fuzzy reactive control of a wheeled mobile robot motion in an unknown environment with obstacles. W.L. Xu, S.K. Tso and Y.H. Fung [1] proposed the use of fuzzy reactive control of a mobile robot incorporating a real/virtual target switching strategy. Navigation control of the robot is realized through fuzzy coordination of all the rules. Sensed ranging and relative target position signals are input to the fuzzy controller. Ranajit Chatterjee and Fumitoshi Matsuno [2] proposed the use of a single side reflex for autonomous navigation of mobile robots in unknown environments. In this work, fuzzy logic based implementation of the single-sided reflex is considered. The use of perceptional symmetry allows perception–action mapping with reduced sensor space dimensions. Simulation and experimental results are presented to show the effectiveness of the proposed strategy in typical obstacle situations. In [3] a fuzzy expert system has been presented for automatic movement control of a platform on a ground with obstacles. System implementation phases are discussed, as well as

J.A.T. Machado et al. (eds.), *Intelligent Engineering Systems and Computational Cybernetics*,

performances of the software tools used for implementation and performances of the implementation.

In this paper the model of the vehicle has two driving wheels (which are attached to both sides of the vehicle) and the angular velocities of the two wheels are independently controlled. The center of the driving wheels is regarded as the gravity center. This model is the simplest and the most suitable for a small-sized and light, battery-driven autonomous vehicle.

First, the vehicle kinematics constraints and kinematics model are analyzed. Then the fuzzy reactive control of a wheeled mobile robot motion in an unknown environment with obstacles is proposed. Output of the fuzzy controller is the angular speed difference between the left and right wheels of the vehicle [1, 2]. Finally, the simulation results show the effectiveness and the validity of the obstacle avoidance behavior in an unknown environment of the proposed fuzzy control strategy.
The paper is organized as follows:

- Section 1: Introduction.
- The modeling of the autonomous wheeled mobile robot is given in Section 2.
- In Section 3 fuzzy control of a wheeled mobile robot motion in an unknown environment with obstacles is proposed.
- In Section 4 the simulation results are illustrated.
- Conclusions are given in Section 5.

2 Modeling of the Autonomous Wheeled Mobile Robots

2.1 Kinematics Constraints

We consider a mechanical system with n generalized coordinate's q subject to m kinematics constraints:

$$A(q)\dot{q} = 0 \tag{1}$$

where: $A \in R^{mxn}$ is a full rank matrix [4]. A large class of mechanical systems, such as wheeled vehicle and mobile robots involve kinematics constraints. In the literature these kinematics constraints can generally be classified as:

- Nonholonomic or
- Holonomic

A mobile robot involving two actuator wheels is considered as a system subject to *nonholonomic constraints.*

2.2 Kinematics Model

Let's consider the kinematics model for an autonomous vehicle. The position of the mobile robot in the plane is shown in Fig. 1.

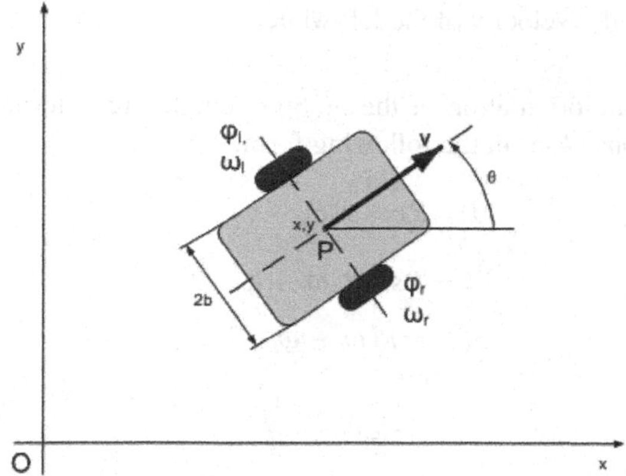

Fig. 1: Position of mobile robot in plane

The inertial-based frame (*Oxy*) is fixed in the plane of motion and the moving frame is attached to the mobile robot. In this paper we will assume that the mobile robots are rigid cart equipped, with non-deformable conventional wheels, and they are moving on a non-deformable horizontal plane.

During the motion: the contact between the wheel and the horizontal plane is reduced to a single point, the wheels are fixed, the plane of each wheel remains vertical, the wheel rotates about its horizontal axle and the orientation of the horizontal axle with respect to the cart can be fixed.

The contact between the wheel of the mobile robots and the non-deformable horizontal plane supposes both the conditions of pure rolling and non-slipping during the motion. This means that the velocity of the contact point between each wheel and the horizontal plane is equal to zero. For low rolling velocities this is a reasonable wheel moving model. The center of the fixed wheel is a fixed point of the cart and b is the distance of the center of the wheel from P.

The rotation angle of the wheel about its horizontal axle is denoted by φ(t) and the radius of the wheel by R. Hence, the position of the wheel is characterized by two constants: b and R and its motion by a time-varying angle:

$\varphi_r(t)$ – the rotation angle of right wheel and

$\varphi_l(t)$ – the rotation angle of left wheel.

The configuration of the mobile robot can be described by five generalized coordinates such as:

$$q = [x, y, \theta, \varphi_r, \varphi_l]^T \tag{2}$$

where: x and y are the two coordinates of the origin P of the moving frame (the geometric center of the mobile robot), θ is the orientation angle of the mobile robot (of the moving frame).

The vehicle velocity v can be found in equation (3):

$$v = R(\omega_r + \omega_l)/2 \tag{3}$$

where:

$\omega_r = \frac{d\varphi_r}{dt}$ – angular velocity of the right wheel,

$\omega_l = \frac{d\varphi_l}{dt}$ – angular velocity of the left wheel.

The position and the orientation of the mobile vehicle are determined by a set of differential equations (4–6) in the following form:

$$\dot{x} = R\cos\theta(\omega_r + \omega_l)/2 \tag{4}$$

$$\dot{y} = R\sin\theta(\omega_r + \omega_l)/2 \tag{5}$$

$$\dot{\theta} = R(\omega_r - \omega_l)/2b \tag{6}$$

Here,

$$\dot{x} = v\cos\theta \tag{7}$$

$$\dot{y} = v\sin\theta \tag{8}$$

Then the matrix form is:

$$\begin{bmatrix} \dot{x} \\ \dot{y} \\ \dot{\theta} \end{bmatrix} = \begin{bmatrix} \cos\theta & 0 \\ \sin\theta & 0 \\ 0 & 1 \end{bmatrix} \begin{bmatrix} v \\ \dot{\theta} \end{bmatrix} \tag{9}$$

Finally, the kinematics model of the vehicle velocity v and the angular velocity $\dot{\theta}$ can be represented by the matrix as follows:

$$\begin{bmatrix} v \\ \dot{\theta} \end{bmatrix} = \begin{bmatrix} R/2 & R/2 \\ R/2b & -R/2b \end{bmatrix} \begin{bmatrix} \omega_r \\ \omega_l \end{bmatrix} \tag{10}$$

In this case, we now consider the mobile robot motion as a nonholonomic mechanical system, where three kinematics constraints exist:

$$\begin{aligned} \dot{x}\sin\theta - \dot{y}\cos\theta &= 0 \\ \dot{x}\cos\theta + \dot{y}\sin\theta &= R\omega_r - b\dot{\theta} \\ \dot{x}\cos\theta + \dot{y}\sin\theta &= R\omega_l + b\dot{\theta} \end{aligned} \tag{11}$$

The constraints can be written in the form (1), where matrix $A \in R^{m\times n}$ $(m = 3, n = 5)$ can be described as:

$$A = \begin{bmatrix} \sin\theta & -\cos\theta & 0 & 0 & 0 \\ \cos\theta & \sin\theta & b & -R & 0 \\ \cos\theta & \sin\theta & -b & 0 & -R \end{bmatrix}$$

In this paper the angular velocities of the two wheels of the mobile robot are independently controlled.

3 Fuzzy Control of a Mobile Robot Motion in an Unknown Environment with Obstacles

3.1 Advantages of Fuzzy Logic Control

In control engineering there are a number of methods to make mobile robot motions intelligent, they can be classified into: fuzzy methods, neural networks and genetic algorithms (or a combination of these). The capabilities of learning, parallel processing, suitability for complex dynamic systems, adaptation and evolution of these methods will be applied to intelligent autonomous mobile robot controller design.

Since the beginning of the 1980s many control architectures for autonomous mobile robots have been proposed in the literature. Fuzzy logic is an attractive technique for mobile robot control problems. Fuzzy controllers are based on three well-known stages: the fuzzification stage – this stage transforms crisp input values from a process into fuzzy sets, the rule base stage – consists of a set of linguistic description rules in the form of "IF-THEN" and the defuzzification stage – transforms the fuzzy sets in the output space into crisp control signals.

Fuzzy controllers have been implemented in many experimental cases and in industrial applications because these controllers have advantages such as: easy implementation, suitability for complex dynamic systems, fast response to the mobile robot navigation, high flexibility and a robust nature. When fuzzy controllers are designed, human knowledge and a set of heuristic rules are implemented without the use of mathematical models.

3.2 Reactive Navigation Strategy of Collision-Free Motion in an Unknown Environment with Obstacles

In this paper fuzzy control is applied to the navigation of the autonomous mobile robot in an unknown environment with obstacles [1, 2]. We supposed that the autonomous mobile robot has two wheels driven independently and groups of ultrasonic sensors to detect obstacles in the front, to the right and to the left of the vehicle. The reactive strategies are based on ultrasonic sensory information and only the relative interactions between the mobile robot and the unknown environment have to be assessed. In this case, a structural modeling of the environment is unnecessary.

If the vehicle is moving towards the target and the ultrasonic sensors detect an obstacle, an avoiding strategy is necessary [5]. While the mobile robot is moving it is important to compromise between avoiding the obstacles and moving towards to the target position.

Suppose that an autonomous mobile robot realizes a collision-free motion in an unknown environment with obstacles and is expected to reach a target position. In moving in the absence of obstacles in an unknown environment, the mobile robot is steered towards the target position (Fig. 2).

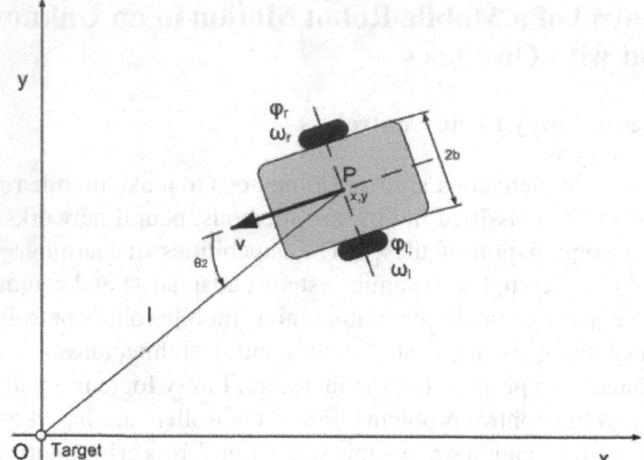

Fig. 2: Definition of target angle θ_2 and target distance l

With obstacles present in the unknown environment, the mobile robot reacts based on both the sensed information of the obstacles and the relative position of the target (Fig. 3) [3].

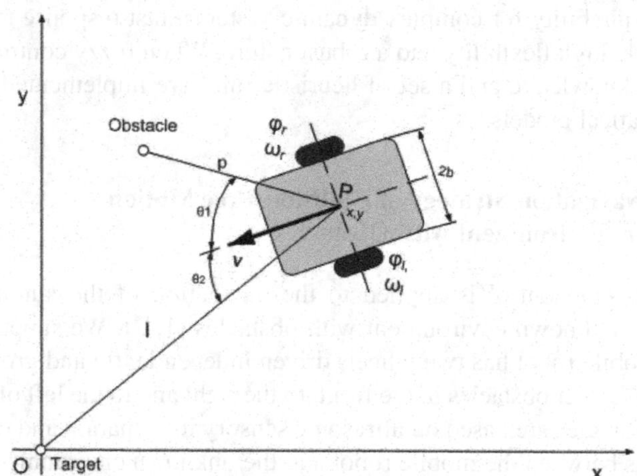

Fig. 3: Definition of obstacle angle θ_1 and obstacle distance p

In moving towards the target and avoiding obstacles, the mobile robot changes its orientation and velocity. When the obstacle in an unknown environment is very close, the mobile robot slows down and rapidly changes its orientation. The navigation strategy is to come as near to the target position as possible while avoiding collision with the obstacles in an unknown environment. The intelligent mobile robot reactive behavior is formulated in fuzzy rules.

3.3 Fuzzy Implementation of Obstacle Avoidance

Fuzzy-logic-based control is applied to realize a mobile robot motion in an unknown environment with obstacles. Inputs to the fuzzy controller are: the obstacle distances p, the obstacle orientation θ_1 (which is the angle between the robot moving direction and the line connecting the robot center with the obstacle), the target distances l, the target orientation θ_2 (which is the angle between the robot moving direction and the line connecting the robot center with the target). Output of the fuzzy controller is the angular speed difference between the left and right wheels (wheel angular speed correction) of the vehicle: $\Delta\omega = \omega_r - \omega_l$. The block diagram of the fuzzy inference system is presented in Fig. 4.

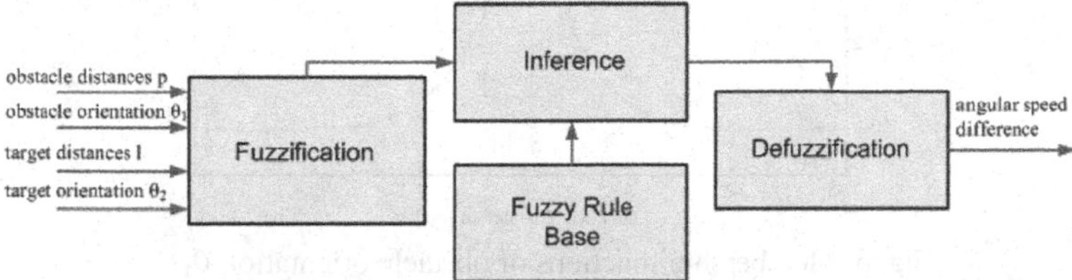

Fig. 4: Block diagram of the fuzzy inference system

The obstacle orientation θ_1 and the target orientation θ_2 are determined by the obstacle/target position and the robot position in a world coordinate system, respectively. The obstacle orientation θ_1 and the target orientation θ_2 are defined as positive when the obstacle/target is located to the right of the robot moving direction; otherwise, the obstacle orientation θ_1 and the target orientation θ_2 are negative [1].

For the proposed fuzzy controller the input variables for the obstacle distances p are simply expressed using two linguistic labels *near* and *far* ($p \in [0, 3\text{ m}]$). Figure 5 shows the suitable Gaussian membership functions.

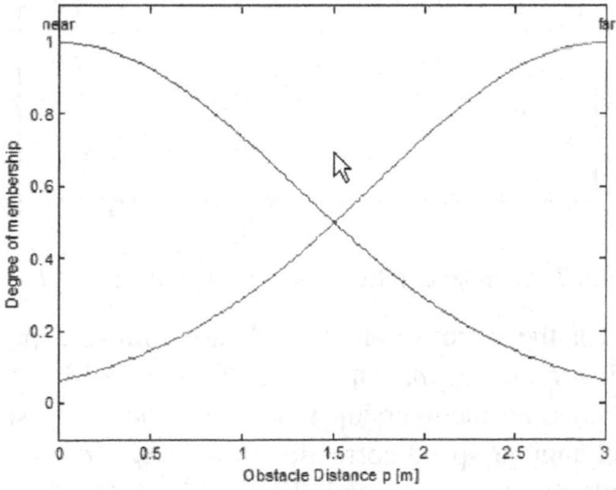

Fig. 5: Membership functions of obstacle distances p

The input variables for the obstacle orientation θ_1 are simply expressed using two linguistic labels *left* and *right* ($\theta_1 \in [-3.14, 3.14\ \text{rad}]$). Figure 6 shows the suitable Gaussian membership functions.

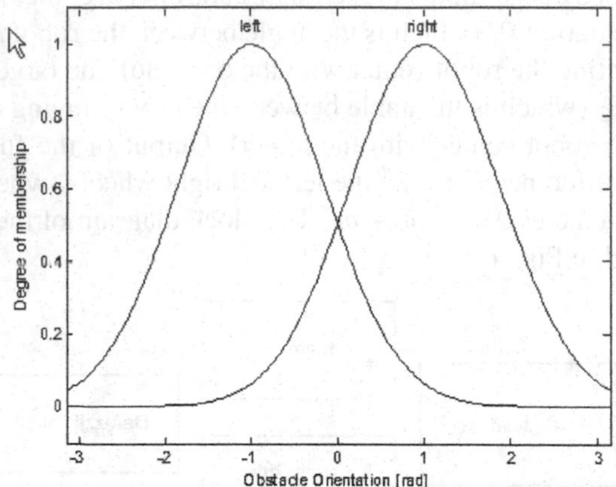

Fig. 6: Membership functions of obstacle orientation θ_1

The input variables for the target distances l are simply expressed using two linguistic labels *near* and *far* ($l \in [0, 3\ \text{m}]$). Figure 7 shows the suitable Gaussian membership functions.

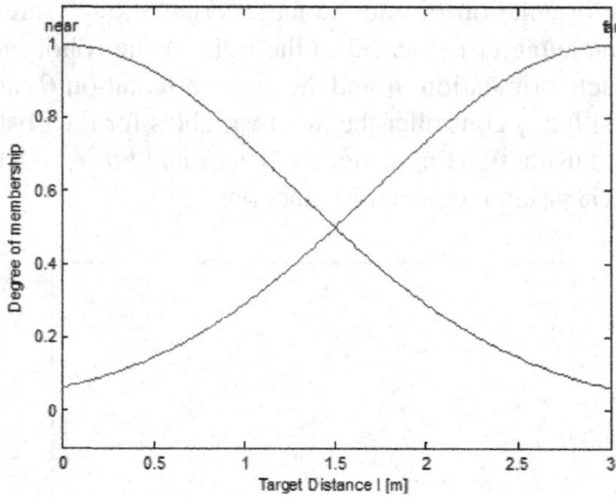

Fig. 7: Membership functions of target distances l

The input variables for the target orientation θ_2 are simply expressed using three linguistic labels *left*, *targetdirection* and *right* ($\theta_2 \in [-3.14, 3.14\ \text{rad}]$). Figure 8 shows the suitable Gaussian membership functions. The fuzzy sets for the output variables – the wheel angular speed correction $\Delta\omega = \omega_r - \omega_l$ (*turn-right*, *zero* and *turn-left*) of the mobile robot – are shown in Fig. 9. The output variables are normalized between: $\Delta\omega \in [-20, 20\ \text{rad/s}]$.

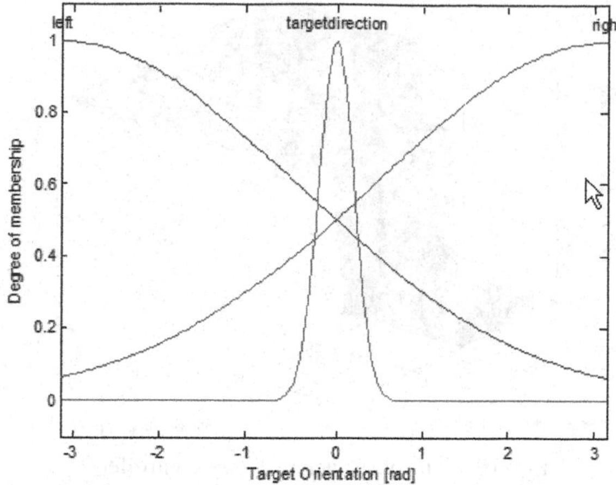

Fig. 8: Membership functions of target orientation θ_2

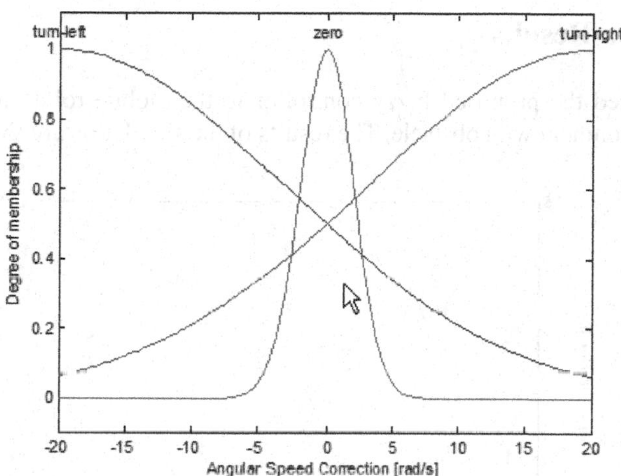

Fig. 9: Membership functions of the angular speed difference between the left and right wheels f

The rule-base for mobile robot fuzzy reaction are:

R1: If θ_2 is right then $\Delta\omega$ is *turn-right*

R2: If θ_2 is left then $\Delta\omega$ is *turn-left*

R3: If p is *near* and l is *far* and θ_1 is left then $\Delta\omega$ is *turn-right*

R4: If p is near and l is far and θ_1 is *right* then $\Delta\omega$ is *turn-left*

R5: If θ_2 is targetdirection then $\Delta\omega$ is *zero*

R6: If p is *far* and θ_2 is targetdirection then $\Delta\omega$ is *zero*

In the present implementation of the fuzzy controller the Center of Area method of defuzzification is used. Control surface of the proposed fuzzy controller as a function of the inputs (the target distances l and obstacle distances p) is shown in Fig. 10.

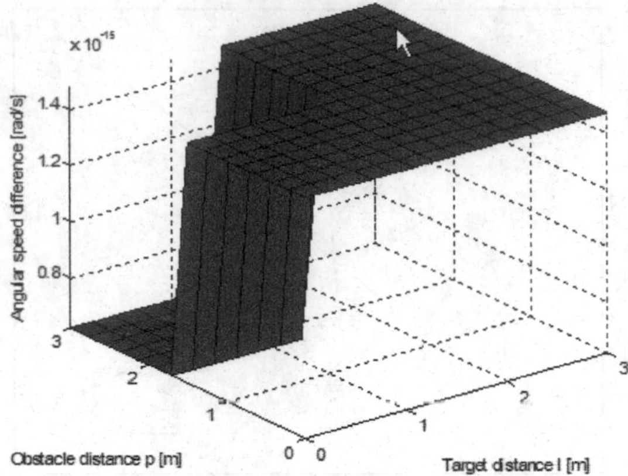

Fig. 10: Control surface of fuzzy controller

4 Simulation Results

Now, we applied the proposed fuzzy controller to the mobile robot moving in an unknown environment with obstacle. The results of the simulation are shown in Figs. 11–13.

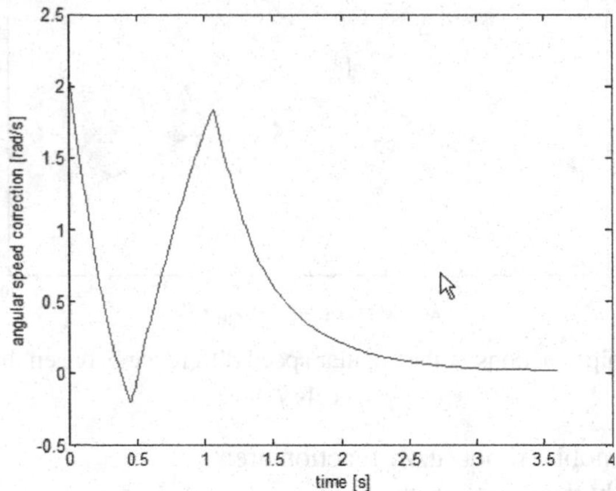

Fig. 11: Wheel angular speed correction

The mobile robot compares the nearness of the obstacles of two sides. Figure 12 shows the obstacle avoidance mobile robot paths of the right side and Fig. 13 shows the obstacle avoidance mobile robot paths of the left side. The fuzzy reactive controller is powerful in view of the short reaction time and rapid decision-making of the obstacle avoidance process.

The simulation results show the effectiveness and the validity of the obstacle avoidance behavior in an unknown environment of the proposed control strategy.

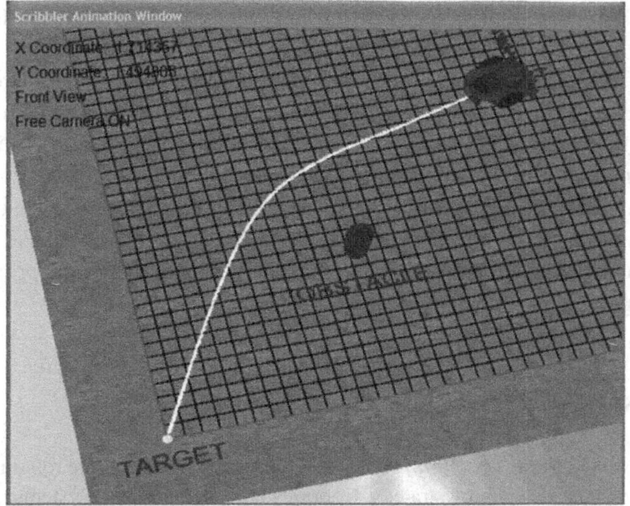

Fig. 12: Obstacle avoidance trajectory of the right side

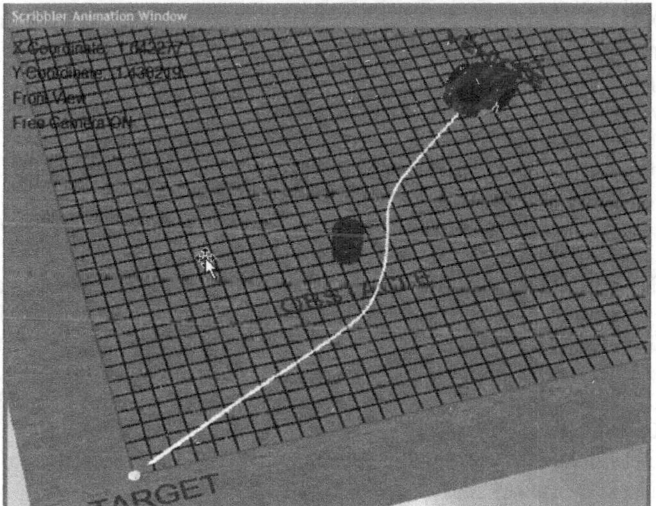

Fig. 13: Obstacle avoidance trajectory of the left side

5 Conclusions

This paper gives the fuzzy reactive control of a wheeled mobile robot motion in an unknown environment with obstacles. The model of the vehicle has two driving wheels and the angular velocities of the two wheels are independently controlled. When the vehicle is moving towards the target and the sensors detect an obstacle, an avoiding strategy is necessary. We proposed fuzzy reactive navigation strategy of collision-free motion in an unknown environment with obstacles. Output of the fuzzy controller is the angular speed difference between the left and right wheels of the vehicle.

First, the vehicle kinematics constraints and kinematics model are analyzed. Then the fuzzy reactive control of a wheeled mobile robot motion in an unknown environment with obstacles is proposed. Finally, the simulation results are illustrated. The implemented controller is computationally simple. The fuzzy reactive controller is powerful in view of the short reaction time and rapid decision-making of the obstacle avoidance process. The simulation results show the effectiveness and the validity of the obstacle avoidance behavior in an unknown environment of the proposed fuzzy control strategy.

References

1. Xu WL, Tso SK, Fung YH (1998) Fuzzy reactive control of a mobile robot incorporating a real/virtual target switching strategy. Robotics and Autonomous Systems, 23:171–186
2. Chatterjee R, Matsuno F (2001) Use of single side reflex for autonomous navigation of mobile robots in unknown environments. Robotics and Autonomous Systems 35:77–96
3. Saletic D, Popovic U (2006) Fuzzy Expert System for Automatic Movement Control of a Platform on a Ground with Obstacles. In: Proceedings of the YuINFO 2006, Kopaonik, Serbia, pp 1–6
4. Wanga J, Zhub X, Oyac M, Sud C-Y (2006) Robust motion tracking control of partially nonholonomic mechanical systems. Robotics and Autonomous Systems 35:332–341
5. Maaref H, Baret C (2002) Sensor based navigation of a mobile robot in an indoor environment. Robotics and Autonomous Systems 38:1–18
6. Mester Gy (2006) Introduction to Control of Mobile Robots. In: Proceedings of the YuINFO 2006, Kopaonik, Serbia, pp 1–4
7. Mester Gy (2006) Modeling the Control Strategies of Wheeled Mobile Robots. In: Proceedings of the Kandó Konferencia 2006, Budapest, Hungary, pp 1–4
8. Mester Gy (2006) Intelligent Mobile Robot Controller Design. In: Proceedings of the Intelligent Engineering Systems, INES 2006, London, pp 282–286
9. Mester Gy (2006) Motion Control of Wheeled Mobile Robots. In: Proceedings of the SISY 2006, Subotica, Serbia, pp 119–130
10. Matijevics I (2006) Autonom rendszerek architektúrája. In: Proceedings of the Kandó Konferencia 2006, Budapest, Hungary, pp 4–6

Local and Remote Laboratories in the Education of Robot Architectures

István Matijevics

University of Szeged, Institute of Informatics, Árpád tér 2, H-6720 Szeged, Hungary
mistvan@inf.u-szeged.hu

Summary. This paper presents a general overview of education of robot hardware architecture and software. A well supplied microcontroller laboratory is needed for mobile robot development. The paper examines the software and hardware used for microcontroller/robotics laboratory. The work presents a project to enhance the microcontroller/robotics education of informatics and electronics engineering students using local and remote microcontroller/ microprocessor laboratory. Solutions over the internet open the possibilities for the distance learning.

1 Introduction

Nowadays and in the future all mobile robots will include one or more microprocessors/ microcontrollers. The use of microprocessors/microcontrollers in a hardware design improves the design's capabilities as well as the design's implementation. Owing to this it is important that students of informatics, electronics and mechatronics have undergraduate experience with microcontrollers. These requirements present the university with a challenge in terms of sufficient laboratory establishment. Other important factors include the large number of students in the education process, the sometimes limited laboratory space and the financial resources of the universities [1–3, 5].

In an effective way we can use existing laboratories of informatics (LAN with PC-s) together with other techniques from the web (internet) and microcontroller trainer boards. Solutions over the internet open the possibilities for the distance learning. Open source distance learning software gives for the lecturers and students the chance for the distance administration, literature access, rapid material renovation, the use of tests. The main standing-point in creating laboratories is low level investment, using existing pieces of equipment together with the improvement of educational effectiveness. Including trainer boards and internet in the laboratories does not change the old functions of the informatics laboratories.

This paper presents a project to enhance the microcontroller/robotics education of informatics and electronics engineering students. Some courses provide preliminary knowledge for students who selected microcontroller/microprocessor based classes. For example the course in Architecture gives them hardware and assembly programming background.

In some obligatory and eligible courses students involve the design and development of microcontroller based technologies, for example in Robotics, Autonomous systems and

J.A.T. Machado et al. (eds.), *Intelligent Engineering Systems and Computational Cybernetics*,

Mechatronics. These courses include both lecture and laboratory components. In some cases in other courses students do interdisciplinary projects, or diploma works, also using microcontroller applications.

There are two aspects to teaching mobile robot architecture course, first the software (programming) and second the hardware (interfacing). Unlike in programming and architecture courses, in robot construction course students must understand deeply the connections between the software and hardware, so low level (assembly) programming, memory fetches, cycles per instructions, stacks, internal registers, accumulator, program counter, flags, timers, I/O lines etc.

Interfacing techniques require some physical and also electrical knowledge from students in microcontroller – external equipment connections. Students must know some electrical laws: common grounds, voltage/current limitations, noise shielding, timing (delaying) problems. Other courses deal with physical/electrical questions, but the experience is, a microcontroller course needs to be reminded of them.

We do not forget the mechanical interfacing aspects, this field is always imperfect in educational process of students.

2 Laboratory Types

The simplest laboratory is the classic informatics laboratory with 10 to 25 PC computers, set up with microcontroller simulation programs in network scheme. It is true that this solution is budget-priced, there are no problems with real physical and electrical factors. Often some mobile robot navigation software for the presentations are used. There are few pieces of simulation software with the ability for the simulation of the peripheries. If some characteristic interfaces are included in the program, student can observe the simulated ports and interfaces behavior. The great problem with the simulators is that the code is correct, so the simulation results are correct, but because it is not working in real time, for the real activities is response time of code too slow or but too fast. After the programming of real system they will not work correctly.

A better solution is the trainer board supported laboratory, old, classical laboratory with PC-s is expanded with evaluation (trainer) boards. Printed circuit board with microcontroller and necessary hardware gives for the student theoretical and practical knowledge, and testing the solutions (program codes) parallel on simulator and real system. Application of mechanical parts for mobile robots (drives and motors, platforms, sensors and actuators) guided the students to the realization of a full, usable mobile robot.

Adding the web properties to the second type of laboratory opens new capacities in mobile robot architecture courses. This type of laboratory opens the distance learning manner, all instruments, trainer boards, software equipments are controlled and observed through the net. So the experiments are flexible, the setup process is fast and satisfy all distance requirements.

3 Equipment Configuration for Microcontroller Laboratory

The optimal number of workplaces is 10 to 15, one lecturer (instructor) in the laboratory can supervise the students' work efficiently. The primary role of this laboratory is the teaching and presentation of the microcontroller theory, I/O interfacing and programming. All workplaces are supplied with the same PC hosts, trainer boards and ICD (In Circuit Debugger) interfaces.

Laboratory equipment, which is used in few fields must have some special as well as universal parameters:

flexibility in programming (Assembly or C), debugging capacity, internal operation showing while executing a program, ability of interfacing with other systems, standardized equipments, ability of interfacing with other external systems, peripheries and robustness.

4 Hardware Type of Laboratory Equipments

After an in-depth analysis of some microcontroller families we chose the PIC type of controllers. There are a large number of very good and operative development pieces of equipment with PIC controllers on the market. Of a high account of parameter comparison we have selected a complex system, PIC16F877 Trainer board from Chipcad [11]. This board is supplied with required peripheries for the microcontroller course needs, these I/O pieces of equipment are also parts of mobile robots. Microchip [10] developed a complex software package MPLAB for program development, testing and programming.

The connection between the PC host and trainer board is via the Microchips ICD2 (In circuit debugger) (Fig. 1).

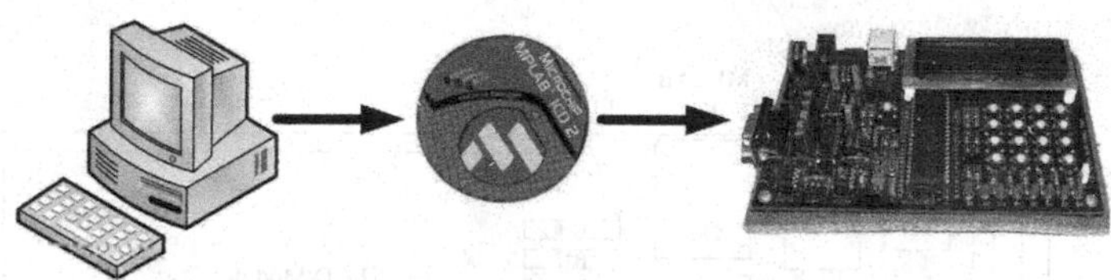

Fig. 1: The entire development system with PC host and trainer board connecting through Microchips in-circuit debugger ICD2 [10]

PIC16F877 trainer board [9, 11] is a single board microcomputer based on PIC16F877 (and other type of PIC) microcontrollers. This is a consummate developmental tool in education, industry and other fields, also it is able to fill a part in industrial applications (Fig. 2).

Some technical characteristics from the board are (Fig. 2) [11]:

General purpose I/O ports, interrupt logic, timers, A/D converter, USART, bus and capture/compare/PWM module.

Main peripheries of board are (Fig. 3):

4 x 4 matrix keyboard, LCD module, 8 LED diodes, RS232 interface bus, serial EEPROM, analog input, PWM output, analog output, PIC port expansion, ICD connector and LDR-key connector.

Microchip ICD2 board is a low cost in-circuit emulator (ICE) to develop and debug programs and transfer code from PC host to the trainer board (Fig. 4).

Features:

USB and RS-232 interface to host, PC real-time execution, MPLAB IDE compatible, built-in voltage/short circuit monitor, firmware upgradeable from PC/web download, totally enclosed diagnostic LED's, Read/Write program and data memory of microcontroller

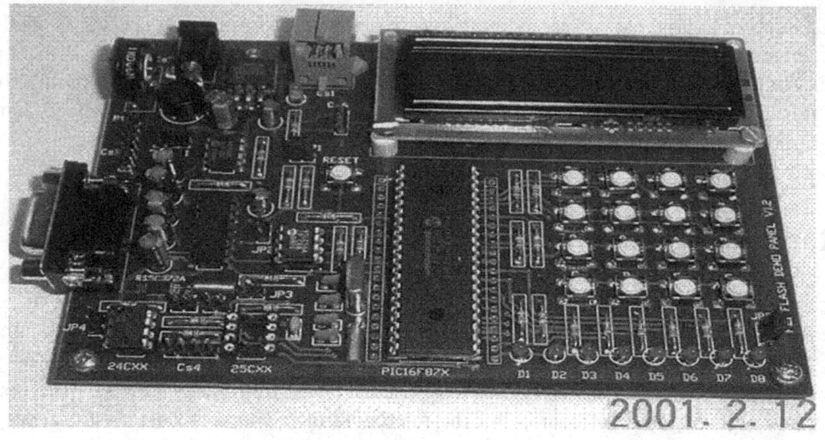

Fig. 2: PIC16F877 trainer board [11] with 16F877 microcontroller

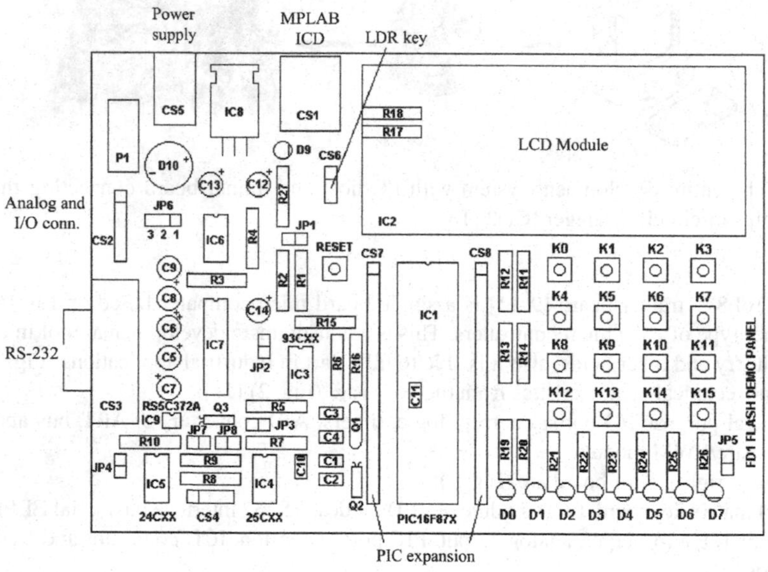

Fig. 3: PIC16F877 Trainer panel blocks

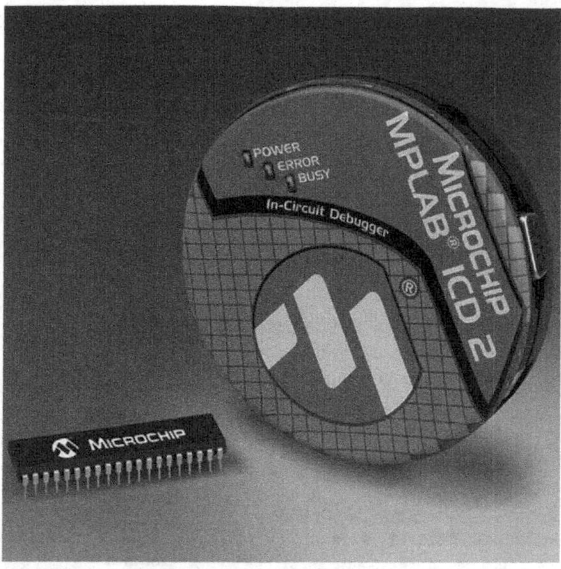

Fig. 4: Microchip ICD2

5 MPLAB - Software Environment [10]

"MPLAB Integrated Development Environment (IDE) is a free, integrated toolset for the development of embedded applications employing Microchip's PICmicro® and dsPIC® microcontrollers. MPLAB IDE runs as a 32-bit application on MS Windows®, is easy to use and includes a host of free software components for fast application development and supercharged debugging. MPLAB IDE also serves as a single, unified graphical user interface for additional Microchip and third party software and hardware development tools. Moving between tools is a snap, and upgrading from the free simulator to MPLAB ICD 2 or the MPLAB ICE emulator is done in a flash because MPLAB IDE has the same user interface for all tools."

MPLAB IDE software package is a full, operative Windows-based off-line and on-line development-system. On Fig. 5 is a classical software design cycle.

For the software design process MPLAB Project Manager gives a complete background (Fig. 6).

After compiling the source code, debugging and linking there are two ways to running the program:

Off-line and

On-line.

In off-line mode the integrated Simulator assures the software realization, in on-line mode the software is sent directly to the microcontroller (development board) and works with real CPU, memories and I/O peripheries.

Main MPLAB IDE components are [10]:

Programmer's text editor, MPLAB SIM, high speed software simulator for PICmicro and dsPIC MCUs with peripheral simulation, complex stimulus injection and register logging, full featured debugger, graphical project manager, visual device initializer (VDI), version

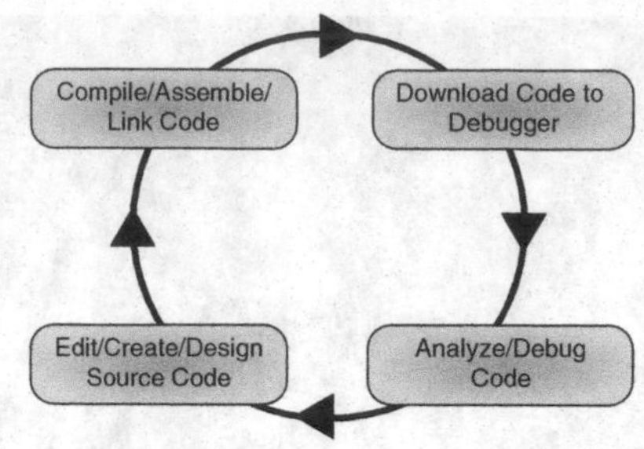

Fig. 5: The design cycle

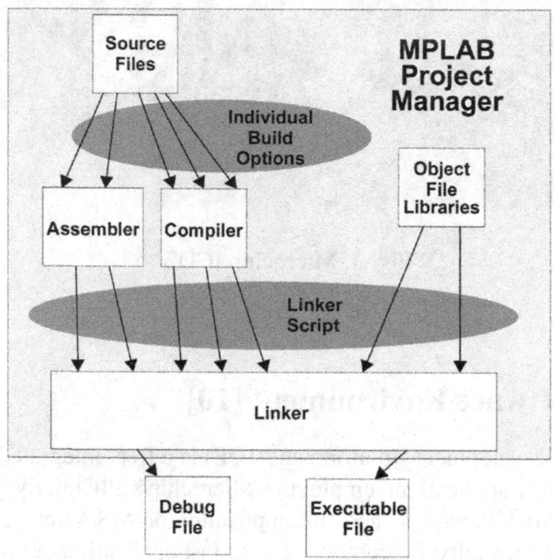

Fig. 6: MPLAB IDE Project Manager

control support for MS source safe, CVS, PVCS, subversion Application Maestro™, configurable firmware module library, MPASM™ macro assembler with MPLINK™ linker and MPLIB™ librarian MPLAB, ASM30 assembler, MPLAB LINK30 and utilities for PIC24 and dsPIC devices, PROCMD command line programmer for MPLAB PM3 and PRO MATE® II Visual PROCMD for simplified GUI control of MPLAB, PM3 and PRO MATE® II Application Maestro, dsPIC Motor Control plug-in, CMX Scheduler viewer plug-in CMX Tiny+ RTOS viewer plug-in.

6 The Microcontroller Laboratory Role

The main topics of Microcontroller Laboratory are to prepare students for the self-supporting activities in the construction and designing of robot hardware and software (and other devices with embedded systems). The program of the microcontroller course includes [4]:

Hardware design engineering and Software (program) design engineering.

Hardware topics consists:

CPU, memory and I/O hardware (digital and analog) functioning and interfacing.

Software topics consists:

instructions, mathematical operations, interfacing concepts, counter/timer, regulators (PID etc.).

The hardware and software interfacing concepts are pointing to the robotics problems:

engine control/management, communication between the robot and PC, communication between two robots, sensor interfacing (sonar, infrared measuring system etc.), navigation problems.

7 The Laboratory Structure

Educational activities are divided into two parts:

compulsory (basics of microcontroller technology) and voluntary.

The compulsory part of the course includes all microcontroller hardware and software knowledge for robotics. The voluntary part of the subject depends on the student, if he has an interdisciplinary project, so together with the compulsory knowledge and with the knowledge from the special field he makes the special system, improves his basic knowledge.

8 Advanced Course – Learning the Architecture of Mobile Robots

The goal of the elementary microcontroller course is to guide students from fundamental electrical, sensor-interfacing, programming techniques to mobile robot building.

The navigation of mobile robots [7, 8] is solved with microcontrollers and microprocessors together with numerous of sensors and actuators. These micro devices have different capacities, they handle more or less in and outgoing signals, have various memory capacities and operating speeds.

The tasks of the robots may vary significantly from one to another, they range from fairly simple tasks to the execution of quite complicated tasks. This fact determines the architecture and complexity used for the robot's navigation.

The trainer board reviewed in [11] is suitable for the building and analyzing some types of sensors and actuators.

Within a mobile robot the architecture applied may be divided in the following way:

a microcontroller for the execution of all tasks (Fig. 7), one-level architecture with several microcomputers.

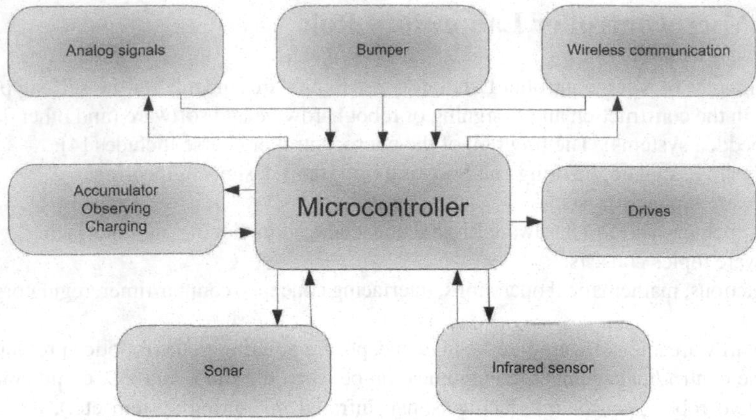

Fig. 7: Simple architecture – one microcontroller

Fig. 8: Lab-in-a-box for embedded applications [6]

9 Distance Education

Remote operation and control of the Mobile Robot Architecture Laboratory opens huge potentials in distance learning. Educational institutes independent from geographical limitations (distributed laboratories) will integrate material and knowledge potentials in virtual but very realistic form to create a whole, in the common educational space. The interactive video link connects two or more laboratories, so the instructor from onelaboratory guides all students

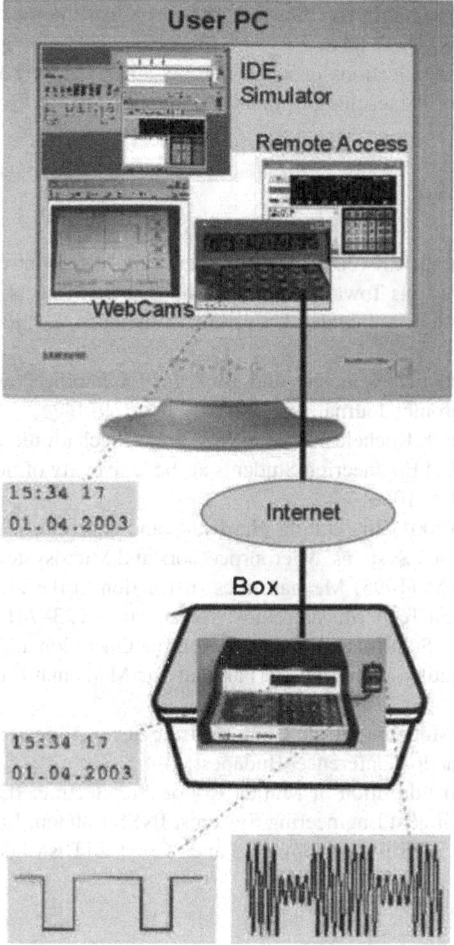

Fig. 9: Graphical user interface [6]

in the common educational space. Students from far workstations can operate through the internet with remote laboratory equipment 24 hours, 7 days a week.

Remote Microcontroller Laboratory [6] have a 'laboratory in a box' for embedded applications (Fig. 8). All integrated equipments are remotely controlled.

The whole remote laboratory is presented through the graphical user interface (Fig. 9).

10 Conclusions

This chapter describes the combined microcontroller/robotics laboratory for several courses in the teaching process of informatics students. The modern laboratory construction was created in accordance with the requirements of the Bologna process demands.

Teaching microcontrollers for robotics applications is feasible for compulsory courses as well as voluntary courses. These laboratory equipments are also appropriate for other microcontroller applications. Applications of internet tools allow building very operative remote controlled laboratories for the teaching of mobile robot architecture.

References

1. Azad AKM, Lakkaraju VK (2003) Development of a Microcontroller Laboratory Facility for Directing Students Towards Application Oriented Projects. American Society for Engineering Education, Valparaiso University, Valparaiso, Sectional Conference, April 4–5, 145–151
2. Kern A (1996) Didactical Concepts and Tools for Explaining Power Electronics Applications. Microelectronics Journal, Vol. 27, Nos 2-3, 139–147
3. Giurgiutiu V, Lyons J, Rocheleau D, Liu W (2005) Mechatronics/Microcontroller Education for Mechanical Engineering Students at the University of South Carolina. Mechatronics, Vol. 15, 1025–1036
4. Al-Dhaher AHG, (2001) Integrating Hardware and Software for the Development of Microcontroller-Based Systems. Microprocessors and Microsystems, Vol. 25, 317–328
5. Ume, Timmerman M (1995) Mechatronics Instruction in the Mechanical Engineering Curriculum at Gorgia Tech. Mechatronics, Vol. 5, No. 7, 723–741
6. Bahring H, Keller J, Schiffmann W (2006) Remote Operation and Control of Computer Engineering Laboratory Experiments. Fakultat fur Mathematik und Informatik 58084 Hagen, Germany
7. Mester Gy (2006) Modeling of the Control Strategies of Wheeled Mobile Robots. Proceedings of The Kando Conference, Budapest, 1–4
8. Matijevics I (2006) Education of Mobile Robot Architecture. IEEE 10th International Conference on Intelligent Engineering Systems, INES London, June 26–28, 180–183.
9. Szego J (2002) PIC16F877 Development Panel. ChipCad Distribution, www.chipcad.hu, Budapest
10. www.microchip.com
11. www.chipcad.hu

Force–Impedance Control of a Six-dof Parallel Manipulator

António M. Lopes and Fernando G. Almeida

IDMEC–Faculdade de Engenharia da Universidade do Porto, Rua Dr. Roberto Frias
4200-465 Porto–Portugal
aml@fe.up.pt, fga@fe.up.pt

Summary. An acceleration based force-impedance controller is presented in this paper. The proposed control strategy is applied to a six-dof parallel robotic mini-manipulator: the Robotic Controlled Impedance Device (RCID).

The control strategy involves three cascade controllers: an inner acceleration controller, built as a set of six Single-Input-Single-Output (SISO) acceleration controllers (one per manipulator axis), an impedance task-space controller, and an outer force controller. The proposed control strategy enables two kinds of manipulator behaviour: force limited impedance control and position limited force control. The type of behaviour only depends on the chosen manipulator trajectories.

The RCID may be used as a force-impedance controlled auxiliary device, coupled in series with a position controlled industrial robot, or as a stand-alone force feedback display, that may be used as a master manipulator in master-slave telemanipulated systems or as a haptic device interacting with virtual environments.

Experimental results of the RCID used as an auxiliary device working coupled to an industrial manipulator are presented.

1 Introduction

It is well known that tasks requiring a strong interaction between the manipulator and its environment necessitate some kind of force control. Polishing, grinding, and deburring industrial tasks [1, 2], medical surgery [3], telemanipulated systems with force feedback capability [4], and haptic devices interacting with virtual environments [5] are a few practical examples. In robotic industrial tasks force control can be chosen passive or active. In passive control, the end-effector of the robot is equipped with an auxiliary mechanical device, typically composed of springs and dampers. In active force control a suitable control strategy and force feedback are required. Active force control is far more flexible because controller parameters can be easily changed according to tasks needs. This paper is organized as follows. Section 2 presents the proposed acceleration based force-impedance controller for a one-dof system, assuming a perfect inner acceleration-tracking controller. Section 3 presents a short description of the RCID parallel robotic mini-manipulator. In Section 4 the acceleration based force-impedance controller outlined in Section 2 is applied to the RCID. Experimental results are presented in Section 5. Conclusions are drawn in Section 6.

J.A.T. Machado et al. (eds.), *Intelligent Engineering Systems and Computational Cybernetics,*

2 Acceleration Based Force–Impedance Control

The control objective of an impedance controller is to impose, along each direction of the task space, a dynamic relation between the manipulator end-effector position error and the force of interaction with the environment, the desired impedance [6–8]. Usually the desired impedance is chosen linear and of second order, as in a mass-spring-damper system. In order to fulfil the task requirements, the user chooses a desired end-effector impedance that may be expressed by (1):

$$\mathbf{M}_d(\ddot{\mathbf{x}} - \ddot{\mathbf{x}}_d) + \mathbf{B}_d(\dot{\mathbf{x}} - \dot{\mathbf{x}}_d) + \mathbf{K}_d(\mathbf{x} - \mathbf{x}_d) = -\mathbf{f}_e \tag{1}$$

where $\mathbf{M}_d$, $\mathbf{B}_d$ and $\mathbf{K}_d$ are constant, diagonal and positive definite matrices, representing the desired inertia, damping, and stiffness system matrices in task space. Vectors $\mathbf{x}$ and $\mathbf{x}_d$ represent the actual and the desired end-effector positions, and $\mathbf{f}_e$ represents the generalised force the environment exerts upon the end-effector. If the manipulator is able to follow an acceleration reference given by

$$\ddot{\mathbf{x}}_r = \ddot{\mathbf{x}}_d + \mathbf{M}_d^{-1} \cdot [-\mathbf{f}_e + \mathbf{B}_d(\dot{\mathbf{x}}_d - \dot{\mathbf{x}}) + \mathbf{K}_d(\mathbf{x}_d - \mathbf{x})] \tag{2}$$

it will behave as described by (1). So, $\ddot{\mathbf{x}}_r$ is the reference signal for an inner loop acceleration-tracking controller, which linearizes and decouples the manipulator nonlinear dynamics.

Figure 1 shows a block diagram of a conventional impedance controller. The transfer function matrix $\mathbf{G}_{aa}(\mathbf{s})$ represents the non-ideal behaviour of the inner loop acceleration controller, such as finite bandwidth and incomplete decoupling. At low frequency, up to a maximum meaningful value, the acceleration controller must ensure that $\mathbf{G}_{aa}(\mathbf{s})$ may be taken as the identity matrix, $\mathbf{I}$.

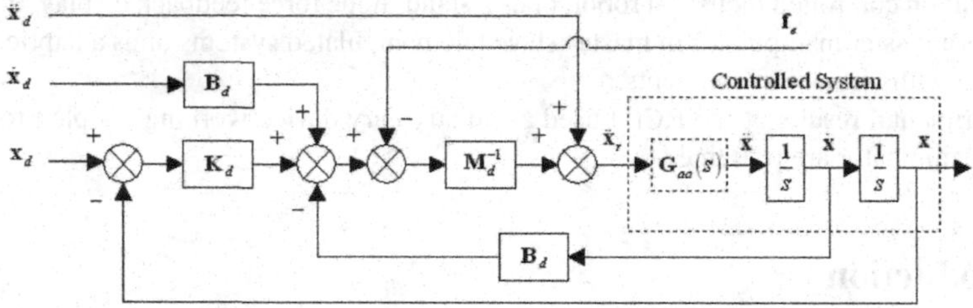

Fig. 1: Block diagram of an impedance controller

Force-impedance control combines the robustness properties of impedance control with the ability to follow position and force references of hybrid control [9–11]. Without loss of generality, the proposed acceleration based force-impedance controller will be first presented for a single-dof system, and assuming a perfect inner loop acceleration-tracking controller [12].

The proposed controller has three cascaded control loops (Fig. 3): an inner acceleration-tracking controller, an impedance controller that generates the acceleration reference, $\ddot{x}_r$, to the inner loop, and a nonlinear force controller used as an external control loop. The external force controller modifies the desired position trajectory, x_d, in order to limit the contact force to a specified maximum value [13].

For each task space direction the user must specify a, possible time variable, force reference, f_d, in addition to the desired position trajectory, x_d, and desired impedance, M_d, B_d, K_d. The force reference has the meaning of a limiting value to the force the end-effector may apply to the environment. The static force-displacement relation imposed by the controller is presented in Fig. 2 . Positive force values are obtained with a "pushing" environment and the negative values with a "pulling" one. The contact force saturation behaviour is obtained by the use of limited integrators on the force control loop.

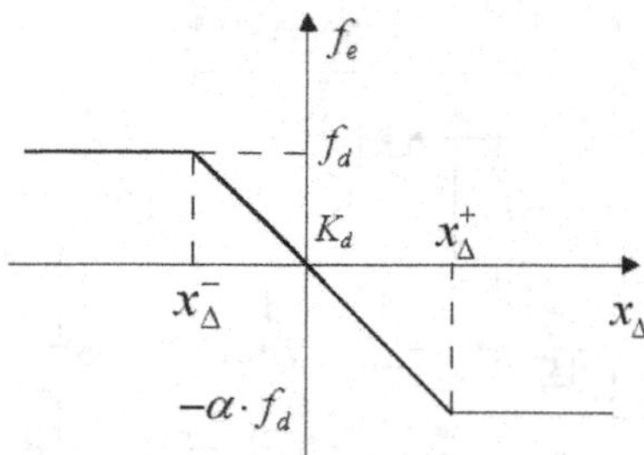

Fig. 2: Block diagram of an impedance controller

While the manipulator is not interacting with the environment the controller ensures reference position tracking. After contact is set up the controller behaviour may be interpreted in two different ways: as a force limited impedance controller or as a position limited force controller.

If contact conditions are planed in such a way that the force reference (limit) is not attained, the manipulator is impedance controlled. If environment modelling errors result in excessive force, the contact force is limited to the reference value. If contact conditions are planed in order to ensure that the force reference is attainable, the manipulator is force controlled. When environment modelling errors result in excessive displacement, manipulator position is limited to its reference value. This manipulator behaviour ensures a high degree of robustness to environment uncertainty.

Figure 3 shows the acceleration based force-impedance controlled system under a contact situation with a purely elastic environment. Its contact surface is positioned at x_e and presents a stiffness K_e. When the force control loop is in action the end-effector position, on the Laplace domain, is described by (3).

$$\begin{aligned} X(s) = {} & \frac{\left(M_d s^2 + B_d s + K_d\right) s}{s\left(M_d s^2 + B_d s + K_d\right) + K_e s + K_F K_e} X_d(s) + \\ & \frac{K_F}{s\left(M_d s^2 + B_d s + K_d\right) + K_e s + K_F K_e} F_d(s) + \\ & \frac{K_F K_e + s K_e}{s\left(M_d s^2 + B_d s + K_d\right) + K_e s + K_F K_e} X_e(s) \end{aligned} \tag{3}$$

If the Routh stability criteria is applied to (3), a maximum value of the force control loop integral gain, K_F, is obtained as

$$K_F < \frac{B_d}{M_d} < \frac{B_d\left(K_d + K_e\right)}{M_d K_e} \tag{4}$$

So, K_F may be chosen in a manner that is independent of the environment stiffness K_e, increasing the controller robustness to environment modelling uncertainty. Obviously, the non-ideal behaviour of the inner loop acceleration controller, such as finite bandwidth and incomplete decoupling, may induce instability, even if (4) is satisfied. Nevertheless, it can be shown that a careful acceleration-tracking controller design may guaranty the stability of the acceleration based force-impedance controlled device in a wide range variation of the desired impedance parameters [13].

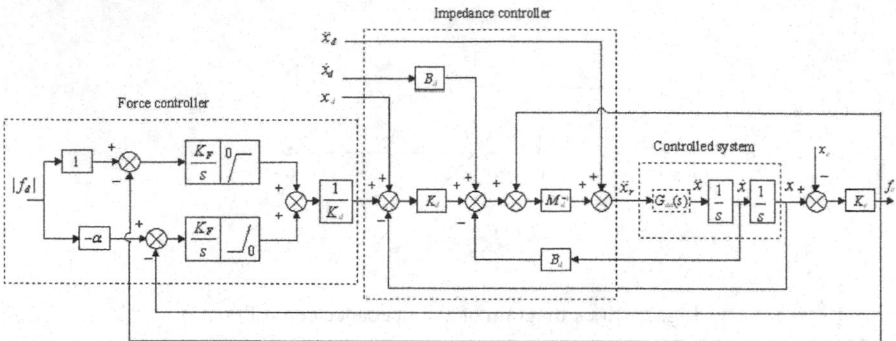

Fig. 3: Acceleration based force-impedance controller

3 The Robotic Controlled Impedance Device (RCID)

The RCID is a six-dof parallel mini-manipulator (Fig. 4). Parallel manipulators are well known for their high dynamic performances and low positioning errors [14]. In the last few years parallel manipulators have attracted great attention from researchers involved with robot manipulators, robotic end effectors, robotic devices for high-precision robotic tasks, machine-tools, simulators, and haptic devices [5].

The mechanical structure of the RCID comprises a fixed (base) platform $\{B\}$ and a moving (payload) platform $\{P\}$, linked together by six independent, identical, open kinematic chains (Fig. 4). Each chain comprises two links: the first link (linear actuator) is always normal to the base and has a variable length, l_i, with one of its ends fixed to the base and the other one attached, by a universal joint, to the second link; the second link (arm) has a fixed length, L, and is attached to the payload platform by a spherical joint. Points B_i and P_i are the connecting points to the base and payload platforms. They are located at the vertices of two semi-regular hexagons, inscribed in circumferences of radius r_b and r_p, that are coplanar with the base and payload platforms.

For kinematic modelling purposes frames $\{P\}$ and $\{B\}$ are attached to the payload and base platforms, respectively. Thus the generalized position of frame $\{P\}$ relative to frame $\{B\}$ may be represented by the vector:

$$
{}^{B}\mathbf{x}_{P}\big|_{B|E} = \begin{bmatrix} x_P \; y_P \; z_P \; \psi_P \; \theta_P \; \varphi_P \end{bmatrix}^T = \\
\begin{bmatrix} {}^{B}\mathbf{x}^T_{P(pos)}\big|_B \;\; {}^{B}\mathbf{x}^T_{P(o)}\big|_E \end{bmatrix}^T \tag{5}
$$

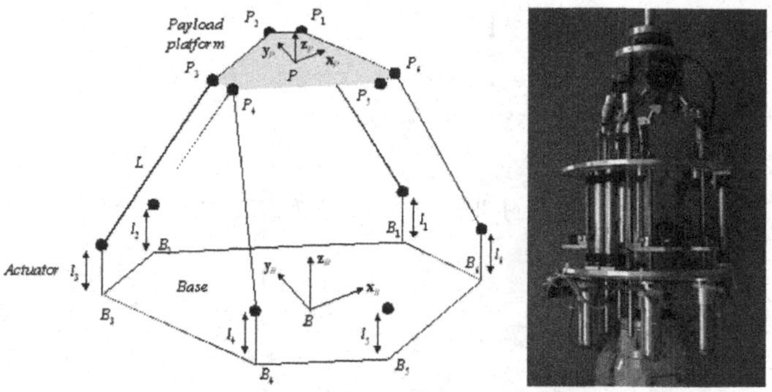

Fig. 4: Schematic view of the mechanical structure (left) and a photography of the RCID (right)

where ${}^{B}\mathbf{x}_{P(pos)}|_{B} = \left[x_P \; y_P \; z_P \right]^T$ is the position of the origin of frame $\{P\}$ relative to frame $\{B\}$, and ${}^{B}\mathbf{x}_{P(o)}|_{E} = \left[\psi_P \; \theta_P \; \varphi_P \right]^T$ defines an Euler angle system representing orientation of frame $\{P\}$ relative to $\{B\}$. Details on kinematics and dynamics of the RCID may be found in [15, 16], for example.

4 Acceleration Based Force–Impedance Control of the RCID

A task space version of the single-dof acceleration based force-impedance controller outlined in Section 2 is now used to control the RCID mini–manipulator.

As the outer force–impedance controller is designed in task space, and the inner acceleration controller of the RCID is built as a set of six SISO acceleration controllers (one per manipulator axis), the force-impedance controller must generate six acceleration references, one for each manipulator axis. Each axis acceleration controller performs acceleration tracking. Figure 5 shows the generation of these acceleration references, $\ddot{\mathbf{l}}_r$.

The RCID payload platform generalized position error $\Delta^{B}\mathbf{x}_{P}|_{B|E}$ is calculated as the difference between the desired position vector ${}^{B}\mathbf{x}_{P}|_{B|E(d)}$ and the actual (measured) position vector, ${}^{B}\mathbf{x}_{P}|_{B|E(m)}$:

$$\Delta^{B}\mathbf{x}_{P}|_{B|E} = \left[\Delta^{B}\mathbf{x}^{T}_{P(pos)}|_{B} \; \Delta^{B}\mathbf{x}^{T}_{P(o)}|_{E} \right]^T = \left[\left({}^{B}\mathbf{x}_{P(pos)}|_{B(d)} - {}^{B}\mathbf{x}_{P(pos)}|_{B(m)} \right)^T \left({}^{B}\mathbf{x}_{P(o)}|_{E(d)} - {}^{B}\mathbf{x}_{P(o)}|_{E(m)} \right)^T \right]^T \tag{6}$$

Vector $\Delta^{B}\mathbf{x}_{P(pos)}|_{B}$ is the position error expressed in Cartesian coordinates. It can be seen as the position of the origin of a frame $\{D\}$ (which represents the desired generalized position) relative to frame $\{P\}$ (witch represents the payload platform measured generalized position). Vector $\Delta^{B}\mathbf{x}_{P(o)}|_{E}$ is the orientation error expressed in Euler angles.

Block $\aleph$, in Fig. 5, is an "operator" that must do all the necessary computations in order to express the generalized position error vector in frame $\{P\}$, $\Delta^{B}\mathbf{x}_{P}|_{P}$.

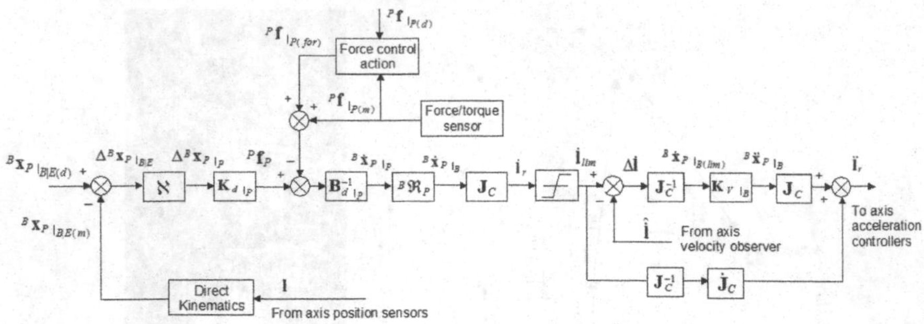

Fig. 5: Acceleration references for the six SISO acceleration controllers (one per manipulator axis)

First, the position error, $\Delta^B \mathbf{x}_{P(pos)}|_B$, is expressed in frame $\{P\}$. This is straightforward since

$$\Delta^B \mathbf{x}_{P(pos)}|_P = {}^B\mathbf{R}_P^T \cdot \Delta^B \mathbf{x}_{P(pos)}|_B \tag{7}$$

where ${}^B\mathbf{R}_P$ is the orientation matrix of frame $\{P\}$ relative to $\{B\}$.

Second, the orientation error is also expressed in frame $\{P\}$. This is done using the vector/rotation angle approach, explained as follows.

If we consider vectors ${}^B\mathbf{x}_P|_{B|E(d)}$ and ${}^B\mathbf{x}_P|_{B|E(m)}$ as

$${}^B\mathbf{x}_P|_{B|E(d)} = \begin{bmatrix} x_d \; y_d \; z_d \; \psi_d \; \theta_d \; \varphi_d \end{bmatrix}^T \tag{8}$$

$${}^B\mathbf{x}_P|_{B|E(m)} = \begin{bmatrix} x_m \; y_m \; z_m \; \psi_m \; \theta_m \; \varphi_m \end{bmatrix}^T \tag{9}$$

where orientation is expressed in Euler angles, then the rotation matrices ${}^B\mathbf{R}_D$ and ${}^B\mathbf{R}_P$ may be easily computed, and ${}^P\mathbf{R}_D = {}^B\mathbf{R}_P^T \cdot {}^B\mathbf{R}_D$ is the rotation matrix of frame $\{D\}$ relative to frame $\{P\}$. This means that frame $\{D\}$, initially parallel to $\{P\}$, rotates an angle θ_{PD} around an axis $\mathbf{r}_{PD}$. As $\mathbf{r}_{PD}$ is expressed in frame $\{P\}$, $\theta_{PD} \cdot \mathbf{r}_{PD}$ allows the computation of the rotation angles of $\{D\}$ around each axis of frame $\{P\}$, supposing all the three rotations are simultaneous.

It can be shown that [17]

$$\sin \theta_{PD} = \pm \frac{1}{2} \sqrt{({}^P r_{D_{32}} - {}^P r_{D_{23}})^2 + ({}^P r_{D_{13}} - {}^P r_{D_{31}})^2 + ({}^P r_{D_{21}} - {}^P r_{D_{12}})^2} \tag{10}$$

$$\mathbf{r}_{PD} = \frac{1}{2 \sin \theta_{PD}} \begin{bmatrix} {}^P r_{D_{32}} - {}^P r_{D_{23}} \; {}^P r_{D_{13}} - {}^P r_{D_{31}} \; {}^P r_{D_{21}} - {}^P r_{D_{12}} \end{bmatrix}^T \tag{11}$$

where ${}^P r_{D_{ij}}, i,j = 1,...,3$, are the elements of matrix ${}^P\mathbf{R}_D$. A good choice of the controller sampling rate usually leads to $|\theta_{PD}|$ very small, and therefore

$${}^P\mathbf{R}_D \approx \begin{bmatrix} 1 & -\delta z & \delta y \\ \delta z & 1 & -\delta x \\ -\delta y & \delta x & 1 \end{bmatrix} \tag{12}$$

where δx, δy, δz are the rotation angles of $\{D\}$ around the x, y and z axis of $\{D\}$. Thus we may write:

$$\Delta^{B}\mathbf{x}_{P(o)}\,|_{P} = \begin{cases} \frac{1}{2}\cdot\left[{}^{P}r_{D_{32}} - {}^{P}r_{D_{23}}\ {}^{P}r_{D_{13}} - {}^{P}r_{D_{31}}\ {}^{P}r_{D_{21}} - {}^{P}r_{D_{12}}\right]^{T}, & if\ \theta_{PD} \le \theta_{th} \\ \theta_{PD}\cdot\mathbf{r}_{PD} &, \ if\ \theta_{PD} > \theta_{th} \end{cases} \tag{13}$$

where θ_{th} is a threshold angle.

Now, the product of the 6 x 6 desired stiffness matrix and the generalized position error expressed in frame $\{P\}$, $\mathbf{K}_{d}|_{P}\cdot\Delta^{B}\mathbf{x}_{P}|_{P}$ gives the generalized force applied and expressed in frame $\{P\}$. This signal is added to the force control action, ${}^{P}\mathbf{f}\,|_{P(for)}$, and the feedback interaction force, ${}^{P}\mathbf{f}\,|_{P(m)}$ (see Fig. 5).

Reference Cartesian velocity is calculated using the desired damping matrix:

$${}^{B}\dot{\mathbf{x}}_{P}\,|_{P} = \mathbf{B}_{d}^{-1}|_{P}\cdot\left({}^{P}\mathbf{f}\,|_{P} - {}^{P}\mathbf{f}\,|_{P(for)} - {}^{P}\mathbf{f}\,|_{P(m)}\right) \tag{14}$$

$${}^{B}\dot{\mathbf{x}}_{P}\,|_{B} = {}^{B}\Re_{P}\cdot{}^{B}\dot{\mathbf{x}}_{P}\,|_{P} \tag{15}$$

with

$${}^{B}\Re_{P} = \begin{bmatrix} {}^{B}\mathbf{R}_{P} & 0 \\ 0 & {}^{B}\mathbf{R}_{P} \end{bmatrix} \tag{16}$$

At this point, the axis velocity references should be computed, and limited to the maximum allowable actuators speed, v_{max}:

$$\dot{\mathbf{l}}_{r} = \mathbf{J}_{C}\cdot{}^{B}\dot{\mathbf{x}}_{P}\,|_{B} \tag{17}$$

$$\dot{\mathbf{l}}_{lim} = minval\left\{\left[factor_{1}\ \cdots\ factor_{6}\right]^{T}\right\}\cdot\dot{\mathbf{l}}_{r} \tag{18}$$

$$factor_{i} = minval\left\{1,\ \left|\frac{v_{max}}{\dot{l}_{r_{i}}}\right|\right\} \tag{19}$$

$\mathbf{J}_{C}$ represents the RCID kinematic jacobian matrix.

The joint velocity error may be computed as $\Delta\dot{\mathbf{l}} = \dot{\mathbf{l}}_{lim} - \hat{\dot{\mathbf{l}}}$, where $\hat{\dot{\mathbf{l}}}$ is the joint estimated velocity; axis position and acceleration are measured and velocity is estimated (a linear velocity observer is used). This error must be expressed in Cartesian space to generate the acceleration reference (see Fig. 5):

$${}^{B}\ddot{\mathbf{x}}_{P}\,|_{B} = \mathbf{K}_{V}\,|_{B}\cdot\Delta^{B}\dot{\mathbf{x}}_{P}\,|_{B(lim)} \tag{20}$$

$$\mathbf{K}_{V}\,|_{B} = {}^{B}\Re_{P}\cdot\mathbf{B}_{d}\,|_{P}\cdot\mathbf{M}_{d}^{-1}|_{P}\cdot{}^{B}\Re_{P}^{T} \tag{21}$$

where $\mathbf{M}_{d}|_{P}$ is the desired inertia matrix.

Finally we have the axis acceleration reference

$$\ddot{\mathbf{l}}_{r} = \mathbf{J}_{C}\cdot{}^{B}\ddot{\mathbf{x}}_{P}\,|_{B} + \dot{\mathbf{J}}_{C}\cdot\mathbf{J}_{C}^{-1}\cdot\dot{\mathbf{l}}_{lim} \tag{22}$$

The term $\dot{\mathbf{J}}_{C}\cdot\mathbf{J}_{C}^{-1}\cdot\dot{\mathbf{l}}_{lim}$ is hard to compute. However, it can be shown that this component is negligible when added to $\mathbf{J}_{C}\cdot{}^{B}\ddot{\mathbf{x}}_{P}\,|_{B}$, and needs not to be considered, reducing computer burden.

5 Experimental Results

The RCID is powered by six DC rotary motors. A ball-screw based transmission converts motor rotation to actuator vertical translation. Linear position and acceleration of each actuator are measured, as well as Cartesian forces and torques applied to the payload platform. The controller is implemented on a PC. It is coded in C language and runs under the HyperKernel™real-time system, on a Windows environment. The inner loop acceleration controller uses a 1 kHz sampling frequency and achieves a closed loop bandwidth of 400 rad/s. The force-impedance controller also runs with a 1 kHz sampling frequency and includes a velocity observer having measured acceleration and position as input variables.

Experimental results were obtained for the RCID working coupled in series to an industrial manipulator: the latter one performs the large amplitude movements while the RCID is only used for the small and high bandwidth movements needed for force-impedance control (Fig. 6). This strategy improves the performance of current commercial industrial robots increasing its flexibility when executing tasks that require some kind of force control.

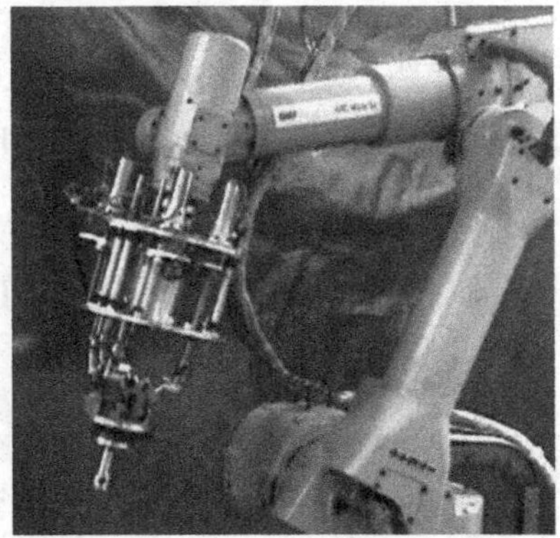

Fig. 6: Serial combination of the RCID and an industrial manipulator

The first experiment is a contour following operation (Fig. 7(a)). Contour following exerting appropriate contact forces may be necessary in industrial tasks like polishing, grinding, and deburring. Current industrial manipulators use some kind of passive force control (passive compliance) while executing this class of tasks. Therefore, special purpose mechanical devices like springs and dampers are attached between the manipulator flange and the robot tool, and/or passive devices are used to hold the work piece, in order to achieve the necessary compliance and keep the contact forces small enough.

This experience shows that a contour following task exerting a desired contact force against the work piece may be performed using the RCID as an active auxiliary force-impedance controlled device.

The manipulator was programmed to follow the surface of an object, at a speed of 10 mm/s, with a small contact force. The object was displaced afterwards and no corrections were done to the program. If this program is rerun a large contact force does result, as shown in Fig. 7(b). However the RCID may be effectively used to limit this force to some desired value (8N on the experiment).

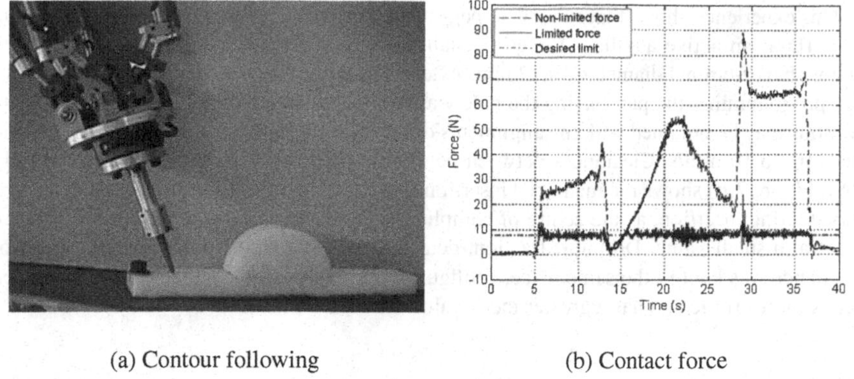

(a) Contour following

(b) Contact force

Fig. 7: Contour following experimental task

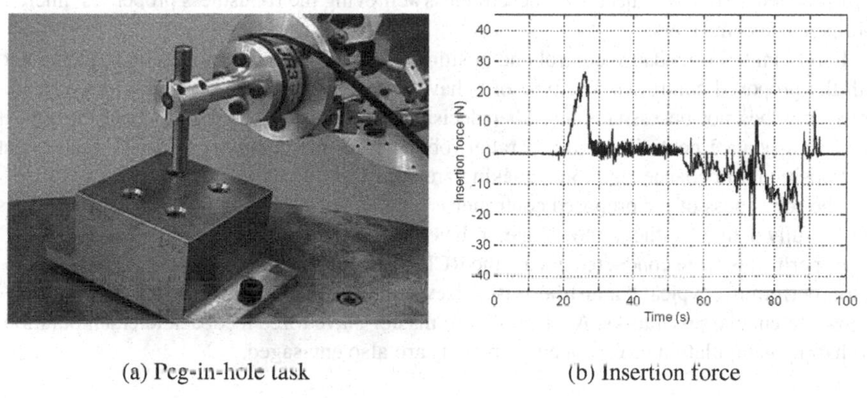

(a) Peg-in-hole task

(b) Insertion force

Fig. 8: Peg-in-hole task experimental task

The second experiment is a peg-in-hole task as shown in Fig. 8(a). The peg-in-hole task may be used as a benchmark for assembly like industrial tasks. The peg-in-hole robotic task is hard to execute because small angular and/or axial misalignments between peg and hole may lead to excessive insertion forces and, consequently, insertion failure. To solve this problem a variety of approaches have been used that are generally based on adding some passive compliance between the robot carrying the peg and the part having the hole. Typically, assembly robotic tasks are executed by special purpose robots, namely the SCARA robots, or industrial robots equipped with special purpose auxiliary passive devices such as the Remote Centre of Compliance (RCC) devices. The RCC devices correct for misalignment errors encountered when parts are mated during assembly operations, but using RCC like devices has some limitations. For example, the insertion of different length pegs necessitates different RCC like devices, thus bringing no flexibility to the assembly system. On the other hand, the RCID device may be used as an active RCC device, the user having the flexibility to specify different impedance matrices (and imposing the centre of compliance) according to tasks needs.

This experience shows that successful peg-in-hole insertion tasks may be performed using the RCID as an active auxiliary force-impedance controlled device. In this experiment peg and hole have nominal diameters of 12 mm. After careful alignment between peg and hole for robot programming, the part having the hole was displaced. When the same program is rerun insertions may be obtained with misalignments of up to 2 mm (hole chamfer size) and 3.5^{o}and speeds up to 10 mm/s. Clearances between peg and hole down to 20 μm were successfully tested. Figure 7(b) shows the measured insertion force during a insertion-extraction cycle at 1 mm/s. During insertion, as the centre of compliance is placed at the peg end, insertion force is kept at a small value. During extraction, detected by force sign reversal, as the centre of compliance was kept in the same place, configuring an unfavourable situation, the extraction force is more erratic and has a greater mean value than the insertion one.

6 Conclusions

The acceleration based force-impedance controller presented in this paper enables the follow up of position and force trajectories, nevertheless achieving the robustness properties inherent to impedance controllers.

Force limited impedance control and position limited force control may be implemented with the proposed controller. The type of behaviour is only dependent on the chosen trajectories and does not necessitate any other decision mechanism. The experimental results of a force-impedance controlled 6-dof parallel robotic mini-manipulator, the RCID, show that good position, impedance, and force tracking are obtained.

The robustness of the proposed controller is also enhanced by the fact that its parameters may be fully defined without knowledge of the environment stiffness.

Experiments were conducted having the RCID coupled in series with an industrial manipulator, performing typical "industrial" tasks. Nevertheless, the RCID has potential applicability in different classes of tasks. Applicability to master-slave force feedback telemanipulation, and haptic manipulation of virtual environments are also envisaged.

References

1. Kazerooni H, Sheridan T B, Houpt P K (1986) Robust compliant motion for manipulators: part I-II. IEEE Journal of Robotics and Automation 2:83–105
2. De Schutter J, Van Brussel H (1988) Compliant robot motion: part I-II. The International Journal of Robotics Research 7:3–33
3. Ho S C, Hibberd R D, Cobb J, Davies B L (1995) Force Control for Robotic Surgery. Proceedings of the International Conference on Advanced Robotics. Sant Feliu de Guixols, spain.
4. Abbott J J, Okamura A M (2006) Stable forbidden-region virtual fixtures for bilateral telemanipulation. ASME Journal of Dynamic Systems, Measurement, and Control 128:53–64
5. Constantinescu D, Salcudean S E, Croft E A (2005) Haptic rendering of rigid contacts using impulsive and penalty forces. IEEE Transactions on Robotics 21:309–323
6. Hogan N (1985) Impedance control: an approach to manipulation: part I-III. ASME Journal of Dynamic Systems, Measurement, and Control 107:1–24

7. Khatib O (1987) A unified approach for motion and force control of robot manipulators: the operational space formulation. IEEE Journal of Robotics and Automation 3:43–53
8. De Schutter J, Bruyninckx H, Zhu W, Spong M (1997) Force control: a bird's eye view. In: Siciliano B, Valavanis K P (eds) Control Problems in Robotics and Automation. Springer, London
9. Anderson R J, Spong M W Hybrid impedance control of robotic manipulators. IEEE Journal of Robotics and Automation 4:549–555
10. Chiaverini S, Sciavicco L (1993) The parallel approach to force/position control of robotic manipulators. IEEE Transactions on Robotics and Automation 9:361–373
11. Seraji H, Colbaugh R (1997) Force tracking in impedance control. The International Journal of Robotics Research 16:97–117
12. Almeida F, Lopes A, Abreu P (1999) Force-impedance control: a new control strategy of robotic manipulators. In Kaynak O, Tosunoglu S, Marcelo Ang Jr (eds) Recent Advances in Mechatronics. Springer, Singapore
13. Lopes A, Almeida F (2006) Acceleration Based Force-Impedance Control. Proceedings of the 25th IASTED International Conference on Modelling, Identification, and Control (MIC 2006), Lanzarote, Spain
14. Chablat D, Wenger P, Majou F, Merlet J-P (2004) An interval analysis based study for the design and the comparison of three-degrees-of-freedom parallel kinematic machines. The International Journal of Robotics Research 615–624
15. Merlet J-P, Gosselin C (1991) Nouvelle Architecture pour un Manipulateur Parallèle à Six Degrés de Liberté. Mechanism and Machine Theory 26:77–90
16. Mouly N (1993) Développement d'une Famille de Robots Parallèles à Motorisation Électrique. Thèse de Doctorat, École des Mines de Paris, Sophia, France
17. Paul R P (1981) Robot Manipulators: Mathematics, Programming. MIT Press, Cambridge, MA

Robotic Manipulators with Vibrations: Short Time Fourier Transform of Fractional Spectra

Miguel F. M. Lima[1], J. A. Tenreiro Machado[2], and Manuel Crisóstomo[3]

[1] Dept. of Electrotechnical Engineering, Superior School of Technology, Polytechnic Institute of Viseu, 3504-510 Viseu, Portugal
lima@mail.estv.ipv.pt

[2] Dept. of Electrotechnical Engineering, Institute of Engineering, Polytechnic Institute of Porto, 4200-072 Porto, Portugal
jtm@isep.ipp.pt

[3] Instituto de Sistemas e Robótica, Department of Electrical and Computer Engineering, University of Coimbra, Polo II, 3030-290 Coimbra, Portugal
mcris@isr.uc.pt

Summary. This paper analyzes the signals captured during impacts and vibrations of a mechanical manipulator. In order to acquire and study the signals an experimental setup is implemented. The signals are treated through signal processing tools such as the fast Fourier transform and the short time Fourier transform. The results show that the Fourier spectrum of several signals presents a non integer behavior. The experimental study provides valuable results that can assist in the design of a control system to deal with the unwanted effects of vibrations.

1 Introduction

In practice the robotic manipulators present some degree of unwanted vibrations. In fact, the advent of lightweight arm manipulators, mainly in the aerospace industry, where weight is an important issue, leads to the problem of intense vibrations. On the other hand, robots interacting with the environment often generate impacts that propagate through the mechanical structure and produce also vibrations.

Motivated by the problem of vibrations, this paper studies the robotic signals captured during an impact phase of the manipulator. The study is done in a fractional calculus (FC) perspective. In order to analyze the phenomena involved an acquisition system was developed. The manipulator motion produces vibrations, either from the structural modes or from end-effector impacts. The instrumentation system acquires signals from multiple sensors that capture the axis positions, mass accelerations, forces and moments and electrical currents in the motors. Afterwards, an analysis package, running off-line, reads the data recorded by the acquisition system and examines them. Bearing these ideas in mind, this paper is organized as follows. Section 2 addresses the motivation for this work. Section 3 describes briefly the

J.A.T. Machado et al. (eds.), *Intelligent Engineering Systems and Computational Cybernetics,*

robotic system enhanced with the instrumentation setup. Section 4 presents the experimental results. Finally, Section 5 draws the main conclusions and points out future work.

2 Motivation

Reference [15] mentions several techniques for reducing vibrations and its implementation either at the robot manufacturing stage or at the operational stage. Briefly, the techniques can be enumerate as: (i) conventional compensation, (ii) structural damping or passive vibration absorption, (iii) control based on the direct measurement of the absolute position of the gripper, (iv) control schemes using the direct measurement of the modal response, (v) active control, driving out energy of the vibration modes, (vi) use a micromanipulator at the endpoint of the larger manipulator and (vii) adjustment of the manipulator command inputs so that vibrations are eliminated.

The work presented here is a step towards the implementation of the sixth technique. In recent years the use of micro/macro robotic manipulators has been proposed for space applications and nuclear waste cleanup. Several authors have studied this technique [16], namely [4, 11] that adopted a command filtering approach in order to position the micromanipulator. Also, [4, 6] used inertial damping techniques taking advantage of a micro manipulator located at the end of a flexible link.

The experiments described in this paper use a macro manipulator, with a low bandwidth, that is compensated through a much faster micromanipulator inserted at the robot endpoint. In this perspective, in order to eliminate or reduce the effect of the vibration is fundamental to study variables, for implementing an adequate control of the macro/micro system.

Bearing these ideas in mind, is presented a study of the robotic signals, in a FC perspective. In fact, the study of feedback fractional order systems has been receiving considerable attention [9, 10] due to the facts that many physical systems are well characterized by fractional-order models [14]. With the success in the synthesis of real noninteger differentiator and the emergence of new electrical circuit element called "fractance" [3], and fractional-order controllers [2], have been designed and applied to control a variety of dynamical processes [13]. Therefore the study presented here can assist in the design of the control system to be used.

3 Experimental Platform

The developed experimental platform has two main parts: the hardware and the software components [7]. The hardware architecture is shown in Fig. 1. Essentially it is made up of a robot manipulator, a Personal Computer (PC) and an interface electronic system. The interface box is inserted between the robot arm and the robot controller, in order to acquire the internal robot signals; nevertheless, the interface captures also external signals, such as those arising from accelerometers and force/torque sensors, and controls the external micro-arm. The modules are made up of electronic cards specifically designed for this work. The function of the modules is to adapt the signals and to isolate galvanically the robot's electronic equipment from the rest of the hardware required by the experiments.

The software package runs in a Pentium 4, 3.0 GHz PC and, from the user's point of view, consists of two applications:

- The acquisition application is a real time program responsible for acquiring and recording the robot signals.

- The analysis package runs off-line and handles the recorded data. This program allows several signal processing algorithms such as, Fourier transform (FT), correlation, time synchronization, etc.

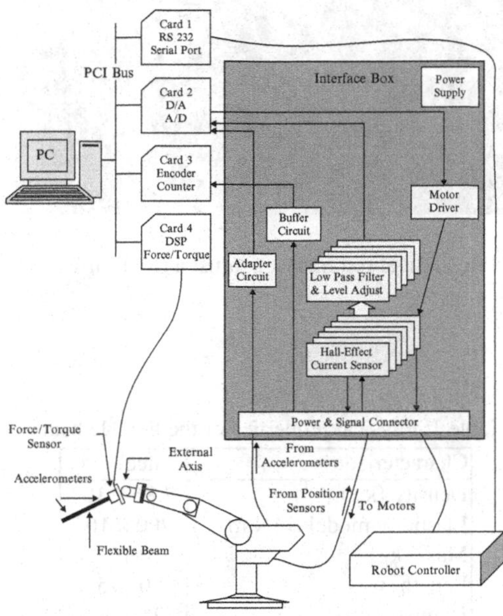

Fig. 1: Block diagram of the hardware architecture

4 Experimental Results

In the experiment is used a steel rod flexible link. To test impacts, the link consists of a long, thin, round, flexible steel rod clamped to the end-effector of the manipulator. The robot motion is programmed in a way such that the rod moves against a rigid surface. Figure 2 depicts the robot with the flexible link and the impact surface. The physical properties of the flexible beam are shown in Table 1.

During the motion of the manipulator the clamped rod is moved by the robot against a rigid surface. An impact occurs and several signals are recorded with a sampling frequency of $f_s = 500$ Hz. The signals come from different sensors, such as accelerometers, force and torque sensor, position encoders and current sensors.

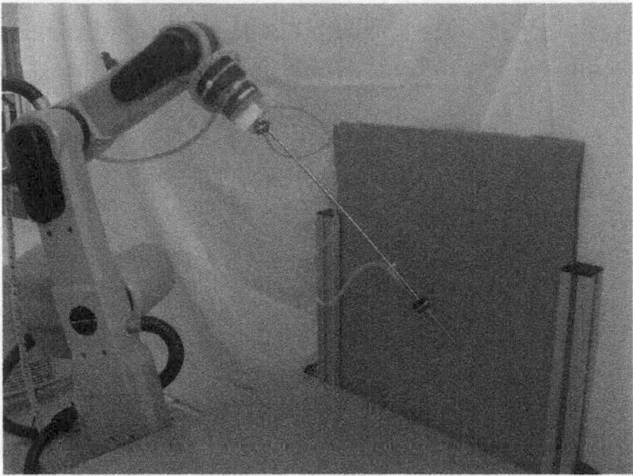

Fig. 2: Steel rod impact against a rigid surface.

Table 1: Physical properties of the flexible beam

Characteristic	Steel rod
Density (kg m^{-3})	7.86×10^3
Elasticity modulus (N m^{-2})	200×10^9
Mass (kg)	0.107
Length (m)	0.475
Diameter (m)	5.75×10^{-3}

4.1 Time Domain

A typical time evolution of some variables is shown in Figs. 3–4 corresponding to: (*i*) the impact of the rod on a rigid surface and (*ii*) without impact [8].

In this example, the signals present clearly a strong variation at the instant of the impact that occurs, approximately, at $t = 4$ sec. Consequently, the effect of the impact force, shown in Fig. 4 (left), is reflected in the current required by the robot motors (see Fig. 3, left). Figure 4 (right) shows the accelerations at the rod free-end (accelerometer 1), where the impact occurs, and at the rod clamped-end (accelerometer 2). The amplitudes of the accelerometers signals are higher near the rod impact side. The two signals are superimposed in Fig. 4 (right). The first acceleration peak (accelerometer 1), due to the impact, corresponds to the rigid surface (case *i*) while the second peak corresponds to the care of no impact (case *ii*).

4.2 Fourier Transform

In order to study the behavior of the signal FT, a trendline can be superimposed over the spectrum. The trendline is based on a power law approximation:

$$|\mathscr{F}\{f(t)\}| \approx c\omega^m \tag{1}$$

where $\mathscr{F}$ is the signal FT, $c \in \Re$ is a constant that depends on the amplitude, ω is the frequency and $m \in \Re$ is the slope.

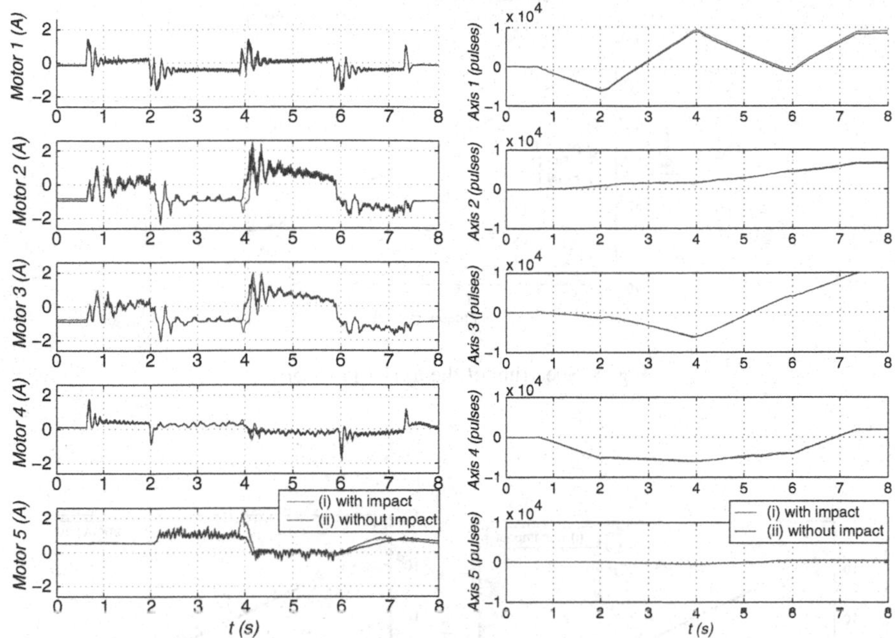

Fig. 3: Electrical currents of robot axis motors (left) and robot axis positions (right)

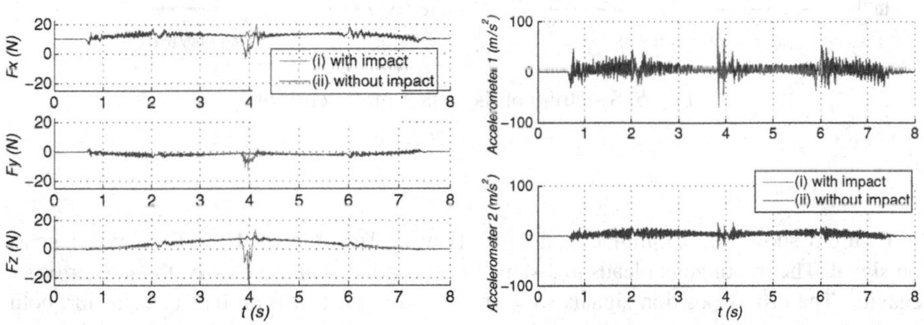

Fig. 4: Forces at the gripper sensor (left) and rod accelerations (right)

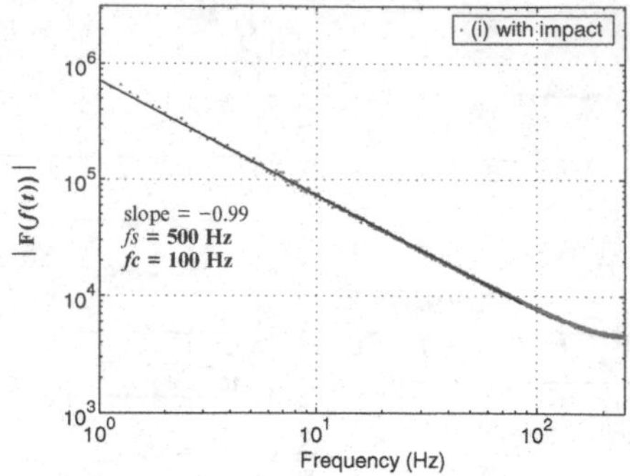

Fig. 5: Spectrum of the axis 1 position

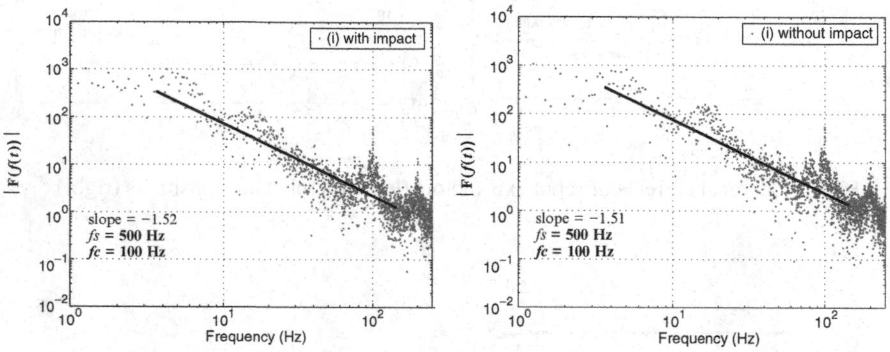

Fig. 6: Spectrum of the axis 3 motor current

Figure 5 shows the amplitude of the Fast Fourier Transform (FFT) of the axis 1 position signal. The trendline (1) leads to a slope $m = -0.99$ revealing, clearly, the integer order behavior. The others position signals were studied, revealing also an integer behavior, both under impact and no impact conditions. Figure 6 shows the amplitude of the FFT of the electrical current for the axis 3 motor. The spectrum was also approximated by trendlines in a frequency range larger than one decade. These trendlines (Fig. 6) have slopes of $m = -1.52$ and $m = -1.51$ under impact (i) and without impact (ii) conditions, respectively. The lines present a fractional order behavior in both cases. Figure 7 depicts the amplitude of the FFT of the electrical current for the axis 4 motor. Here the trendlines present slopes that vary slightly (slopes $m = -1.58$ with impact and $m = -1.64$ without impact) but, in both cases, continue to reveal a fractional order behavior. The others axis motor currents were studied, as well. Some of them, for a limited frequency range, present also fractional order behavior while others have a complicated spectrum difficult to approximate by one trendline. According to the robot manufacturer specifications the loop control of therobot has a cycle time of $t_c = 10$ mSec. This

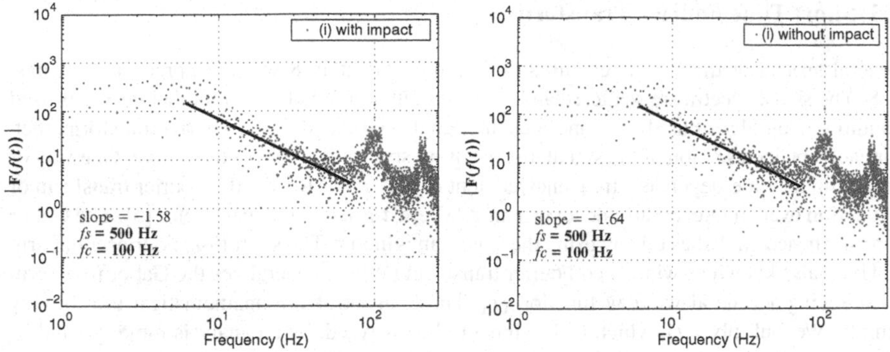

Fig. 7: Spectrum of the axis 4 motor current

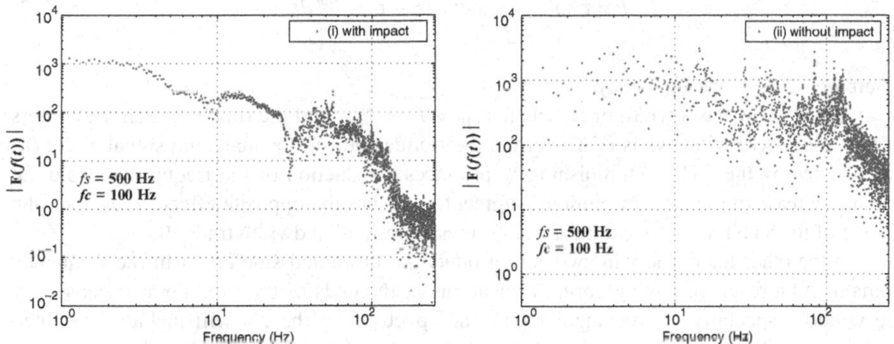

Fig. 8: F_z force spectrum with impact (left) and acceleration spectrum of the rod free-end without impact (right)

fact is observed approximately at the fundamental ($f_c = 100$ Hz) and multiple harmonics in all spectra of motor currents (Figs. 6 and 7). Figure 8 (left) shows, as example, the spectrum of the F_z force. This spectrum is not so well defined in a large frequency range. All force/moments spectra present identical behavior and, therefore, it is difficult to define accurately the behavior of the signals. Finally, Fig. 8 (right) depicts the spectrum of the signal captured from the accelerometer 1 located at the rod free-end of the beam.

Like the spectrum from the other accelerometer, this spectrum is spread and complicated. Therefore, it is difficult to define accurately the slope of the signal and consequently its behavior in terms of integer or fractional system.

4.3 Short Time Fourier Transform

Several spectra of the signals captured during approximately 8 sec are represented in Figs. 5–8. The signal spectra are scattered and, therefore, in order to obtain smoother curves is used a multiwindow algorithm. If we time slice the signals and calculate the Fourier transform, then for each obtained section of the signal, the resulting spectrum is a smoother curve. One way of obtaining the time-dependent frequency content of a signal is to take the Fourier transform of a function over an interval around an instant τ, where τ is a variable parameter [5]. The Gabor Transform accomplishes this by using the Gaussian window. The short time Fourier transform (STFT), also known as windowed Fourier transform (WFT), generalizes the Gabor transform by allowing a general window function [1]. The concept of this mathematical tool is very simple. We multiply $x(t)$, which is the signal to be analyzed, by an analysis moving window $x(t)g(t-\tau)$, and then we compute the Fourier transform of the windowed signal $x(t)g(t-\tau)$. Each FT gives a frequency domain 'slice' associated with the time value at the window centre. Actually, windowing the signal improves local spectral estimates [12]. Considering the window function centered at time τ, the STFT is represented analytically by:

$$F_w(\tau, \omega) = \int_{-\infty}^{+\infty} x(t)w(t-\tau)e^{-j\omega t}\,dt \tag{2}$$

where $\omega = 2\pi f$ is the frequency.

Each window has a width of t_w, where $t_{min} < t_w < t_{max}$ and the time between the centers of two successive windows is δ. Therefore, the frequencies of the analyzing signal $f < 1/t_w$ are rejected by the STFT. Diminishing t_w produces a reduction of the frequency resolution and an increase in the time resolution. Augmenting t_w has the opposite effect. Therefore, the choice of the STFT window entails a well-known duration-bandwidth tradeoff.

On the other hand, the window can introduce an unwanted side effect in the frequency domain. As a result of having abrupt truncations at the ends of the time domain caused by the window, specially the rectangular one, the spectrum of the FT will include unwanted "side lobes". This gives rise to an oscillatory behavior in the FT results called the Gibbs Phenomenon [12]. In order to reduce this unwanted effect, generally is used a weighting window function that attenuate signals at their discontinuities. For that reason there are several others popular windows normally adopted in the STFT, namely, Hanning, Hamming, Gaussian, Blackman, etc.

In this line of thought, several experiments were developed with different windows shapes. Due to space limitations we are only depicting the more relevant features. In Fig. 9 are shown the spectra versus time of the axis 4 motor current, using a STFT with rectangular, Hamming, Gaussian and Hanning window, respectively. The STFT with the rectangular window presents clearly the Gibbs Phenomenon. The tests show that the STFT with the Hamming window present also this phenomenon, though to a smaller extend. The STFTs with the Hanning, Blackman and Gaussian windows present the best behavior, with slightly differences. Therefore, in the experiments described we adopt the rectangular, Hamming, Gaussian and Hanning windows. The adequate time parameters of the window, namely the width (t_w) and the interspacing time (δ) for the STFT were determined by trial and error. Figure 10 depicts the amplitude of the FFT of the electrical current for the axis 4 motor under impact condition using a Gaussian window for $\tau = 2$ sec. In this example the trendline presents a slope of approximately $m = -1.89$. According with the position in time of the moving window, the slope of the trendline will varies revealing different spectral component along the acquisition time.

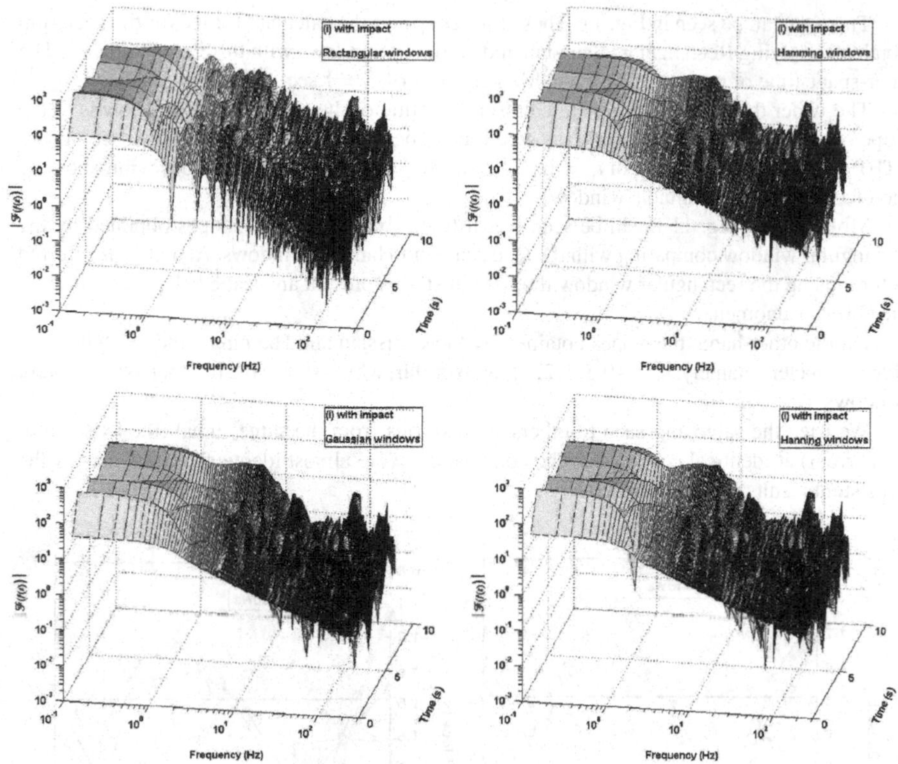

Fig. 9: STFT of axis 4 motor current using the window {rectangular, Hamming, Gaussian, Hanning}, for $\delta = 1$ (sec) and $t_w = 1$ (sec)

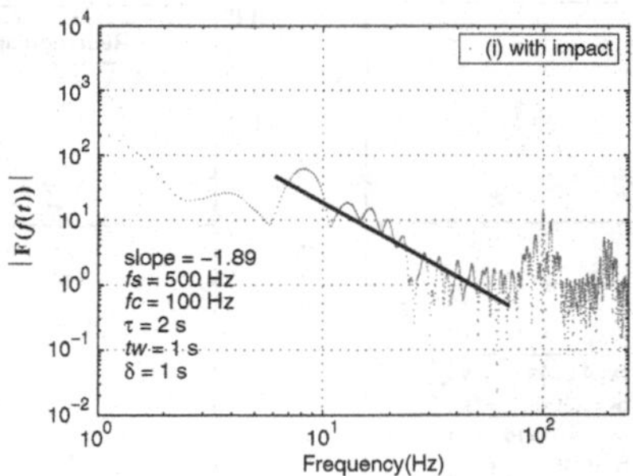

Fig. 10: Spectrum of the axis 4 motor current using a Gaussian window

This fact can be seen in Fig. 11. These figures shows the calculated slope for the spectrum obtained applying Rectangular, Gaussian and Hanning windows with different widths t_w. The inter-space time of two successive windows consists of $\delta = 1$ sec.

The wider the window of time, the closer the value of the slope of each window is to the slope calculated by the classical FT, as would be expected. From Fig. 11 it can be seen that a STFT using a window length of $t_w = 0.25$ sec presents a kind of an erratic behavior, showing therefore that is an unsuitable window.

Moreover, in Fig. 11 it can be seen the different behavior of the slopes obtained by the rectangular window comparing with the Gaussian and Hanning windows. Actually, as referred before, using the rectangular window the several slopes present an erratic behavior caused by the Gibbs phenomena.

On the other hand, the slopes obtained by the Gaussian and Hanning windows with suitable parameters, namely, $t_w = \{0.5, 1, 2, 4\}$, are similar, which shows the best behavior of these windows.

We tried the same approach to others acquisitions from the same signal (the axis 4 motor current) at identical conditions. The conclusions were almost identical, which reveals the consistent results presented here.

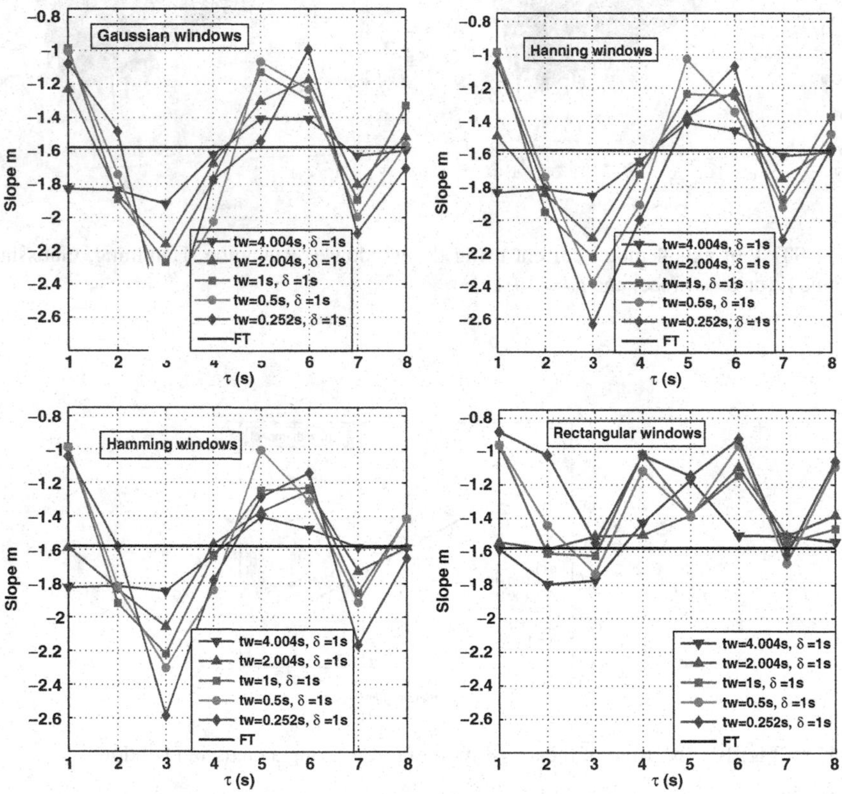

Fig. 11: Slopes of the spectrum of the axis 4 motor current under impact condition using {Gaussian, Hanning, Hamming, rectangular} windows with $t_w = \{0.25, 0.5, 1, 2, 4\}$ (sec) and $\delta = 1$ (sec)

5 Conclusions

In this paper an experimental study of several robot signals was presented. The study was done in a FC perspective and the use of the STFT confirms the fractional nature of the signals whose behavior was analyzed by the classical FT also. This study provides useful information that can assist in the design of a control system to be used in eliminating or reducing the effect of vibrations.

The next stage of development of the software and hardware apparatus is to reduce the vibrations and its effect upon the robot structure. In this line of thought, is under development a micromanipulator, with a higher frequency response than the main manipulator, mounted at the end-effector and actively counter-acting the undesirable dynamics.

References

1. Allen R L, Mills D W (2004) Signal Analysis. Wiley-Interscience, New York
2. Barbosa R S, Tenreiro Machado J A, Ferreira I M (2005) Tuning of pid controllers based on Bode's ideal transfer function. In: Nonlinear Dynamics, pp 305–321. Kluwer, Netherlands.
3. Bohannan G W (2002) Analog realization of a fractional control element – revisited. In: mechatronics.ece.usu.edu/foc/cdc02tw/cdrom/aditional/FOC_Proposal_Bohannan.pdf
4. Cannon D W, Magee D P, Book W J, Lew J Y (1996) Experimental study on micro/macro manipulator vibration control. In: Proceedings of the IEEE International Conference on Robotics & Automation. Minneapolis, Minnesota, USA.
5. Gabor D (1946) Theory of communication. J. IEE, 93:429–457, 1946.
6. Lew J Y, Trudnowski D J, Evans M S, Bennett D W (1995) Micro-manipulator motion control to suppress macro-manipulator structural vibrations. In: Proceedings of the IEEE International Conference on Robotics & Automation, Nagoya, Japan, pp 3116–3120.
7. Lima M F M, Tenreiro Machado J A, Crisóstomo M (2005) Experimental set-up for vibration and impact analysis in robotics. WSEAS Transactions on Systems 4(5):569–576
8. Lima M F M, Tenreiro Machado J A, Crisóstomo M (2006) Fractional order Fourier spectra in robotic manipulators with vibrations. In: Second IFAC Workshop on Fractional Differentiation and Its Applications, Porto, Portugal, pp 386–391
9. Tenreiro Machado J A (1997) Analysis and design of fractional-order digital control systems. Journal of Systems Analysis-Modelling-Simulation 27:107–122
10. Tenreiro Machado J A (2003) A probabilistic interpretation of the fractional-order differentiation. Journal of Fractional Calculus & Applied Analysis 6:73–80
11. Magee D P, Book W J (1995) Filtering micro-manipulator wrist commands to prevent flexible base motion. In: Proceedings of American Control Conference
12. Oppenheim A V, Schafer R W, Buck J R (1989) Discrete-Time Signal Processing. Prentice-Hall, upper Saddle River, NJ
13. Oustaloup A, Moreau X, Nouillant M (1997) From fractal robustness to non-integer approach in vibration insulation: the crone suspension. In: Proceedings of the 36th Conference on Decision & Control. San Diego, California, USA.
14. Podlubny I (2002) Geometrical and physical interpretation of fractional integration and fractional differentiation. Journal of Fractional Calculus & Applied Analysis 5(4): 357–366

15. Singer N C, Seering W P (1988) Using acausal shaping techniques to reduce robot vibration. In: Proceedings of the IEEE Internaltional Conference on Robotics & Automation. Philadelphia PA, USA.
16. Yoshikawa T, Hosoda K, Doi T, Murakami H (1993) Quasi-static trajectory tracking control of flexible manipulator by macro-micro manipulator system. In: Proceedings of the IEEE International Conference on Robotics & Automation, Atlanta, GA, USA, pp 210–215

Part II

Artificial Intelligence

Classifying Membrane Proteins in the Proteome by Using Artificial Neural Networks Based on the Preferential Parameters of Amino Acids

Subrata K. Bose[1], Antony Browne[2], Hassan Kazemian[3], and Kenneth White[4]

[1] MRC Clinical Sciences Centre, Faculty of Medicine, Imperial College London, UK
subrata.bose@ic.ac.uk

[2] Department of Computing, School of Engineering and Physical Sciences, University of Surrey, Guildford, UK
a.browne@eim.surrey.ac.uk

[3] Intelligent Systems Research Centre, Department of Computing, Communication Technology and Mathematics, London Metropolitan University, London, UK
h.kazemian@londonmet.ac.uk

[4] Institute for Health Research and Policy, Departments of Health and Human Sciences, London Metropolitan University, London, UK
kenneth.white@londonmet.ac.uk

Summary. Membrane proteins (MPs) are large set of biological macromolecules that play a fundamental role in physiology and pathophysiology for survival. From a pharma-economical perspective, though it is the fact that MPs constitute ∼75% of possible targets for novel drugs but MPs are one of the most understudied groups of proteins in biochemical research. This is mainly because of the technical difficulties of obtaining structural information about transmembrane regions (these are small sequences that crossways the bilayer lipid membrane). It is quite useful to predict the location of transmembrane segments down the sequence, since these are the elementary structural building blocks defining their topology. There have been several attempts over the last 20 years to develop tools for predicting membrane-spanning regions but current tools are far away from achieving a considerable reliability in prediction. This study aims to exploit the knowledge and current understanding in the field of artificial neural networks (ANNs) in particular data representation through the development of a system to identify and predict membrane-spanning regions by analysing primary amino acids sequence. In this paper we present a novel neural network (NNs) architecture and algorithms for predicting membrane spanning regions from primary amino acids sequences by using their preference parameters.

1 Introduction

Proteins are fundamental macromolecules of life that constitute about half of a living cell's dry weight [1]. They are polymers of one or more unclefted chains of amino acids ranging typically from 200–300 to few thousands. The focus of this research is one key property of

J.A.T. Machado et al. (eds.), *Intelligent Engineering Systems and Computational Cybernetics,*

proteins, whether or not they are what biochemists call integral membrane proteins. From a biochemical point of view proteins can be classified as being either soluble, such as the proteins found in blood or in the liquid compartments of cells (cytosol), or bound to cell membranes.

Many of the proteins, which coat the surface of cells, actually have part of the protein anchored in the cell membrane and protrude into the interior of the cell. The part of the protein, which makes contact with the cell membrane, is called the transmembrane domain or the membrane spanning region. Long before the advent of genomics, it was realized that the parts of proteins which have to contact membranes tend to be made of fat-loving (hydrophobic) amino acids since the membranes of cells are largely made of fat. It is now well established that a majority of transmembrane domains share a common structure of α-helical arrangement of 21–26 hydrophobic amino acids. Transmembrane proteins(TM) are one of the most understudied groups of proteins in biochemical research because of the technical difficulties of obtaining structural information about transmembrane regions.

Three-dimensional structures of proteins derived by X-ray crystallography have been determined for about 15,000 proteins, but only about 80 of these are transmembrane proteins (despite the fact that TM proteins may account for about 30% of the proteome). While it is of use to be able to identify membrane spanning regions from simple analysis of sequence, there is biochemical evidence to suggest that there are functionally distinct sub-groups of membrane spanning regions. Hence it would be of great value to develop reliable ways of identifying sub-types of membrane spanning regions. In particular, many TM proteins act as receptors for hormones. Adrenaline, insulin and others bind to these receptor proteins, and the act of binding sends a signal inside the cell to trigger a biological response. Thus, these TM proteins on the cell surface provide an interface between the cell and its environment.

In this paper, we will describe the construction of a novel neural network (NNs) architecture, the application of the algorithm to new datasets, and the analysis of the results using both statistical techniques and theoretical analysis to determine their significance and robustness of the developed model.

2 System and Methods

Artificial Neural Networks were encouraged by neuro-biological theory involving to the behavior of the brain as a network of units called neurons. An artificial Neural Networks can be defined as a parallel, distributed information-processing component or physical cellular systems that functions in a manner that resembles some operation of a biological neuron and can acquire, store and utilize experiential knowledge [2]. In other word NNs are a collection of several elements that can process information in parallel (and in connection) is a network of artificial neurons.

ANNs could be used as a very useful tool for biological data mining and classification, for their adaptive nature [3–5]. Due to the architecture of the neural networks – learning from examples makes this technique interesting for problem like secondary protein structure prediction because of the lack of experience but availability of the training data. Though several applications of ANNs are previously used in membrane protein prediction but most produced as output a classification for the amino acid in the centre of the input window. To improve the accuracy of those ANNs, majority of the designer has put effort on network's algorithms, but the elementary concern, that is, amino acid encoding and representations for input into the NNs have not been thoroughly studied yet.

2.1 Selection of Neural Technique

The first concern in designing the prediction model was the type of neural technique to be adopted. Here in this model we have chosen feedforward multilayer perceptrons (MLPs). MLPs differs from the Single Layer Perceptron by having an additional layer of neurons between the input and output layer, known as the hidden layer, which vastly increases the learning power of the MLP. By setting sufficient number of hidden units and accurate weights and biases, from a compact region of input space it can approximate to arbitrary accuracy any continuous function. The multi-layered feed-forward neural network used here consists of an input layer, a single hidden layer and an output layer and training was done by the back-propagation algorithm (Fig. 1).

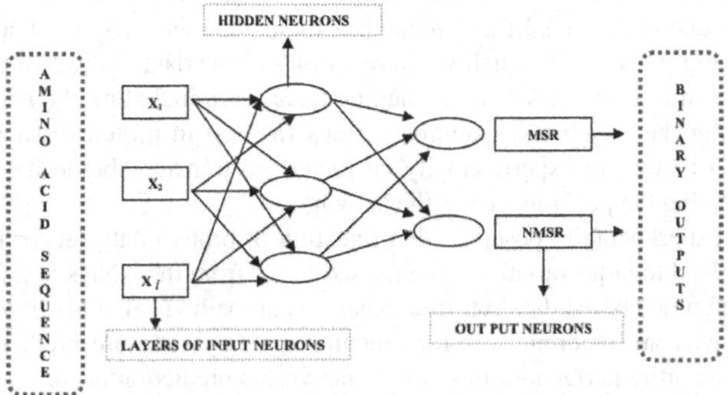

Fig. 1: Neural Network Architecture Used, the input is propensity values of amino acids and output is binary – 0 for nonmembrane spanning region (NMSR) and 1 for membrane spanning regions (MSR)

Feed forward nets used here have a distinctive layered architecture where neurons or nodes are connected to one or more other nodes by real-valued weights but not with same layers node. Here, the first layer takes input signal creates linear combinations of these inputs. The reason of this combination is to present a set of intermediate activation variables and one variable is connected with one hidden unit. Hidden layer vastly increases the learning power of the MLPs. Finally, these values were sent through the output-unit activation function to give put values.

2.2 Training and Testing

In simple term training proceeded in the following way. First, the weights and biases in the network are initialized, to small random values. A training pattern is then applied to the input units and the activation of neurons in the hidden layer is calculated. The outputs produced by these neurons via the transfer function are then fed on to neurons in the following layer. This forward pass process is repeated at each layer until an output signal from neurons in the output layer is obtained. The difference between the actual and desired output values is measured, and the network model connection strengths are changed so that the outputs produced by the network become closer to the desired outputs.

The NN model was trained by the combination of changing the learning rate within a range, along with changing the number of hidden units and a random starting weight initialization. During the training the validation dataset was used to check the progress of training after regular intervals. Once the validation accuracy rate decreased the training stopped and the best result obtained is saved. Initially we designed the experiment with a threshold value of 0.5 to get the most suited experimental parameters for the particular algorithm. Without changing the parameter of the best NN model, we have changed the threshold values ranging from 0.1 to 0.9.

2.3 Data Used

The lack of desired performance of the membrane spanning regions prediction methods from their primary amino acid sequences and some time overrating the accuracy of their methods could be mainly due to the lack of non-redundant membrane datasets which are used in the learning and tuning process. The quality and reliability of the data are two important factors before the selection of data set for any modelling techniques including the neural networks. Perhaps predicting the membrane spanning regions (MSRs) in proteins using ANNs techniques required trustworthy experimentally characterized transmembrane data for training, validating and testing the performance of the network.

It has been experimentally observed that selection of protein data set for test and training purpose can lead to large variation, so unbiased result from the ANNs largely depends on membrane and nonmembrane data set. In another recent methodical study, it was found that larger data sets play an important role for protein secondary structure prediction with neural network, noteworthy perfection, in terms of network's prediction accuracy was achieved (62.6–67.0%) for larger data sets as compared with performance on smaller databases. Fortunately, in recent years there has been a tremendous augmentation of protein sequences data, which is set down in specific well curated databases as a comprehensive source of information on proteins. One of the most comprehensive central repositories of protein sequence knowledgebase is UniProtKB/Swiss-Prot and it (Release 48.8 of 10-Jan-06) contains 205,780 sequence entries, comprising 74,898,419 amino acids abstracted from 138,273 references. In this knowledgebase, 8191 entries are found with a search containing 'TRANSMEM' keyword.

There are limited reports about the transmembrane topology prediction methodologies based on reliable datasets and all most all of the methodologies used a small number of datasets. So selection of a reliable set of examples amino acid sequences from Swiss-Prot database that construct the membrane protein was really a difficult task. To overcome this limitation, an effort to use larger well-characterized reliable datasets for both training a testing has been made. We used two protein datasets with known biochemical characterizations of membrane topology which is assembled by Möller and TMPDB Release 6.2 [6–9].

Möller [6] considered C-terminal fusion, antibody binding and x-ray diffraction methods as benchmarking experiment criteria for a source for transmembrane annotation. Möller's initial data set was based on SWISS-PROT release 39, but the current SWISSPROT database (Release 48.8) grown significantly. Personal communication has been made to get an updated membrane protein dataset and generously Apweiler's group given us the updated data set.

In TMPDB curators used X-ray crystallography, NMR, gene fusion technique, substituted cysteine accessibility method and Asp (N)-linked glycosylation method as an experimental confirmation for assortment of the data set.

From these two sets of data, we have identified the common protein by searching the unique ID name and those are excluded from our experimental database. In this experiment we have used 707 MPs.

To train the ANNs with negative sequence, the remaining non-transmembrane sequences of the Möller's and TMPDB [6–9] dataset have not been used because the left over sequences are not inevitably non-membrane protein and in future, some sequences on the same entry might be annotated as transmembrane protein. DB-NTMR dataset [10] have been chosen which is freely available at http://biophysics.biol.uoa.gr/DB-NTMR.

2.4 Distribution of the Data

From the final data set about 75% of the data was randomly taken for training the network, and remaining 25% of the data were selected as the test data. To detect when the over-fitting starts and to prevent the over-fitting problem, the training set was further partitioned into two disjoint subsets: training and validation dataset. The size of the validation data was about 25% of the training data.

2.5 Amino Acid Encoding

A predicament of using ANNs to analyse primary amino acid sequence is that most advanced ANNs are incapable of recognising non-numerical character strings. So conversion of the amino acid sequences into real number in order to get numerical input vectors is an important critical step of constructing novel neural network architecture. In this study we have used a propensity for each amino acid [11]. Propensity can be described as a numerical value to calculate the possibility that a given amino acid to be in a transmembrane region.

3 Results

The algorithm produces a confusion matrix (two-by-two) which is used to critically analyse the performances of all the algorithms. Here in this study three parameters – classification rate (%), true positive ratio or hit rate (TPr), false alarm rate or false positive ratio (FPr) are used for comparison of the algorithms (Table 1). We have started with minimum number of hidden layer and increased it gradually to investigate the effect of no of hidden layer to the model and dataset.

Initially we designed the experiment with a threshold value of 0.5 to get the best experimental parameters for the algorithm. The best network give performance at a classification rate of 82.6541 (Fig 2(a) and 2(b)) with 58 hidden layers and 0.5 decision threshold (Table 1 and Fig. 3).

We analysed the effect of decision threshold boundaries on the NNs performances (Fig. 3) it is clear that with the lowest decision threshold boundary (0.1) the network's performance is poor and the performance increases at every increment of the threshold boundary at a rate of 0.1 and it continues till 0.5 and after any further increase it starts to decrease.

The NN performance was least with the highest decision boundary threshold. So it can be suggested that for designing ANNs to solve biological data mining problem, the decision threshold plays an important role.

Without changing any parameter of this NN model, we have changed the threshold values ranging from 0.1 to 0.9 to calculate the receiver operating characteristic (ROC) curve to asses the robustness of the algorithms.

Table 1: Results generated by the model. The classification Rate (CR) of best NN architecture with different Decision Boundary Threshold (DBR)

DBR	Confusion matrix				False alarm rate	Hit rate	CR (%)
	TP	FN	FP	TN			
0.1	371	629	22	892	0.024070	0.371	65.9875
0.2	542	458	45	869	0.049234	0.542	73.7200
0.3	673	327	80	834	0.087527	0.673	78.7356
0.4	766	234	109	805	0.119256	0.766	82.0794
0.5	836	164	168	746	0.183807	0.836	82.6541
0.6	894	106	249	665	0.272429	0.894	81.4525
0.7	937	63	325	589	0.355580	0.937	79.7283
0.8	967	33	434	480	0.474836	0.967	75.6008
0.9	985	15	583	331	0.637856	0.985	68.7565

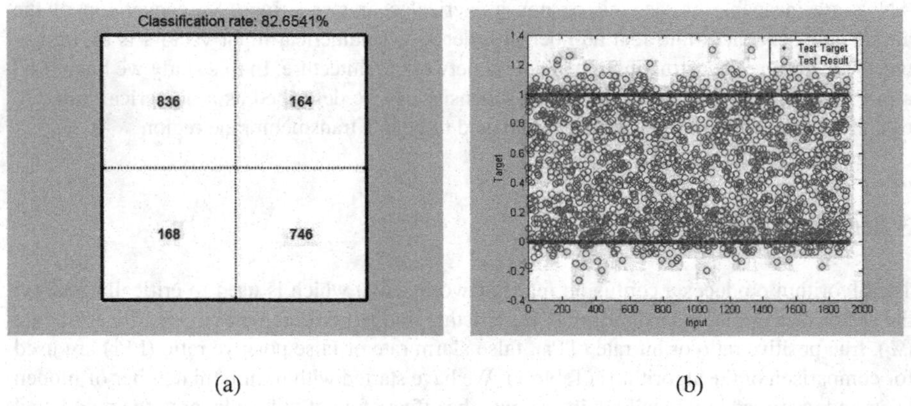

Fig. 2: (a) Two-by-two confusion matrix (also called a contingency table) and (b) is Test target vs Test results produced by best network

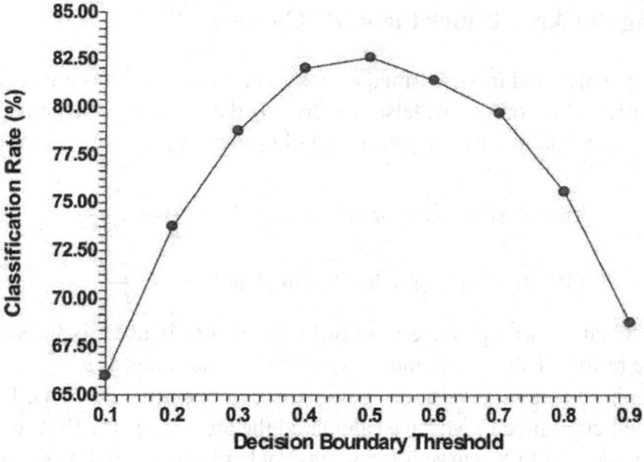

Fig. 3: Effects of decision boundary thresholds on classification performance

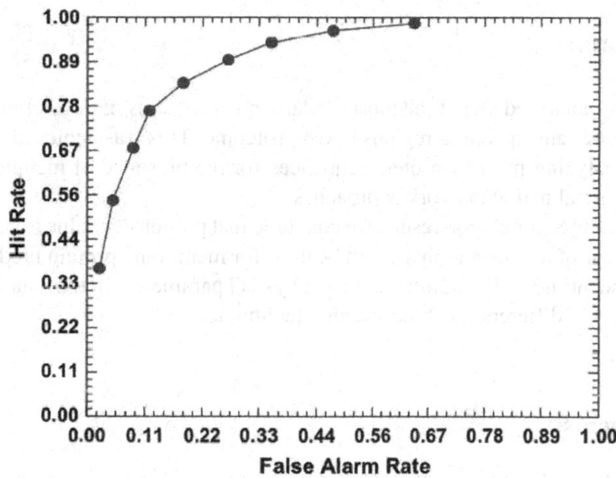

Fig. 4: Area under the ROC curve is 0.5347062, where maximum hit rate is 0.985 and FA rate at max of hit rate is 0.6378556, minimum and maximum FA rate is 2.407002E-02 and 0.6378556 respectively

3.1 Calculating the Area Under the ROC Curve

If the instance is positive and it is classified as positive, it is counted as a true positive; if it is classified as negative, it is counted as false negative. If the instance is negative and classified as negative it is, it is counted as true negative; if it classified as positive, it is counted as a false positive.

$$\text{True Positive ratio or Hit Rate } (HR) = \frac{TP}{TP+FN}$$

$$\text{False Positive ratio or False Alarm Rate } (FA) = \frac{TN}{TN+FP}$$

To plot a ROC curve, true positive ratio or hit rate (HR) is used as the vertical axis (Y) and false positive ratio or False alarm rate (FΛ) the horizontal one (Y).

An ROC graph depicts relative trade – offs between benefits and costs. For quantitative analysis of the prediction model, we have calculated the area under the ROC curve using nine points. The area under the ROC curve is 0.5347062, where maximum hit rate is 0.985 and FA rate at max of hit rate is 0.6378556.

4 Conclusions

In this research we analysed one of the most fundamental problems of developing ANNs tools for predicting membrane-spanning regions in the proteome. This was achieved by developing techniques for analyzing primary protein sequences for the presence of membrane spanning regions using artificial neural network approaches.

From the computer simulation results we conclude that propensity value is a good physico-chemical parameter of amino acid that could be used for membrane protein prediction. Future research will concentrate on including more bio-physical parameters in the data representation and applying them in different machine learning techniques.

Acknowledgments

The authors thank London Metropolitan University, for the financial support to this work.

References

1. Parris N, Onwulata C (1995) Food proteins and interactions. In: Meyers RA (ed) Molecular Biology and Biotechnology: A comprehensive Desk Reference. VCH Publishers, Cambridge, pp. 320–323.
2. Zurada, JM (1992) Introduction to Artificial Neural Systems. PWS Publishing Company, Boston, MA.
3. Browne A, Hudson BD, Whitley DC, Ford MG, Picton P (2004) Biological data mining with neural networks: Implementation and application of a flexible decision tree extraction algorithm to genomic problem domains. Neurocomputing, 57, 275–293. Elsevier - ISSN: 0925-2312.

4. Bose S, Browne A, Kazemian H, White K (2003) Knowledge Discovery in Bioinformatics using Neural Networks. In: 6th International Conference on Computer and Information Technology (ICCIT) 2003, Dhaka, Bangladesh. ISBN: 984-584-005-1.
5. Bose S, Kazemian H, White, K, Browne, A (1995) Use of neural networks to predict and analyse membrane proteins in the proteome. BMC Bioinformatics 6 (Suppl 3): P3.
6. Möller S, Kriventseva E, Apweiler R (2000) A collection of well characterized integral membrane proteins. Bioinformatics, 16, 1159–1160.
7. Shimizu T, Nakai K (1994) Construction of a membrane protein database and an evaluation of several prediction methods of transmembrane segments. In: Miyano S, Akutsu T, Imai H, Gotoh O and Takagi T (eds), Proceedings of Genome Informatics Workshop, Universal Academy Press, Tokyo, pp. 148–149.
8. Kihara D, Shimizu T, Kanehisa M (1998) Prediction of membrane proteins based on classification of transmembrane segments. Protein Eng., 11, 961–970.
9. Ikeda M, Arai M, Lao DM, Shimizu T (2002) Transmembrane topology prediction methods: A re-assessment and improvement by a consensus method using a dataset of experimentally-characterized transmembrane topologies. In Silico Biol., 2, 19–33.
10. Pasquier C, Hamodrakas SJ (1999) An hierarchical artificial neural network system for the classification of transmembrane proteins. Protein Eng., 12(8), 631–634.
11. Pasquier C, Promponas VJ, Palaios GA, Hamodrakas JS, Hamodrakas SJ (1999) A novel method for predicting transmembrane segments in proteins based on a statistical analysis of the SwissProt database: The PRED-TMR algorithm. Protein Eng., 12(5), 381–385.

8. Hirokawa T, Boon-Chieng S, Mitaku S (1998) SOSUI: classification and secondary structure prediction system for membrane proteins. Bioinformatics 14:378–379

Multi-Channel Complex Non-linear Microstatistic Filters: Structure and Design

Dušan Kocur, Jozef Krajňák, and Stanislav Marchevský

Department of Electronics and Multimedia Communications, Technical University of Košice, Park Komenského 13, 041 20 Košice, Slovak Republic
dusan.kocur@tuke.sk, jozi.krajnak@tuke.sk, stanislav.marchevsky@tuke.sk

Summary. In this article, the structure and design procedure of a new time-invariant complex-valued multi-channel non-linear microstatistic filter (C-M-CMF) will be proposed. The C-M-CMFs belong to a group of minimum mean-square estimators based on the estimation of desired signals by using a linear combination of vector elements obtained by a threshold decomposition of the input signals of the filter. The C-M-CMF represents a modification of multi-channel real-valued conventional microstatistic filter (M-CMF) originally developed in [4, 10, 11] for the purpose of design of a non-linear multi-user receiver for code division multiple access transmission systems (CDMA). The C-M-CMF theory developed in this paper can be applied for the design of new complex-valued receivers for multi-carrier code division multiple access transmission systems (MC-CDMA).

1 Introduction

In the last decades, the great attention in the field of digital filters has been devoted to development of non-linear filters. The motivation for that work has been the intention to overcome the limitation of linear filters in system modelling and signal estimation. It is well known, that the most real-world problems are intrinsically non-linear and they can be modelled as linear ones only within limited range of values to be processed. On the other hand, it follows from the estimation theory that the optimum estimators of non-Gaussian signals are non-linear estimators. Therefore, there are many practical applications where linear estimators cannot provide acceptable quality of signal processing, and currently, where non-linear estimators could provide possible solutions. Here, Volterra filters, Hammerstein filters, non-linear Wiener filters, non-linear Kalman filters, stack filters, order statistic filters, estimators based on neural networks and conventional piece-wise linear filters can be given as the examples of non-linear estimators [3].

Single-channel conventional microstatistic filters (CMF) originally introduced in [5] represents another promising class of non-linear estimators. They belong to a group of minimum mean-square estimators based on the estimation of a desired signal by using a linear combination of vector elements obtained by a threshold decomposition of the input signals of the filter. It can be shown (e.g. [9, 12]) that the CMFs represent a sub-set of piece-wise linear filters. In [6, 7], a new class of microstatistic filters referred to as Volterra microstatistic filters has been introduced. The output of Volterra microstatistic filters is given by a non-linear

J.A.T. Machado et al. (eds.), *Intelligent Engineering Systems and Computational Cybernetics,*

combination of vector elements obtained by a threshold decomposition of the input signals of the filter. In this case, the non-linear combination of threshold decomposer outputs is done by well-known Volterra filters [3, 6, 7].

The first application of CMFs has been proposed in [4, 10, 11]. Here, the M-CMFs have been used with success in the structure of non-linear single-stage multi-user receiver for CDMA. It is expected, that the multi-user receivers for MC-CDMA could be another promising application of microstatistic filters. The expected success of this application can be emphasized by the fact that multi-carrier transmission systems based on orthogonal frequency division multiplex are highly sensitive to non-linear amplification, which requires large back off in the transmitter amplifier and, as a consequence, inefficient use of power amplifiers [2]. Here, multi-user receivers based on microstatistic filters could be applied for non-linear distortion effect suppressing what can result in increasing efficiency of transmitter amplifiers and decreasing bit error rate of transmission systems.

MC-CDMA multi-user receivers are based on complex-valued signal processing. Because the C-M-CMFs have not been yet proposed, a basic theory of C-M-CMFs will be introduced in this article, where the main emphasis will be put on the C-M-CMF structure and design procedure.

2 C-M-CMF Structure

A block scheme of the C-M-CMF is given in the Fig. 1. Here, M, $y^{(i)}(n)$ and $\hat{d}^{(k)}(n)$ are the number of the input and output channels, the i-th input complex-valued signal and the k-th output complex-valued signal of the C-M-CMF, respectively. It can be seen from this figure that the C-M-CMF consists of M complex-valued decomposers (TD) and the set of complex-valued multi-channel Wiener filter (WF).

Because the input signals to the TD are complex-valued, a simple real-valued threshold decomposer (R-TD) applied in the M-CMF [5] cannot be used for their decomposition directly. In order to develop a suitable TD let us assume, that the complex-valued signal $y^{(i)}(n)$ is expressed in the form

$$y^{(i)}(n) = Y^{(i)}(n)e^{j\phi(i,n)} \tag{1}$$

where

$$Y^{(i)}(n) = |y^{(i)}(n)| \tag{2}$$

and

$$\phi(i,n) = arg\left[y^{(i)}(n)\right] \tag{3}$$

Generally, the performance of the i-th TD (TD_i) of C-M-CMF can be described as

$$D_C^{(i)}[y^{(i)}(n)] = \left[y^{(i,1)}(n)\dots y^{(i,L)}(n)\right]^{\mathsf{T}} \tag{4}$$

In this expression, $D_C^{(i)}[.]$ represents the complex-valued threshold decomposition of the signal $y^{(i)}(n)$ due to TD_i into a set of the L signals $y^{(i,j)}(n)$ and the superscript T signifies transposition. Analogous to R-TD, it is expected, that the expression

$$y^{(i)}(n) = \sum_{j=1}^{L} y^{(i,j)}(n) \tag{5}$$

is valid for TD, too.

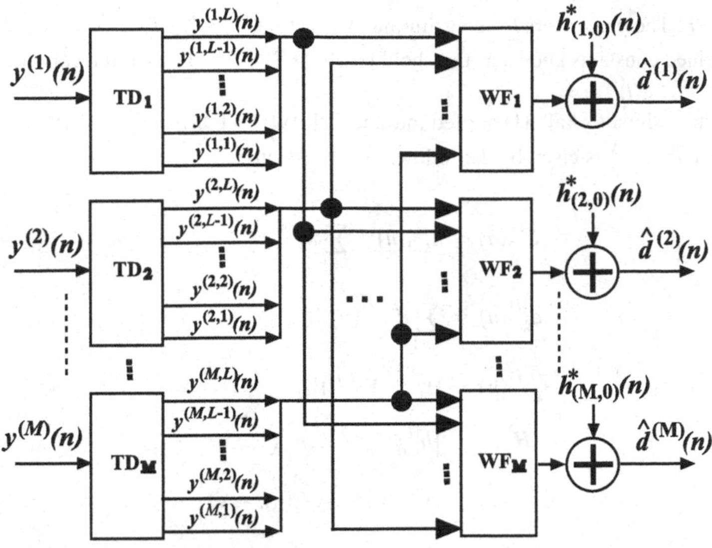

Fig. 1: Complex-valued multi-channel non-linear microstatistic filter

An intention of non-linear filter applications is e.g. to model a non-linear systems, where the non-linearity is related especially to the signal magnitude of signals (e.g. non-linearity of satellite communication systems [1]). On the other hand, in many applications of complex-valued signal processing (e.g. in communication engineering), signal arguments $\phi(i,n)$ are uniformly distributed over the interval $< -\pi, \pi >$. Therefore, a decomposition of $\phi(i,n)$ does not result in any clear benefit. With regard to this considerations and taking into account (5), the following decomposition of the complex-valued signals can be proposed:

$$\begin{aligned} y^{(i,j)}(n) &= D_C^{(i)}[y^{(i)}(n)] = D_R^{(i)}[Y^{(i)}(n)]e^{j\phi(i,n)} = \\ &= \left[Y^{(i,1)}(n)\dots Y^{(i,L)}(n)\right]^{\mathsf{T}} e^{j\phi(i,n)} = \\ &= \left[y^{(i,1)}(n)\dots y^{(i,L)}(n)\right]^{\mathsf{T}} \end{aligned} \tag{6}$$

where

$$y^{(i,j)}(n) = Y^{(i,j)}(n)e^{j\phi(i,n)} \tag{7}$$

In (6), $D_R^{(i)}[.]$ represents the real-valued threshold decomposition of the positive real-valued signal $Y^{(i)}(n)$ into a set of the real-valued signals given by

$$\begin{aligned} Y^{(i,j)}(n) &= D_R^{(i,j)}\left[Y^{(i)}(n)\right] = \\ &= \begin{cases} 0 & \text{if} \quad Y^{(i)}(n) < l_{j-1}^{(i)} \\ Y^{(i)}(n) - l_{j-1}^{(i)} & \text{if} \quad l_{j-1}^{(i)} \le Y^{(i)}(n) < l_j^{(i)} \\ l_j^{(i)} - l_{j-1}^{(i)} & \text{if} \quad l_j^{(i)} \le Y^{(i)}(n) \end{cases} \end{aligned} \tag{8}$$

for $1 \leq j \leq L$. The parameters $l_j^{(i)}$ constituting the vector $\boldsymbol{L}^{(i)} = \left[l_1^{(i)} l_2^{(i)} \ldots l_L^{(i)}\right]^\mathsf{T}$ are the positive real-valued constants known as threshold levels of TD_i. The threshold levels are confined as $0 < l_1^{(i)} < \ldots < l_L^{(i)} = \infty$.

The output signals of all TD are fed into the k-th WF (WF_k). Then, the k-th output of the C-M-CMF $\left(\hat{d}^{(k)}(n)\right)$ is given by the following expressions:

$$\hat{d}^{(k)}(n) = h_{k,0}^*(n) + \sum_{i=1}^{M} \hat{d}_k^{(i)}(n) \tag{9}$$

$$\hat{d}_k^{(i)}(n) = \sum_{j=1}^{L} d_k^{(i,j)}(n) \tag{10}$$

$$\hat{d}_k^{(i,j)}(n) = \boldsymbol{H}_k^{(i,j)^\mathsf{H}} \boldsymbol{Y}^{(i,j)}(n) \tag{11}$$

$$\boldsymbol{H}_k^{(i,j)} = \left[h_{k,0}^{(i,j)} h_{k,1}^{(i,j)} \ldots h_{k,N}^{(i,j)}\right]^\mathsf{T} \tag{12}$$

$$\boldsymbol{Y}^{(i,j)}(n) = \left[y^{(i,j)}(n) \ldots y^{(i,j)}(n-N)\right]^\mathsf{T} \tag{13}$$

where the asterisk denotes complex conjugation, the superscript H signifies Hermitian transposition, the sequence $h_{k,l}^{(i,j)^*}$ represents the part of impulse response of the WF_k fed by signal $y^{(i,j)}(n)$. The constant term $h_{k,0}^*$ is applied in the C-M-CMF structure in order to obtain an unbiased C-M-CMF output.

Now, let us define the block vector $\boldsymbol{H}_k^{(i)}$ containing the vectors $\boldsymbol{H}_k^{(i,j)}$ and the block vector $\boldsymbol{Y}^{(i)}(n)$ containing the vectors $\boldsymbol{Y}^{(i,j)}(n)$ as follows

$$\boldsymbol{H}_k^{(i)} = \left[\boldsymbol{H}_k^{(i,1)^\mathsf{T}} \ldots \boldsymbol{H}_k^{(i,L)^\mathsf{T}}\right]^\mathsf{T} \tag{14}$$

$$\boldsymbol{Y}^{(i)}(n) = \left[\boldsymbol{Y}^{(i,1)^\mathsf{T}}(n) \ldots \boldsymbol{Y}^{(i,L)^\mathsf{T}}(n)\right]^\mathsf{T} \tag{15}$$

Then by using (14) and (15), the expression (10) can be obtained in this form

$$\hat{d}_k^{(i)}(n) = \boldsymbol{H}_k^{(i)^\mathsf{H}} \boldsymbol{Y}^{(i)}(n) \tag{16}$$

Finally, let us define the vector $\boldsymbol{H}_k$ and the vector $\boldsymbol{Y}(n)$ as follows

$$\boldsymbol{H}_k = \left[h_{k,0} \quad \boldsymbol{H}_k^{(1)^\mathsf{T}} \ldots \boldsymbol{H}_k^{(M)^\mathsf{T}}\right]^\mathsf{T} \tag{17}$$

$$\boldsymbol{Y}(n) = \left[1 \quad \boldsymbol{Y}^{(1)^\mathsf{T}}(n) \ldots \boldsymbol{Y}^{(M)^\mathsf{T}}(n)\right]^\mathsf{T} \tag{18}$$

Then by using (17), (18) and (19), the k-th output of the C-M-CMF $\hat{d}_k(n)$ is then given by

$$\hat{d}_k(n) = \boldsymbol{H}_k^\mathsf{H} \boldsymbol{Y}(n) \tag{19}$$

In this expression, $\boldsymbol{H}_k^\mathsf{H}$ represents the impulse response of WF_k. It can be seen from (19) that the C-M-CMF responses are given by a linear combination of vector elements obtained by the decomposition of the input signals of the filter. With regard to that fact, the C-M-CMF responses are still linear functions with respect to the C-M-CMF coefficients.

3 C-M-CMF Design Procedure

Let us assume that the input signals of the time-invariant C-M-CMF $y^{(i)}(n)$ and the desired signals $d^{(k)}(n)$ are stationary random processes, then $\boldsymbol{H}_k(n) = \boldsymbol{H}_k$. Because C-M-CMF is a minimum mean square estimator, the set of parameters of the optimum time-invariant M-CMF given by $\boldsymbol{L} = \left[\boldsymbol{L}^{(1)^{\mathsf{T}}} \ldots \boldsymbol{L}^{(M)^{\mathsf{T}}}\right]^{\mathsf{T}}$ and $\boldsymbol{H}_k$ is obtained as the solution that minimizes the cost functions

$$MSE(\boldsymbol{H}_k,\boldsymbol{L}) = E\left[e^{(k)}(n)e^{(k)^*}(n)\right] = E\left[|e^{(k)}(n)|^2\right] \tag{20}$$

where

$$e^{(k)}(n) = d^{(k)}(n) - \hat{d}^{(k)}(n) = d^{(k)}(n) - \boldsymbol{H}_k^{\mathsf{H}}\boldsymbol{Y}(n) \tag{21}$$

$$e^{(k)^*}(n) = d^{(k)^*}(n) - \hat{d}^{(k)^*}(n) = d^{(k)^*}(n) - \boldsymbol{Y}^{\mathsf{H}}(n)\boldsymbol{H}_k \tag{22}$$

and $E[.]$ denotes the expectation operator.

Before $MSE(\boldsymbol{H}_k,\boldsymbol{L})$ minimization, it is necessary to determine the dimensions of $\boldsymbol{L}$ and $\boldsymbol{H}_k$ vectors. Generally, the dimension of $\boldsymbol{L}$ and $\boldsymbol{H}_k$ will be obtained as the results of trade-off between expected performance properties of the final filter application and its computational complexity.

Under the condition that the dimensions of $\boldsymbol{L}$ and $\boldsymbol{H}_k$ are defined, the design procedure of the M-CMF is based on an iteration process, where one iteration consists of three basic steps [8].

In the first step, the vector $\boldsymbol{L}$ is estimated. For that purpose, Scanning Method (SC-M), Genetic Algorithm Based Method (GA-M) and Method of Cumulative Distribution Function (CDF-M) can be applied [8]. Then, based on $\boldsymbol{L}$ estimation, the coefficients of the optimum WFs are computed (the second step) as follows

$$\boldsymbol{H}_k^{opt} = \boldsymbol{R}^{-1}\boldsymbol{P}_k \tag{23}$$

where

$$\boldsymbol{R} = E\left[\boldsymbol{Y}(n)\boldsymbol{Y}^{\mathsf{H}}(n)\right] \tag{24}$$

$$\boldsymbol{P}_k = E\left[d^{(k)}(n)\boldsymbol{Y}(n)\right] \tag{25}$$

$\boldsymbol{R}$ is correlation matrix of vector sequence $\boldsymbol{Y}(n)$ and $\boldsymbol{P}_k$ is the cross-correlation vector of the desired signal $d^{(k)}(n)$ and vector sequence $\boldsymbol{Y}(n)$. Under the condition that $y^{(i)}(n)$ and $d^{(k)}(n)$ are stationary random processes, $\boldsymbol{R}$ and $\boldsymbol{P}_k$ can be estimated e.g. by using a training sequence.

As the last step of the iteration, the evaluation of the cost functions of the C-M-CMF (mean square error given by (20)) for the estimated vector $\boldsymbol{L}$ and $\boldsymbol{H}_k$ is made. If the values of the cost functions are the minimum once or if they are acceptable from the application point of view, the iteration process is stopped and the values $\boldsymbol{L}$ and $\boldsymbol{H}_k$ providing the best values of the cost function are declared as the optimum (or sub-optimum) parameters of C-M-CMF. If the obtained values of the cost functions are not acceptable, the next iteration of the design procedure has to be started.

Under the condition that $\boldsymbol{H}_k^{opt}$, $\boldsymbol{R}$ and $\boldsymbol{P}_k$ are computed according to (23) - (25), then the mean-square error corresponding to optimum C-M-CMF is given by

$$\begin{aligned} MSE(\boldsymbol{H}_k^{opt},\boldsymbol{L}) &= \sigma_d^2 - \boldsymbol{P}_k^{\mathsf{H}}\boldsymbol{H}_k^{opt} - \boldsymbol{H}_k^{opt^{\mathsf{H}}}\boldsymbol{P}_k + \boldsymbol{H}_k^{opt^{\mathsf{H}}}\boldsymbol{R}\boldsymbol{H}_k^{opt} = \\ &= \sigma_d^2 - \boldsymbol{P}_k^{\mathsf{H}}\boldsymbol{R}^{-1}\boldsymbol{P}_k \end{aligned} \tag{26}$$

where

$$\sigma_d^2 = E\left[d^{(k)}(n)d^{(k)^*}(n)\right] \tag{27}$$

4 Conclusions

In this paper, the structure and design procedure of the new time-invariant C-M-CMF has been proposed. For the C-M-CMF structure, the new decomposer of complex-valued signals based on threshold decomposition of the signal magnitude has been introduced. The design procedure of the optimum C-M-CMF consists in modification of the design procedure of the M-CMF reflecting requirements of complex-valued signal processing. Our motivation for this work has been to develop the theoretical base for the design of non-linear multi-user receivers for MC-CDMA based on C-M-CMF. The development of this C-M-CMF application will be the core of the next research of ours.

Acknowledgments

This work has been funded by Ministry of Education of Slovak Republic under the project COST 289 *"Spectrum and Power Efficient Broadband Communications"*, project COST 297 *"High Altitude Platforms for Communications and Other Services"* and project *"VEGA 1/4088/07"*.

References

1. McCarthy J R (1999) Spread Spectrum Communications over Nonlinear Satellite Channels, Institute for Telecommunication Research, University of South Australia, A thesis submitted for the degree of Doctor of Philosophy
2. Hanzo L, Münster M, Choi B J, Keller T (2003) OFDM and MC-CDMA for Broadband Multi-User Communications, WLANs and Broadcasting, Willey, England
3. Kocur D (2001) Adaptive Volterra Digital Filters (in Slovak), Elfa s.r.o. Košice
4. Kocur D, Marchevský S, Čížová J (2003) Microstatistic Multi-User Detection Receiver, Proceedings of ICCC 2003, Siófok, Hungary, 363–366
5. Arce G R (1992) Microstatistics in Signal Decomposition and the Optimal Filtering Problem, IEEE Transactions on Signal Processing, 40(11):2669–2682
6. Kocur D, Drutarovský M, Marchevský S (1996) A New Class of Nonlinear Filters: Microstatistic Volterra Filters, Radioengineering, 5(1):19–24
7. Kocur D, Drutarovský M, Marchevský S (1995) Microstatistic Volterra Filters, Proceedings of the 40th Internationales Wissenschaftliches Kolloquium, 376–381
8. Kocur D, Krajňák J, Marchevský S (2005) Sub-Optimum MSF-MUD for CDMA Systems, 2nd COST 289 Workshop on Special Topics on 4G Technologies, 65–71
9. Kocur D, Hendel I (2005) PWL vs. Conventional Microstatistic Digital Filters, Acta Electrotechnica et Infromatica, 5(1):22–26
10. Kocur D, Čížová J, Marchevský S (2003) Adaptive Multi-Channel Microstatistic Filters, 48th Internationales Wissenschaftliches Kolloquium, Technische Universit *ä*t Ilmenau, Germany, 91–92

11. Kocur D, Čížová J, Marchevský S (2004) Microstatistic Multi-User Detection Receiver, Journal of Advanced Computational Intelligence and Intelligent Informatics (JACI3), 8(5):482–487
12. Heredia E A, Arce G A (1996) A Piecewise Linear System Modeling Based on a Continuous Threshold Decomposition, in IEEE Transactions on Signal Processing, 44(6):1440–1453

Legal Ontologies and Loopholes in the Law

Sandra Lovrenčić[1], Ivorka Jurenec Tomac[2], and Blaženka Mavrek[3]

[1] Faculty of organization and informatics, Pavlinska 2, Varaždin, Croatia
sandra.lovrencic@foi.hr
[2] Konzum d.d., I. Kukuljevića 14, Varaždin, Croatia
ivorka.jurenec@vz.t-com.hr
[3] Varaždin County Court, B. Radić 2, Varaždin, Croatia

Summary. The use of ontologies is today widely spread across many different domains. The main effort today is, with the development of Semantic Web, to make them available across the Internet community with the purpose of reuse. The legal domain has also been explored concerning ontologies, both on the general as on the sub-domain level. In this paper are explored problems of formal ontology development regarding areas in specific legislation acts that are understated or unequally described across the act – popularly said: loopholes in the law. An example of such a problematic act is shown. For ontology implementation, a well-known tool, Protégé, is used. The ontology is made in formal way, using PAL – Protégé Axiom Language, for expressing constraints, where needed. Ontology is evaluated using known evaluation methods.

1 Introduction

Ontologies are nowadays widely used. Various domains are already described, or are planning to be described, using ontologies. Taxonomies, that were used before to represent domain concepts and their hierarchy, are in present time supplemented with ontologies that have constraints expressed in some formal language. With several development methods and many languages and tools that exist ([7–10, 15–17, 26]) one can at first think that the only problem is to choose appropriately among all those options. The main factor when making such decisions is the domain that will be represented with ontology and the way in which its knowledge is expressed. Knowledge can be gathered from a number of sources in different ways, which again depends on the domain.

If a legal domain is considered and the aim is to create ontologies of specific legislation acts, then the knowledge should be gathered strictly from them. It is presumed that such an act comprises all facts and regulations that can be questioned, since it actually represents a law that must be obeyed. But, in many cases certain things stay understated or are not worked out in the same detail across the entire act. Of course, some of them are worked out in detail in legal acts at lower level, but some are not. In concrete cases in court they are then subjected to "individual interpretation".

Ontology development of those acts can therefore be problematic, which will be shown on the Family Legislation Act example.

J.A.T. Machado et al. (eds.), *Intelligent Engineering Systems and Computational Cybernetics,*

2 Ontologies

2.1 Basic Notions

First definition of ontology that can be found in almost all papers concerning ontologies is the one of T. Gruber, saying it is "a formal explicit specification of a shared conceptualization for a domain of interest", meaning that it "has to be specified in a language that comes with formal semantics" [19]. Ontologies are commonly distinguished as representation, general, top-level or upper-level, domain, task, domain task, method or application ontologies [17]. Any domain ontology gives a good insight into all its concepts and hierarchical structure, but also defines concept attributes and all restrictions that are incorporated into specific domain, using various formalisms.

To formally develop a specific ontology there are a number of languages and language definitions, for example: Description Logic (DL) [3], Frame logic (F-logic) [2], Resource Description Framework (RDF) and RDF Schema [22], OWL – Web Ontology Language (`http://www.w3.org/2004/OWL/`).

Also, there are a number of tools for building ontologies and some well known are:
OntoStudio (`http://www.ontoprise.de/`, formerly OntoEdit),
OilEd (`http://oiled.man.ac.uk/`),
Protégé (`http://protege.stanford.edu`),
Ontolingua (`http://www.ksl.stanford.edu/software/ontolingua/`).
OWL, based on DL, along with the term Semantic Web, becomes more and more important, because ontologies should ensure semantic on the Web. Various factor comparisons of a number of tools and languages (including Protégé) can be found in [8, 17, 23] and also in the numerous reports from OntoWeb project [15] and its OntoWeb Portal (`http://www.ontoweb.org/`) that is now followed with the Semantic Web project on a broader level (`http://knowledgeweb.semanticweb.org`).

Along with Semantic Web, main concerns today regarding ontologies are their reusability and evaluation. Reusability means that ontologies should be made in a way so that they can be applied, with minor changes/additions, to similar domains. A number of evaluation methods using various factors are already proposed, as can be seen in [4, 10, 11, 13, 16, 17, 20].

2.2 Legal Ontologies

The law domain is also "attractive" for ontology development for more than fifteen years. According to [5], ontologies have already been modeled for "traffic regulations, tax – criminal – and administrative law; international treaties on trade, and safety at sea". For example, in [25] is presented ontology for financial fraud prevention that aims at compiling for several languages and merging different European laws. Along with law rules, it should contain information about companies and suspected fraudsters. A general ontology that is considered in [21] is dealing with "causation in fact" and tries to define under what conditions causation occurs, which can be applied to all laws. Law ontologies also try to help in determining liability in complex legal cases [28]. Furthermore, in [5] and [27] is discussed a core ontology of the law domain, that was being developed more than ten years. Based on it, ontology libraries that are intended for reusability could be developed. The resulting LRI – Core ontology has five major portions: physical, mental and abstract classes, roles and occurrences [6].

Ontology development for individual legislation acts should clarify relations among their specific concepts and explicitly represent regulations as constrains on concept attributes. It is presumed that such an act comprises all facts and regulations that can be questioned, since it actually represents a law that must be obeyed and therefore must cover all possible situations involving major act concepts. But, in many cases certain regulations stay understated, which is commonly known as loopholes in the law. In concrete cases in court they are then subjected to "individual interpretation". The question is how such situations can be reflected in ontology.

3 Legal Ontology Development

In this unit at first will be described a tool that was used for ontology development, Protégé. Since legislation act regulations (constraints) will be expressed as axioms in formal logic, Protégé's formal language will be described in detail. Then, the concrete example will be presented.

3.1 Protégé

Classes and Attributes

Protégé (`http://protege.stanford.edu`), as an ontology development tool, exists almost two decades and today is widely used. It is made in Java, it's free and open source. Today it has various features, some delivered in standard distribution and other accessible by many plug-ins. Comparisons of Protégé and other ontology modeling tools can be found in [15, 17, 23].

In Protégé ontology can be developed using OWL or Frames interface. Concepts are organized into class hierarchy and have their attributes. Attribute values can be restrained, but constraints can also be expressed in formal logic using PAL – Protégé Axiom Language.

Therefore, ontology modeling in Protégé includes [24]:

- Determining the domain and the ontology scope
- Considering a reuse of existing ontologies
- Enumerating important terms in the ontology
- Defining classes in the ontology – domain concepts
- Arranging classes in a subclass-super class hierarchy
- Defining slots – properties of classes
- Defining facets on slots – features and attributes
- Filling in values for slots and instances

Axioms

Constraint development in Protégé is enabled with PAL – Protégé Axiom Language [18, 31], created as a plug-in that is delivered together with the tool. It is developed for writing logical constraints and queries about frames of a knowledge base (classes, instances, slots and facets). PAL works with model-checking, making "closed-world" assumptions, so it can be used only for constraints on existing knowledge and not for asserting new ones. It has constraint-checking engine that can be run against the knowledge base to find frames that violate constraints.

A constraint in PAL includes [18, 31] a set of variable range definitions and a logical statement that must hold on those variables. The syntax is a variant of Knowledge Interchange Format (KIF) [12], but, although it supports KIF connectives, it does not support all constants and predicates. A constraint statement is a sequence of sentences that use a number of predefined predicates and functions, ontology and other predicates linked with logical connectives (<=>, =>, *and*, *or*, *not*) in a prefix notation. Every variable in the constraint has to be quantified with universal (*forall*) or existential quantifier (*exists*).

PAL variable range definition must be in the form:
(defrange var type [additional information]), where is
- *defrange* – reserved word for denoting a variable range definition
- *var* – the name of the variable, begins with *?* for local and with *%* for global variables
- *type* – the type of the variable, can be *:SET, :FRAME, :SYMBOL, :STRING, :INTEGER, :NUMBER*
- *[additional information]* – depends on the type of the variable, for example *:FRAME [classname][slotname]*

Syntax of predicate and function assertions is the same:
(pred-name/func-name symbol-1 symbol-2 symbol-3 ...), where is
- *pred-name/func-name* – the name of the predicate or function
- *symbol-n* – slot value, class or instance

Connectives and not are used as follows:
(not assertion)
(connective-name assertion assertion), where is
- *assertion* – predicate assertion described above
- *connective-name* – *<=>, =>, and, or*

Quantifier syntax is in the form:
(forall ?name full-assertion)
(exists ?name full-assertion), where is
- *?name* – the name of the variable being quantified
- *full-assertion* – assertion with connectives and not

3.2 Family Legislation Act Example

As a private civil law, family law has not been of such interest as other, more public laws, as crime or trade law. But it is omnipresent and its regulations are concerning all people in some points of their lives. Let the ontology being developed be aimed at easier understanding of the Family Legislation Act (FLA) [29, 30] by the very same people influenced by it. For that reason, some usually used concepts will be omitted. This means that, for example, *Mother* and *Father* won't be subclasses of the possible class *Role* (omitted), but simply subclasses of class *Person*. In general, Croatian Family Law includes the same considerations as in other countries. As an example, the part of it covering motherhood and parenthood will be used. Figure 1 shows a part of class hierarchy of the ontology.

Each class has a set of important attributes. For example, some of important attributes for class *Mother* are: *gender*, *alive*, *capability* and *became_ m*. *Gender* is set only to be female. Attribute *alive* is also important in some aspects of motherhood and fatherhood. *Capability* is referring on law term for possible person incapability of some legal actions. Attribute *became_m* shows possible manners of becoming a mother. This attribute is set separately for *Mother* and *Father* (*became_f*) classes, because it is commonly presumed that a mother is a person that gave a birth to a child and a father is recognized automatically only if he is a husband of a mother – if not, he has to admit fatherhood and a mother or court have to give consent to his admission [29, 30]. Possible values for attribute *became_f* is shown in Fig. 2.

As mentioned before, it is possible that some regulations stay understated or are not worked out in the same detail across the entire act. Example that will be represented is that of a motherhood or fatherhood disputing. The disputation in the cases of conception with medical help will not be included. In articles 76 and 77 of Family Legislation Act [29, 30] are given regulations about disputing motherhood by a mother herself or other woman, respectively, and in article 82 about disputing fatherhood of a man that once admitted it. In articles 79 and 81 [29, 30] are given regulations about disputing fatherhood by mother's husband or mother herself, respectively. Both last articles also state: "If a husband/mother lacks capability completely or partially, but without permission to undertake any actions considering personal conditions, lawsuit for disputing fatherhood can be undertaken by her/his guardian with prior consent from Social Care Center." This part is missing from the first three mentioned articles.

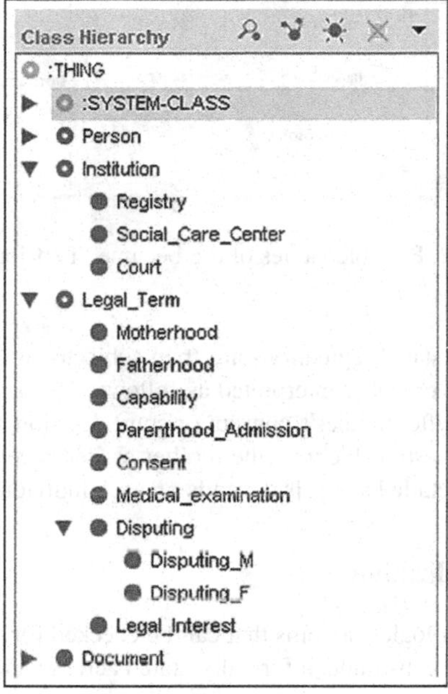

Fig. 1: Part of FLA Class Hierarchy

Explanation of an expert and coauthor that was involved in ontology development is that it is very rarely (and not once in her practice), that a mother herself or other woman disputes the motherhood. Today very rarely child replacements in hospitals or false entries into birth register occur. Also, a men will rarely admit fatherhood if he is not completely convinced that he is a father. Since this is very rare, the possibility of such a person being incapable is almost impossible. This is a reason that those articles aren't worked out in the same detail as most of those about fatherhood.

Also, there is no article about disputing motherhood by a father – a mother is having more priority because she gave birth. Considering disputing of fatherhood by another man, it is explicitly stated in article 83 [29, 30] that he can dispute the fatherhood of a person that admitted it. But, it is not stated whether he can dispute the fatherhood of a mother's husband. Again, the interpretation is that this is very rare, mothers are usually the ones that dispute the fatherhood of their husbands. Another example is that a mother can dispute a fatherhood of her husband [29, 30] but it is not stated whether she can dispute a fatherhood of a person that admitted it.

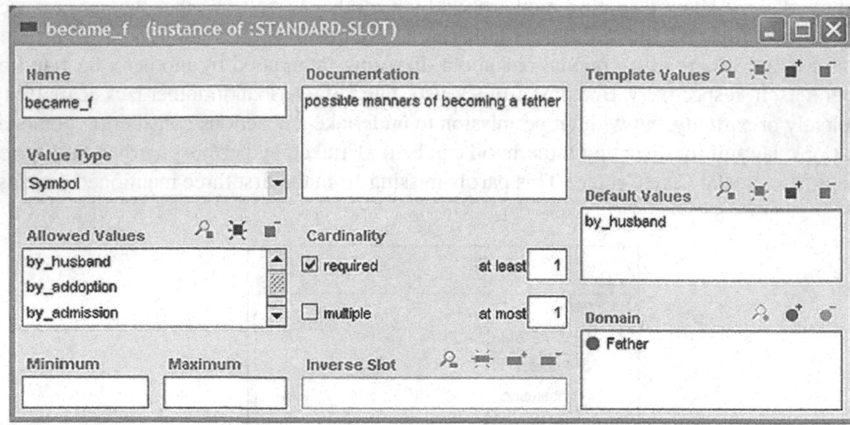

Fig. 2: Possible values of the became_ f attribute

Such and similar understated regulations are then subjected to individual interpretation. The last example should be probably interpreted as follows: The person that admitted fatherhood became a father only after mother's consent or court decision. Court decision cannot be disputed and it would be questionable why the mother earlier gave consent and now tries to dispute fatherhood. More detailed analysis depends on each individual case.

3.3 Regulations Formalization

Expressing regulations with logical axioms that can be checked for consistency ensures clear and precise definition of all constraints informally stated across legislation acts. For that purpose, continuing with the example, a class *Disputation* is created as a subclass of a class *Legal_ Term*. Each disputation has a person that tries to dispute motherhood or fatherhood, which is represented as an attribute called *disputant*. Several conditions must be fulfilled, so that a lawsuit for disputation can be made. According to relevant articles, those are [29, 30]:

- Disputant's capability, which denotes what kind of person disputant can be
- Time passed from finding out facts that caused disputation (not later than six months of the findings or six months from the child birth, depending on the disputation and the disputant)
- Child's age (not after seven years of child's age for all but the child itself – the child can dispute until 25th birthday)
- Demand for establishing disputant's motherhood/fatherhood if she/he is not the person legally acknowledged as such or acting for such a person

Each of those conditions could be expressed with one constraint including all possible disputants – if all of them would be included in all articles. But, since it is explained that the part about disputant's capability isn't included in all articles considering disputation, then there are several possibilities:

- To use the interpretation that such a condition can be applied equally to the articles from which it is omitted
- To create subclasses for disputing motherhood and fatherhood separately – but still there are articles for each of those that does and does not include the statement about capability
- To create subclasses for each disputant – that would certainly be too detailed and still would not solve the problem
- To make a combination of second and third proposal – a very consuming solution

In consultation with the expert it is decided that first solution will be applied. Every person in the case of incapability will be given a guardian, according to other legislative acts [1]. Generally said, a disputant must be guardian if the person that disputes is incapable or a child. The constraint is therefore formally stated as follows:

```
(defrange ?Disputing :FRAME Disputing)
(defrange ?Person :FRAME Person)
(forall ?Disputing (forall ?Person
 (or (=> (= (capability ?Person) capable)
         (or (= (disputant ?Disputing) Mother)
             (= (disputant ?Disputing) Father)
             (= (disputant ?Disputing) Non_Parent)))
     (=> (or (= (capability ?Person) incapable)
             (= (capability ?Person) child))
         (= (disputant ?Disputing) Guardian)))))
```

When considering who can dispute motherhood or fatherhood to whom, some loopholes in the law exist as stated in examples above. In the case when a man that considers himself as a father dispute fatherhood of a person that admitted fatherhood, the question is whether he can dispute the fatherhood of a mother's husband. Since he can dispute only with demand that his own fatherhood must be established, he can dispute both. After adding the attribute *disputation_ object* that can have values of *motherhood* and *fatherhood*, the constraint will be formally expressed as:

```
(defrange ?Disputing :FRAME Disputing)
(defrange ?Father :FRAME Father)
(defrange ?Non_Parent :FRAME Person)
(forall ?Disputing (forall ?Father (exists ?Non_Parent
(=> (or (= (became_f ?Father) by_husband)
        (= (became_f ?Father) by_admission))
    (and (= (disputant ?Disputing) Non_Parent)
         (= (gender ?Non_Parent) M)
         (= (disputation_object ?Disputing) fatherhood))))))
```

The possible view on the last example, when a mother can dispute a fatherhood of her husband but it is not stated whether she can dispute fatherhood of a person that admitted it, is already explained above. In the case that she could dispute both, statement similar to above would regulate it, but instead of male *Non_ parent* disputant would be *Mother*. If the conclusion is that she cannot dispute the fatherhood of the man that admitted it, this should be incorporated in the statement using negation of the existence of mother in such a case. The third option, not to resolve the problem, would have to be accentuated.

4 Ontology Evaluation

Each ontology should be evaluated for purpose of verification that it defines the domain correctly. Family Legislation Act Ontology is evaluated using general criteria to evaluate ontologies proposed and described in detail in [13, 14, 17]. They are: consistency, completeness, conciseness, expandability and sensitiveness.

Consistency is checked by verifying that all ontology definitions are individually and inferentially consistent, which is not the case if they can lead to contradictory conclusions. Furthermore, to verify individual consistency, it is necessary to confirm that there is no contradiction in interpretation of formal and informal definition regarding the real world and also to confirm that they both have the same meaning. The expert has confirmed that informal definition of domain concepts is correct and that it has the same meaning as formal definition, which also verifies the consistency of formal definition. Inferential consistency is ensured by attribute facets and formal axioms and checked with individual cases.

Completeness of ontologies is still a problem and it is difficult to prove. But, from the incompleteness of an individual definition, which can be proved, the incompleteness of ontology can be deduced. The domain expert confirmed that everything that is regulated by law is either explicitly stated (in the scope of individual definitions) or can be inferred. Although the example is given in a very narrow domain, the problem are, of course, loopholes in the law. Suggested solutions are described above, but, since they represent an "individual interpretation", a mutual agreement on higher level should be achieved.

Conciseness can be verified by checking whether the ontology doesn't have any unnecessary definitions and explicit or implicit redundancies between definitions. In expert's opinion, developed ontology doesn't have any unnecessary or redundant definitions. For example, as already explained, the general class *Role* is omitted, because it wasn't necessary for this specific ontology.

Expandability (for example, on other parts of Family Legislation Act) still isn't checked. The authors presume that some problems in that part can be expected due to loopholes in the law.

Sensitiveness can be checked together with expandability, and same problems are expected in that area, too.

5 Conclusions

Ontologies in legal domain already cover many sub-domains and core ontology is almost fully developed. Still, there is a number of sub-domains that are not yet completely explored and a number of questions that need to be answered. Law is in many countries based on cases and many, if not all, legislative acts have some parts that are understated. This is normal, because, as expert coauthoring in this paper says, the law is built upon practice. There are certain parts of some regulations that are left to individual interpretation. Those are mostly cases that occur very rarely, if not never, but are theoretically possible.

In all domains, ontology development is a task that tries to comprise as much knowledge as can be found. It is most undesirable to left some parts of the domain undefined. Some understated regulations can be resolved by using most often presumed "individual interpretations", because they are not very conflictive. The solution can be found using comparison with similar articles or in a higher-level legislative act. On the other hand, some regulative interpretations can very differ from case to case, so a detailed description of the problem should be available to the ontology users. Some delicate situations, after careful consideration, may better be left unsolved, if the damage is considered to be less in that way.

The main conclusion is that there still doesn't exist final solution how to interpret legal act understatements, which leads to the problem of formally defining all parts of specific legal domain. Additional knowledge resources, including other experiences in this area and advices from domain experts can be considered as a possible solution.

References

1. Alinčić M, Dika M, Hrabar D, Jelavić M, Korać A (2003) Obiteljski zakon – novine, dvojbe i perspektive (Family Legislation Act – novelties, doubts and perspectives). Narodne novine d.d., Zagreb
2. Angele J, Lausen G (2004) Ontologies in F-logic. In: Staab S, Studer R (eds) Handbook on Ontologies. Springer, Berlin
3. Baader F, Horrocks I, Sattler U (2004) Description Logics. In: Staab S, Studer R (eds) Handbook on Ontologies, Springer, Berlin
4. Brank J, Grobelnik M, Mladenič D (2005) A Survey of Ontology Evaluation Techniques. Conference on Data Mining and Data Warehouses, Ljubljana. http://kt.ijs.si/dunja/slkdd2005/Papers/BrankEvaluationSiKDD2005.pdf, 20.12.2005.
5. Breuker J (2003) Managing Legal Domains: In Search of a Core Ontology for Law. In: Proceedings of the Workshop on Knowledge Management and the Semantic Web at KCAP-2003. ftp://ftp-sop.inria.fr/acacia/proceedings/2003/kcap-kmsw/kcap2003-kmsw-Breuker.pdf, 10.01.2006.
6. Breuker J (2004) Epistemology and Ontology in Core Ontologies: FOLaw and LRI-Core, Two Core Ontologies for Law. In: Proceedings of EKAW Workshop on Core Ontologies. http://ftp.informatik.rwth-aachen.de/Publications/CEUR-WS/Vol-118/paper2.pdf, 10.01.2006.
7. Corcho O, Fernández-López M, Gómez-Pérez A (2003) Methodologies, tools and languages for building ontologies. Where is their meeting point? Data and Knowledge Engineering 46:41–64
8. Denny M (2002) Ontology Building: A Survey of Editing Tools. http://www.xml.com/pub/a/2002/11/06/ontologies.html, 23.09.2004.
9. Duineveld A J, Stoter R, Weiden M R, Kenepa B, Benjamins V R (1999) Wonder Tools? A Comparative Study of Ontological Engineering Tools. http://hcs.science.uva.nl/wondertools/html/paper.htm, 24.09.2004.

10. Fernández-López M et al. (2002) A Survey on Methodologies for Developing, Maintaining, Integrating, Evaluating and Reengineering Ontologies. http://www.aifb.uni-karlsruhe.de/WBS/ysu/publications/OntoWeb_Del_1-4.pdf, 05.02.2005.
11. Gangemi A, Catenacci C, Ciaramita M, Lehmann J (2005) Ontology Evaluation and Validation. Technical Report. http://www.loa-cnr.it/Files/OntoEval4OntoDev_Final.pdf, 20.12.2005.
12. Genesereth M R (1995) Knowledge Interchange Format Specification. http://logic.stanford.edu/kif/specification.html, 20.10.2004.
13. Gómez-Pérez A (2001) Evaluation of Ontologies. International Journal of Intelligent Systems 16:391–409
14. Gómez-Pérez A (2004) Ontology Evaluation. In: Staab S, Studer R (eds) Handbook on ontologies. Springer, Berlin
15. Gómez-Pérez A et al (2002) OntoWeb Deliverable 1.3: A Survey on Ontology Tools. http://ontoweb.org/About/Deliverables/D13_v1-0.zip/, 17.05.2004.
16. Gómez-Pérez A et al (2002) Technical Roadmap D.1.1.2. http://ontoweb.org/About/Deliverables/D1.1.2_v1_0.zip, 17.05.2004.
17. Gómez-Pérez A, Fernández-López M, Corcho O (2004) Ontological Engineering. Springer, London
18. Grosso W (2001) The Protégé Axiom Language: Overall Design Considerations. http://protege.stanford.edu/plug-ins/paltabs/OverallDesignConsiderations.zip, 11.10.2004
19. Gruber T (1993) A Translation Approach to Portable Ontology Specifications. Knowledge Acquisition 5(2):199–220
20. Hartmann J et al (2005) D1.2.3 Methods for Ontology Evaluation. http://www.starlab.vub.ac.be/research/projects/knowledgeweb/KWeb-Del-1.2.3-Revised-v1.3.1.pdf, 20.12.2005.
21. Lehman J, Breuker J, Brouwer B (2003) Causation in AI& Law. ICAIL 2003 Workshop on Legal Ontologies & Web-based Legal Information Management. http://lri.jur.uva.nl/~winkels/LegOnt2003/Lehmann.pdf, 10.01.2006.
22. McBride B (2004) The Resource Description Framework (RDF) and Its Vocabulary Description Language RDFS. In: Staab S, Studer R (eds) Handbook on Ontologies. Springer, Berlin
23. Mizoguchi R (2004) Ontology Engineering Environments. In: Taab S, Studer R (eds) Handbook on Ontologies. Springer, Berlin
24. Noy N F, McGuinness D L (2000) Ontology Development 101: A Guide to Creating Your First Ontology. http://smi-web.stanford.edu/pubs/SMI_Abstracts/SMI-2001-0880.html, 23.09.2004.
25. Schafer B, Kingston J, Vandenberghe W (2003) Ontology Modelling in the Legal Domain – Realism Without Revisionism. In: Proceedings of the KI2003 Workshop on Reference Ontologies and Application Ontologies. http://ceur-ws.org/Vol-94/ki03rao_vandenberghe_al.pdf, 10.01.2006.
26. Su X, Ilebrekke L (2002) A Comparative Study of Ontology Languages and Tools. In: Proceedings of the 14th International Conference on Advanced Information Systems Engineering. Lecture Notes in Computer Science 2348:761–765
27. Valente A, Breukner J (1996) Towards Principled Core Ontologies. In: Proceedings of Tenth Knowledge Acquisition for Knowledge-Based Systems Workshop. http://ksi.cpsc.ucalgary.ca/KAW/KAW96/valente/doc.html, 10.01.2006.

28. van Laarschot R, van Steenbergen W, Stuckenschmidt H, Lodder A R, van Harmelen F (2005) The Legal Concepts and the Layman's Terms. In: Proceedings of the 18th Annual Conference on Legal Knowledge and Information Systems. IOS Press, Amsterdam Berlin Oxford Tokyo Washington DC
29. van Laarschot R, van Steenbergen W, Stuckenschmidt H, Lodder A R, van Harmelen F (2003) Obiteljski zakon. (Family Legislation Act) Narodne novine 116(1583)
30. van Laarschot R, van Steenbergen W, Stuckenschmidt H, Lodder A R, van Harmelen F (2004) Zakon o izmjenama i dopunama obiteljskog zakona. (Family Legislation Act Amendments Law) Narodne novine 17(0484)
31. van Laarschot R, van Steenbergen W, Stuckenschmidt H, Lodder A R, van Harmelen F (2001) PAL Documentation, http://protege.stanford.edu/plug-ins/paltabs/pal-documentation/index.html, 11.10.2004

Part III

Computational Intelligence

Extracting and Exploiting Linguistic Information from a Fuzzy Process Model for Fed-Batch Fermentation Control

Andri Riid[1] and Ennu Rüstern[2]

[1] Department of Computer Control, Tallinn University of Technology, Ehitajate tee 5, Tallinn 19086, Estonia
andri@dcc.ttu.ee

[2] Department of Computer Control, Tallinn University of Technology, Ehitajate tee 5, Tallinn 19086, Estonia
Ennu.Rystern@dcc.ttu.ee

Summary. A class of fuzzy process models can be subjected to linguistic inversion, a technique that has great potential for use in process control but the inversion technique itself needs refinement before it can be regarded as a viable controller design method. The key elements of the procedure that is under observation here are the reliability-oriented modelling algorithm and the automatic inversion technique. When the developed approach is applied to the fed-batch fermentation benchmark, good control performance is observed, confirming the validity of our assumptions. It should be noted, however, that the approach has its natural limits because it fails to explore the solution space exhaustively.

1 Introduction

Many applications of fuzzy logic use it as the underlying logic system for expert systems to reason about data. Fuzzy logic systems are able to capture the inexact and approximate nature of human knowledge and have therefore found use as a man-machine interface to make accumulated human experience available for the applications in different fields of automated decision-making, including control. Extraction of linguistic information from fuzzy models constructed on the basis of raw identification data, which would give us insight into the behavior of the observed system and reveal causal relationships between system variables is much rarer, though – mostly because data-driven fuzzy modelling today is primarily a black-box phenomenon; further emphasized by the fact that fuzzy system structure commonly used belongs to the class of first order Takagi–Sugeno (TS) systems [3], with inherent poor interpretation values. The problem that fuzzy models are generally non-transparent to interpretation and that interpretation of extracted fuzzy rules can lead to invalid conclusions has been treated in the past [12] by developing transparent fuzzy modelling techniques that maintain transparency of the model throughout the training process, thus ensuring that combined meaning of fuzzy rules and fuzzy membership functions (MFs) of the final model is consistent with the observed numerical behavior of the model.

J.A.T. Machado et al. (eds.), *Intelligent Engineering Systems and Computational Cybernetics*,

This, however, is only a partial solution to the problem because transparency is a measure of model's internal consistency and does not give any guarantee that the model is consistent with training data (in this context, external consistency). The primary measure of external consistency is the approximation error but, in particular, difficulties arise with neural networks inspired learning algorithms [8] (rather typical choice in fuzzy modelling) that rely on global learning techniques driven by numerical approximation error when the parameters of a fuzzy model are determined on the basis of scarce data. In such situation the missing rules are obtained by drawing conclusions through the extrapolation of existing data samples, often resulting in fuzzy rules that are unrealistic or simply untrue for the given application. For example, least squares estimation [3], commonly used for consequent parameter identification, has such properties.

What makes the issue even more critical is that scarcity of training data is actually very typical situation in applications because it is usually difficult to get good coverage of the input space by training data when the number of input variables is larger than two, moreover, certain regions of inputs space may be physically impossible to cover (an antecedent "if sun is bright and rain is heavy" is an example of this). In conclusion, if reliability of the fuzzy rules is not guaranteed in model development, any attempt of putting linguistic information to some practical use is bound to fail.

In this work the latter problem is approached. We present a combined approach for modelling a biochemical process and (consequent) controller extraction by model inversion. Furthermore, we are able to demonstrate that the performance of a controller obtained this way is good for the fed-batch fermentation benchmark.

2 Linguistic Inversion for Process Control

Arguably, the ultimate goal of controller design is to derive the inverse model of the process. In theory, the use of an inverse model possesses the advantages of open-loop control, i.e. inherent stability and perfect control with zero error. In practice, however, it is not guaranteed if the inverse configuration actually exists or if it is physically realizable.

Global inversion of the model, where all states become the outputs of the inverted model and the output of the original system becomes the state variable (Fig. 1) has normally non-unique solution and must be given by a family of solutions. In case of partial inversion, only one of the states (x_1 in Fig. 1) of the original system becomes the output of the inverted model and other states together with the original output are the inputs of the inverted model.

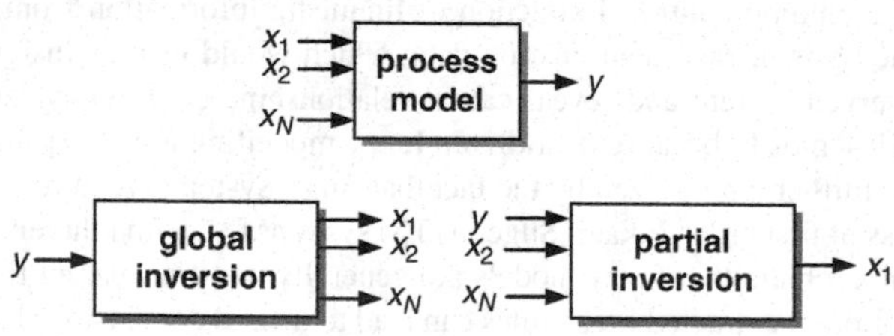

Fig. 1: Process model and its inversions

The main convenience of partial inversion is that it has an unique solution if the original model is strictly monotone in respect to the inverted state (it can also be more easily embedded into the control system than the globally inverted model). Numerical identification of the inverse model that is the common way for obtaining them (techniques have been developed by neural network research), however, may be computationally expensive, requiring many training epochs/samples to converge. The issue of invertibility is also not very well handled with automatic generation of the inverted model. Fuzzy systems, then, give a different approach angle to the problem because unlike neural networks (a) they can be interpreted in linguistic terms; (b) if transparent [12], their parameters can be interpreted in terms of their influence to the input-output relationship and therefore allow approximate linguistic inversion [9–11] as well as exact analytical inversion [7].

Linguistic inversion (causality inversion), what we concentrate on in current paper, is obtained through the exchange of antecedent and consequent variables in fuzzy rules. Consider a three-input fuzzy system from [7], given in Table 1.

Table 1: Rule base of the sample model

	x_1 is small	x_1 is medium	x_1 is large
x_2 is low AND x_3 is low	Zero	Low	Medium
x_2 is low AND x_3 is high	Low	Medium	High
x_2 is high AND x_3 is low	Low	Medium	High
x_2 is high AND x_3 is high	Medium	Medium	High

The inversion procedure may have three possible results marked in Table 2: (i) The input configuration is unique. This is the ideal case. (ii) The input configuration is non-unique, meaning that the rule base is non-invertible. The approximate solution is to choose the input configuration with the lowest control energy (in linguistic sense). (iii) There are no inputs that allow one-step transition to the desired output.

Table 2: Inverted rule base

x_2, x_3	y is zero	y is small	y is medium	y is high
x_2 is low AND x_3 is low	Small (i)	Medium (i)	Large (i)	Large(iii)
x_2 is low AND x_3 is high	Small (iii)	Small (i)	Medium (i)	Large (i)
x_2 is high AND x_3 is low	Small (iii)	Small(i)	Medium (i)	Large (i)
x_2 is high AND x_3 is high	Small (iii)	Small (iii)	Small (ii)	Large (i)

The reason why the latter situation occurs is twofold. First, the number of linguistic labels ("low", "high", etc.) given for x_1 and y is not equal, consequently, the expected number of rules in the inverted model is 16, whereas the number of rules in the original model is 12. The second reason is that the original system may not simply allow one-step transition to the desired output from the given state. The approximate but necessarily not the optimal solution is to choose the "nearest" output (again in linguistic sense).

Things become even worse when the rule base of the model is incomplete (there are less than 12 rules in the original model) or if the model is of 0-th order TS type (output MFs are not shared among the rules).

In such situations, it would be still possible to invert the model if the specification for desired closed control loop behavior exists. Assume that our goal is to achieve closed loop mapping that is depicted in Fig. 2.

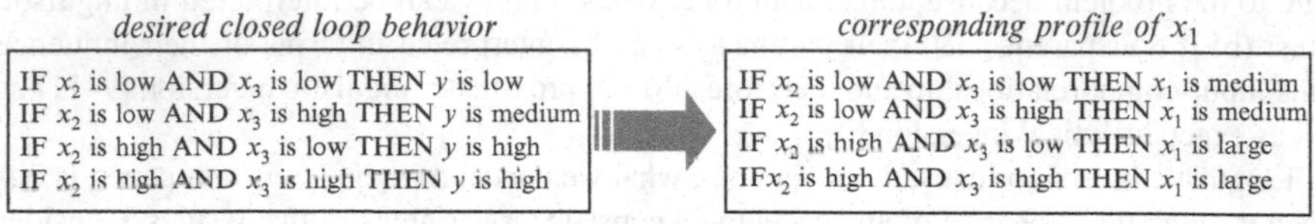

Fig. 2: Closed loop behavior-oriented model inversion

From the rules of the model that have linguistically identical premises in terms of x_2 and x_3 (i.e. rows in Table 1) only one that corresponds to the desired label of y is selected and inverted (Fig. 2). The resulting fuzzy logic system can be embedded into the control loop (Fig. 3).

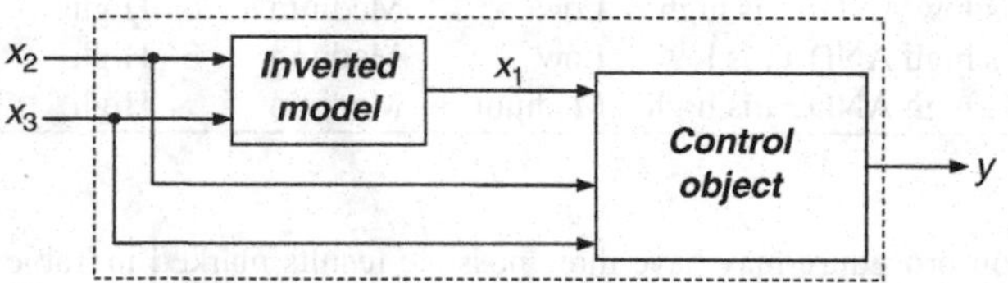

Fig. 3: Inverted model in closed loop

Clearly, there are advantages with the proposed approach – the inverted model has S_y (the number of linguistic labels of y) times less rules than the full-scale partial linguistic inversion and if the closed-loop specification is realistic we do not encounter (ii) and (iii) type rules. The main disadvantage is that if the specification for closed-loop behavior is changed, the model must be re-inverted again. However, it will be shown in Section 4 that on certain conditions it is possible to automate the inversion procedure.

Because of the approximate nature of linguistic inversion (the inversion is inexact by default), there will always be a difference between the specified closed loop behavior and the one we are able to achieve with the inverted model. Modelling error, which is usually nonzero, also contributes to this discrepancy but if the plant is robust enough it is still possible to produce satisfactory control results as will be shown in Section 5.

3 Plant Description

In 1996, Industrial Control Center of University of Webminster held a Modelling and Control Competition. The process simulation software that was available for the participants, has become a benchmark in subsequent years and has been utilized in several instances [1, 2, 4], including contributions [5, 13] of our own. Due to its challenging characteristics this application is in use here once again as a test bed for the experiments of model inversion inspired control techniques.

The plant is supplied as a program that simulates a fed-batch fermentation process producing a secondary metabolite as the product. The microorganism in this process needs two substrates (s_1 and s_2) for growth and production and the process has two inputs, f_1 and f_2 in terms of substrate feed rates. There are five measurements that are x – biomass, s_1 – concentration of substrate 1, s_2 – concentration of substrate 2, p – product concentration and, V – volume. Maximum feed rates and the volume of the fermentor are limited ($f_{\max} = 50, V_{\max} = 4{,}000$). It is believed that the nominal profile as provided with the assignment

$$\begin{cases} f_1 = 10 + \frac{25}{(1+e^{5-0.1t})} \\ f_2 = 3.5 + \frac{3.5}{(1+e^{10-0.15t})} \end{cases} \tag{1}$$

is not good enough for the production. The optimal feed pattern should be investigated in order to improve the process productivity. The productivity criterion J may be selected as: $J = pV/T$, where T is the duration of fermentation. Other process environment variables such as temperature and pH are assumed to be constant (at theirs optimum).

The model representing the plant is not given explicitly (to make the exercise comparable to working on a practical plant) but supplied as a black box with a set of specified inputs and has the features (the initial state varies randomly within a subspace for each batch, the parameters of the model vary within specified limits, a set of non-measurable disturbances, a set of constraints) that make it impossible to obtain the same output series for each batch (also more realistic to the practical situation).

4 Modelling and Inverting the Fermentation Process

The first step in controller design is the generation of the 0-th order Takagi–Sugeno model of the fermentation process having the following structure

$$\begin{array}{c} \text{IF } x \text{ is } A_{1r} \text{ AND } s_1 \text{ is } A_{2r} \text{ AND } s_2 \text{ is } A_{3r} \\ \text{THEN } \Delta x = b_{1r} \text{ AND } \Delta p = b_{2r}, \end{array} \tag{2}$$

where A_{ir} denote the linguistic labels of the input variables and b_{jr} are the consequent singleton values of output variables – growth of biomass (Δx) and growth of the product (Δp), associated with the r-th rule ($i = 1,\ldots,3, j = 1,\ldots,2, r = 1,\ldots,R$).

0-th order TS systems possess computationally inexpensive inference algorithm that gives numerical relationship between the system (2) variables

$$y_j = \frac{\sum_{r=1}^{R} \tau_r b_{jr}}{\sum_{r=1}^{R} \tau_r}, \tau_r = \bigcap_{i=1}^{N} \mu_{ir}(x_i) \tag{3}$$

where τ_r is the activation degree of the r-th rule and μ_{ir} denotes the MF of the i-th input variable (representing A_{ir}) associated with the same rule.

In current application our primary concern is model ability to reveal valid causal relationships between system variables and therefore both transparency and reliability of the model are important requirements. According to [12], transparency can be preserved by imposing certain restrictions to MF parameters as follows.

In order to satisfy input transparency conditions we require triangular input MFs (given by parameters $a_i^s, b_i^s, c_i^s, s = 1, \ldots, S_i$, see Fig. 4) to form a partition (4) – i.e. $a_i^s = b_i^{s-1}, c_i^s = b_i^{s+1}$. Note that with (4), $\forall r (\sum_{r=1}^{R} \tau_r = 1)$.

$$\forall x_i \in X_i : \sum_{i=1}^{S} \mu_i^s(x_i) = 1 \tag{4}$$

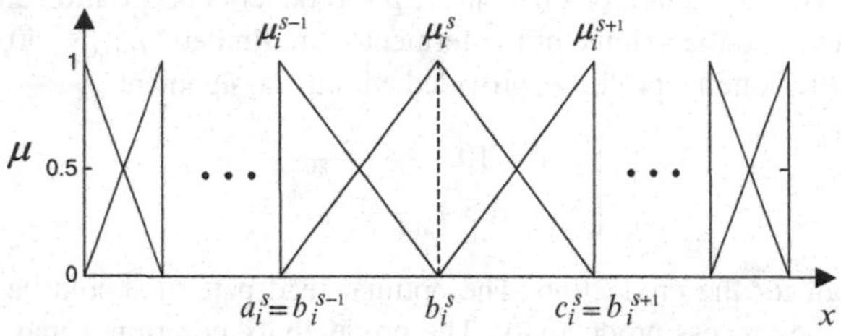

Fig. 4: Input partition style, employed in current paper

The problem of reliability is addressed through the modelling algorithm that we use. First of all, for the identification of the input partition no "real" algorithm is used. Given $S_i (i = 1, \ldots, N)$ input MFs per i-th input variable these are distributed logarithmically along the x-axis:

$$b_i^s = x_i^{min} + (x_i^{max} - x_i^{min})(\frac{s-1}{S_i - 1})^m, s = 1, \ldots, S_i, \tag{5}$$

where b_i^s is the core of the observed MF and $m > 0$ is the exponent. It is easy to see that with $0 < m < 1$ higher values of x_i are modelled with greater resolution, with $m = 1$, we have uniform distribution, and with $m > 1$, smaller quantities of x_i are modelled with greater resolution (as in Fig. 5).

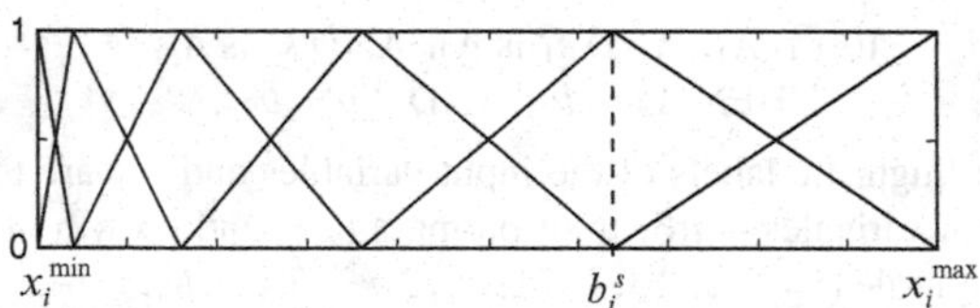

Fig. 5: Logarithmic input partition ($m = 2$)

Given such input partition, it is possible to consider up to $R_{max} = S_1 \cdot S_2 \cdot \cdots \cdot S_N$ rules (all possible antecedent combinations). However, each potential rule is checked for validity before adding it to the model by comparing $max(\tau_r(k)), (k = 1, \ldots, K)$ to some pre-specified threshold value (in current project the threshold value equals 0.5^N) and added only if the former exceeds that. Moreover, consequent parameters of all M output variables of the r-th rule are computed simultaneously with the creation of rules using the method of Nozaki et al. [6]

$$b_{jr} = \frac{\sum_{k=1}^{K} (\tau_r(k))^{\alpha} y_j(k)}{\sum_{k=1}^{K} (\tau_r(k))^{\alpha}}, j = 1, 2, \quad (6)$$

where α is the parameter that influences model accuracy in terms of RMSE (it is reported in [6] that $\alpha = 10$ provides best results in ideal environment and that it should be smaller if data is bad, for current application best results were obtained with $\alpha = 2$). Thus, reliability of the model is addressed in two ways – rule validity routine helps to filter out the rules that are not really evidenced by available training data (and keeps the model compact as well) and as the consequent parameters for the given rule are computed as the weighted average of relevant (measured by rule activation degree τ_r in (6)) output samples, it gives the model interpolating rather than extrapolating character.

Once the model is obtained, the extracted rules are grouped into S_x subsets (S_x is the number of MFs defined for biomass and consequently, each subgroup contains the rules that have the same label of x) of which one rule is selected and inverted. As shown in [13], the selection and inversion task can be performed automatically. All numerical values of (b_{1r}, b_{2r}) corresponding to the rules of given subgroup are fed into respective inputs of the decision making mechanism (Fig. 6) that computes the corresponding pair (τ_1, τ_2). Of the subset of rules the one with maximum τ_1 or τ_2 (whichever happens to be larger) is declared the winner and inverted – rewritten in the format

$$\text{IF } x \text{ is } A_{1r} \text{ THEN } s_1 = p_{1r} \text{ AND } s_2 = p_{2r}, \quad (7)$$

where $p_{1r} = \text{core}(A_{2r})$, $p_{2r} = \text{core}(A_{3r})$. The resulting controller or process manager consisting of S_x rules will then be embedded into the control system, where PI controllers take care of the substrate flow regulation (Fig. 7).

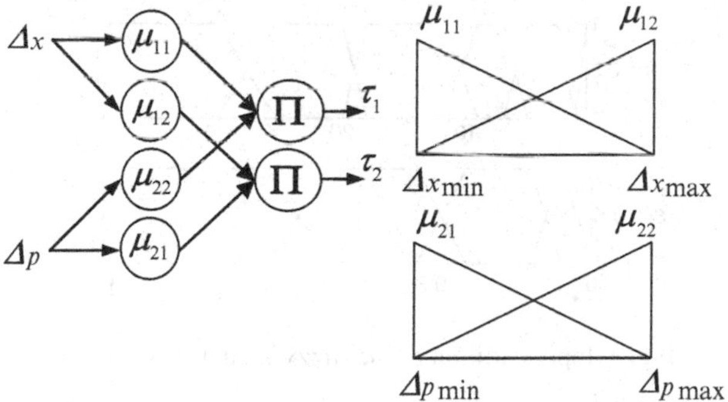

Fig. 6: Decision making mechanism and its parameters

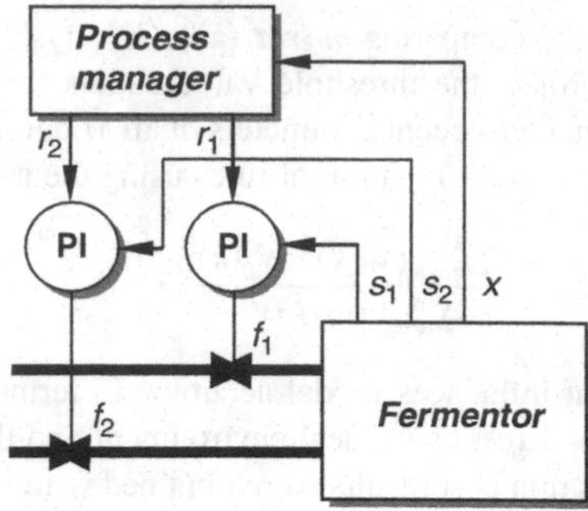

Fig. 7: The control system

5 Results

First, the combined effectiveness of the automatic inversion mechanism and Nozaki's algorithm is tested as the latter is applied to the input partition depicted in Fig. 8 that belongs to one of the successful controllers from [5]. Application of (6) results in a 38-rule (so few of 140 potential rules actually pass to the model through the rule validation filter) model that has root-mean-squared-error (RMSE) 10.75 and 0.18 for Δp and Δx, respectively. In the next step the model is inverted automatically into a four-rule process manager. Averaged J (J_{av}) over ten process runs ($1.32 \cdot 10^5$) is 59% higher than the one corresponding to nominal profiles and 29% higher than the number reported in [5], which really shows that proposed approach is more effective than manual inversion and modelling approach taken in [5].

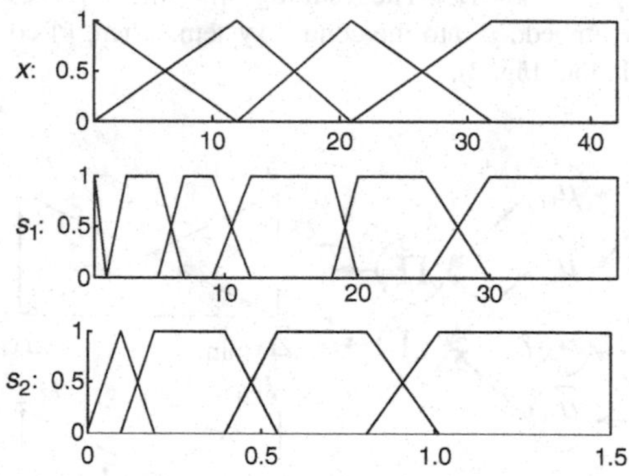

Fig. 8: Input partition of the fuzzy model (from [6])

However, it is not always the case that the input partition is readily available. In such cases using the partitioning style (5) makes sense. We also take into account that in present application the concentration of biomass is regarded as the scheduling variable – its partition divides the process into separate stages for which we need to choose appropriate concentrations of substrates – so it is sufficient to consider few (four or five) uniformly distributed fuzzy sets for x. The substrate concentrations, however, need to be modelled with greater resolution, to provide rich information content. Smaller substrate concentrations are more important (it is unlikely to maintain higher concentrations in the second phase of the process as the volume V grows), therefore in this case the exponent $m = 2$. As it is impossible to predict optimal partition beforehand, several combinations of S_i were tried out. Table 3 contains the results of modelling and control (the figures are the average values of 10 process runs).

Table 3: Results

Case	RMSE($\Delta x/\Delta p$)	Partition	R	J_{av}	p_{av}	T_{av}
1	0.242/12.88	4-8-8	47	$1.10 \cdot 10^5$	3255	117
2	0.259/9.663	5-8-8	51	$1.22 \cdot 10^5$	3012	97
3	0.274/11.46	4-9-9	56	$1.25 \cdot 10^5$	3227	102
4	0.268/10.30	5-9-9	62	$1.31 \cdot 10^5$	3063	92
5	0.225/8.71	4-10-10	66	$1.13 \cdot 10^5$	3411	119
6	0.225/8.59	5-10-10	73	$1.23 \cdot 10^5$	3263	105

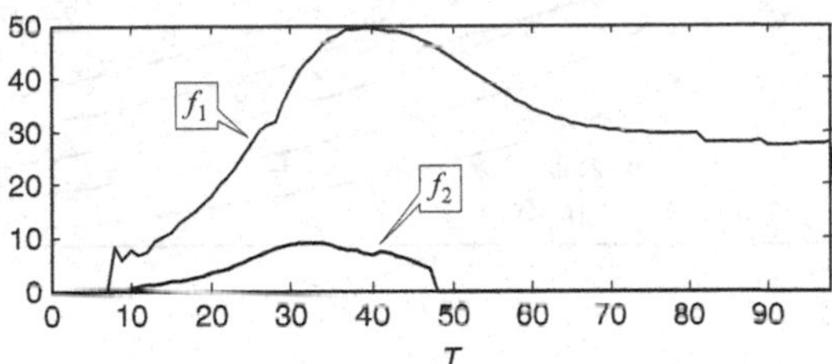

Fig. 9: Averaged feed profiles

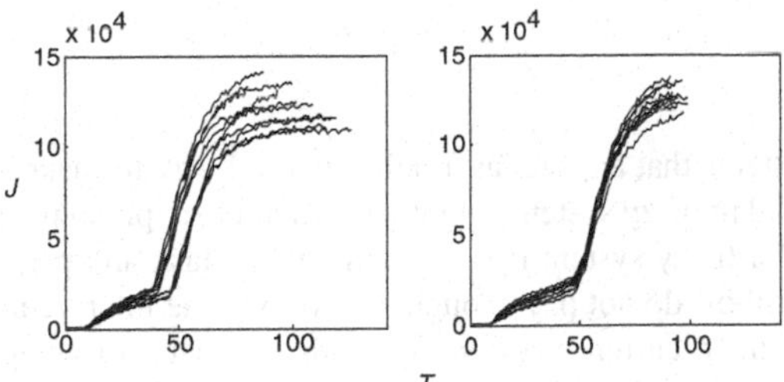

Fig. 10: Control system performance over 10 process runs. Supervisory control system (left), averaged feed profiles (right)

There are yet additional ways for improving both the performance and consistency of control. First, the problem with the supervisory control system depicted in Fig. 7 is the high variance of results – e.g. process time (case 3) varies from 80 to 124 time units and product concentration from 2,700 to 3,400 during different runs. In order to provide more stability, we use averaged feed profiles (Fig. 9) of 5 best process runs of tested 10 computed by the supervisory control system, which lead to much more consistent (as can be seen from Fig. 10) and further improved ($J_{av} = 1.28 \cdot 10^5, T = 94 \pm 4, p_{av} = 3023(2855 - 3130)$ results.

Secondly, from Fig. 9 it is obvious that feed rate in the first quarter of the process is rather small, which is ultimately the reason why it takes quite long time fill up the fermentor. The process can be sped up by manipulating the value of p_{21} – the setpoint of s_2 corresponding to the first quarter of the process – which currently is equal to 0.0235. When this is done in the range of $0.02 - 1.5$, it appears that what we have got here is a very simple yet effective knob to influence process productivity and duration (Fig. 11). It might be asked, why importance of the value of p_{21} was not discovered during the modelling and inversion procedure in first place, but the reason is really simple – nominal feed profiles do not produce very high concentrations of the substrate in the first quarter of the process, thus there was simply nothing to discover.

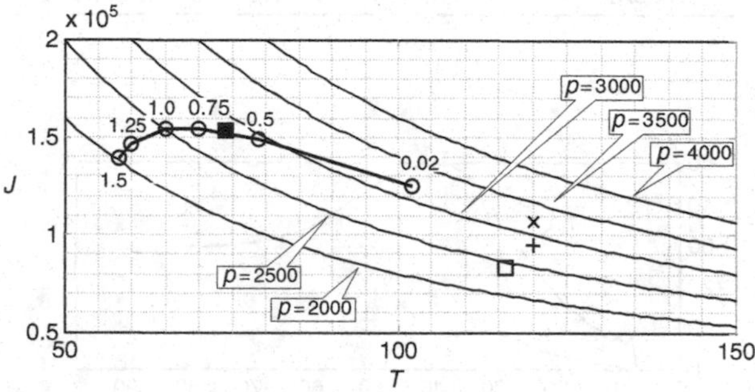

Fig. 11: Productivity vs. process duration. Comparison: ○ – current results (with different values of p_{21}, which are indicated in the figure), ■ – Liang [2], + – DISOPE [1], × – Zhang [4], □ – nominal feed profiles.

6 Conclusions

Inversion of fuzzy systems that appears as a rather natural way to make use of the linguistic knowledge accumulated in fuzzy systems, is rather difficult to apply in practice. Partly because full-scale inversion of a fuzzy system is simply unrealistic task and partly because available fuzzy modelling algorithms do not pay enough attention to the interpretability and reliability properties of the model. In current work we have shown that narrowing the task to the inversion of the desired profile of specified variable, automation of the inversion procedure and application of the modelling algorithm that enforces transparency and reliability properties of the model, substantially simplifies this task and pays back in terms of control performance.

When assessing the performance of the benchmark fed-batch fermentation process controller that was designed using the proposed methodology, it is of interest that other known attempts at the same benchmark that we quoted here for comparison purposes violate at least one of the premises of the original assignment, e.g. [1,4] ignore the fermentor volume limitation and manage to obtain their results only at the expense of that. The application in [2] on the other hand makes explicit use of equations hidden in the black-box model and has therefore access to supposedly forbidden information.

Acknowledgements

This work was partially supported by Estonian Science Foundation, grant no. 6837.

References

1. Becerra V M, Roberts P D (1998) Application of a novel optimal control algorithm to a benchmark fed-batch fermentation process. Trans Inst MC 20(1):11–18
2. Liang J, Chen Y (2003) Optimization of a fed-batch fermentation process control competition problem using the NEOS server. Proc Inst Mech Engrs Part I: J Syst Contr Eng 217(5):427–432
3. Takagi T, Sugeno M (1985) Fuzzy identification of systems and its applications to modeling and control. IEEE Trans Syst Man Cybern SMC-15(1):116–132
4. Zhang H, Lennox B (2003) Multi-way optimal contol of a benchmark fed-batch fermentation process. Trans Inst MC 25(5):403–417
5. Riid A, Rüstern E (2000) Supervisory fed-batch fermentation control on the basis of linguistically interpretable fuzzy models. Proc Estonian Acad Sci Eng 6(2):96–112
6. Nozaki K, Ishibuchi H, Tanaka H (1997) A simple but powerful heuristic method for generating fuzzy rules from numerical data. Journal of Fuzzy Sets and Systems 86:251–270
7. Baranyi P, Bavelaar I M, Babuška R, Kóczy L T, Titli A, Verbruggen H B (1998) A method to invert a linguistic fuzzy model. Int J Syst Sci 29(7):711–721
8. Jang J-S R, Sun C-T, Mizutani E (1996) Neuro-Fuzzy and Soft Computing: A Computational Approach to Learning and Machine Intelligence. Prentice-Hall, Upper Saddle River, NJ
9. Fantuzzi C (1994) Linguistic Rule Synthesis of a Fuzzy Logic Controller. Proceedings of IECON'94, pp 1354–1358
10. Raymond C, Boverie S, Titli A (1995) Fuzzy multivariable control design from the fuzzy system model. In: Proceedings of the IFSA World Congress, Sao-Paulo, Brazil, pp 509–511 citeseer.ist.psu.edu/484086.html
11. Braae M, Rutherford D A (1979) Theoretical and linguistic aspects of the fuzzy logic controller. Automatica 15(5):553–557
12. Riid A, Rüstern E (2003) Transparent Fuzzy Systems in Modeling and Control. In: Casillas J, Cordon O, Herrera F, Magdalena L (eds) Interpretability Issues in Fuzzy Modeling. Springer, Heidelberg, pp 452–476
13. Riid A, Rüstern E (2006) Automatic Linguistic Inversion of a Fuzzy Model for Fed-Batch Fermentation Control. In: Proceedings of 2006 International Conference on Intelligent Engineering Systems, London, pp 129–134

Transformations and Selection Methods in Document Clustering

Kristóf Csorba[1] and István Vajk[2]

[1] Department of Automation and Applied Informatics, Budapest University of Technology and Economics, Goldmann Gy. tér 3, Budapest 1111, Hungary
kristof@aut.bme.hu

[2] Department of Automation and Applied Informatics, Budapest University of Technology and Economics, Goldmann Gy. tér 3, Budapest 1111, Hungary
vajk@aut.bme.hu

Summary. Document clustering is an important and widely researched part of information retrieval. It aims to assign natural language document to various categories based on some criteria. In this case, this criteria is the topic of the document, which means, that the goal is to identify the topic of documents and group the similar ones together. As there are many clustering methods and noise filtering techniques to support this procedure, this paper focuses on the composition of such transformations and on the comparison of the configurations built from a subset of these transformations techniques as tiles of the whole procedure. Altogether five tile methods (term filtering, frequency quantizing, singular value decomposition (SVD), term clustering (for double clustering) and document clustering of course) are used. These are compared based on the maximal achieved F-measure and time consumption to find the best composition.

1 Introduction

Topic based document clustering aims to separate a huge amounts of documents according to their topic. As there isn't an exact definition for topic, there are many ways to identify it. A frequently used approach is based on common terms in the documents: two documents are said to be of similar topic if they share many terms. That means, documents with many common terms should get into the same category after the clustering. This model considers documents as bags of terms, the order of the terms is not used. As even humans often disagree how to order a set of documents into two groups, there are many ways for defining the optimal solution. The reason for this is the hierarchic and overlapping nature of topics.

Vector space models [1] create a feature space to characterize the documents and comparisons are performed after a transformation into this space. The first step is usually the creation of the term document matrix, which contains the relevance of each term for each document. Many weighting schemes are used to estimate this relevance. In this paper we are going to use occurrence numbers, which means the number of occurrences of a term (represented by a row) in a document (represented by a column). This matrix is created as a result of parsing of the documents.

J.A.T. Machado et al. (eds.), *Intelligent Engineering Systems and Computational Cybernetics,*

One could say that as the (unsupervised) clustering of the documents is equivalent to the clustering of the column vectors of the term document matrix, a clustering method could be applied directly to this matrix to get the solution. The most important problem is caused by the terms with similar meaning: even documents with very similar topic tend to contain different words (like synonyms), which makes their column vectors too different: without further knowledge on the nature of terms, the differences caused by synonyms and completely different terms cannot be distinguished: The distance between "car" and "auto" is the same, as the distance of "newspaper" and "mountain". This fact leads to the need of further examinations in the term document matrix.

In this paper we examine the usefulness of some transformations in the feature space in respect to the support of document clustering. In Section 2 we provide a brief description of these transformations as "tiles" of the clustering procedure. These include term filtering (Subsection 2.1), term frequency quantizing (Subsection 2.2), singular value decomposition for dimensionality reduction (Subsection 2.3), term clustering (the key of double clustering, Subsection 2.4) and document clustering of course (Subsection 2.5). After describing the transformations, the best configuration is selected (Section 3) and as the system can be very sensitive to the correct parameter settings, the most important parameters are examined in Section 4. Finally conclusions are made in Section 5.

Table 1: The most important matrices and scalars used in this paper

Notation	Interpretation
X	*n*-by-*m* term document matrix
Y	*t*-by-*m* term-cluster document matrix
T	*t*-by-*kt* feature matrix of terms after SVD
n	Number of used terms
m	Number of documents
kt	Number of term features after SVD
t	Number of term-clusters
kd	Number of document features after SVD
d	Number of document-clusters

2 Tiles for the Document Clustering

An overview of the notations is presented in Table 1, which contains the most important matrices and scalars used in this paper.

The process of unsupervised document clustering starts in this case by generating the n-by-m term document matrix $\mathbf{X}$. The element $\mathbf{X}_{i,j}$ is the occurrence number of the i-th term in the j-th document.

In the following the possible tile methods used in the experiments are described in detail.

2.1 Term Filtering

Term filtering is an essential way to reduce the dimensionality of the feature space. Although there are many much more sophisticated term filtering techniques, we use a relatively simple and fast method. There are many terms in a document, which do not help us in identifying the topic. These can be very rare terms occurring only for instance in 1% of the documents, or they are too frequent, occurring for example 99% of the documents. The former are removed based on their frequency (rate of documents they appear in), and the later ones based on a stopword list. Stopword removal is a very common step in natural language processing, which removes words like "the", "a", "with", etc. from the documents. In this case, after stopword removal a frequency based filtering was applied to all the documents: all terms occurring rarer than a given limit in average are removed. (Other possibilities would be for instance techniques based on information gain [2] or part of speech tagging.)

Given the term document occurrence number matrix $\mathbf{X}$, using a norm defined by the sum of occurrences

$$\|\mathbf{v}\| = \sum_{i=1}^{n} \mathbf{v}_i \tag{1}$$

the term filter removes terms (row vectors of $\mathbf{X}$) i for which

$$\|(\mathbf{X})_{i,\bullet}\| < MinFrequency \tag{2}$$

The remaining rows build the $\mathbf{X}$ used in the ongoing processes. As this filtering step is very important due to time consumption reasons, all the tested configurations in our experiments contain this tile as the first step.

2.2 Quantizing Filter

The related literature presents many solutions for the calculation of the values in the term document matrix. The most common group of techniques uses some kind of TFIDF scheme, which means that some function of the term frequency (TF) and the inverse document frequency (IDF) is employed. Although people often simply multiply TF and IDF, there are much more sophisticated functions like the Okapi weighting and the Pivoted normalization weighting [1]. The reason why we use simply the occurrence number of a term in the given document is the employment of this quantizing filter based on the following assumption: Considering the nature of terms, the exact number of occurrences of a given term in a document is relative unimportant from the topic's point of view. One can define frequency categories like "never", "rare", "frequent", but a deeper level of granularity makes only the comparison of the documents harder. (The usefulness of more allowed values is examined as an important parameter of the procedure in Section 4).

Based on this assumption, a frequency quantizing filter can be employed to remove unnecessary diversity. The quantizing procedure assigns new values to the elements of $\mathbf{X}$ based on a quantizing list $\mathbf{q}$, which contains the possible quantized values, like $[0,1,2]$ for example.

$$quantize(x) = \max_i \{\mathbf{q}_i \le x\} \tag{3}$$

$$\mathbf{X}_{i,j} = quantize(\mathbf{X}_{i,j}) \tag{4}$$

Bypass mode: As this tile may be turned off, in "bypass" mode the quantizing is not performed and the original values are left unchanged.

2.3 Singular Value Decomposition

The singular value decomposition [3] is a matrix-analytical method to search for a space, where a given linear transformation (usually given by a rectangular matrix) is a scaling along the base vectors. More precisely if $\mathbf{X}$ is an n-by-m matrix, SVD has a result in the form

$$\mathbf{X} = \mathbf{USV}^T \tag{5}$$

where $\mathbf{S}$ is the r-by-r diagonal matrix of the singular values of $\mathbf{X}$ (r is the rank of $\mathbf{X}$) and $\mathbf{U}$ and $\mathbf{V}$ have orthonormal columns. SVD of $\mathbf{X}$ is very close to the eigenvalue decomposition of $\mathbf{X}^T\mathbf{X}$ into the form $\mathbf{VLV}^T$, where $\mathbf{V}$ contains the eigenvectors in its columns and $\mathbf{L}$ the corresponding eigenvalues. In fact, $\mathbf{U}$ consists of the eigenvectors of $\mathbf{XX}^T$ and $\mathbf{V}$ of the eigenvectors of $\mathbf{X}^T\mathbf{X}$. $\mathbf{L}$ is equivalent to $\mathbf{S}^2$ in this case.

In a statistical sense, singular value decomposition is an effective method to reduce the dimensionality of observations by preserving as much information as possible. The r-th principal component is a probabilistic variable, which has maximal deviation among all probabilistic variables which

- Have the form $\mathbf{b}^T\mathbf{X}$, where $\mathbf{b} \in \mathbb{R}^p$ and $\|\mathbf{b}\| = 1$
- Are uncorrelated with the first $r-1$ principal components

If the row vectors of $\mathbf{U}$ and $\mathbf{V}$ are taken geometrically as coordinates in the m dimensional space, the terms and the documents became points in a vector space. In this "feature space" the coordinates are linear combinations of documents (or terms respectively). The diagonal of $\mathbf{S}$ serves as a scaling of the axes in this space reflecting the contribution of the dimensions in the overall similarity structure. As the singular values in the diagonal of $\mathbf{S}$ are provided in descending order, one can easily select the most significant components.

The SVD step can be used to generate a reduced dimensionality representation of both the terms and the documents. For a compressed term representation $\mathbf{T}$ can be retrieved as follows:

$$\mathbf{T} = \mathbf{U}_{\bullet,1..kt}\mathbf{S}_{1..kt,1..kt} \tag{6}$$

In the feature space terms with similar occurrence behaviors are located near each other while words related to different topics (and used in different documents) are farer in the sense of cosine distance. (The cosine distance of two vectors $\mathbf{A}$ and $\mathbf{B}$ is defined as $distance = 1 - \cos\varphi$, where $\cos\varphi = \frac{\mathbf{AB}}{|\mathbf{A}||\mathbf{B}|}$.)

A similar compressed representation of the documents can be retrieved from the $\mathbf{V}$ matrix for the feature space reduction of the document vectors. This tile is referred to as "Document SVD".

In the next step, $\mathbf{T}$ can be used as a basis of the term clustering instead of the row vectors of $\mathbf{X}$.

An example for the result of the principal component analysis can be seen in Fig. 1, where large dots and triangles represent terms occurring in only one of the document classes and small dots stand for common terms.

Bypass mode: The SVD produces a lower dimensional approximation of the original matrix. If this transformation is turned off ("bypass" mode), the original $\mathbf{X}$ is used to calculate the $\mathbf{T}$ matrix.

2.4 Term Clustering

Term clustering (the key of double clustering [4]) means the clustering of the terms before document clustering to enablea clustering of the documents based on a term-cluster document

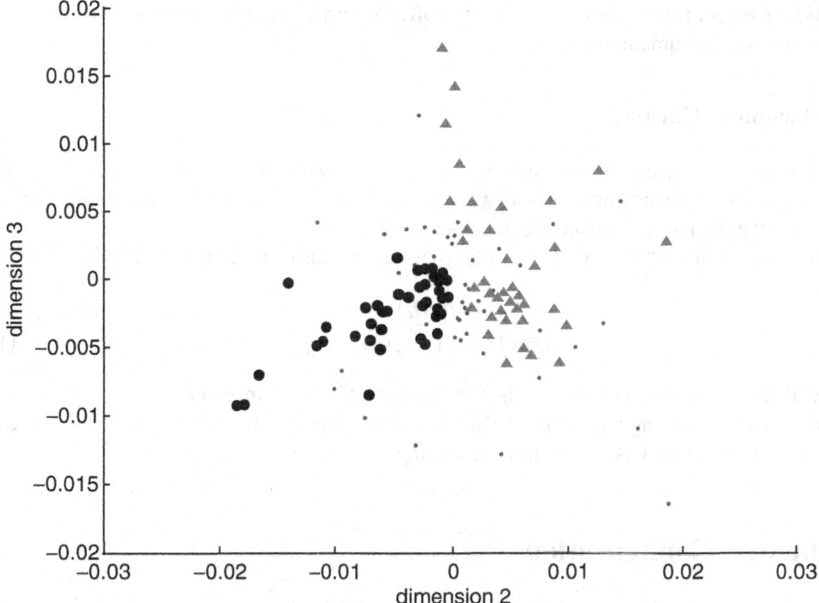

Fig. 1: Terms in the space of two useful components after SVD

matrix instead of the term document matrix. Terms are assigned to clusters and row vectors of the term document matrix belonging to the same term-cluster are merged together.

Term clustering can be performed based on the **T** matrix retrieved from a previous SVD step or directly based on **X** (if the SVD tile is turned off). The process based on the former way is as follows:

$$j_i = \alpha(\mathbf{T}_{i,\bullet}) \tag{7}$$

$$\alpha^{-1}(j) = \{\mathbf{T}_{i,\bullet} | \alpha(\mathbf{T}_{i,\bullet}) = j\} \tag{8}$$

The function $\alpha(\mathbf{T}_{i,\bullet})$ returns the index of the cluster the given term (represented by $\mathbf{T}_{i,\bullet}$) is assigned to. $\alpha^{-1}(j)$ returns the set of terms assigned to the $j-th$ term cluster. In our experiments the k-means algorithm was applied as the α function. The merging of the terms in a cluster is the simple addition of their representing row vectors:

$$\mathbf{Y}_{j,\bullet} = \sum_{\mathbf{v} \in \alpha^{-1}(j)} \mathbf{v} \tag{9}$$

Y is the term-cluster document matrix containing term-clusters instead of terms as row vectors. The documents are represented as column vectors of **Y**. If SVD is not employed to reduce document feature dimensionality, $\mathbf{D}_i = \mathbf{Y}^T_{\bullet,i}$. The values of **Y** show, how frequent the terms of a term-cluster are among the documents. **Y** can be a basis of the document clustering in the next step instead of the **X** term document matrix.

Bypass mode: If term clustering is turned off, the term document matrix is used instead of the term-cluster document matrix.

2.5 Document Clustering

Please note, that according to Subsection 2.3, a singular value decomposition may be applied to the term-cluster document matrix as well, which means a further low-dimensional approximation of the matrix, just before the document clustering.

The document clustering step is similar to the term clustering as follows:

$$j_i = \beta(\mathbf{D}_i) \tag{10}$$

$$\beta^{-1}(j) = \{\mathbf{D}_i \mid \beta(\mathbf{T}_i) = j\} \tag{11}$$

As this step is essential to our goals, this tile method is used in every tested configurations, just like the term filtering mentioned earlier. In our experiments the β function is the k-means algorithm just as in the case of the term clustering.

3 The Best Configuration

To measure the capabilities of the tiles a pipe architecture (Fig. 2) was built, where most tiles can be activated or deactivated independently.

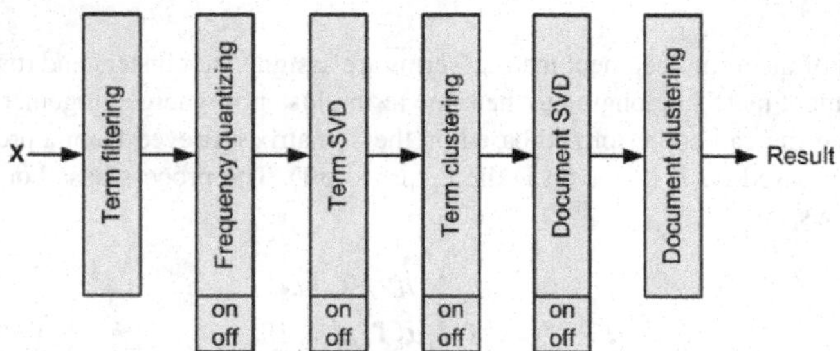

Fig. 2: Comparison of configurations

As the testing data set, a part of the 20Newsgroups corpus ([5]) was selected: 250 documents from the "auto" and "graphics" section each. To find the best configuration, the behavior of the configurations were simulated and compared based on the widely used performance measure F-measure:

$$F = \frac{2 \cdot recall \cdot precision}{recall + precision} \tag{12}$$

F-measure is the weighted harmonic mean of precision and recall, also known as F_1 measure, because recall and precision are weighted evenly. If a cluster is trying to capture the documents of a document class, recall is the fraction of all documents from that class that are returned in the cluster. Precision is the rate of relevant documents in the set of all documents returned by the given cluster.

The results of the method comparison are shown in Fig. 3. "Q" stands for the activated quantizing filter, "tSVD" for the term dimensionality reduction with SVD, "tClust" for term clustering and "dSVD" for the document dimensionality reduction (with SVD). The maximal achieved F-measure is taken over a wide range of parameter settings examined.

Based on the measurements the configuration resulting in the best performance contained frequency quantizing, term SVD and term clustering. (Document SVD was not employed in this case.)

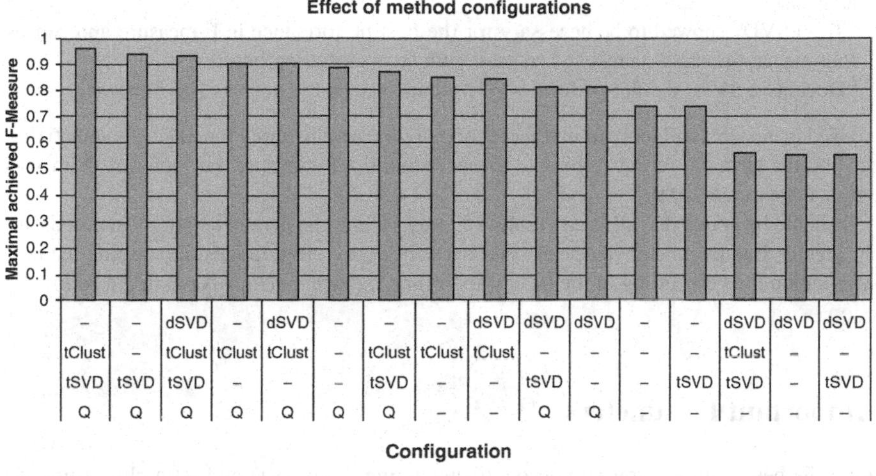

Fig. 3: Comparison of configurations

The second most important aspect of performance is the time consumption. Figure 4 shows, that the best configuration above is relative good in this aspect as well. Configurations with less active components are faster of course, but the little additional time is paying in the F-measure aspect.

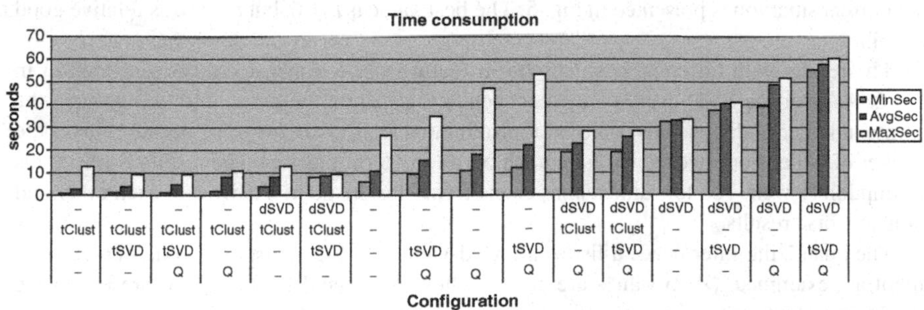

Fig. 4: Time consumption of configurations

In general, we can observe the followings:

- The quantizing filter is essential for F-measures over 0.9. It is active in every configuration with good clustering capabilities. This means that the assumption, that the exact number of occurrences is not important, seems to be correct. The additional time need caused by this step seems not to be relevant.
- The term clustering (double clustering) is important for the good clustering performance as well, but its most important rule is to decrease the time consumption: all the fastest configurations contain term clustering. (Good clustering capability can be achieved without it as well.)
- "Term SVD" showed to be necessary for the best performance in F-measure and causes minimal additional time need when used with term clustering before it.
- "Document SVD" did not perform very well and seemed to decrease performance.

Used in its own (with only term filtering and document clustering together) the quantizing filter and the term clustering were able to overcome 0.8 F-measure, but none of them was enough alone to reach 0.9.

It should be noted, that all these results are only valid if the parameters of the methods are set correctly. Bad parameter values can ruin even the best configurations. Examining the best configuration the best setting of the most important parameters are analyzed in the following section.

4 Important Parameters

Taking the best configuration into account (quantizing, term SVD and term clustering) the followings are the most important parameters:

- **q** is the list of valid values after quantizing the term occurrence numbers.
- *kt* is the new dimensionality of the term feature vectors produced by term SVD.
- *t* is the number of term-clusters created by the term clustering.

The *t* number of term clusters has to be set intuitively to a value, which enables the separation of enough subtopics to distinguish between the terms of the document classes (standing for the two topics "auto" and "graphics"). This value usually moves between 5 and 12 clusters. The current situation is presented in Fig. 5. The best value is $t = 6$, but $t = 9$ was relative good as well.

The results with different *kt* values (term feature space dimensionality) can be seen in Fig. 6. As after the SVD every component tends to separate the terms into two groups, the optimal number of dimensions could be approximated as $kt_{opt} = \lceil log_2(t) \rceil$, because this is the number of dimensions needed to distinguish between *t* term-clusters. The results confirm this assumption as well. Too few dimensions cause term-clusters get too narrow to each other and leads to worse results.

The quantizing filter needs a list of allowed values. In Fig. 7 lists with one non-zero element are examined. (Zero values are always allowed to stand for terms not present in the document.) The best setting is clearly the list $[0; 1]$ which means a binary term document occurrence matrix. If we examine quantizing lists with two non-zero elements, performance is shown in Fig. 8. None of the settings showed an ability to overcome the performance of the $[0; 1]$ list. This is an interesting result as it states the uselessness of more than one non-zero quantizing values.

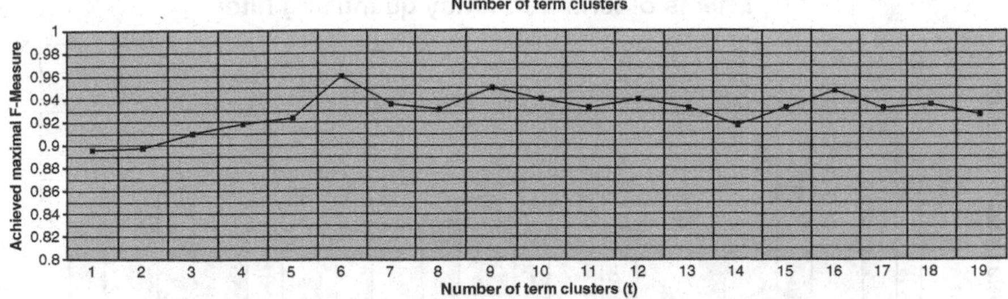

Fig. 5: Number of term clusters

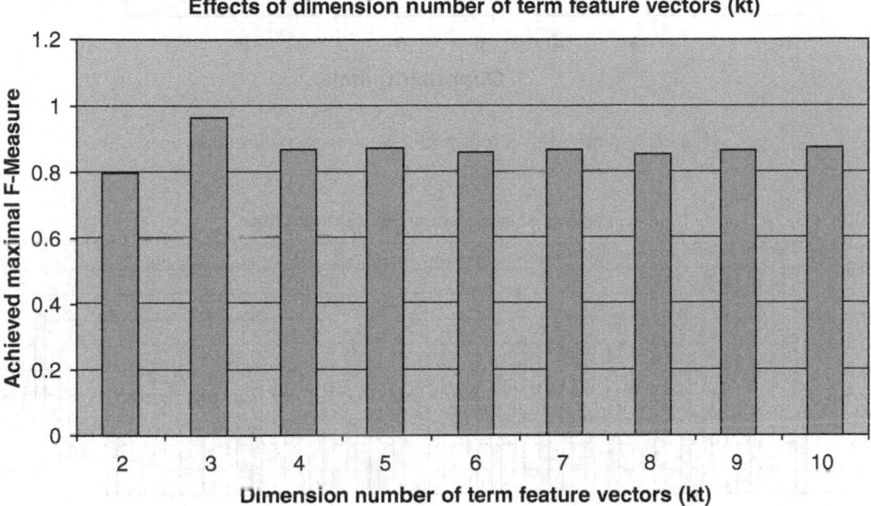

Fig. 6: Dimension number of term feature space

5 Conclusions

Using the best configuration, our system could reach 0.96 F-measure, which is very promising. About the tiles of the clustering system, the following conclusions can be made:

- Various settings of the quantizing filter were thought to bring more interesting results, but finally the pure binary quantizing was the best. But the binary quantizing was of essential importance.
- SVD was already known to be effective. The performance together with the other components was shown to be even better in the term-clustering phase, but not in the feature space of the documents, where it could not improve the performance.
- The term clustering was a very effective component: it provides not only strong dimensionality reduction and acceleration, but is effective against the problem of synonyms as well.

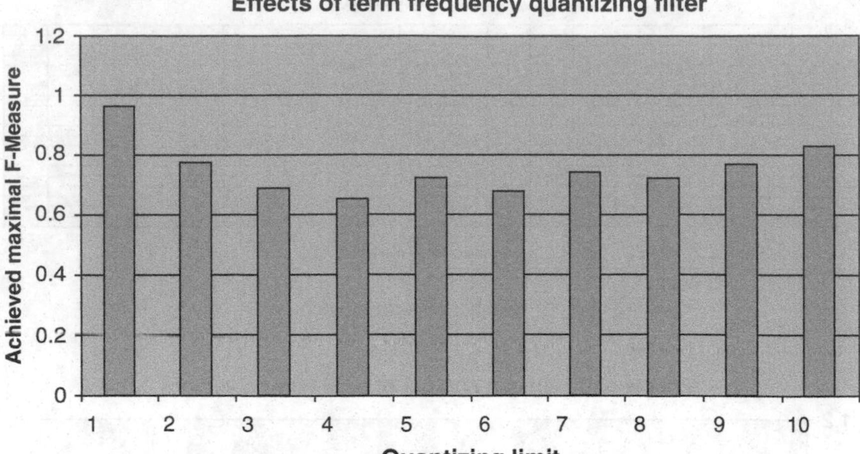

Fig. 7: Quantizing vectors with one non-zero elements

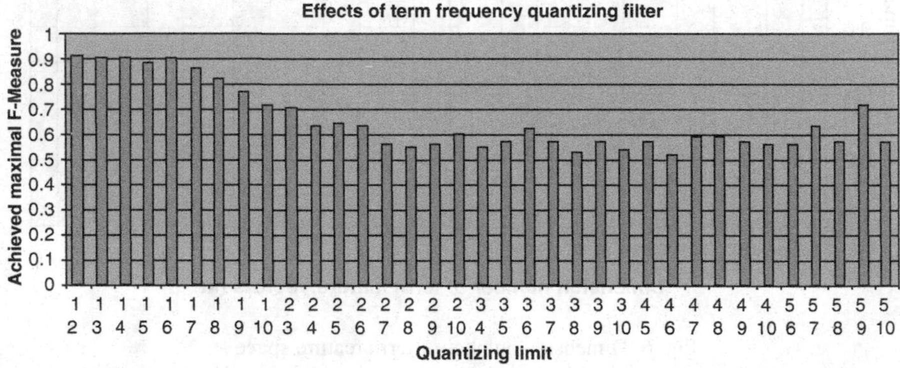

Fig. 8: Quantizing vectors with two non-zero elements

The setting of the parameters influence the result very strongly, thus the selection of the optimal values is important.

Acknowledgments

This work has been funded by the Hungarian Sciences for control research and the Hungarian Research Fund (grant number T042741).

References

1. Singhal A (2001) Modern information retrieval: A brief overview. IEEE Data Engineering Bulletin 24(4):35–43
2. Li L, Chou W (2002) Improving latent semantic indexing based classifier with information gain. Technical Report May 16
3. Furnas G, Deerwester S, Dumais S T, Landauer T K, Harshman R, Streeter L A, Lochbaum K E (1988) Information retrieval using a singular value decomposition model of latent semantic structure. In: Chiaramella Y (ed), Proceedings of the 11th Annual International ACM SIGIR Conference on Research and Development in Information Retrieval (Grenoble, France), pp 465–480, ACM
4. Slonim N, Tishby N (2000) Document clustering using word clusters via the information bottleneck method. In: Proceedings of the 23rd Annual International ACM SIGIR Conference on Research and Development in Information Retrieval, Clustering, Athens, Greece, pp 208–215
5. Lang K (1995) Newsweeder: Learning to filter netnews. In: ICML, Tahoe City, California, USA pp 331–339

F-Logic Data and Knowledge Reasoning in the Semantic Web Context

Ana Meštrović[1] and Mirko Čubrilo[2]

[1] Faculty of Philosophy, Department of Information Sciences, Omladinska 14, 51000 Rijeka, Croatia
akaic@ffri.hr

[2] Faculty of Organization and Informatics, Department of the Theoretical and Applied Foundations of Information Sciences, Pavlinska 2, 42000 Varaždin, Croatia
mirko.cubrilo@foi.hr

Summary. The paper addresses problems of data and knowledge reasoning in the domain of Semantic Web. The objective of this research was to explore object-oriented logic programming languages in the light of Semantic Web concept. It will be shown that logical formalisms integrating the deductive component and the object-oriented approach and featuring the second-order syntax and the first-order semantics provide an adequate Web data representation and reasoning. The paper describes elements relevant for Web data representation and integration for chosen data domain.

1 Introduction

Semantic Web assumes a new generation of the Web. A new element that emerged in the representation of knowledge within the Semantic Web is the meta level of knowledge containing a formal description of data on the Web pages and representing an extension of data model at semantic level [4]. The concept of Semantic Web has been particularly topical in recent years and it has currently been in the production stage. The Semantic Web data and Web data in general are featured by the object-oriented approach. Besides that, when handling such data some other particularities should be considered (semi-structured data, representation of data schema, etc.). In that light it is necessary to choose an adequate language for data and knowledge reasoning. The language should be semantically rich and formal enough to enable reasoning in the Semantic Web context.

Web data reasoning underlines the problem of schema-less and heterogeneous data reasoning. Generally, the problem of knowledge representation, data modeling and data manipulation, has been one of the central problems in the development of the Semantic Web. Numerous papers and researches have addressed the development and analysis of the language and data structure used for representation of Semantic Web data [1,2,4]. Previous researches resulted in languages and data structures that enable formal representation of data on the Web, such as XML, XML Schema, RDF, RDF Schema, OWL and other ontological languages. On the other hand, language development projects have been initiated in order to enable data manipulation, querying and bounding data together on the Web pages [1,2,4,16].

J.A.T. Machado et al. (eds.), *Intelligent Engineering Systems and Computational Cybernetics,*

In this paper it is interesting to explore the potentials of existing logical formalisms, logical languages and deductive databases. The F-logic is a formalism that supports the object-oriented approach and the deductive component, thus being a proper choice for the development of Semantic Web concepts. Some of the papers have considered theoretical potentials of such languages [3, 5, 9, 20], however, some practical solutions have seen the light as well [6, 7, 19]. This paper discusses potentials of the F-logic in the domain of the Semantic Web data modeling. Also, the potentials of XSB and Flora-2 systems have been analyzed. Flora-2 system is an implementation of F-logic integrated in the XSB system.

There is a short introduction into F-logic given in the second section. The third section presents special data features on the Web. In the fourth section different approaches for Web data manipulation are explained. In section five the integration process is described. The sixth section describes the development of the reasoning system in the Semantic Web Context. Finally some possible improvements are discussed and some future work plans are presented.

2 F-Logic

2.1 Introduction to F-logic

The frame logic (abbr., F-logic) is a formalism connecting object-oriented approach, frame languages and logical reasoning. The theoretical approach on which F-logic is based was first presented in 1995 in paper [10]. The paper offers definitions of the F-logic syntax and semantics as well as the entire theoretical base of F-logic. Shortcomings in previous attempts of connecting logic programming and object-oriented approach inspired the introduction of F-logic in logic programming. The authors of F-logic extend the classical predicate calculus aiming to define a logic that would enable inferring in object-oriented databases. At syntax level, F-logic is extended with a set of additional symbols, while at semantics level formulas of F-logic assume meaning where implementation the basic concepts of object-oriented approach. is possible. Defined in such way, the F-logic retained some quality properties of the classical predicate logic in terms of defining derivation rule that is analogous to the resolution with unification procedure as in the classical logic.

One of the F-logic implementations was achieved in system Flora-2 [18, 19]. The Flora-2 system as part of XSB system joins three logical formalisms: F-logic, HiLog and Transaction logic. This paper considers F-logic properties, but elements of HiLog and Transaction logic are also used for reasoning with Web data. HiLog enables processing of meta data [5]. Transaction logic supports the manipulation of procedural knowledge in the scope of declarative programming. Higher order logic formalisms enable more natural data modeling, meta-querying and representation of knowledge in general. Besides, the design of Flora-2 system was influenced by some requirements made by the Semantic Web. Recently, system designers, who are also authors of theoretical formalisms, recognize the potential of implementing F-logic in the Semantic Web and they extend the existing system in that direction. In [20] it is stated that a system has been designed in line with various tasks performing and automating of the Semantic Web, ranging from management at meta level to information integration and intelligent agents. These resulted in a flexible system that combines a system of rules and object-oriented paradigm.

2.2 Semantic Web Requirements Realized in F-Logic

Languages of the Semantic Web, that are declared as standard by W3C[1] organization, i.e., XML, RDF, OWL, assume hierarchical data representation and presentation of knowledge through concepts of classes, objects and attributes. Another major feature of the Semantic Web data is expressing semantics through data scheme. Besides, data on the Web have a range of specific characteristics such as incompleteness of data, meaning that the chosen language should be flexible enough to enable representation of such data. The following three sections present as to how basic features of the Semantic Web data can be realized in F-logic, i.e., in Flora-2 system.

Implementation of object-oriented approach

F-logic is structured so that it supports basic concepts of object-oriented programming. Object-oriented languages comprise expressive constructs that describe data in a more natural way so that an object-oriented database is also a more natural environment for the Web data in comparison to a relational database. F-logic concepts are prominent in respect to requirements of the Semantic Web data modeling and reasoning in Flora-2 system.

Basic formulas of the F-logic show objects and their respective attributes or methods `(object[attribute=>value])` and appurtenance of an object to a class `(object : class)` as well as appurtenance of a subclass to a superclass `(subclass : superclass)`.

Other important concepts of the object data representation refer to classes, methods, types, inheritance, encapsulation and method overloading. All these elements are supported through a set of expressive elements in F-logic and implemented in Flora-2 system. In Table 1 there is a short description of some object-oriented concepts given.

Table 1: Implementation of OO concepts in F-logic

	Realization in F- logic
Types	In F-logic it is possible to specify object types. Specification of object types is expressed through signature expressions.
Inheritance	**The F-logic support inheritance. The concept of inheritance is derived as inheriting of superior classes methods. Problems with inheritance as well as possible paradoxes have been considered at theoretical level.**
Encapsulation	**Encapsulation was applied by introduction of modular programming and concepts 'public' and 'private' into the Flora-2 system.**

[1] W3C – World Wide Web Consortium

Data scheme representations and meta queries

F-logic includes two levels of data representation: data level and schema level. Schema level is also called meta level. These two levels are distinguished in terms of being data expression or signature expression. Various types of designations are used at schema and data level so that it is always possible to make a distinction between levels.

Languages containing meta level must provide a query facility for making queries on meta structure. It should be noted that, at the level of queries and data analysis in general, F-logic query language provides query facility for both data and schema without a different query languages being defined for schema and data respectively.

The second order syntax, as implemented in HiLog, actually results with the fact that a variable can take predicate place, so the following queries are legal in HiLog: `?- X(p)` or `?- p(X), X, X(Y,X)`. In combination with the F-logic, the second order syntax provides queries facility on data schema. In that way retrieval of any data referring to methods of certain classes and objects is possible. It is also possible to check hierarchical structure and interrelations among objects and classes. For example, query returning names of all attributes (noninheritable in this particular case) for a class conference is defined as follows:

```
?-conference[Methd=»_];
conference[Methd=»_].
```

In that light Flora-2 system is considerably simpler than some other systems as it is quite common to define a separate language for querying data scheme, which language cannot be combined with query language intended for data.

Meta queries are extremely important for accomplishing reasoning in the Semantic Web context because querying the data meta level is required for deriving new knowledge. Some languages used for the representation of meta data are XML Schema and RDF Schema. The language to be used for manipulating such data should enable the representation of data schema.

Meta programming supported in Flora-2 system will enable development of data translation system and integration system. Classic first order logic and languages based on predicate calculus have less expressive power for the accomplishment of the Semantic Web and therefore languages providing reasoning within higher order logic represent a better choice.

Hierarchical data structure and incomplete data representation

The F-logic assumes hierarchical data structure, which is a usual data structure on the Web. The F-logic, as an object-oriented language, assumes hierarchical data model drawn up through definition of superclasses and subclasses. The appurtenance of a subclass to a superclass in F-logic is specified using the so called 'isa' statements with denotation `subclass::superclass`; also, appurtenance of an object to a class is designated with the formula `object:class`.

Representation of the Web data assumes linking of several heterogeneous data sources. Furthermore, this implies an incompleteness of data since each of the given data sources can be additionally updated. Additional updates should be integrable into existing data schema. Flora-2 system and F-logic assume partial knowledge on data and therefore Flora-2 system data models can always accommodate new knowledge elements. First, the relation between subclass and superclass, which is expressed as subclass::superclass, can always be updated with a class that would be nested between these two classes without disturbing the defined hierarchical structure. Also, appurtenance of an object to a class can be extended with some additional class. Knowledge of an object is considered as incomplete. This actually reflects the situation on the Web since some additional object attribute can emerge anytime during the process of data integration from heterogeneous knowledge sources. Statements on object can be extended without disturbing initial data model. At any times it is possible to add new attribute values as well as to add new attributes to an object. Here it should be noted that traditional object-oriented languages and database languages do not support such flexibility when additional modification of data scheme is required.

3 Problem of Schema-Less and Heterogeneous Data on the Web

The main goal of Semantic Web is to provide automated reasoning with Web data and generating new knowledge. The idea of that approach is to have semantic annotation and some formal structure of Web data. Still, data on the Web are mostly unstructured or semi-structured and have no scheme nor semantic definition. Besides that, Web data and knowledge reasoning involves data from different, heterogeneous Web sources. In order to provide Web data reasoning it is necessary to translate all data from different Web sources into unique language and into unique schema. Therefore it is necessary first to define data schema, then to translate data into unique language, and finally to integrate all data from different sources on schema and data level. Adding structure and semantic to data would provide reasoning as in Semantic Web context.

In this section some of the main Web data characteristic and problems are described. The first problem with enabling automated Web data reasoning is that most often data has no structure defined. In Table 1 there is a list of three main types of Web data according to their structure.

Table 2: Web data structures

TYPE OF STRUCTURE	DATA ON THE WEB
Unstructured	Text on HTMl pages.
Semistructured	XML or RDF data format
Structured	XML or RDF data format with XML Schema or RDF Schema; OWL format

Data with no schema defined, unstructured data is data represented as a text. It is a usual case that Web page offers only text with no structure. In order to perform formal reasoning with text data it is necessary to transform text into formal database. Sometimes text can be partially structured using HTML format, but this form of data also needs transformation into more formal shape. Semi-structured data can have schema, but it is not formally given. It is the case with XML or RDF data. In that case data schema should be derived from data. Besides that, those representation formats can have schema defined as XML Schema or RDF Schema.

Another important Web data characteristic and problem is data heterogeneity. Generally, different data sources may have different structures, they may vary in syntax and semantics and may differ as to the query language features, organizations, availability and may contain data with different schemata. Most frequently, this is the issue of reasoning with the data from the Web pages where data are presented in HTML, XML, RDF or some other format.

If data schema is defined, it is possible that each of sources has defined a different schema. Schemata may differ in terms of syntax and semantic properties and expressiveness. Different schemata normally have different attribute names as well, which will be later referenced to the same data. Data such as dates may be recorded in various formats. At structure level, it may happen that data in one Web source is expressed in quite a different way than data in another one. It would mean that within a single data source several data could constitute the value of one single attribute, while in another data source the very same data may be represented as the value for several different attributes, hence broken down.

Besides, inconsistencies may occur in terms of different sources offering different information for the same entity at data level. Representation of data on the Web most commonly includes only partial characteristics of some entity. These are some of the inconsistencies that must be considered while reasoning about Web data.

According to the two described problems, reasoning with Web data can be realized through two main phases. The first phase is schema definition and data translation and the second phase is data integration. These two phases will be described in the next two sections.

4 Data Translation in F-Logic

In this approach F-logic is chosen as a basic language for data representation. Web data has to be transformed into database in F-logic. All data should be limited to a certain domain in order to develop a system that can automatically transform Web data into F-logic database.

After having chosen the adequate language for representation of the Semantic Web data, the procedure of translating data into chosen basic language can be performed. Paper [8, 14] describes translation procedures in Flora-2 system and basic implementation elements of the system for reasoning in the Semantic Web context. Generally, a wrapper is introduced for solving translation problems. The wrapper is a system that will enable translation of all data into a defined language and it normally functions within a data integration system. In the XSB system, it is possible to develop quite complex wrappers since the XSB system is featured by the implementation of various grammars [11–13]. However, if Web data are involved, sometimes it is not necessary to develop complex systems for data translation, primarily because the F-logic naturally corresponds to semi-structured data [1, 2]. An object-oriented approach to application development of the Semantic Web is described in paper [6], however in this research the FLORID system was used, whose meta programming features are somewhat poorer as compared to the Flora-2 system.

As explained in the previous section, there are many different data forms and structures on the Web. Data can be given as a text, semi-structured data or structured data, but there is a very little amount of structured data. Depending on data structure, there are few different approaches in data translation. Text data should be translated into F-logic formalism by defining a schema and providing information extraction. Semi-structured data have to be transformed into F-logic using schema definition that can be provided automatically or manually (depending on the situation). For structured data schema can be automatically translated into F-logic. In both cases, first, it is necessary to define or generate a target schema. The target schema will be given as an object-oriented data model in the F-logic. After that it is possible to translate data into F-logic database.

4.1 Text Information Extraction

HTML pages contain mostly text data. Text data are given without any structure. There are different strategies defined for transforming text into formal database. Generally, different techniques of semantic analysis are used to analyze the text. Different methods of parsing are defined and used. In a case of limited domain, information extraction is an advisable technique. It is the approach described in this section. The whole process of information extraction for limited domain can be performed in Flora-2 system. In order to have a picture of whole system, a short preview of information extraction in Flora-2 is given. It is described just as a conceptual model, the process in detail is described in [15].

First it is necessary to define a semantic background for a chosen domain. Semantic background is the domain knowledge that should be captured in: the semantic dictionary, categories, phrases and data schema that is called output templates. All that knowledge can be represented in F-logic [15]. Each word from the chosen domain has to be associated with an appropriate semantic category. Semantic categories can be naturally represented in F-logic as classes and subclasses. Each of these categories consists of semantic subcategories. The hierarchical model of semantic categories can be implemented in Flora-2 system. In the approach described here, semantic categories are at the same time used for generating data schema. Also, semantic categories can be represented as a lattice. There is exactly one class that is a subclass of all classes and there is exactly one class that is a superclass of all classes given by semantic categories. These two classes represent the minimal and the maximal category. With class inclusion as a partial order relation, this set of classes forms a lattice. Lattice can be used for determining sentence semantic category.

After formally defining domain knowledge, text information extraction can be performed in through three more phases. The simplified model of text information extraction is shown in Fig. 1 with four main phases highlighted.

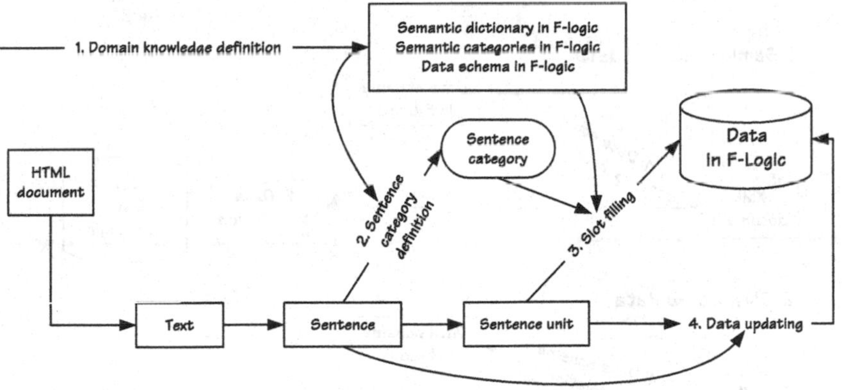

Fig. 1: Text information extraction in F-logic

Text analysis starts with determining a semantic category for each sentence. First, text is divided into sentence. Each word from sentence can be associated with one semantic category of the chosen domain. According to semantic categories of the words in a sentence, it is possible to define the dominant semantic category of a sentence using lattices. After that, each sentence should be decomposed into semantic units. The semantic unit is a part of the sentence that contains necessary information to fill attribute values represented in data schema. Using dictionary, semantic categories and data schema, objects are generated and relevant words from the text are transformed into attribute values.

Some output classes can have missing data because they contain relative terms. The incomplete data of relative terms are updated with missing values. Missing categories are updated with the information gathered from the preceding and succeeding semantic units in a sentence using resolving rules specified in F-logic.

The final result of a whole process is database in F-logic with important data extracted from the input text.

4.2 Semistructured and Structured Data Translation

Semi-structured data is described as Schema-less or self-describing [1]. Semi-structured data is characterized by the lack of any formal scheme, but it has some implicit structure. Data on the Web can be represented as semi-structured data, in XML or RDF format. The idea of reasoning with semi-structured data implies defining data schema that can capture whole data and connecting data from different Web sources. F-logic provides a natural environment for semi-structured data representation and formal structure generation for that data. In some cases data schema is given together with Web data as XML Schema or RDF Schema for example. In those cases there is no need for data schema generation, but it is necessary to transform the existing schema into F-logic format. After data schema is generated in F-logic, data can be extracted into F-logic database.

In Fig. 2 there are two possible situations of Web data transformation into F-logic database given. First situation is for the semi-structured data where no data schema is given. For example, that can be data in XML format. The second situation describes transformation Web data together with data schema. For example, data in XML format with XML Schema given.

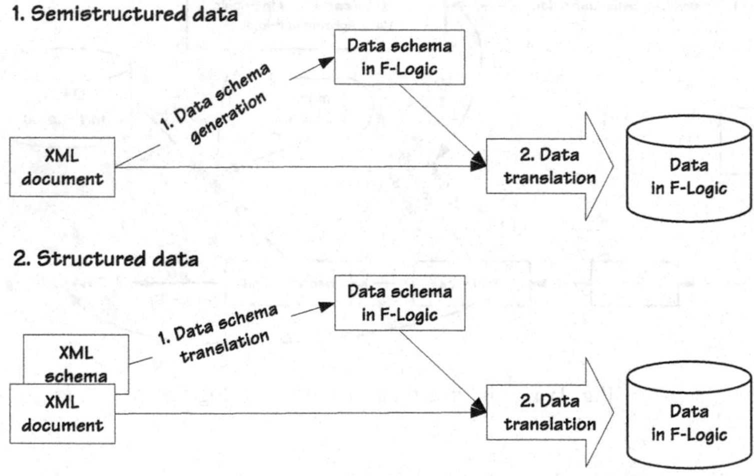

Fig. 2: Web data translation conceptual models

The data schema generation process assumes recognition of data hierarchical structure, classes, objects, attributes and methods and definition of such given schema in the F-logic. If data schemata are not given, it is necessary to query data on data source side in order to generate the data schema. In fact, according to the current situation in the Semantic Web development a part of data is considered to be only semi structured meaning that the schema is not explicitly given, rather, it must be generated from data themselves. Paper [20] offers two possible approaches to querying translating data schema: schema-level meta-querying and instance-level meta-querying. The first approach assumes direct access to data schema, while the latter assumes data schema query on the basis of instances when the schema is not given. In the Flora-2 system it is possible to query the database at meta-level as well as offers features for manipulation with data schemata.

After data schema is defined or translated, data should be translated into F-logic. Detailed data translation is described in Section 6.

5 Data Integration in F-Logic

Web data integration system is a part of a more complex system, which represents a Semantic Web application, however it represents a separate part, and it can be considered independently of the application. It is very important to analyze data properties when developing an integration system conceptual model. Therefore this section presents some preliminary considerations in respect to the Web data structure followed by the theoretical approach in the development of data integration system.

After translating data into unique language, it is possible to integrate data from different Web sources resolving the inconsistencies. The integration of data from heterogeneous knowledge sources represents a consolidation of heterogeneous data aiming to generate new knowledge that cannot be derived from single data sources. Researching in that sense has been topical for quite a long time, numerous papers were produced [7, 11–13, 16] and various strategies and approaches were defined. The integration process in the domain of the Semantic Web has been intensively analyzed during recent years and some of the defined strategies for the data integration systems have been implemented. However, some specific properties of Web data must be considered here. The mediator system model has been generally accepted. The mediator is a system that supervises data integration through distribution of users' queries to heterogeneous knowledge sources and then integrates obtained data into a single unit. One of the possible approaches is to use existing logic programming languages. The basic idea was to develop an integration system similar to the mediator systems. The implementation of such a system is generally rather complex due to a variety of data sources.

In general, the integration process is performed in two separate parts, i.e. at two levels. Further, the process is performed in several steps. It is important to consider the data consistency in this step. Since data from various sources are involved in integration, some inconsistencies may occur. After resolving the data inconsistency problem, a complete knowledge database is obtained.

Figure 3 shows the basic data integration system model involving an application for the Semantic Web. User's queries are forwarded to unrelated data source on the Web, while the final objective is to generate a response based on the integrated data. In order to generate a reply to a query, data should be translated into the F-logic and integrated both at the schema and data level. In Flora-2 system modular programming is assumed so that a separate module has been developed for each part of the system.

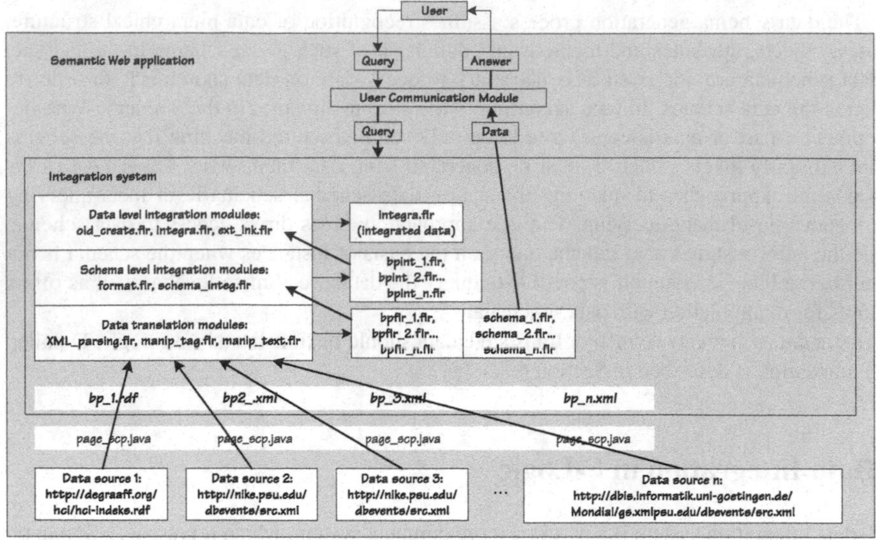

Fig. 3: Conceptual integration system model

6 Application Development

This section addresses the particular Web data integration process for the Semantic Web application intended for searching data on conferences.

6.1 Data Translation

Development of the integration system starts with analysis of data from the problem domain. In this particular case these are data on scientific conferences and data in respect to conferences. It is important that such data contain at least a minimum meta-level, i.e. at least tags indicating what kind of data are dealt with.

The XML and RDF data contain tags although in this particular case, when considering conferences, there are neither XML, nor RDF data schemata so that the data schema is defined within the application design. The schema can be reached in several ways. A target schema only referring to a part of the entire application data is defined. These are data on conferences gathered from independent knowledge sources. The target schema is defined in advance on the basis of schemata and data offered by independent knowledge sources. These are mainly basic data on conferences, which will be used to create the unique database. In respect to data model, advantages of the object-oriented model should be emphasized in comparison to the relational data models when data in XML or RDF format are involved.

Given n of independent heterogeneous knowledge sources containing data in XML, RDF and OWL format, let data schemata are (possibly) given along with data.

Schema translation rules can be defined within the Flora-2 system. One of the ways of translating XML and RDF data into the Flora-2 system language is the extraction of attributes and their values from the document on the basis of data recorded in data schema. In order to recognize and extract required attributes, schema of each database must be known in advance. Database schema must be previously translated into the F-logic. So, it is possible to extract attributes given in schema based on defined schemata: schema_1, schema_2, ..., schema_n for each particular XML document.

An algorithm has been defined for reading all tags from the document. The algorithm recognizes which of the read tags refer to object and attributes respectively. The recognition of objects is based on reading objects and attributes recorded in schema. All values are recorded in format id[Att~>Value][2] in the Flora-2 system local databases. Names of attributes and methods whose values are to be drawn out from the XML document are extracted from the defined target schema. The idea is to separate data on classes and attributes from a given schema using meta-queries. For that purpose meta-queries on data schema are used, which data schema can be defined in a separate model, for example schema_geo.flr. Meta-queries on given schema can provide names of the classes and attributes or methods. For further attributes extraction procedure a predicate value of itemtag/1 is defined, whose values are class names, and predicate att/2, which contains data on attributes of given class for each class. Values of predicates itemtag/1 and att/2 for the given target schema can be generated using the following meta-queries:

```
?-['C:\\magrad\\flr\\schema_geo.flr'>>sh].
itemtag(Itemtag):-Itemtag[_X=>_Y]@sh.
att(Itemtag,Att):-Itemtag[Att=>_Y]@sh.
att(Itemtag,Att):-Itemtag[Att=>>_Y]@sh.
att(Itemtag,Att):-Itemtag[Att=>_Y]@sh.
att(Itemtag,Att):-Itemtag[Att=>>_Y]@sh.
```

These clauses are implemented in translating data on countries and continents from the XML document into the F-logic. However, the translation of data about conferences located in independent, separate documents requires further slight modification of the procedure and consideration of the database ordinal number, which number is used to identify each database. All of n schemata are considered here, i. e., attributes defined for this local schema is searched for each database. The rule used for generating data in the F-logic can be simplified as follows:

```
X[Att->Value]@SchemaN :-\verb
X[Att =>_Y]@SchemaN,
tag(Line, Tag),
Tag is Att,
extract_tag(Line,Att,Value).
```

However, the entire algorithm is somewhat more complex and defined through predicate read_file/1, whose parameter is the name of XML document to be translated and predicate read_att/3. On the basis of the given parameters, which are text lines, class and object names and identifier, predicate read_att/3 generates and records all attribute values for a given object into the new document. Predicate extract_tag/3, which recognizes required tag and its value in a given text line, is defined for the purpose of extracting tags from the given document and it records the given text in the F-logic. All these predicates are defined in the XMPparsing.flr module. Besides, rules for manipulation with tags are defined in a separate module manip_tag.flr. However, module manip_text.flr is required as well. The above procedure results in n databases recorded in the F-logic. However, attribute names in these databases are not always uniform and translated into the target schema. Translation of XML data into the F-logic is shown in Fig. 4.

---

[2] ~> refers to one of tags: ->, *->, ->>, *->>

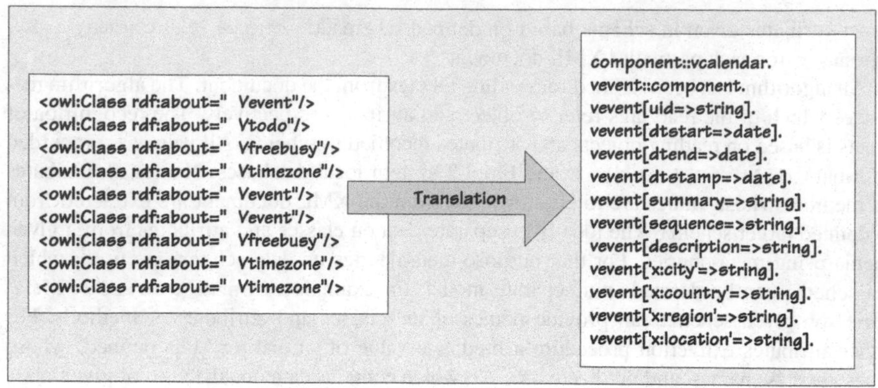

Fig. 4: Translation of data in XML format into F-logic

6.2 Schema Level Integration

The next step is the schema level integration. Since local schemata may not necessarily conform in attribute names, target schema will unify their names. An intervention by the application designer is required here in order to make sense in translating each schema into the target schema. The translation assumes linking of each single attribute for local schema to the relevant attribute in the target schema. It is necessary to define a predicate that would link target and local schema attributes in each database `att_integ/3`, where the first attribute represents the name of the local attribute, while the second one represents the name of the same attribute in the target schema and the third argument indicates the name of attribute `Att_loc` location in particular local database:

```
att_integ(Att_loc,Att_goal,N).
```

All data in local databases require predicate `att_integ/3` arguments to be defined. Further, several clauses must be defined for the process of translating attribute and method names. Clauses for translating names at data level are defined separately from defining clauses for translating names at schema level. However, in this particular case it is not necessary to apply data translation clauses because eventually all data will conform to the target schema. Classes are defined for translation of both scalar type and set type methods, noninheritable and inheritable, and attribute names. Clause format defined for scalar noninheritable attributes is as follows:

```
X[Att_new->Y]@bp_n:-
at_int(Att_new,Att_old,N),
X[Att_old->Y]@bp_n.
```
[3]

Rules for translating set type and inheritable attributes are defined in analogous way. It is also necessary to somehow coordinate names of the classes and objects. In this particular case the situation is not complex because all objects will belong to class ‘event’, but in more complex situations it is possible to define rules that would resolve this segment of translation. Besides, it is possible to define analogous clauses at schema level for extracting attributes and methods in format:

[3] Analogously for `-»`, `*->`, `*-»`

```
X[Att_new=>Y]@bp_n:-
at_int(Att_new,Att_old,N),
XAtt_old=>Y@bp_n.4
```

Another problem may occur and that is data format. In fact, there may be different records of some data in different schemata or databases. There are no default data formats in Flora-2 so these should be defined in advance at the design level. All modifications of attribute records may be resolved by applying additional rules definitions. This will assure a unique record of attributes in local databases. It is also possible that several data are contained in one database or XML document as a value for a single attribute. For example:

```
<description>
Heraklion, Crete, Greece.http://www.esws2004.org/
</description>
```

In this case it is necessary to define additional rules to be used for data extraction. Integration rules at the schema level are located in module `schema_integ.flr`. However, an additional module, `format.flr`, is required for data formatting.

The process of data translation and integration at the schema level may be described through several activities that are distinguished as the process key activities. Table 3 shows and describes such activities as well as module resolving each of the steps required in implementation.

We have results form the first part of the translation procedure: `schema_1`, `schema_2`, ..., `schema_n`. Based on defined schemata, it is possible to extract attributes given in schema for each single document it.

Table 3: Schema integration process activities

ACTIVITY	MODULE	RESULT
Definition of target schema	-	schema.flr
Data collection	page_scp.java	bp_1.xml bp_2.xml ... bp_*n*.xml
Generation data schema	It is necessary to define data schema for each source.	schema_1.flr schema_2.flr ... schema_*n*.flr
Translation of data into F-logic	XMLparsing.flr manip_tag.flr manip_text.flr	bpflr_1.flr bpflr_2.flr ... bpflr_*n*.flr
Data conversion	format.flr	(Result is obtained after translation into a target schema.)
Integration at schema level	schema_integ.flr	*n* databases: bpint_1.flr bpint_2.flr ... bpint_*n*.flr

[4] Analogously for =»

Furthermore, it is necessary to define an algorithm in Flora-2 system that would read all tags from a document and recognize which of the tags refer to objects and attributes respectively. Then the algorithm reads values of the tags referring to attributes of given objects. All these values are recorded in the Flora-2 system local databases in format `id[Att~>Value]`.[5] It is also necessary to extract those attributes and method names from the defined target schema, whose values are to be extracted from an XML document. In general, the idea is to extract data on classes and attributes from a given schema using meta-queries.

Data model in F-logic corresponds to XML data model on the basis of equalizing tag values in XML document with attribute values in schemata given in F-logic and tag names with attribute names in F-logic. From XML document and defined schema recorded in F-logic it is possible to uniquely generate a database in F-logic. It has been demonstrated that a list of attributes for each entity and class may be obtained for a defined database schema. It is also possible to obtain data on defined classes from schema. Since the F-logic has the second order syntax it is possible to make queries at meta-level, which queries return names of the attributes, methods and classes as values.

6.3 Data level integration

In the second part, after all the databases have been consolidated to a unique schema, it is necessary to integrate the data. It is necessary to solve the problem of possible inconsistent data that may occur. For the Web data integration, the optimistic i.e. naive strategy has been chosen. Such a choice makes sense since data are expected to be from different conferences, which have to be consolidated into a single database. The pessimistic strategy is out of question because in this case it is the union of data that is aimed rather than intersection of data from the source. Such a procedure actually results in database, which is the union of all data from all independent knowledge sources. In the second part of the integration it is necessary to exclude all inconsistent data. The chosen approach was to reject data that would lead to an inconsistent state, but regardless of this, some data should be considered, which is defined by additional rules. However, some other approaches are practicable for solving the inconsistent databases issue.

If several sources contain same data, it is necessary to coordinate knowledge coming from such sources. For solving conflicts in data coming from various sources, a new predicate could be defined, which would refer to those attributes that are not allowed to appear in several different values for the same attribute. However, in F-logic there are attributes that may have several different values. It occurs when a set type method is given. In such a situation occurrence of several different values for one single method, i.e. attribute, does not represent an inconsistent situation. These two cases should be distinguished, and the difference between the set type and scalar type methods is defined at the schema level. However, along with these two cases, there is the third situation where it is also necessary to define reasoning rules. For an attribute that can assume only one value it is possible to have different values within various sources, but yet these values do not cause a conflict situation, i.e., we do not want to reject them. For example, conference name is the attribute that can assume only one value, but in various data sources the value of this attribute may differ if the name is abridged or quoted as an acronym. This is only one of the situations where we allow a scalar attribute to assume several values. On the other hand, conference opening date is the attribute that can have only one value and if several different data for that attribute occur an intervention is needed in order to solve the conflict situation. In this case the pessimistic approach was chosen when dealing

[5] `~> refers to one of tags: ->, *->, ->>, *->>`

with scalar attributes that are allowed to assume only one value, meaning that all conflict data will be automatically rejected.

Sometimes it is possible to choose some other approach as well. If, for example, there is some verified index of source reliability, a hierarchy of sources could be established and the conflict situations will be solved on the basis of source reliability index. It is also possible to offer user with several different solutions, but this may not be the best strategy in all situations. In Web environment, it seems that there is more sense in rejecting data that disturb the database consistency. If a reliable data source is found afterwards, it may possibly be included later, which in this particular case would be the specific conference Web page.

Each of three possible approaches required adequate clauses to be defined for solving a conflict situation:

1. Solving a conflict situation for scalar attributes that do not allow any conflicts: predicate `conflict/4` is defined for scalar attributes, which predicate checks if there are two different attribute values C and D for given object A and attribute B. If different values exist, it is necessary to remove all these values from the database in order to preserve the database consistency. Attribute `conflict/4` is defined for noninheritable and inheritable data methods as follows:

```
conflict(A,B,C,D):-A[B->C]@M1,
A[B->D]@M2, C\=D, unique(A,B).
conflict(A,B,C,D):-A[B*->C]@M1,
A[B*->D]@M2, C\=D, unique (A,B).
```

Predicate `unique/2` indicates that attribute B is uniquely defined for object A and it is an application designer which rules that no conflicts will be allowed, i.e., linking of ambiguous solutions such as in dates.

2. Solving conflict situation for scalar type attributes where conflicts can be allowed: if the intention is to offer a user with all of the mentioned solutions, a predicate defining all such values could be created.

```
notconflict (A,B,C,D):-A[B->C]@M1,
A[B->D]@M2, C\=D, notunique (A,B).
notconflict (A,B,C,D):-A[B*->C]@M1,
A[B*->D]@M2, C\=D, notunique (A,B).
```

Predicate `notunique/2` indicates that the attribute B is ambiguously defined for object A. It is defined by the application designer if no conflicts are allowed, i.e. linking ambiguous solutions, which is, for example, acceptable in contents description.

3. Solving conflict situation for set type attributes: for set type expressions the following situation is allowed:

```
A[B->>C]@M1,A[B->>D]@M2, C\=D.
A[B*->>C]@M1,A[B*->>D]@M2, C\=D.
```

In such a case no intervention is required in terms of additional rules and clauses because database remains consistent. Intervention is necessary only if some of the data is not correct.

Table 4 summarizes activities in the second part of the integration process assuming integration at data level. The final result of the second part of the integration process is an integral database `bp_conf.flr` containing data on conferences. However, this database contains only some basic data on events. For any other queries the program returns to the data sources and generates the required answers.

Table 4: Data integration process activities

ACTIVITY	MODULE	RESULT
Generating OID for data	oid_create.flr	bpflr_1.flr bpflr_2.flr ... bpflr_*n*.flr
Data integration	integra.flr	bp_conf.flr
URLs extraction-required for further queries	ext_lnk.flr	schema_1.flr schema_2.flr ... schema_*n*.flr
Translation of data into F-logic	XMLparsing.flr manip_tag.flr manip_text.flr	Document containing URLs: adr.txt

7 Conclusions

The objective of this paper was to demonstrate that object-oriented logic formalisms and logic formalisms with second order syntax represent a solid theoretical and conceptual basis for data and knowledge reasoning in the Semantic Web context. HiLog and Flora-2 programming languages have been selected for the data and schema modeling, translation and integration system development. F-logic, as logic formalism, has been developed so that it contains all relevant elements of the object-oriented database and thus it represents an adequate language for data representation in the Semantic Web environment. Rules for translating schemas can be defined in Flora-2 system since this system contains a deductive component and therefore it provides not only data representation but data manipulation as well.

Languages containing data meta-level must provide a query facility for making queries on data meta-structure. The F-logic combined with HiLog, as implemented in Flora-2 system, provides query facility for data scheme as well as all possible combinations on data schema and on data. In that way retrieval of any data referring to methods of certain classes and objects is possible. It is also possible to check the hierarchical structure and interrelations among objects and classes. In that light Flora-2 system is considerably simpler than some other systems as it is quite common to define a separate language for querying data scheme, which language cannot be combined with query language intended for data.

Another conclusion is that logic programming languages can be used for deriving a required model. Implementation of existing systems such as Flora-2 system is possible and there is no need for developing new languages to be used for the development of mediator systems. Logic programming languages featuring the second order syntax can be extended by means of amalgams and annotations [11–13], thus making the development of mediator systems possible. Further researches in that sense are possible in terms of exploring development of more general applications and systems for manipulation with the Semantic Web data, searching heterogeneous data sources and generating new knowledge.

References

1. Abiteboul S, Buneman P, Szciu D (2000) Data on the web. From Relations to Semistructured Data and XML. Morgan Kaufmann, San Francisco, CA
2. Abiteboul S (1997) Querying Semi-Structured Data. In: Proceedings of the 6th International Conference on Database Theory. Springer, London, 1–18
3. Angele J, Lausen G (2003) Otologies in F-logic. Ontoprise GmbH
4. Berners-Lee T, J. Hendler J, Lassila O (2001). The Semantic Web. Scientific American 284(5):34–43
5. Chen W, Kifer M, Warren D S (1993) A Foundation for Higher-Order Logic Programming. Journal of Logic Programming 15:187–230
6. Frohn J, Himmeriider R, Kandzia P, Lausen G, Schlepphorst C (1997) FLORID: A Prototype Form F-logic. In: Proceedings of the 13th International Conference on Data Engineering, Birmingham, UK, 583
7. Hendler J, Berners-Lee T, Miller E (2002) Integrating Applications on the Semantic Web. Journal of the Institute of Electrical Engineers of Japan 122(10):676–680
8. Kaić A (2005) Conceptual Modeling and Implementation of Semantic Web Using F-logic (in Croatian), MA thesis, Zagreb University, Faculty of Organization and Informatics Varaždin, Varaždin, Croatia
9. Kifer M, Lara R, Polleres A, Zhao C, Keller U, Lausen H, Fensel D (2004) A Logical Framework for Web Service Discovery. In: ISWC 2004 Workshop on Semantic Web Services: Preparing to Meet the World of Business Applications. Hiroshima, Japan.
10. Kifer M, Lausen G, Wu J (1995) Logical Foundations of Object-Oriented and Frame-Besed Languages. Journal of the ACM 42(4):741–843
11. Lovrenčić A, Čubrilo M (1999) Amalgamation of Heterogeneous Data Sources Using Amalgamated Annotated HiLog. In: Proceedings of the 3rd IEEE Conference on Intelligent Engineering Systems, INES'99. Stara Lesna, Slovakia.
12. Lovrenčić A (1999) Knowledge Base Amalgamation Using Higher-Order Logic-Based Language HiLog. Journal of Information and Organizational Sciences 23(2):133–147
13. Lovrenčić A (2003) Logic Programming Languages for Development of Systems for Integration of Heterogenous Knowledge Sources (in Croatian), Ph. D. thesis, Zagreb University, Faculty of Organization and Informatics Varaždin, Varaždin, Croatia
14. Meštrović A, Cubrilo M (2006) Semantic Web data integration using F-logic, in the proceedings of International Conference on Intelligent Engineering Systems, INES 2006, London
15. Ludäscher B, Himmeroder R, Lausen G, May M, Schlepphorst C (1998) Managing Semistructured Data with Florid: A Deductive Object-Oriented Perspective. Information Systems 23(8):589–612
16. Vdovjak R, Houben G (2001) RDF-based architecture for semantic integration of heterogeneous information sources. In: International Workshop on Information Integration on the Web, Rio de Janeiro, RJ, Brasil
17. Yang G, Kifer M, Zhao C (2003) FLORA-2: A Rule-Based Knowledge Representation and Inference Infrastructure for the Semantic Web. In: Second International Conference on Ontologies, Databases and Applications of Semantics (ODBASE). Catania, Italy
18. Yang G, Kifer M, Zhao C (2003) FLORA-2: User's manual, Version 0.92, Department of Computer Science, Stony Brook University, http://flora.sourceforge.net/

19. Yang G, Kifer M (2003) Inheritance and Rule in Object-Oriented Semantic Web Langugaes. In: Second International Workshop on Rules and Rule Markup Languages. Sanibel Island, Florida, USA.
20. Yang G, Kifer M (2003) Reasoning About Anonymous Resources and Meta Statements on the Semantic Web. Journal of Data Semantics 1:69–97

Study on Knowledge and Decision Making

Dana Klimešová[1,2]

[1] Czech University of Life Sciences, Prague, Faculty of Economics and Management, Department of Information Engineering, Kamýcká 129, 165 21, Praha 6 – Suchdol, Czech Republic
klimesova@pef.czu.cz

[2] Institute of Information Theory and Automation, Prague, Academy of Sciences of the Czech Republic, Pod vodárenskou věží 4, 182 08 Praha 8, Czech Republic
klimes@utia.cas.cz

Summary. The paper deals with knowledge transformation process on the background of geospatial data modelling and discusses the possibilities of the context use as a reflection of the system of understanding. The problem of the relation to the decision support system is addressed and GIS as a tool dealing with all phases of knowledge structure. The paper shows that with the development of the Web services architecture there is a clear trend towards GIS becoming more open, robust and interoperable.

1 Introduction

1.1 Geo-Information

The past decade has registered a significant increase in the quality and quantity of geospatial information from various sources. Consequently, the quest for knowledge from the massive geospatial information for scientific, commercial and decision-making activities is great. While conventional spatial statistics methods remain their power and popularity in numerous studies, many new techniques and approaches have appeared in response to the newly available geospatial data and their possibilities due spatially interconnections process and gain knowledge [2]. A GIS makes it possible to link, or integrate, information that is difficult to associate through any other means. Thus, a GIS can use combinations of mapped variables to build and analyze new variables. variety of analysis and modelling approaches have been brought into geospatial domain such as network modelling, connectivity, stream and spread cumulative functions application, agent-based modelling, qualitative and in the last time also fuzzy reasoning. These techniques are efficient and effective in discovering hidden structures, patterns and associations within the geospatial data [3,4]. On the other hand, emerging visualization and interaction technologies provide a powerful tool for obtaining additional insights into geospatial information for spatial analysis and modelling process. There has been an increasing convergence of the analytical reasoning and visualization towards creation and discovery of geospatial knowledge for real world applications and support its transfer. There are many factors that complicate knowledge transfer, including [5,6]:

- Geography
- Language

J.A.T. Machado et al. (eds.), *Intelligent Engineering Systems and Computational Cybernetics*,

- Areas of expertise
- Internal conflicts (e.g. professional territoriality)
- Generational differences
- Union-management relations
- Incentives

The use of visual representations to transfer knowledge (knowledge visualization) To support the knowledge transfer process, it needs:

- Identifying the key knowledge holders within the organization
- Motivating them to share
- Designing a sharing mechanism to facilitate the transfer
- Executing the transfer plan
- Measuring to ensure the transfer
- Applying the knowledge transferred

1.2 Access to Data

Thanks to Internet and Mobile Internet GIS, Mobile Web Map Services and Mobile Web Analytical Services that facilitate the acquiring of data it is much easier to monitor and map temporal states of the objects and phenomena.

Mobile Internet GIS and Mobile Web Map Services, it is a new solution that offers mobile Internet access to the data including theirs updated versions, provides the transmission of maps that are composed under the demands and the transfer of map attributes and provides the access to the raster data, orthophotomaps and vector data.

Web Analytical Services make possible to process and analyze spatial data through Internet and mobile internet. This system increases the access to information for the management in the field of forestry, agriculture, regional development, environment protection and others. Mentioned technologies (Mobile Internet GIS, Mobile Web Map Services, Mobile Web Analytical Services) can significantly increase the processing effect because they make possible:

- To increase the speed of finding information (phase of collection)
- To increase the ability to analyse the large files of information (phase of analysis)
- To increase the effect of the context evaluation (top analysis)
- To increase the effect of the presentation (distribution)

2 Decision Support Systems & GIS

A Decision Support System (DSS) is an interactive computer-based system or subsystem intended to help decision makers use communications technologies, data, documents, knowledge and/or models to identify and solve problems, complete decision process tasks, and make decisions [2]. DSS refers to an academic field of research that involves designing and studying DSS in their context of use. In general, DSS are a class of computerized information system that support decision-making activities.

The DSS and their construction are based on the models. These models are an approximation of reality, which are dependent on the subjective interpretation of the knowledge. It means that new observations may lead to a refinement, modification, or completion of the already constructed model. On the other hand, the models may guide further acquisition of knowledge and the knowledge is the base for decision support [7]. Moreover, besides knowledge modelling also knowledge representation is very important field of DSS.

A mathematical model is an equation, inequality, or system of equations or inequalities, which represents certain aspects of the physical system modelled. Models of this type are used extensively in the physical sciences, engineering, business, and economics [8].

Geographic information system (GIS) provides essential marketing and customer intelligence solutions that lead to better business decisions. Geography is a framework for organizing our global knowledge and GIS is a technology for being able to create, manage, publish and disseminate this knowledge for all of society. With GIS, it is possible to analyse:

- Site selection and location analysis
- Customer segmentation, profiling, and prospecting
- Demographics and customer spending trends
- Potential new markets

GIS allows visualizing and interpreting data in ways simply not possible in the rows and columns of a spreadsheet. GIS can help your business save time and money, while improving access to information and realizing a tangible return on your GIS investment [9].

3 Knowledge & GIS

Organizations around the world spread their information technology investments by integrating GIS technology. GIS gives the tools to be able to:

- Make informed decisions
- Know where, when, why, and how to take action
- Share knowledge with others
- Help students understand real-world problems using data analysis
- Prepare students for GIS jobs available in business and government
- Share information across multiple disciplines and promote a holistic approach to learning

Using GIS is about sharing what you know and setting new courses that will sustain our world in the years to come [10]. Standards and interoperability are extensively important elements in our overall software development and support efforts.

GIS technology provides essential information tools for many levels of society. As developers, you need to be able to:

- Develop applications using the language of your choice
- Deploy applications on a variety of platforms
- Access and manipulate GIS data in multiple formats

GIS plays a significant part in the day-to-day functions of information gathering agencies; the way this information is distributed to other agencies and organizations, and how it is disseminated to the public. Across government, agencies are integrating GIS software solutions as a central component in building a strong GIS. By integrating GIS with government processes, organisations can:

- Create an information base that shares information resources, reduces data redundancy, and increases data accuracy
- Perform joint project analysis and provide decision support

- Streamline processes to increase efficiency, automate tasks, and save time and money

With the development of the Web services architecture [6, 12] there is a clear trend towards IT including GIS becoming more open, robust and interoperable. Web has a unique ability to integrate diverse data through shared location and specially GIS Web services offer real potential for meeting the demands of users and will bring significant benefit to knowledge-based society. Web provides universal and rapid access to information at a scale that has never been seen before and GIS technology has become easier to use and more accessible and make possible to think about large context of processed data.

GIS and DSS allow disparate data sets to be brought together to create a complete picture of a situation. GIS and DSS technology has the specialized tools focused on [7]:

- Knowledge Identification
- Knowledge Sharing/ Dissemination
- Knowledge Acquisition
- Knowledge Preservation
- Knowledge Development
- Knowledge Utilization

The focus on single process steps allows the structuring of the management process; detection of problems in this process and detection of problems, which interfere with the overall Knowledge Management process, is simplified.

GIS provides essential information tools for many levels of society. IT professionals need those tools to be able to:

- Coordinate and communicate key concepts between departments within an organization
- Share crucial information across organizational boundaries
- Manage and maintain a central spatial data infrastructure, often within a service-oriented architecture (SOA)

4 Context & Knowledge

The contextual modelling, as mentioned over, deals with different types of context information. The attempt to specify the context sources [9–11] follows:

- Context as the reflection of object or phenomena using different interpretation through the system of cognition:
  – Perception
  – Conception
  – Interpretation
- Context as the reflection of selected facts is concerned with validity of statements and the system of argumentation:
  – Identification
  – Analysis – coordination
  – Synthesis – decision
- Context as the reflection when hypothesis stays instead of experience in the system of abduction – instinct based context:
  – Recognition of patterns
  – Coordination by intuition
  – Judgement due to synthetic inference

- Context as the reflection concerning validity of statement using knowledge generating system – knowledge based context:
  - Abduction – iconic analogy to experience
  - Deduction – model of ideal world
  - Induction
- Context as the reflection of internal/external learning processes through the system of quality – learning based context:
  - Abduction
  - Deduction
  - Induction

Using context it is possible to derive new quality of information that can be effectively utilized in geoinformation analysis.

5 Conclusions

The contribution deals with more abstract level for reflection and understanding of the various modelling processes in geo-information processing. Due to this fact the modelling tool has been introduced as a formal framework reflecting the context in various representational levels. Due to this argumentation the understanding of the model as related to an actual context represents perfectly this idea. In this article this general view on the way of building models is presented as a formalized modelling tool and the capacity is illustrated due to the aspects of argumentation and learning.

Acknowledgments

This work is supported by the Project: Information and knowledge support of strategic control MSM 6046070904.

References

1. Aamodt A, Nygard M (1995) Different roles and mutual dependencies of data, information and knowledge. Data & Knowledge Engineering 16:191–222
2. Power DJ (2002) Decision Support Systems: Concepts and Resources for Managers, Quorum Books, Westport, CT
3. Decker S, Erdmann M, Fensel D, Studer R (1999) Ontobroker: Ontology Based Access to Distributed and Semi-Structured Information. In: Meersman R et al. (eds), DS-8 Semantic Issues in Multimedia Systems, Kluwer, Boston, MA
4. O'Leary D (1998) Knowledge Management Systems: Converting and Connecting IEEE Intelligent Systems, 13(3), May/June 1998, pp 30–33
5. Bernbom G (2001) Information Alchemy: The Art and Science of Knowledge Management, EDUCAUSE Leadership Series No 3. Jossey-Bass, San Francisco, CA. Graham, Ricci
6. Fensel D, Decker S, Erdmann M, Studer R (1998) Ontobroker: Transforming the WWW into a Knowledge Base. In: Proceedings of the 11th Workshop on Knowledge Acquisition Modeling and Management, Banff, Canada, April 18–23

7. Musen MA (1993) An Overview of Knowledge Acquisition. In: David JM et al. (eds), Second Generation Expert Systems, Springer Verlag, New York
8. Studer R, Decker S, Fense D, Staab S (2000) Situation and Prospective of Knowledge Engineering. In: Cuena J, Demazeau Y, Garcia A, Treur J (eds), Knowledge Engineering and Agent Technology. IOS Series on Frontiers in Artificial Intelligence and Applications, IOS Press
9. Worboys MF (1995) GIS – a computing perspective. Taylor and Francis, London
10. Klien E, Lutz M, Kuhn W (2006) Ontology-based discovery of geographic information services. Computers, Environment and Urban Systems 30(1), January, 102–123
11. Peuqeut DJ (2002) Representations of Space and Time, Guilford Press, New York
12. Mitra P, Kersten M, Wiederhold G (2000) Graph-Oriented Model for Articulation of Ontology Interdependencies. In: Proceedings of the 7th International Conference on Extending Database Technology (EDBT), Konstanz, Germany

CNMO: Towards the Construction of a Communication Network Modelling Ontology

Muhammad Azizur Rahman[1], Algirdas Pakstas[1], and Frank Zhigang Wang[2]

[1] Department of Computing, Communications Technology and Mathematics, London Metropolitan University, England
mam406@londonmet.ac.uk, a.pakstas@londonmet.ac.uk
[2] Cranfield University, Bedfordshire MK43 0AL, UK
f.wang@Cranfield.ac.uk

Summary. Ontologies that explicitly identify objects, properties, and relationships in specific domains are essential for collaboration that involves sharing of data, knowledge or resources. A communications network modelling ontology (CNMO) has been designed to represent a network model as well as aspects related to its development and actual network operation. Network nodes/sites, link, traffic sources, protocols as well as aspects of the modeling/simulation scenario and operational aspects are defined with their formal representation. A CNMO may be beneficial for various network design/simulation/research communities due to the uniform representation of network models. This ontology is designed using terminology and concepts from various network modeling, simulation and topology generation tools.

1 Introduction

Modern communications networks are complex systems consisting from many components which are complex systems by itself. Planning and design of such systems often involves use of various tools which may help to the designers or researchers to make decisions or get insights into behaviour of certain network [sub] systems. There are many tools created for different purposes by various relevant communities which are using sometimes different terminology than other communities for the same/similar concepts or objects. Authors are too aware about such situation due to their involvement into a project aiming to integrate various network modeling, simulation, etc. tools [1].

Development of ontology is one of the ways to serve the purposes of knowledge sharing and reuse as it represents fundamental concepts of a domain. A *communications network modelling ontology (CNMO)* may be beneficial for designers and researchers due to uniform representation of the models and the ways these models are developed and used. The purposes of this paper are (a) to rise an issue regarding need for CNMO, (b) to discuss issues related to the development of CNMO, (c) to suggest principles for development of the CNMO, and (d) to design elements of the CNMO in a formal way (formal ontology).

J.A.T. Machado et al. (eds.), *Intelligent Engineering Systems and Computational Cybernetics,*

1.1 Need for CNMO

Communications networks planning and design, as a branch of the human's activity, is still pretty much in the stage of an "art" or "handicraft" rather than "engineering" [2, 3]. The construction of CNMO is an effort to address the need for consistent descriptions of the most common concepts as well as components of network models using common vocabularies. Many tools surveyed by the authors have various incompatibilities among the network concepts and models on which they rely. Such situation (a) makes it difficult for end users of the tools to decide which tool is suitable for particular network design/research project, (b) makes it difficult to produce the fare benchmarks for the tools, and (c) produces confusing stream of publications where terms are used differently in different context, etc. Thus, while there are projects aiming to build shared ontologies for various engineering areas [4, 5, 6] the situation with CNMO is currently pretty much similar to one existed before activity towards development of the OSI/ISO standards for communications networks have started at the end of 1970s.

1.2 Issues Related to the Development of CNMO

There are a few aspects related to the network design/research/exploitation processes: (a) formal descriptions of the network configurations (topology, component parameters, traffic flows, functions, etc.) are useful for network monitoring and management purposes as inputs for setting targets; (b) the same formal descriptions can be used as inputs to the tools for the purposes of analysing performance, planning changes, etc.; (c) description of the of possible "situations" in the networks such as link's failure/restoring, wireless bit error rates, etc. can be useful as input for network design and simulation tools; (d) documenting of the system maintenance process can be based on the formal descriptions of the involved components in the network as well as events which are related to this component (consequences of failure, related repairing activities, costs, etc.).

Thus, it is obvious that all above mentioned processes are using overlapping parts of the knowledge regarding communications networks. Development of the CNMO may help to clarify and identify concepts and terminology for the related processes.

The rest of the paper is organized as follows. Section 2 describes principles for development of the CNMO including terminology, symbols and axioms. In Section 3, the method of designing the CNMO is illustrated step by step. Related work is discussed in Section 4. Finally, conclusions are presented in Section 5.

2 Principles for Development of the CNMO: Terminology, Symbols and Axioms

Ontology design involves the specification of concepts and relationships that exist in the domain, besides their definitions, properties and constrains [7–9]. Although methodologies for building an ontology are not mature enough, there are some methodologies available. Authors mostly followed methodologies described in [10–13]. At first the problems concerned with the domain have been identified. Then axioms based on the use of first order logic have been defined. This ontology is designed based on the models represented by different network modelling, simulation, topology generation and discovery tools developed by different communities.

2.1 Terminology Used in the CNMO Design

The terminology used in the CNMO design is based on definitions presented in [10, 14–19]:

Taxonomy: It is a hierarchy of textitconcepts in which each link is an **is-a** link or a **part-of** link. An ontology is a taxonomy of concepts in which each concept is clearly defined. Taxonomy is also similar to ontology. Formally, an ontology consists of *terms*, their *definitions*, and *axioms* relating them; terms are typically organized in a taxonomy.

Concepts: The *classes* of ontology are known as concepts [10].

Slots: Properties of each concept describing various features and attributes of the concepts are referred to slots. It is also called role or property [10].

Terminology: Terminology is a theory of labels of concepts. The labels of concepts are named after coming to an agreement on them.

Vocabulary: A set of words where each word indicates some concepts. Vocabulary is language-dependent.

Ontology: An ontology has been defined in several ways by different ontologists [14–17]. A concise explanation from Tom Gruber and the Knowledge Systems Laboratory at Stanford University [14] is as follows: *"An ontology is an explicit specification of a conceptualisation"*. In the "knowledge base" community, ontology is defined as *"a system of primitive vocabulary/concepts used for building artificial systems"* [15].

Formal ontology: Axiomatic description of an ontology. It can answer questions on the *capability* of ontology.

Axiom: Declaratively and rigorously represented knowledge, which has to be accepted without proof. In predicate logic case, a formal inference engine is implicitly assumed to exist.

2.2 Symbols Used in the CNMO Design

The following symbols are used for the configuration of CNMO. The links between slots or concepts are represented by arrow indicating a part of the relation. The concept/class is represented by rectangle and slot/attribute is represented by ellipse.

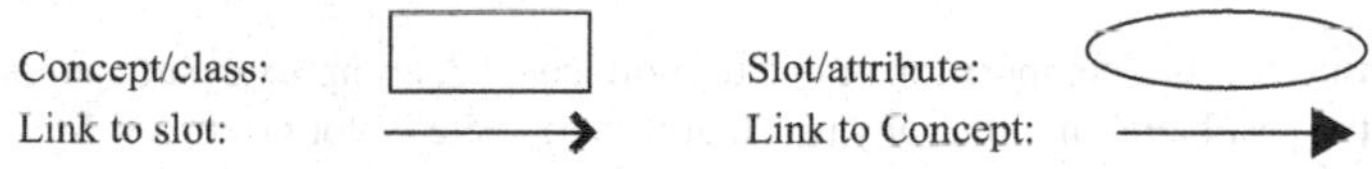

The following symbols are used to represent the axioms designed for the CNMO:

Symbol	Meaning	Symbol	Meaning	Symbol	Meaning
$\forall$	For all	$\rightarrow$	Implication	$\vee$	Or
$\exists$	There exist	$\leftrightarrow$	Equivalent	$\wedge$	And
$\neg$	Not	$\subseteq$	Belong to	$\sqcup$	Union

2.3 Axioms Applied in the CNMO Design

This section introduces the axioms which will be used in the design of CNMO. A *part* is a component of the artefact being designed. The *concept* of part introduced here represents a *physical identity* of the artefact, software components and services. The *structure* of a part is defined in terms of the *hierarchy* of its component parts. The relationship between a part and its components is captured by the *predicate* **partOf**. Between two parts x and y, partOf(x,y) means that x is a part/component (subpart) of y. The following two axioms states that a part cannot be a component of itself and it is never the case that a part is a component of another part which in turn is a component of the first part. This shows that the relation partOf is *non-reflexive* and *anti-symmetric*:

$$(\forall x)\neg \text{partOf}(x,x) \tag{1}$$

$$(\forall x,y)\text{partOf}(y,x) \rightarrow \neg \text{partOf}(x,y) \tag{2}$$

The relation partOf is *transitive*; that is, if a component of another part that is components of a third part then the first part is a component of the third part.

$$(\forall x,y,z)\text{partOf}(z,y) \wedge \text{partOf}(y,x) \rightarrow \text{partOf}(z,x) \tag{3}$$

A part can be a (sub) component of another part. But since each part has a unique ID (its name), it cannot be a sub-component of two of more distinct parts that are not components of each other.

$$(\forall x,y,z)(\text{partOf}(x,y) \wedge \text{partOf}(x,z) \rightarrow y) \rightarrow x \vee \text{partOf}(y,z) \vee \text{partOf}(z,y) \tag{4}$$

The above four axioms guarantee that the part structure is in the form of a *forest* consisting of one or more *trees* of parts. Parts are classified into two types, depending upon the partOf relationship it has with the other parts in the hierarchy. The two types are: *primitive* and *composite*.

- A primitive part is a part that cannot be further subdivided into components. These types of parts exist at the lowest level of the artefact decomposition hierarchy. Therefore, a primitive part cannot have sub-parts.

  $$(\forall x)\mathit{primitive}(x) \rightarrow (\neg\exists y)\text{partOf}(y,x) \tag{5}$$

  Primitive parts serve as a connection between the design stage and the manufacturing stage.

- A composite part is a composition of one or more parts. A composite part cannot be a leaf node on the part hierarchy; thus any part that is composite is not primitive.

  $$(\forall x)\mathit{composite}(x) \rightarrow \neg \mathit{primitive}(x) \tag{6}$$

Most composite parts are *assemblies* which are composed of at least two or more parts.

$$(\forall x)\mathit{assembly}(x) \leftrightarrow (\exists y,z)\text{partOf}(y,x) \wedge \text{partOf}(z,x) \wedge y \neq z \tag{7}$$

Sometimes a designer may need to find out the direct component of a part. A part is a *direct component* of another part if there is no middle part between the two in the product hierarchy.

$$(\forall y,z)\mathit{direct}_{\text{partOf}}(y,z) \leftrightarrow$$
$$\leftrightarrow \text{partOf}(y,z) \wedge (\neg(\exists x)\text{partOf}(y,x) \wedge \text{partOf}(x,z) \tag{8}$$

That is, y is a direct part of z if y is a component of z and there is no x such that y is a part of x and x is a part of z. If y is a part of x then x is the whole of the y

$$(\forall x,y)\text{partOf}(y,x) \leftrightarrow \mathit{wholeOf}(x,y) \tag{9}$$

Classes are disjoint if they cannot have any instance in common:

$$(\forall x,y)\mathit{disjoint}(x,y) \rightarrow (\neg\exists z)\text{partOf}(z,x) \wedge \text{partOf}(z,y) \tag{10}$$

3 Design of CNMO

3.1 Structure of CNMO

An ontology can be divided into several component ontologies [20]. Thus, for constructing of the CNMO, component ontologies are designed separately and then an unified CNMO is produced. All of the vocabularies are taken from the network modelling, simulation, topology generation and discovery tools [2, 21–23] selected by the authors for integration purposes [1]. There are various languages and tools for constructing ontologies [24]. Additionally to the concepts which are used in various tools related to the target domain there are also aspects which are purely dedicated to Human-Computer Interaction issues. Thus, the differences in vizualisation of the model components such as nodes, links, etc. in various tools has to be addressed.

3.2 Communications Network Node Ontology

A node is an active device connected to a network and is described by the following slots (Fig. 1): NodeNumber, NodeName, NodeType, NodeNetAddress, NodeCostMonth (monthly cost), NodeDelay, TrafficIn (incoming traffic), TrafficOut (outgoing traffic) and NodeUtili (node utilization). Additionally there are classes for node vizualization (NodeVizCoord and NodeVizualisation) and location (Node GeoCoord for geographic location, longitude and latitude coordinates). A NodeVizCoord (Fig. 2) can be described by the natural spatial attributes such as XCoord, YCoord and ZCoord. A need for NodeVizualization ontology arises due to the fact that different network modelling and simulation tools represent their outputs in different formats and styles. For instance, a node can be associated to various icons, colours, shapes and labels. So the attributes of the classes/concepts for the node can be NodeIcon, NodeColor, NodeShape, NodeLabel (Fig. 3). A NodeGeoCoord (Fig. 4) just describes geographic coordinates of the node.

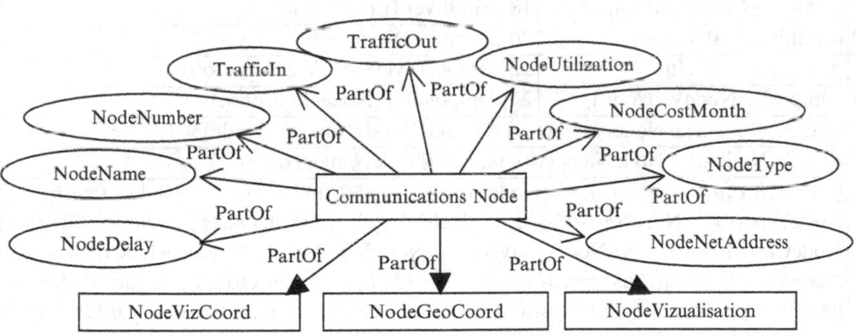

Fig. 1: Node ontology

The relevant axioms are shown in Tables 1, 2, 3 and 4. For instance, **Axiom 27** (Table 1) shows the concept: node consists of all parts. This approach to represent axioms is similar to that presented in [9]. Slots can have different facets describing the value type, cardinality (allowed values and the number of the values), and other features of the values the slot can take. This type of approach to represent the slot type and other facets is applied in [10].

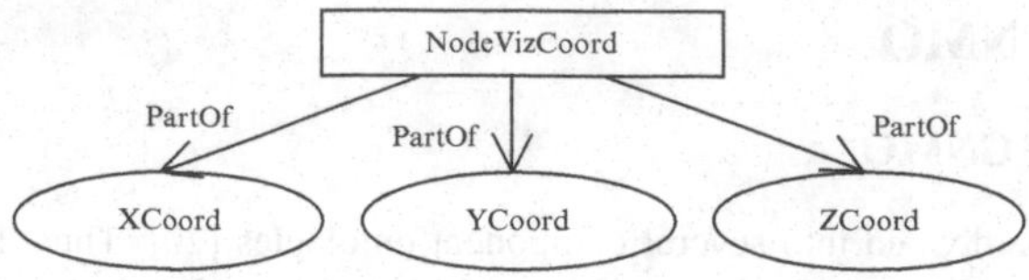

Fig. 2: NodeVizCoord ontology

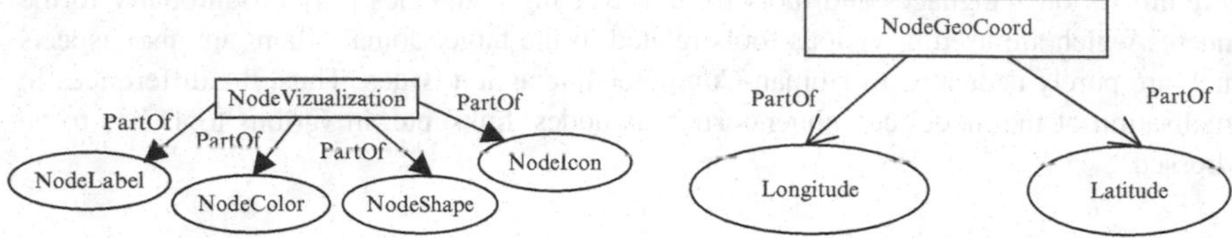

Fig. 3: NodeVizualization ontology

Fig. 4: Geographic coordinate ontology

Table 1: Axioms for node ontology

ID	Axioms	ID	Axioms
1	partOf(NodeNumber, node)	2	partOf(NodeName, node)
3	partOf(NodeType, node)	4	partOf(NodeNetAddress, node)
5	partOf(NodeCostMonth, Node)	6	partOf(TrafficOut, Node)
7	partOf(TrafficIn, Node)	8	partOf(NodeUtili, Node)
9	partOf(NodeDelay, Node)	10	partOf(NodeVizualization, node)
11	partOf(NodeVizCoord, node)	12	partOf(NodeGeoCoord, node)
13	primitive(NodeNumber)	14	primitive(NodeName)
15	primitive(NodeType)	16	primitive(NodeNetAddress)
17	primitive(NodeCostMonth)	18	primitive(TrafficOut)
19	primitive(NodeDelay)	20	primitive(NodeUtili)
21	primitive(TrafficIn)	22	compositive(NodeVizualization)
23	composite(NodeVizCoord)	24	compositive(NodeGeoCoord)
25	$(\forall x, \exists y)\text{Node}(x) \wedge \text{NodeNumber}(y) \wedge \text{partOf(NodeNumber, Node)} \vee T$		
26	$(\forall x, \exists y)\text{Node}(x) \wedge \text{NodeName}(y) \wedge \text{part of (NodeName, Node)} \vee T$		
27	$(\forall x, \exists a,b,c,d,e,f,g,h,I,j,k,l)\,\text{Node}(x) \wedge \text{NodeNumber}(a) \wedge \text{NodeName}(b) \wedge \text{NodeType}(c) \wedge \text{NodeNetAddss}(d) \wedge \text{TrafficIn}(e) \wedge \text{TrafficOut}(f) \wedge \text{NodeUtili}(g) \wedge \text{NodeCostMonth}(h) \wedge \text{NodeDelay}(i) \wedge \text{NodeVizcord}(j) \wedge \text{NodeGeoCoord}(k) \wedge \text{NodeVizualization}(l) \wedge \text{partOf}(a,x) \wedge \text{partOf}(b,x) \wedge \text{partOf}(c,x) \wedge \text{partOf}(d,x) \wedge \text{partOf}(e,x) \wedge \text{partOf}(f,x) \wedge \text{partOf}(g,x) \wedge \text{partOf}(h,x) \wedge \text{partOf}(i,x) \wedge \text{partOf}(j,x) \wedge \text{partOf}(k,x) \wedge \text{partOf}(l,x)$		

Table 2: Axioms for NodeVizCoord ontology

ID	Axioms	ID	Axioms
1	partOf(Xcoord, NodeVizCoord)	2	partOf(Ycoord, NodeVizCoord)
3	partOf(Zcoord, NodeVizCoord)	4	primitive(Xcoord)
5	primitive(Ycoord)	6	primitive(Zcoord)

Table 3: Axioms for NodeVizualizaion ontology

ID	Axioms	ID	Axioms
1	partOf (NodeLabel, NodeVizualization)	2	partOf (NodeColor, NodeVizualization)
3	partOf (NodeShape, NodeVizualization)	4	partOf (NodeIcon, NodeVizualization)
5	primitive(NodeLabel)	6	primitive(NodeColor)
7	primitive(NodeShape)	8	primitive(NodeIcon)

Table 4: Axioms for NodeGeoCord ontology

ID	Axioms	ID	Axioms
1	partOf(Longitude, NodeGeoCoord)	2	partOf (Latitude, NodeGeoCoord)
3	primitive(Longitude)	4	primitive(Longitude)

A value-type facet (see example Table 5) describes what types of values can fill in the slot. The most common value types are *string, number, enumerated*. Some systems distinguish only between single cardinality (allowing at most one value) and multiple cardinality (allowing any number of values).

Table 5: Concept: Node

Slots	Value type	Cardinality	Other facets Allowed value
NodeNumber	Number	Single	Classes = Node
NodeName	String	Single	Classes = Node
NodeType	String	Single	Classes = Node
NodeNetAddress	String	Single	Classes = Node
TrafficIn	Number	Single	Classes = Node
TrafficOut	Number	Single	Classes = Node
NodeCostMonth	Number	Single	Classes = Node
NodeUtili	Number	Single	Classes = Node
NodeDelay	Number	Single	Classes = Node
Xcoord	Number	Single	Classes = NodeVizCoord
Ycoord	Number	Single	Classes = NodeVizCoord
Zcoord	Number	Single	Classes = NodeVizCoord
Longitude	Number	Single	Classes = NodeGeoCoord
Latitude	Number	Single	Classes = NodeGeoCoord
NodeLabel	String	Single	Classes = NodeVizualization
NodeColor	String	Single	Classes = NodeVizualization
NodeShape	Enumerated	Single	Classes = NodeVizualization circle, rectangle, square,
NodeIcon	String	Single	Classes = NodeVizualization

3.3 Communications Network Link Ontology

A link is a communications channel which is transferring data between nodes and it is described by the following slots which are mostly self-explanatory (Fig. 5): LinkStartNode, LinkEndNode, LinkCapacity (in bits per second), LinkMaxThroughput (observed throughput, in bits per second), LinkUtilization (observed/predicted, in percent), LinkType (e.g. T1 as used in some tools), LinkTechnology (e.g. optical fibre), Duplexity (*simplex, half-duplex* or *full-duplex*), LinkDelay (for propagation plus transmission delay in milliseconds), LinkLength (physical distance between the two end nodes in meters), LinkReliability, BitErrorRate, LinkQueueSize (buffer size, number of packets), LinkQueueAlgorithm (e.g. DROPTAIL, RED etc. as used in ns-2) and LinkCostMonth. The link may be viewed differently in different tools and this is reflected in the vizualisation concept LinkViz (Fig. 6) which has attributes/slots such as LinkColour (for link colour), LinkOrient (for direction of link in degrees), LinkStyle (bold, thin, thick etc.). The relevant axioms are shown in Tables 6 and 7.

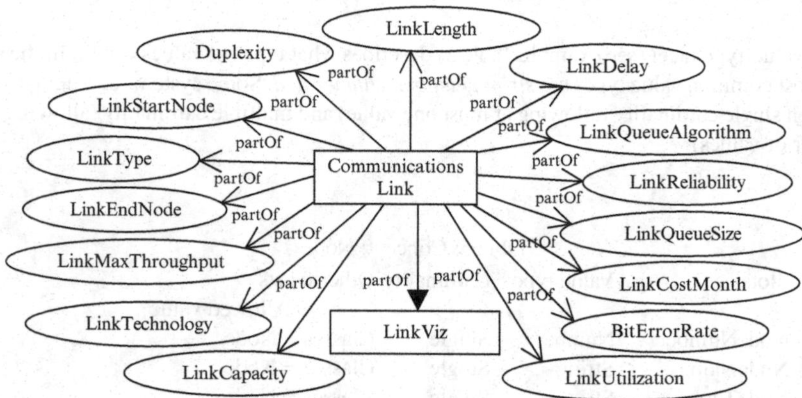

Fig. 5: Link ontology

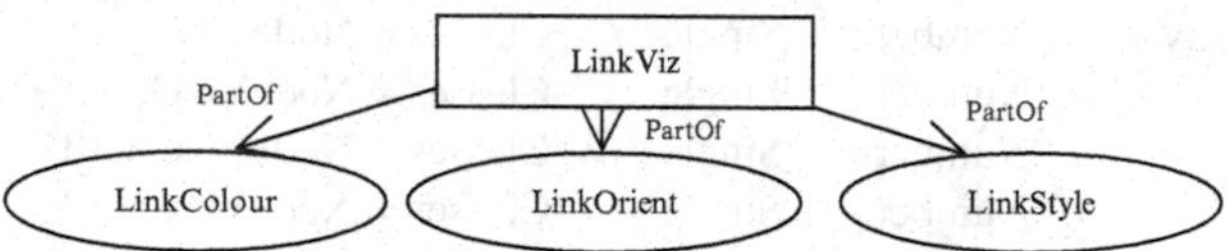

Fig. 6: LinkVizualization ontology

The attributes/slots, concept/class, possible constrains with possible values related to the communications link are shown in Table 8.

3.4 TransportEntity Ontology

This ontology deals with the transport protocols (called agents in network simulator ns-2 [21]) and it is described by the following slots which are mostly self-explanatory (Fig. 7): TPName

Table 6: Axioms for Link ontology

ID	Axioms	ID	Axioms
1	partOf(Duplexity , Link)	2	partOf(LinkStartNode , Link)
3	partOf(LinkEndNode, Link)	4	partOf(LinkMaxThroughput, Link)
5	partOf(LinkType, Link)	6	partOf(LinkTechnology, Link)
7	partOf(LinkCapacity, Link)	8	partOf(LinkLength, Link)
9	partOf(LinkDelay, Link)	10	partOf(LinkQueueAlgorithm, Link)
11	partOf(LinkReliability, Link)	12	partOf(LinkQueueSize, Link)
13	partOf(LinkCostMonth, Link)	14	partOf(BitErrorRate, Link)
15	partOf(LinkUtilization, Link)	16	partOf(LinkViz, Link)
17	primitive(Duplexity)	18	primitive(LinkStartNode)
19	primitive(LinkEndNode)	20	primitive(LinkMaxThroughput)
21	primitive(LinkType)	22	primitive(LinkTechnology)
23	primitive(LinkCapacity)	24	primitive(LinkLength)
25	primitive(LinkDelay)	26	primitive(LinkQueueAlgorithm)
27	primitive(LinkReliability)	28	primitive(LinkQueueSize)
29	primitive(LinkCostMonth)	30	primitive(BitErrorRate)
31	primitive(LinkUtilization)	32	compositive(LinkViz)
33	$(\forall x, \exists y)\text{Link}(x) \wedge \text{Duplexity}(y) \wedge$ partOf(Duplexity, Link) $\vee T$	34	$(\forall x, \exists y)\text{Link}(x) \wedge \text{LinkStartNode}(y) \wedge$ partOf(LinkStartNode, Link) $\vee T$
35	$\forall(x, \exists y)\text{Link}(x) \wedge \text{LinkEndNode}(y) \wedge$ partOf(LinkEndNode, Link) $\vee T$		

Table 7: Axioms for LinkViz ontology

ID	Axioms	ID	Axioms
1	partOf(LinkColor, LinkViz)	2	partOf(LinkOrient, LinkViz)
3	partOf(LinkStyle, LinkViz)	4	primitive(LinkOrient)
5	primitive(LinkColor)	6	primitive(LinkStyle)

(name/label for transport protocol entity), TPType (e.g. TCP, UDP, etc.), TPNode (the node name in where this entity is running), TPPriority (the priority assigned to packet from this entity), FId (e.g. 1, 2, 3, etc. which is IP layer flow id), TPPacketSize, TPWindowInit (the initial size of the congestion window on the slow start TCP), TPMaxCWnd (the upper bound of the congestion window for the TCP connection), TPWindowSize (upper bound on the advertised window for TCP connection). The relevant axioms are shown in Table 9.

The attributes/slots, concept/class, possible constraints with possible values related to the transport entities are shown Table 10.

3.5 TEConnection Ontology

This ontology describes the transport connections between the pairs of transport entities represented by slots Source and Sink (Fig. 8). Table 11 shows the axioms.

The attributes/slots, concept/class, possible constraints with possible values related to TEConnection are shown Table 12.

Table 8: Concept: Link

Slots	Value type	Cardinality	Other facets	Allowed value
LinkStartNode	String	Single	Classes = Link	
LinkEndNode	String	Single	Classes = Link	
LinkCapacity	Number	Single	Classes = Link	
LinkMaxThroughput	Number	Single	Classes = Link	
LinkUtilization	Number	Single	Classes = Link	
LinkType	String	Single	Classes = Link	
LinkTechnology	String	Single	Classes = Link	
Duplexity	Enumerated	Single	Classes = Link	simplex, duplex
LinkDelay	Number	Single	Classes = Link	
LinkLength	Number	Single	Classes = Link	
LinkReliability	Number	Single	Classes = Link	
BitErrorRate	Number	Single	Classes = Link	
LinkQueueSize	Number	Single	Classes = Link	
LinkQueueAlgorithm	String	Single	Classes = Link	
LinkCostMonth	Number	Single	Classes = Link	
LinkColor	String	Single	Classes = LinkViz	
LinkOrient	Alphanumeric	Single	Classes = LinkViz	In degree
LinkStyle	String	Single	Classes = LinkViz	

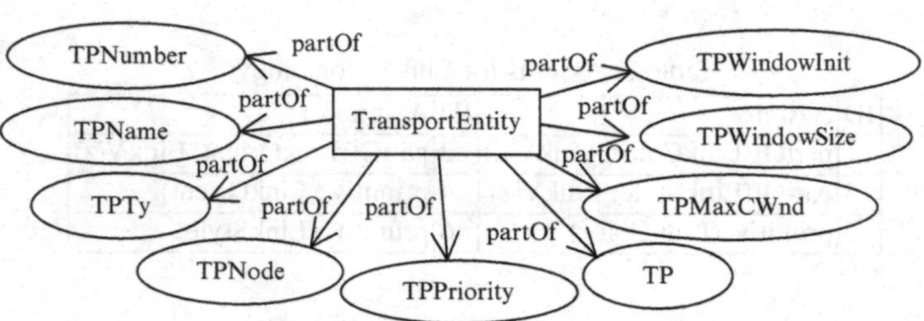

Fig. 7: TransportEntity ontology

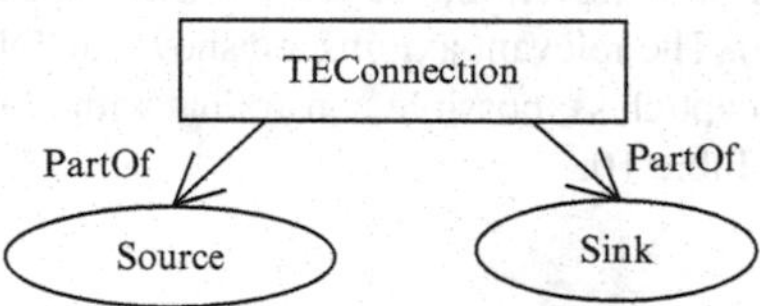

Fig. 8: TEConnection ontology

Table 9: Axioms for TransportEntity ontology

ID	Axioms	ID	Axioms
1	partOf(TPNumber, TrasportEntity)	2	partOf(TPName, TrasportEntity)
3	partOf(TPType, TrasportEntity)	4	partOf(TPNode, TrasportEntity)
5	partOf(TPPrior., TrasportEntity)	6	partOf(TPWindowInit, Trasport Entity)
7	partOf(TPWindowSize, Trasport Entity)	8	partOf(FId, TrasportEntity)
9	primitive(TPNumber)	10	primitive(TPName)
11	primitive(TPType)	12	primitive(TPNode)
13	primitive(TPPriority)	14	primitive(TPWindowInit)
15	primitive(TPWindowSize)	16	primitive(FId)
17	$(\forall x, \exists y)$TrasportEntity(x) $\wedge$ TPNumber(y) $\wedge$ partOf(TPNumber, TrasportEntity) $\vee\, T$	18	$(\forall x, \exists y)$TrasportEntity(x) $\wedge$ TPName(y) $\wedge$ partOf(TPName, TrasportEntity) $\vee\, T$
19	$(\forall x, \exists y)$TrasportEntity(x) $\wedge$ TPType(y) $\wedge$ partOf(TPType, TrasportEntity) $\vee\, T$	20	$(\forall x, \exists y)$TrasportEntity(x) $\wedge$ TPNode(y) $\wedge$ partOf(TPNode, TrasportEntity) $\vee\, T$

Table 10: Concept: TraspotEntity

Slots	Value type	Cardinality	Other facets Allowed Value
TPNumber	Number	Single	Class = TrasportEntity
TPName	String	Single	Class = TrasportEntity
TPType	Enumerated	Single	Class = TrasportEntity TCP, TCPSink, UDP, NULL
TPNode	String	Single	Class = TrasportEntity
TPPriority	String	Single	Class = TrasportEntity
FId	Number	Single	Class = TrasportEntity
TPWindowSize	Number	Single	Class = TrasportEntity
TPWindowInit	Number	Single	Class = TrasportEntity
TPMaxCWnd	Number	Single	Class = TrasportEntity
TPPacketSize	Number	Single	Class = TrasportEntity

Table 11: Axioms for TEConnect ontology

ID	Axioms	ID	Axioms
1	partOf(Source, TEConnection)	2	partOf(Sink, TEConnection)
3	primitive(Source)	4	primitive(Sink)

Table 12: Concept: TEConnection

Slots	Value type	Cardinality	Other facets	Allowed value
Source	String	Single	Class = TEConnect.	
Sink	String	Single	Class = TEConnect.	

3.6 TrafficSource Ontology

TrafficSource ontology is needed to define an application or protocol which is using transport connection and it is described by the following slots which are mostly self-explanatory (Fig. 9): TSNumber (ID number of the traffic source in the model), TSName, TSType (type of the source, e.g. FTP, Telnet etc.), TSAgent (agent attached to the source), TSRoundTrip (the round trip time in seconds), TSInterval (the time interval between sending of two consecutive packets), TSStart (starting time when source starts sending the packets in seconds), TSEnd (corresponding stopping time). TSCastingType (*e.g. unicast, multicast, narrowcast, broadcast*), TSCastingProtocol (e.g. IGMP, EGP, CBT, MTP, etc.). The relevant axioms are shown in Table 13.

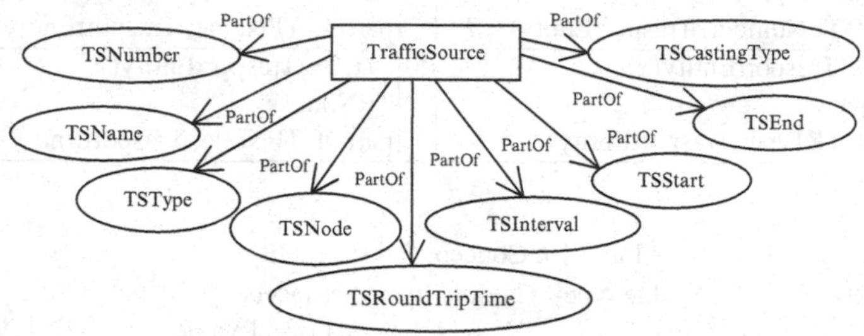

Fig. 9: TrafficSource ontology

Table 13: Axioms for TrafficSource ontology

ID	Axioms	ID	Axioms
1	partOf(TSNumber, TrafficSource)	2	partOf(TSName, TrafficSource)
3	partOf(TSType, TrafficSource)	4	partOf(TSNode, TrafficSource)
5	partOf(TSRoundTripTime, TrafficSour.)	6	partOf(TSInterval, TrafficSource)
7	partOf(TSStart, TrafficSource)	8	partOf(TSEnd, TrafficSource)
9	partOf(TSCastingType, TrafficSource)	10	primitive(TSNumber)
11	primitive(TSName)	12	primitive(TSType)
13	primitive(TSNode)	14	primitive(TSRoundTripTime)
15	primitive(TSInterval)	16	primitive(TSStart)
17	primitive(TSEnd)	18	primitive(TSCastingType)
19	$(\forall x, \exists y)$TrafficSource(x) $\wedge$ TSNumber(y) $\wedge$ partOf(TSNumber, TrafficSource) $\vee T$	20	$(\forall x, \exists y)$TrafficSource(x) $\wedge$ TSName(y) $\wedge$ partOf(TSName, TrafficSou.) $\vee T$
21	$(\forall x, \exists y)$TrafficSource$(x) \wedge$ TSType$(y) \wedge$ partOf(TSType, TrafficSource) $\vee T$	22	$(\forall x, \exists y)$TrafficSource(x) $\wedge$ TSNode(y) $\wedge$ partOf(TSNode, TrafficSource) $\vee T$

The attributes/slots, concept/class, possible constraints with possible values related to the traffic source are shown in Table 14.

Table 14: Concept: TrafficSource

Slots	Value type	Cardinality	Other facets Allowed value
TSNumber	String	Single	Class = TrafficSource
TSName	String	Single	Class = TrafficSource
TSType	Enumerated	Single	Class = TrafficSource CBR, FTP, Telnet, HTTP
TSNode	String	Single	Class = TrafficSource
TSRoundTrip	String	Single	Class = TrafficSource
TSInterval	String	Single	Class = TrafficSource
TSStart	String	Single	Class = TrafficSource
TSEnd	String	Single	Class = TrafficSource
TSCastingType	String	Single	Class = TrafficSource

3.7 ModellingFiles Ontology

This ontology defines the names of the files associated to the network modelling process and it is mostly self-explanatory (Fig. 10).

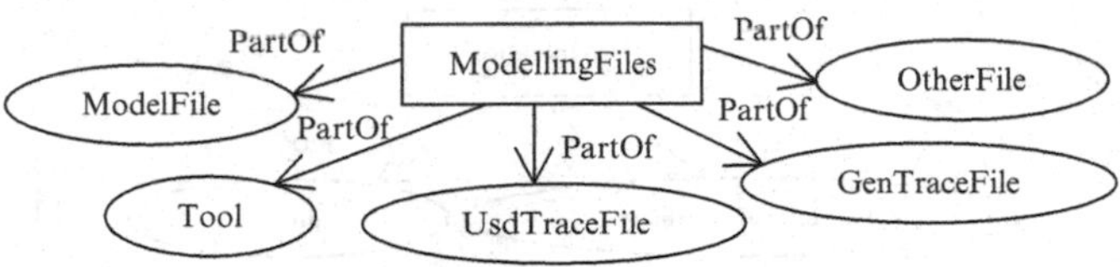

Fig. 10: ModellingFiles ontology

Here are defined slots ModelFile (name of a file containing the network model information), Tool (name of corresponding tool), UsdTraceFile (name of trace file used during the simulation), GenTraceFile (name of trace file generated by the tool), OtherFile contains name of other file (if any). The relevant axioms are shown in Table 15.

Table 15: Axioms for NetSimulation ontology

ID	Axioms	ID	Axioms
1	partOf(ModelFile, ModellingFiles)	2	partOf(Tool, ModellingFiles)
3	partOf(UsdTraceFile, ModellingFiles)	4	partOf(GenTraceFile, ModellingFiles)
5	partOf(OtherFile, ModellingFiles)	6	primitive(ModelFile)
7	primitive(Tool)	8	primitive(UsdTraceFile)
9	primitive(GenTraceFile)	10	primitive(OtherFile)

The attributes/slots, concept/class, possible constraints with possible values related to ModellingFiles are shown Table 16.

Table 16: Concept: ModellingFiles

Slots	Value type	Cardinality	Other facets	Allowed values
ModelFile	String	Single	Class = ModellingFiles	
Tool	Enumerated	Single	Class = ModellingFiles	Ns-2, Delite, Brite, Glomosim
UsdTraceFile	String	Single	Class = ModellingFiles	.tr, .trace
GenTraceFile	String	Single	Class = ModellingFiles	.tr, .trace
OtherFile	String	Single	Class = ModellingFiles	

3.8 NetOperation Ontology

NetOperation ontology is needed to define some outputs/information which may be produced/found after/during the network operation and it is described by the following slots which are mostly self-explanatory (Fig. 11): PacketLossProbability (packet loss probability in percent) and Jitter (delay variation in milliseconds). Relevant axioms are shown in Table 17.

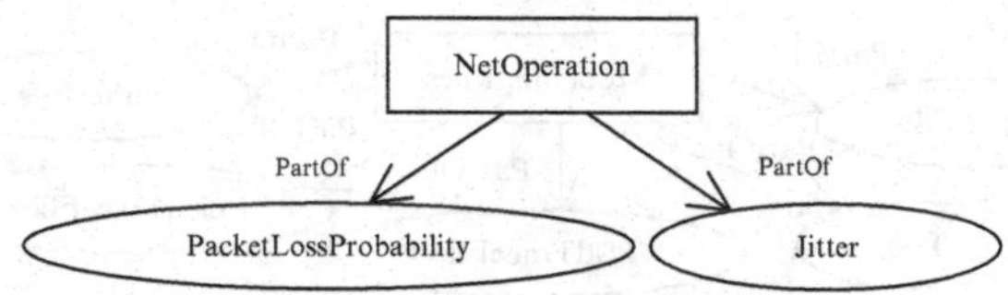

Fig. 11: Network operation ontology

Table 17: Axioms for NetOperation ontology

ID	Axioms	ID	Axioms
1	partOf(PacketLossProbability, NetOperation)	2	partOf(Jitter, NetOperation)
3	primitive(PacketLossProbability)	4	primitive(Jitter)

The attributes/slots, concept/class, possible constraints with possible values related to NetOperation are shown Table 18.

Table 18: Concept: NetOperation

Slots	Value type	Cardinality	Other facets	Allowed Values
PacketLossProbability	Number	Single	Class = NetOperation	
Jitter	Number	Single	Class = NetOperation	

3.9 CNMO as a Whole

A CNMO as a whole is shown in Fig. 12. Node domain is built up with ontologies: Node, NodeVizCood, NodeGeoCoord, and NodeVizualization whose details are shown in previous sections. Link domain consists of ontologies LinkInfor and linkViz. Formally in logic, a CNMO is:

$(\forall x : \text{Network})\ (\exists A : \text{Node})\ (\exists F : \text{Link})\ (\exists H : \text{TransportEntity})(\exists I : \text{TrafficSource})\ (\exists J : \text{NetOperation})\ (\exists K : \text{ModellingFiles})\ (\exists L : \text{TEConnection})\ (\text{partOf}(A,x) \wedge \text{partOf}(F,x) \wedge \text{partOf}(H,x) \wedge \text{partOf}(I,x) \wedge \text{partOf}(J,x) \wedge \text{partOf}(K,x) \wedge \text{partOf}(L,x))$

Formally a node is:

$(\forall A : \text{Node})\ (\exists B : \text{NodeVizCoord})\ (\exists C : \text{NodeGeoCoord})\ (\exists D : \text{NodeVizualization})\ (\text{partOf}(B,A) \wedge \text{partOf}(C,A) \wedge \text{partOf}(D,A))$.

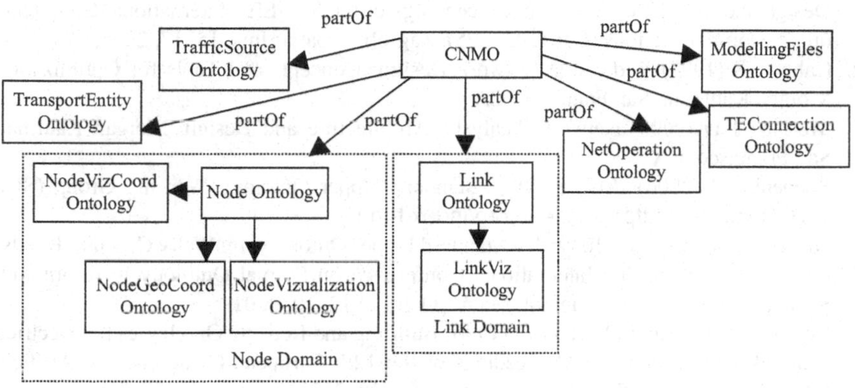

Fig. 12: CNMO as a whole

4 Related Work

Two ontologies are presented in [4, 5]: "Ontology for Communication" and "Ontology for Distributed Computing". The Communication ontology provides definition of terminology related to telephone communications system, radio system, television broadcast etc., but they are pretty general. Basic computing terminology (e.g. computer, CPU, HardDiskDrive, ComputerProcess, etc.) along with a few communications terms (e.g. LAN, DataTransfer, IPAddress, etc.) is defined in the Distributed Computing ontology. This one is basically an ontology for computing system rather than for communications system. Also, a method of developing an ontology for electrical network applications and its reuses have been described in [6]. The CNMO has parts which can be used together with these ontologies in order to create a wider scope. Additionally, the parts of CNMO are related not only to the modelling/research processes but also to the components of the communications network itself.

5 Conclusions

An approach to construction of the CNMO is discussed in this paper. The first-order logic is used for defining the terminology of the CNMO what gives a precise and unambiguous semantics for each term. The precision helps to avoid possible conflicts and different interpretations by different network designers/researchers. A set of axioms definitions and constrains on the terminologies that enable automatic deduction from the design knowledge is developed. The axioms also allow integrity checking of the network model knowledge, i.e. detecting invalid data and avoid updates which may cause conflicts among the data and a model. Furthermore, these common vocabularies can be used to integrate all network modelling, simulation, topology generation and discovery tools together [1].

References

1. Rahman M A, Pakstas A, Wang F Z (2005) An Approach to Integration of Network Design and Simulation Tools. In: Proceedings of the 8th IEEE International Conference on Telecommunications (ConTEL 2005), Zagreb, Croatia, June 15–17, 2005, 173–180
2. Cahn R S (1998) Wide Area Network Design: Concepts and Tools for Optimization. Morgan Kaufman, San Francisco, CA
3. McCabe J D (2003) Network Analysis, Architecture and Design. Morgan Kaufman, San Francisco, CA
4. Schoening J (2006) IEEE P1600.1 Standard Upper Ontology Working Group (SUO WG) Home Page http://suo.ieee.org/index.html
5. Niles I, Pease A (2001) Towards a Standard Upper Ontology. In: Welty C, Smith B (eds), Proceedings of the 2nd International Conference on Formal Ontology in Information Systems (FOIS-2001), Ogunquit, Maine, October 17–19, 2001
6. Bernaras A, Laresgoiti I, Corera J (1996) Building and Reusing Ontologies for Electrical Network Applications. In: Proceedings of the 12th European Conference on Artificial Intelligence (ECAI), Budapest, August 1996, 298–302
7. Gomes Mian P, de Almeida Falbo R (2004) Supporting Ontology Development with ODEd. Journal of the Brazilian Computer Society 9(2):57–76
8. Gruninger M, Lee J (2002) Ontology Applications and Design. Introductory article to a special issue on Ontology Engineering. Communications of the ACM, February 45(2):39–41
9. Gomes Mian P, de Almeida Falbo R (2003) Buidling Ontologies in a Domain Oriented Software Engineering Environment. In: IX Argentine Congress on Computer Science (CACIC 2003), La Plata, Argentina, October 6–10, 2003, 930–941
10. Noy N F, McGuinness D L (2001) Ontology Development 101: A Guide to Creating Your First Ontology. Stanford Knowledge Systems Laboratory Technical Report KSL-01-05 and Stanford Medical Informatics Technical Report SMI-2001-0880, Stanford University, Stanford, CA, USA, March 2001
11. Lin J, Fox M S, Bilgic T (1996) A requirement ontology for engineering design. Concurrent Engineering: Research and Applications 4(4):279–291
12. Uschold M, King M (2005) Towards a Methodology for Building Ontologies. Technical Report, AIAI-TR-183, University of Edinburgh, Edinburgh, July 1995 (Presented at the Workshop on Basic Ontological Issues in Knowledge Sharing, IJCAI95, Montreal) http://www.aiai.ed.ac.uk/~entprise/enterprise/ontology.html

13. Uschold M, Grüninger M (1996) Ontologies: Principles, Methods and Applications. Knowledge Engineering Review 11(2):93–155, June 1996
14. Gruber T R (1992) A Translation Approach to Portable Ontology Specifications. In: Proceedings of the Second Japanese Knowledge Acquisition for Knowledge-based Systems Workshop (JKAW'92), Kobe and Hatoyama, November 1992, 89–108
15. Mizoguchi R (1993) Knowledge Acquisition and Ontology. In: Proceedings of the KB& KS'93, Tokyo, 1993, 121–128
16. Mizoguchi R, Ikeda M (1997) Towards Ontology Engineering. In: Proceedings of the Joint Pacific Asian Conference on Expert Systems/Singapore International Conference on Intelligent Systems (PACES/SPICIS '97), Singapore, February 24–27, 1997, 259–266
17. Gennari J H, Altman R B, Musen M A (1995) Reuse with PROTÉGÉ-II: From Elevators to Ribosomes. In: Proceedings of the 1995 Symposium on Software Reusability, Seattle, Washington, DC, 72–80
18. Mizoguchi R (2005) The Role of Ontological Engineering for AIED Research. In: Computer Science and Information System (ComSIS), Vol. 2, No. 1, Belgrade, Serbia and Montenegro, June 2005, 31–42
19. Jin L, Ikeda M, Mizoguchi R (1996) Ontological Issues on Computer-Based Training. In: PRICAI-96 Workshop on Knowledge-Based Instructional Systems in an Industrial Setting, Cairns, Australia, August 26–30, 1996, 55–66
20. Sunagawa E, Kozaki K, Kitamura Y, Mizoguchi R (2003) Management of Dependency between Two or More Ontologies in an Environment for Distributed Development. In: Proceedings of the International Workshop on Semantic Web Foundations and Application Technologies (SWAFT), Nara, Japan, March 12, 2003
21. Fall K, Varadhan K (2003) The NS Manual. The VINT Project, December 13, 2003 http://www.isi.edu/nsnam/ns/doc/ns_doc.pdf
22. GloMoSim Manual (ver. 1.2) (2001). University of California, Los Angeles, CA, February 7, 2001 http://pcl.cs.ucla.edu/projects/glomosim/GloMoSimManual.html
23. Medina A, Lakhina A, Matta I, Byers J (2001) Brite: Universal Topology Generator from a User's Perspective. Technical Report, BUCS-TR-2001-003, Boston University, Boston, MA, April 12, 2001 http://www.cs.bu.edu/brite/publications/usermanual.pdf
24. Su X, Ilebrekke L (2002) A Comparative Study of Ontology Languages and Tools. In: Proceedings of the 14th International Conference on Advanced Information Systems Engineering, London, May 27–31, 2002, 761–765

Computational Intelligence Approach to Condition Monitoring: Incremental Learning and Its Application

Christina B. Vilakazi and Tshilidzi Marwala

School of Electrical and Information Engineering, University of the Witwatersrand, Private Bag 3, Wits, 2050, South Africa
c.vilakazi@gmail.com, t.marwala@ee.wits.ac.za

Summary. Machine condition monitoring is gaining importance in industry due to the need to increase machine reliability and decrease the possible loss of production due to machine breakdown. Often the data available to build a condition monitoring system does not fully represent the system. It is also often common that the data becomes available in small batches over a period of time. Hence, it is important to build a system that is able to accommodate new data set as it becomes available without compromising the performance of the previously learned data. Two incremental learning algorithm are implemented, the first method uses Fuzzy ARTMAP (FAM) algorithm and the second uses Learn++ algorithm. Experimental results show that both methods can accommodate both new data and new classes.

1 Introduction

Industrial machinery has a high capital cost and its efficient use depends on low operating and maintenance costs. To comply with this requirements, condition monitoring and diagnosis of machinery have become established industry tools [1]. Condition monitoring approaches have produced considerable savings by reducing unplanned outage of machinery, reducing downtime for repair and improving reliability and safety. This is done so that potential problems can be detected and diagnosed early in their development, and corrected by suitable recovery measures before they become severe enough to cause plant breakdown and other serious consequences. Hence, there is a need for development a reliable, fast and automated diagnostic technique allowing relatively unskilled operators to make important decisions without the need for a condition monitoring specialist to examine data and diagnose problems. Various computational intelligence techniques such as neural networks, support vector machines, have been used extensively to the problem of condition monitoring. However, many computational intelligence based methods for fault diagnosis rely heavily on adequate and representative set of training data. In real-life applications it is often common that the available data set is incomplete, inaccurate and changing. It is also often common that the training data set becomes available only in small batches and that some new classes only appear in subsequent data collection stages. Hence, there is a need to update the classifier in an incremental fashion without compromising on the classification performance of previous data. Firstly, the objective of this work is to investigate the effectiveness of the Fuzzy ARTMAP as compared to support vector machine (SVM), multi-layer perceptron (MLP) and extension neural network

J.A.T. Machado et al. (eds.), *Intelligent Engineering Systems and Computational Cybernetics,*

(ENN). Furthermore, this work aims to implement incremental learning system to ensure that the condition monitoring system knowledge base is updated in an incremental fashion without compromising the performance of the classifier on previously learned information.

2 High Voltage Bushing Data

Dissolved gas analysis is one of the most popular chemical techniques that are used in oil-filled equipment. DGA is the most commonly used diagnostic technique for oil-filled machines such as transformers and bushings [2]. DGA is used to detect oil breakdown, moisture presence and partial discharge activity. The gaseous byproduct are produced by degradation of transformer and bushing oil and solid insulation, such as paper and pressboard, which are all made of cellulose. The gases produced from the transformer and bushing operation can be listed as follows [3]:

- Hydrocarbons and hydrogen gases: methane, ethane, ethylene, acetylene and hydrogen
- Carbon oxide: carbon monoxide and carbon dioxide
- Naturally occurring gases: nitrogen and oxygen

The symptoms of faults are classified into four main groups; corona, low energy discharge, high energy discharge and thermal. The quantity and types of gases reflect the nature and extent of the stressed mechanism in the bushing.

3 Background on Classifiers

3.1 Multi-Layer Perceptron

Artificial neural networks are data processing system that learns complex input-output relationships from data. A typical ANN consists of simple processing elements called neurons that are highly interconnected in an architecture that is loosely based on the structure of biological neurons in human brain. There are different types of ANN models; two that are commonly used are MLP network and radial basis function (RBF) network. However, only MLP are implemented in this paper because they have been successfully applied to various condition monitoring applications. A good general introduction to neural networks is provided by Bishop book:bishop.

3.2 Support Vector Machine

Kernel-based classifiers have recently been used as popular and powerful tools for classification, due to their strong theoretical origin from statistical learning theory as well as their high performance in practical applications [5]. SVM classifiers are kernel-based learning algorithms, determining the optimal hyperplane decision boundary in the feature space. In kernel-based algorithms, a kernel transformation process the data in a feature space without the explicit knowledge of a nonlinear mapping from the data space to a feature space. In statistical learning theory, the complexity term of the upper bound of the expected risk is minimized by maximizing the margin of the separating hyperplane. The minimization of the upper bound can be viewed as avoiding the over-fitting problem. The maximization of the margin can be formulated as a quadratic optimization problem so that a global solution can be easily obtained. A detailed description of SVM is found in book:vapnik.

3.3 Extension Neural Network

The Extension Neural Network is a new pattern classification system based on concepts from neural networks and extension theory [6]. The extension theory uses a novel distance measurement for classification processes, and the neural network can embed the salient features of parallel computation power and learning capability. ENN comprises of input layer and the output layer. The input layer nodes receive an input feature pattern and uses a set of weighted parameters to generate an image of the input pattern. There are two connection weights between input nodes and output nodes; one connection represents the lower bound for this classical domain of the features, and the other represents the upper bound.

3.4 Fuzzy ARTMAP

The fuzzy ARTMAP architecture is an auto-organized learning system [7]. This kind of network has supervised training and pertains to the adaptive resonance theory (ART) family; its structure is based on the adaptive resonance theory and is similar to the fuzzy ART network, which employs calculus based on fuzzy logic. This network is composed of two fuzzy ART modules; ART_a and ART_b, interconnected by an inter-ART by an associative memory module. There are three fundamental parameters for the performance and learning of the fuzzy ART network, namely, the chosen parameter, training rate, and vigilance parameterjournal:carpenter. The chosen parameter, ($\alpha > 0$) which acts on the category selection. Training rate, ($\beta\ \varepsilon\ [0,1]$) which controls the velocity or the learning rate of the network adaptation, $\beta = 1$ permits the system to adapt faster while $0 < \beta < 1$ allows the system to adapt slower. Vigilance parameter, ($\rho\ \varepsilon\ [0,1]$) which controls the network resonance. The vigilance parameter is responsible for the number of categories formed. If the vigilance parameter is very large, it produces a good classification, providing the generation of several categories and this means that the network has less errors but has less capacity for generalization. If it is very small, it will generate few categories, which implies that the network will have more capacity for generalization, but more possibility of making mistakes.

4 Comparison of Classifiers

This experiment aims to compare the performance of a batch trained fuzzy ARTMAP in terms of classification accuracy with batch trained MLP, SVM and ENN. An MLP with 12 hidden layer units that uses a tangent activation function is used. A fuzzy ARTMAP with a vigilance parameter of 0.578 determined empirically was used. Support vector machine that uses an RBF kernel function was used. ENN with a learning rate of 0.125 which was determined empirically was implemented. Table 1 shows that the classification performance of MLP, ENN, SVM and fuzzy ARTMAP.

Table 1 shows that FAM gave slightly better results than MLP and SVM. However, ENN gave slightly better results than the fuzzy ARTMAP.

5 Incremental Learning

An incremental learning algorithm is an algorithm that is able to learn additional information from data while retaining previously learned information. The algorithm should not require

Table 1: Comparison of classifiers for bushing fault diagnosis

Classifier	Validation accuracy (%)	Testing accuracy (%)	Time (s)
MLP	98.00	95.87	0.2031
SVM	97.70	95.63	30.7031
ENN	99.00	98.50	0.0156
fuzzy ARTMAP	98.50	97.50	0.5687

access to the original data and must also be able to learn classes that may be introduced by incoming data. The FAM and Learn++ fulfill all these criteria.

The characterization of machine incremental learning applies to three dimensions of machine incremental learning; structural changes, learning parameter adjustments and input data variations. In characterization, let $s = (s_i)_{i=0...n}$, $l = (l_j)_{j=0...n}$ and $d = (d_k)_{k=0...n}$ be families of real numbers. An incremental learning system $I = (s;l;d)$ is a learning system which is parameterized by three families of incremental learning parameters which can be modified during training [8]. The first is the structure parameters ($s = (s_i)_{i=0...n}$), which are, for example, the number of neurons, density of connections, or other parameters which determine structure and functionality of a neural network. Learning parameters ($l = (l_j)_{j=0...n}$) which are, for example, evolutionary or other learning parameters, such as the stepsize. Data-complexity parameters ($d = (d_k)_{k=0...n}$) which can represent any complexity measure of the training data. Accordingly, there are three main forms of incremental learning. Each of them modifies members of one of the parameter families defined above [8]:

- Structure Incremental Learning: The structure or functional capacity of the neural network is changed during learning.
- Learning Parameter Incremental Learning: A selection of learning parameters from $l = (l_j)$ is adapted during learning.
- Data Incremental Learning: The data set or its complexity is increased in stages during learning controlled by parameter changes of $d = (d_k)$.

The incremental learning in the fuzzy ARTMAP is a combination of structure and data incremental learning. As data of different classes is added to the system or more data on existing classes are added, the structure of the system adapts creating a larger number of categories in which to map the output labels. The incremental learning in Learn++ is a data incremental learning because as new data is added new classifiers are build to be added to the existing system. Learning new information without requiring access to previously used data, however, raises "stability-plasticity dilemma". This dilemma indicates that a completely stable classifier maintains the knowledge from previously seen data, but fails to adjust in order to learn new information, while a completely plastic classifier is capable of learning new data but lose prior knowledge.

6 Fuzzy ARTMAP and Incremental Learning

6.1 System Design

The proposed system consists of three stages, data processing, creation of ensemble, classification process with incremental learning. The data preprocessing is done using min-max normalization. The normalization is a requirement for using the FAM, since the FAM complement coding assumed the data is normalized. The population of classifiers is used to introduce classification diversity. Since, a bad team of classifiers may even lead to worse performance than the best single classifier due to the fusion of error decision. A classifier selection process is executed using the correlation measure. As a result, an optimal team of classifier is formed to improve classification accuracy. After the classifier selection, majority voting fusion algorithm is employed for the final decision.

Creation of an Ensemble

The creation of an ensemble of classifiers follows a series of steps. First, an initial population of 12 classifiers with different permutations of the training data are created. This permutation is needed in order to create diversity of the classifiers being added since FAM learns in an instance-based fashion, which makes the order in which the training patterns are received an important factor. From the created population, the best performing classifier of the team is selected based on the classification accuracy. Then the correlation degree between the best classifier and all the members of the population is calculated using (1).

$$\rho_n = \frac{nN^F}{N - N^F - N^R + nN^F} \tag{1}$$

where N^F is the number of samples which are misclassified by all classifier N^R represent the samples which were classified correctly by all classifiers. N is the total number of experiments example. From the unselected classifiers, the classifier with low correlation is selected for fusion. This is repeated until all the unselected classifiers are selected. Lastly, the selected classifiers are fused using the majority voting to give the final decision. The algorithm for this system is shown in Fig. 1.

TRAINING PHASE

1. Create a population of 12 classifiers with different permutation of the input data.
2. Select an appropriate performance measure as the initial evaluation criterion such as accuracy rate, which is the ratio. number of sample classified correctly to the total sample.
3. Find the best performing classifiers to be the first classifier of the ensemble.
4. Calculate the correlation degree between the first classifier and other classifiers, respectively using Equation 1
5. Select the classifier with low correlation for fusion. 6 Repeat 4 and 5 between selected classifiers yet to be selected until all the classifiers are determined.
6. Fuse the individual classifier's prediction using majority voting strategy.

OPERATION PHASE

1. If a new data becomes available during the operation phase.
2. Add new data to the ensemble of classifiers by training the individual classifier and then combine through majority voting.

Fig. 1: The algorithm for the fuzzy ARTMAP system

6.2 Experimental Results and Discussion

The available data was divided into three data set; training, validation and testing data sets. The validation data set is used to evaluate the performance of each classifiers in the initial population and this performance is used to select the classifiers for the ensemble. The testing data set is used to evaluate the performance of the system on unseen data. The first experiment evaluates and compares the incremental capability of a single FAM and the ensemble of FAM for new classes. The last experiment uses an ensemble of FAM so as to compare the performance of FAM with that of Learn++.

Comparison of a single FAM and the ensemble

For this experiment, the training dataset, validation dataset and the testing dataset consisted of 200, 300 and 500 data points from each classes, respectively. The results for different numbers of classifiers are shown in Table 2. When classifier selection was performed it was found that the optimal results is achieved with six classifiers. The best classifier gave an accuracy of 98.5% and 95.2% on the validation and testing dataset, respectively. On the other hand, the ensemble gave a classification accuracy of 100% and 98% on the validation and testing dataset, respectively and this shows that the ensemble of fuzzy ARTMAP performs better than the best fuzzy ARTMAP classifier.

Table 2: Results of the FAM classifiers created to be used in the ensemble for the bushing data

Classifier	Validation accuracy (%)	Correlation (ρ)
1	98.5	Best classifier
2	97.5	0.624
3	97.5	0.6319
4	97.5	0.6239
5	95.0	0.6310
6	95.0	0.6296
7	95.0	0.6359
8	92.5	0.6377
9	92.5	0.6143
10	92.5	0.6178
11	92.5	0.6340
12	92.5	0.6476

Incremental Learning

The incremental learning capability of a single FAM is compared with that of an ensemble of classifiers. For the initial experiment the knowledge of a fourth class is added to the classification system. Using the existing model of the classifier in the ensemble, each classifier is trained on data of this new class. Experimentation is performed on independent test set for this class for both the best classifier and ensemble of the classifiers. Both the best classifier and the ensemble gave a classification accuracy of 100%. Experimentation of the system on the testing data set of the initial three classes is performed, this is to determine how addition of new information affects previously learned information. The classifier accuracy achieved is 94.88% for the best classifier and 98% for the ensemble. This simple experiment shows that the system is able to learn new classes, while still preserving existing knowledge of the system. The small change in the accuracy of the three classes is due to the tradeoff between stability and plasticity of the classifier during training. The learning ability of the system on a further fifth class is shown in Fig. 2.

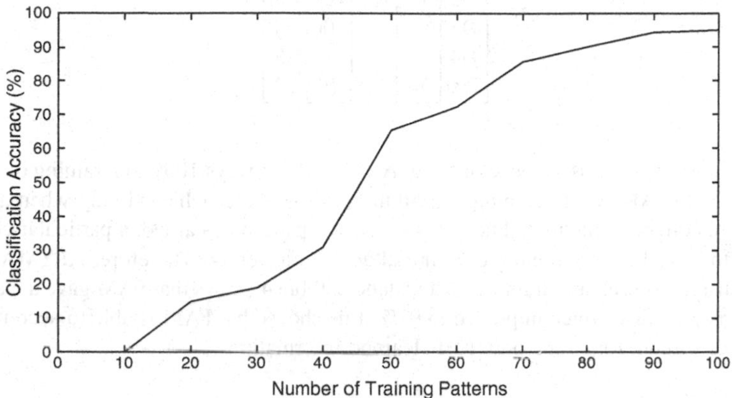

Fig. 2: Incremental learning of the best classifier on the fifth class

The classification accuracy of the best classifier and the ensemble is 95% and 96.33%, respectively. The best classifier gave a classification accuracy of 90.20% while the ensemble gave a classification accuracy of 89.67% on the original test data with three classes. As more training examples are added the ability of the system to correctly classify the data generally increases as shown by the increase in classification accuracy.

Incremental Learning with the Ensemble of FAM

Table 3 shows the distribution of the classes in various databases, D_j $j = 1,\ldots,4$ indicated by the first column.

Table 3: Distribution of the five classes on different databases

	Normal	Corona	Low energy discharge	High energy discharge	Thermal
D1	300	300	0	300	0
D2	20	20	300	0	0
D3	20	20	20	20	300
D4	20	20	20	20	300
Test	300	300	300	300	300

Table 4 shows the results of the ensemble of FAM, the first row shows the training cycles, T_i $i = 1, \ldots, 4$ and the first column indicate the various databases D_j $j = 1, \ldots, 4$.

Table 4: Classification performance of FAM on five classes

	T1	T2	T3	T4
D1	100	100	100	100
D2	–	100	100	100
D3	–	–	100	100
D4	–	–	–	100
Test	48	68	79	93

The classification performance of fuzzy ARTMAP is always 100% on training data, since according to the ARTMAP learning algorithm convergence is achieved only when all training data are correctly classified. Furthermore, once a pattern is learned, a particular cluster is assigned to it, and future training does not alter this clustering. Therefore, ARTMAP never forgets what it has seen as a training data instance. Table 4 shows that FAM gave a classification accuracy of 48% which improved to 93%, this shows that FAM is able to accommodate new classes without forgetting previously learned information.

7 Learn++ and Incremental Learning

7.1 Overview of Learn++

Learn++ is an incremental learning algorithm that uses an ensemble of classifiers that are combined using weighted majority voting [13]. Each database D_j, which is an independent set of data that is to be added to the system consists of a training instances x_i, and corresponding labels y_i. The algorithm specifies a weak learner and an integer T_k, which is the number of classifiers to add to the system for a given database D_j. The system trains each of the T_k classifiers with a different subset of the training data, effectively creating multiple hypotheses for the training data. These multiple hypotheses are combined using weighted majority voting scheme, where the voting weight for each classifier in the system is determined using the performance of that particular classifier on the entire set of training data used for the current increment. More details on Learn++ can be found in [13].

7.2 System Design

The architecture of the proposed condition monitoring framework consists of two major stages after data acquisition, which are; pre-processing and/or feature extraction stage, and classification stage with incremental learning.

7.3 Experimental Results and Discussion

Three experiments are performed to evaluate the effectiveness of Learn++ on the high voltage bushing data. The first experiment evaluates the incremental capability of the algorithm. This is done by evaluating the performance of Learn++ as new data is introduced. The second experiment is performed to evaluate how well Learn++ can accommodate new classes. The last experiment compares the performance of Learn++ with that of a batch trained MLP. This is done to compare the classification rate of Learn++ with that of a strong learner. The format of the tables in the following section was adopted from [13].

For the first experiment, the training data was divided into four databases each with 300 training instances. In each training session, Learn++ is provided with each database and generates 12 classifiers. The base Learner uses an MLP with 30 hidden layer neurons and 100 training cycles. To ensure that the method retains previously seen data, for each training session, the previous database is tested. Table 5 presents the results in which first four rows indicates the classification performance of Learn++ on each of the databases after each training session while the last row shows the generalization capability on the testing data. This demonstrates the performance improvement of Learn++ as inherited from AdaBoost on a single database. Table 5 further shows that classifiers performances on the testing dataset, gradually improved from 65.7% to 95.8% as new databases become available thereby demonstrating incremental learning capability of Learn++. The table further shows that Learn++ does not forget previously learned information, when new data set is introduced. The small change in accuracy is due to the trade-off between stability and plasticity.

Table 5: Classification performance of Learn++ on new data for bushing data

	T1	T2	T3	T4
D1	89.5	85.8	83	86.9
D2	–	91.5	94.2	92.9
D3	–	–	96.5	90.46
D4	–	–	–	98
Test	65.7	79.0	85	95.8

For the second experiment, the training data was divided into four databases. Table 3 shows the distribution of the data in four databases. In each training session, Learn++ is provided with each database and generates a specified number of hypotheses, which is indicated by the number inside the bracket in Table 6. Table 6 shows that the classifiers performance increases from 49% to 95.67% as new classes are introduced in the subsequent training dataset.

Table 6: Classification performance of Learn++ on new classes for bushing data

	T1 (8)	T2 (12)	T3 (15)	T4 (18)
D1	97.00	96.42	93.00	94.58
D2	–	98.78	96.00	92.10
D3	–	–	95.00	95.00
D4	–	–	–	98.00
Test	49	67.00	81.00	95.67

An MLP with the same set of training examples as Learn++ was trained. The trained MLP was tested with the same validation data as Learn++. The training data consisted of all the data in the four databases and an MLP that consists of 12 hidden layer units was trained. The MLP gave a classification rate of 100% tested on the same testing data as Learn++. This shows that the classification accuracy of Learn++ is comparable with that of an MLP trained using batch learning.

8 Comparison of Learn++ and Fuzzy ARTMAP

One disadvantage of Fuzzy ARTMAP is that it is very sensitive to the order of presentation of the training data. Fuzzy ARTMAP is also extremely sensitive to the selection of the vigilance parameter, and finding the optimal value for the vigilance parameters can be quite challenging.

In ensemble approaches that use a voting mechanism for combining classifier outputs, each classifier votes on the class it predicts. The final classification is then determined as the class that receives the highest total vote from all classifiers. Learn++ uses weighted majority voting, where each classifier receives a voting weight based on its training performance. This works well in practice for most applications. In incremental learning problems that involve introduction of new classes, the voting scheme proves to be unfair towards the newly introduced class. Since none of the previously generated classifiers can pick the new class, a relatively large number of new classifiers to recognize the new class are needed, so that their total weight can out-vote the first batch of classifiers on instances of the new class. This in return populates the ensemble with an unnecessarily large number of classifiers. When incrementing the system with new classes, it is best to ensure that the number of classifiers that is added to the system is greater than the number of classifiers added during a previous system increment. It is also better to include some data from classes that have previously been seen. This will ensure that if any pattern is classified into one of the new classes, the votes from the previous classifiers do not 'outvote' votes received from new classifiers. The major disadvantage of Learn++ is that it is computationally expensive. Generally, to allow incremental learning of the classes, 43 classifiers had to be generated for the system, while the same performance was obtained from a single classifier trained in batch mode.

9 Conclusions

The fuzzy ARTMAP gave slightly better results than the MLP, SVM and ENN classifiers. This might be due to the fact that the structure of the fuzzy ARTMAP adapts, creating a larger number of categories in which to map the output labels as the training data becomes available. The ensemble of classifiers gave an improvement on the classification accuracy from a single fuzzy ARTMAP. Two incremental learning algorithm were implemented. The first incremental learning algorithm uses an ensemble of MLP classifier that are combined using the weighted majority voting known as Learn++. The second algorithm uses the fuzzy ARTMAP and to enhance the performance of a single FAM and an ensemble of FAM, two different scenarios were considered. Both the algorithm were able to accommodate both new data and new classes without forgetting previously learned information.

Acknowledgments

The financial assistance of the National Research Foundation (NRF) of South Africa, the Carl and Emily Fuchs Foundation and the Council of Scientific and Industrial Research towards this research is hereby acknowledged. Opinions and conclusions arrived at, are those of the authors and are not necessarily to be attributed to the NRF.

References

1. Tavner PJ, Penman J (1987) Condition Monitoring of Electrical Machines. Wiley, England
2. Saha TK (2003) Review of Modern Diagnostic Techniques for Assessing Insulation Condition in Aged Transformers. IEEE Transactions on Electrical Insulation 10(5):903–917
3. Dhlamini SM, Marwala T (2004) Using SVM, RBF and MLP for Bushings. Proceedings of IEEE Africon 2004:613–617
4. Bishop CM (2003) Neural Networks for Pattern Recognition. Oxford University Press, New York
5. Vapnik VN (1999) The Nature of Statistical Learning Theory. Second edition. Springer, Berlin
6. Wang TK (2003) A Novel Extension Method for Transformer Fault Diagnosis. IEEE Transactions on Power Delivery 18(1):164–169
7. Carpenter GA, Grossberg S, Markuzon N, Reynolds JH, Rosen DB (1992) Fuzzy ARTMAP: A Neural Network Architecture for Incremental Supervised Learning of Analog Multidimensional Maps. IEEE Transactions on Neural Networks 3:698–713
8. Chalup SK (2002) Incremental Learning in Biological and Machine Learning Systems. International Journal of Neural Systems 12(6):90–127
9. Wang EH, Kuh A (1992) A smart algorithm for incremental learning. In: Proceedings of International Joint Conference on Neural Networks, pp 121–126
10. Osorio FS, Amy B (1999) INSS: A Hybrid System for Constructive Machine Learning. Neurocomputing 28:191–205
11. Mitra P, Murthy CA, Pal SK (2000) Data condensation in large databases by incremental learning with support vector machines. In: Proceedings of 15th International Conference on Pattern Recognition, 2000, Vol 2, pp 708–711
12. Li K, Huang HK (2002) Incremental learning proximal support vector machine classifiers. In: Proceedings of International Conference on Machine Learning and Cybernetics, 2002, Vol 3, pp 1635–1637
13. Polikar R, Udpa L, Udpa AS Hanovar V (2001) Learn++: An Incremental Learning Algorithm for Supervised Neural Networks. IEEE Transactions on Systems, Man and Cybernetics 31(4):497–508

An Approach for Characterising Heavy-Tailed Internet Traffic Based on EDF Statistics

Karim Mohammed Rezaul and Vic Grout

Centre for Applied Internet Research (CAIR), University of Wales,
NEWI Plas Coch Campus, Wrexham, UK
karim@cair-uk.org, v.grout@newi.ac.uk

Summary. In this research, statistical analyses of Web traffic were carried out based on the Empirical Distribution Function (EDF) test. Several probability distributions, such as Pareto (simple), extreme value, Weibull (three parameters), exponential, logistic and Pareto (generalized) have been chosen to fit the experimental traffic data (traces), which show an analytical indication of traffic behaviour. The issues of traffic characterisation and performance shown by these models are discussed in terms of the heavy tailedness and fitness of the curves. The aim of the research is to find a suitable analytical, method which can characterise the Web traffic.

1 Introduction

In the last decade, various researchers have shown that worldwide web transfers exhibit characteristics of self-similarity, which can be explained by the heavy-tailedness of the various distributions involved. A fitting algorithm based on the expectation-maximisation is used in [1] for pure traffic statistics as well as queuing studies. Self-similarity and heavy-tailedness are of great importance for network capacity planning purposes, for which researchers are interested in developing analytical methods to analyse the traffic characteristics. Paxson [2] presented a number of analytic models for describing the characteristics of TELNET, NNTP, SMTP, and FTP connections, drawn from wide-area traces collected from different sites. The argument presented in the paper is that, while wide-area traffic cannot be modeled exactly in a statistical sense, simple analytic models with a good approximation can be constructed. If it is possible to develop such models, characteristics of network traffic can be easily approximated. The main characteristic of the Internet is the fact that it is a large-scale, wide-area network for which the importance of measurement and analysis of the Internet is vital. As Internet technology becomes more widespread, and the network becomes more complex, the characteristics of traffic also become more diverse. Therefore, it is essential to develop a reliable model to analyse the network traffic behaviour. When modelling network traffic, network events such as packet and connection arrivals are often modelled as Poisson processes for analytical simplicity because such processes have attractive theoretical properties [3]. However, a number of studies have shown that for both local-area [4, 5] and wide-area [6, 7] network traffic, packet inter-arrivals are clearly not exponentially distributed as expected from the Poisson processes. With the growth of data networks, modelling congestion in telecommunication networks becomes more intricate than before. Multilayer protocols are used at the present time and many complicated traffic statistics are creating greater difficulty in analyzing network traffic. There

J.A.T. Machado et al. (eds.), *Intelligent Engineering Systems and Computational Cybernetics,*

are also a far greater number of applications (such as voice conversation, email, audio, video etc.), each with its own traffic characteristics, and new applications can arise at any time. There are more varieties of network connectivity, architecture, and equipment, and accordingly, different types of traffic flow. Since Web traffic accounts for a large proportion of the traffic on the Internet, understanding the nature of WWW traffic is increasingly important. This chapter is organised as follows. Section 2 explains the probability distribution and EDF statistics where the Anderson–Darling test (A^2) was used for normality (which is considered to be the most powerful goodness-of-fit test for normality) as well as for other distributions. Cramer–von Mises (W^2) and Watson (U^2) tests were also performed for such distributions. Section 3 describes the traffic traces used in this study. Finally the results are presented in Section 4.

2 Probability Distribution and EDF Statistics

In this section we will describe the probability distributions used with the experimental data to find the consistency (e.g. best matching) of traffic behaviour. Empirical distribution function (EDF) test statistics will also be described if the data (traces) are from a particular distribution and best fittings will be made using the method of EDF statistics. Parameters (e.g. location parameter, scale parameter, shape parameter) that influence in fitting the curve will be highlighted to realize their effect.

2.1 Pareto Distribution

The cumulative distribution for the Pareto (simple) random variable is obtained from

$$F(x) = P(X \leq x) = 1 - \left(\frac{\alpha}{x}\right)^{\beta}; \quad \alpha, \beta \geq 0,\ x \geq \alpha$$

where α = Location parameter; β = Shape parameter.

2.2 Tests Based on EDF Statistics

Here we consider tests of fit based on the empirical distribution function (EDF). The EDF is a step function, calculated from the sample, which estimates the population distribution function. EDF statistics are measures of the discrepancy between the EDF and a given distribution function and are used for testing the fit of the sample to the distribution.

Suppose a given random sample of size n is $X_1, \ldots, X_n$ and $X_{(1)} < X_{(2)} < \cdots < X_{(n)}$ be the order statistics; suppose further that the distribution of x is F(x). The empirical distribution function (EDF) is $F_n(x)$ defined by

$$F_n(x) = \frac{\text{number of observations} \leq x}{n}; \qquad -\infty < x < \infty.$$

The calculation is done by using the Probability Integral Transformation (PIT), $Z = F(X)$; when $F(x)$ is the true distribution of x, the new random variable Z is uniformly distributed between 0 and 1. Then Z has the distribution function $F^*(z) = z$, $0 \leq z \leq 1$. The following formulas [8] have been applied for calculating EDF statistics from the Z-values. The formulas involve the Z-values arranged in ascending order, $Z_{(1)} < Z_{(2)} < \cdots < Z_{(n)}$.

$$\left.\begin{aligned} W^2 &= \sum_{i=1}^{n} \{Z_{(i)} - (2i-1)/(2n)\}^2 + 1/(12n) \\ U^2 &= W^2 - n\,(\bar{z} - 0.5)^2\,;\ \text{where } \bar{z} = \text{ sample mean} = \frac{\sum_{i=1}^{n} Z_i}{n} \\ A^2 &= -n - \left(\tfrac{1}{n}\right) \sum_{i=1}^{n} (2i-1) \left[\log Z_{(i)} + \log\left\{1 - Z_{(n+1-i)}\right\}\right] \end{aligned}\right\} \quad (A)$$

Here $\log x$ means $\log_e x$ (natural logarithm).

2.3 Anderson–Darling Test for Normal Distribution

The procedure [8] for performing the Anderson–Darling test (A^2) is

1. Arrange the sample in ascending order, $X_{(1)} \leq \cdots \leq X_{(n)}$
2. Calculate standard values, $Y_{(i)}$

$$Y_{(i)} = \frac{X_{(i)} - \bar{X}}{s}, \text{ where } \bar{X} = \frac{\Sigma X}{n} \text{ and } s = \sqrt{\frac{\Sigma (X - \bar{X})^2}{n-1}} \text{ for } i = 1, \ldots, n \quad (1)$$

3. Calculate $P_{(i)}$ for $i = 1, \ldots, n$

$$P_{(i)} = \Phi(Y_{(i)}) = \int_{-\infty}^{Y_{(i)}} \frac{e^{-\frac{t^2}{2}}}{\sqrt{2\pi}} dt \quad (2)$$

Here $\Phi(y)$ represents the cdf of the standard normal distribution and P_i is the cumulative probability corresponding to the standard score $Y_{(i)}$. P_i can be found either from standard normal tables or using the following equations.

$$P_i = Q_i;\ where\ Q_i = 1 - \frac{1}{2}\left(1 + C_1 y + C_2 y^2 + C_3 y^3 + C_4 y^4\right)^{-4}$$

where $C_1 = 0.196854$, $C_2 = 0.115194$, $C_3 = 0.000344$, $C_4 = 0.019527$.

For $Y_{(i)}$ such that $0 \leq Y_{(i)} < \infty$ define $y = Y_{(i)}$.

$P_i = 1 - Q_{(i)}$; for $Y_{(i)}$ such that $-\infty < Y_{(i)} \leq 0$ define $y = -Y_{(i)}$.

4. Compute the Anderson–Darling statistics

$$A^2 = -n - \left(\frac{1}{n}\right) \sum_{i=1}^{n} (2i-1) \left[\log P_{(i)} + \log\left\{1 - P_{(n+1-i)}\right\}\right] \quad (3)$$

and W^2 and U^2 statistics from equation (A)

5. Compute the modified statistic

$$\begin{aligned} A^* &= A^2\left(1.0 + 0.75/n + 2.25/n^2\right) \quad (4) \\ W^* &= W^2\,(1.0 + 0.5/n) \\ U^* &= U^2\,(1.0 + 0.5/n) \end{aligned}$$

Reject the null hypothesis of normality if A* exceeds 0.631, 0.752, 0.873, 1.035, and 1.159 at levels of significance 0.10, 0.05, 0.025, 0.01, and 0.005 respectively.

2.4 EDF Test for Extreme Value Distribution

The extreme value distribution is

$$F(x) = \exp\left[-\exp\left\{-\frac{(x-\alpha)}{\beta}\right\}\right]; \quad -\infty < x < \infty \tag{5}$$

where $-\infty < \alpha < \infty$ and $\beta > 0$; α = location parameter; β = scale parameter.

Let us consider the both parameters and unknown (case 3). We suppose the parameters will be estimated by maximum likelihood and can be found by the following equations [9]:

$$\hat{\beta} = \sum_j x_j/n - \left\{\sum_j x_j \exp\left(-x_j/\hat{\beta}\right)\right\} / \left\{\sum_j \exp\left(-x_j/\hat{\beta}\right)\right\} \tag{6}$$

and

$$\hat{\alpha} = -\hat{\beta}\log\left\{\sum_j \exp\left(-x_j/\hat{\beta}\right)/n\right\} \tag{7}$$

$\hat{\alpha}$ and $\hat{\beta}$ replaces α and β respectively. Equation (6) is solved iteratively for $\hat{\beta}$, and then (7) can be solved for $\hat{\alpha}$. The procedure is as follows:

1. Parameters are estimated by equations (6) and (7).
2. Calculate $Z_{(i)} = F\left(x_{(i)}\right); for\, i = 1, \ldots\ldots, n$, using equation (5).
3. Use formulas (A) to calculate the EDF statistics.
4. Modification of test statistics [8]

$$W^* = W^2\left(1 + 0.2/\sqrt{n}\right) \tag{8}$$

$$U^* = U^2\left(1 + 0.2/\sqrt{n}\right) \tag{9}$$

$$A^* = A^2\left(1 + 0.2/\sqrt{n}\right) \tag{10}$$

2.5 EDF Test for Three Parameter Weibull Distribution

The three parameter Weibull distribution is

$$F(x;\alpha,\beta,m) = 1 - \exp\left[-\{(x-\alpha)/\beta\}^m\right], \quad x > \alpha, \tag{11}$$

where α = location parameter; β = Scale parameter and m = Shape parameter. β and m are positive constants.

The parameters can be estimated by likelihood equations [10].

$$\frac{1}{m} - \frac{\Sigma\left(x_{(i)}-\alpha\right)^m \log\left(x_{(i)}-\alpha\right)}{\Sigma\left(x_{(i)}-\alpha\right)^m} + \frac{\Sigma\log\left(x_{(i)}-\alpha\right)}{n} = 0 \tag{12}$$

$$\frac{m-1}{m}\Sigma\left(x_{(i)}-\alpha\right)^{-1} - n\frac{\Sigma\left(x_{(i)}-\alpha\right)^{m-1}}{\Sigma\left(x_{(i)}-\alpha\right)^m} = 0 \tag{13}$$

Equations (12) and (13) are solved for and m, then is estimated by the following equation:

$$\beta = \left\{\Sigma\left(x_{(i)}-\alpha\right)^m/n\right\}^{\frac{1}{m}} \tag{14}$$

The transformation $Z_{(i)} = F\left(x_{(i)};\alpha,\beta,m\right)$ has been made for $i = 1, 2, \ldots, n$ by equation (11). Then EDF statistics are calculated by using equation (A). Finally the table provided [10] is entered for critical points, $c = 1/m$. The null hypothesis is rejected at significance level p if the statistic used is greater than the tabulated value given [10] for level p.

2.6 EDF Test for the Exponential Distribution

The exponential distribution, denoted by Exp (α, β), is the distribution

$$F(x;\alpha,\beta) = 1 - \exp\{-(x-\alpha)/\beta\},\ x > \alpha;\ \beta > 0, \tag{15}$$

where α = location parameter, β = scale parameter.

The null hypothesis is H_0: the random sample $X_1,\ldots,X_n$ comes from the distribution Exp(α, β), with α, β unknown. The procedure [8] is as follows:

1. Parameter estimation:

$$\hat{\beta} = n\left(\bar{x} - x_{(1)}\right)/(n-1), \tag{16}$$

where $\bar{x}$ =sample mean,

$$\hat{\alpha} = x_{(1)} - \hat{\beta}/n \tag{17}$$

2.

$$W_i = \left(x_{(i)} - \hat{\alpha}\right)/\hat{\beta} \tag{18}$$

3.

$$Z_i = 1 - \exp(-W_i); \tag{19}$$

4. EDF statistics are calculated from equation (A).
5. Modification [8]

$$W^* = W^2\left(1 + 2.8/n - 3/n^2\right) \tag{20}$$

$$U^* = U^2\left(1 + 2.3/n - 3/n^2\right) \tag{21}$$

$$A^* = A^2\left(1 + 5.4/n - 11/n^2\right) \tag{22}$$

The corresponding values of significance level are given in the table [8] for the modified statistics. The hypothesis is rejected if the modified value is greater than the tabulated value.

2.7 EDF Test for Logistic Distribution

A random sample of n values of x comes from the logistic distribution

$$F(x) = [1 + \exp\{-(x-\alpha)/\beta\}]^{-1},\ -\infty < x < \infty, \tag{23}$$

where α = location parameter; β = scale parameter.

The null hypothesis H_0: the random sample $X_1,\ldots,X_n$ comes from the distribution (23), with α, β unknown.

The test procedure is as follows:

1. Parameter estimation [10]:

The parameters are estimated from the sample by maximum likelihood; the estimate are given by the equations

$$n^{-1}\sum_i\left[1 + \exp\left\{(x_i - \hat{\alpha})/\hat{\beta}\right\}\right]^{-1} = \frac{1}{2} \tag{24}$$

$$n^{-1}\sum_i \left(\frac{x_i-\hat{\alpha}}{\hat{\beta}}\right)\frac{1-\exp\left\{(x_i-\hat{\alpha})/\hat{\beta}\right\}}{1+\exp\left\{(x_i-\hat{\alpha})/\hat{\beta}\right\}}=-1 \tag{25}$$

Equations (24) and (25) are solved iteratively; suitable starting values for $\hat{\alpha}$ and $\hat{\beta}$ are $\alpha_0=\bar{x}$ and $\beta_0=s$, where $\bar{x}$ and s^2 are respectively the sample mean and variance.

2. $W^*=\left(nW^2-0.08\right)/(n-1.0)$, where F(x) is given in (23).
3. EDF statistics are calculated by using the equations (A).
4. Modifications [11]

$$\begin{aligned} W^* &= \left(nW^2-0.08\right)/\left(n-1.0\right) && (26)\\ U^* &= \left(nU^2-0.08\right)/\left(n-1.0\right) && (27)\\ A^* &= A^2\left(1.0+0.25/n\right) && (28)\end{aligned}$$

The hypothesis is rejected if the modified value is greater than the tabulated value for the corresponding significance level.

2.8 EDF Test for Generalized Pareto Distribution

The generalized Pareto distribution (GPD) has the following distribution function [12]:

$$F(x)=1-(1-kx/a)^{1/k}, \tag{29}$$

where $a=$ Positive scale parameter and $k=$ Shape parameter. The density function is

$$f(x)=(1/a)\ (1-kx/a)^{(1-k)/k} \tag{30}$$

$$0\le x<\infty \text{ for } k\le 0 \text{ and } 0\le x\le a/k \text{ for } k>0$$

The mean and variance are $\mu=a/(1+k)$ and $\sigma^2=a^2/\{(1+k)^2(1+2k)\}$; thus the variance of the GPD is finite only for $k>-0.5$. For the special values $k=0$ and 1, the GPD becomes the exponential and uniform distributions respectively [12]. The name generalized Pareto was given by Pickands [13]; the distribution is sometimes called simply Pareto when $k<0$. In this case GPD has a long tail to the right [12]. Suppose that $X_1,\ldots,X_n$ is a given random sample from the GPD given in equation (29), and let $x_{(1)}\le x_{(2)}\le\cdots\le x_{(n)}$ be the order statistics. The estimation is done by maximum likelihood. The log-likelihood is given [12] by

$$\begin{aligned} L(a,k) &= -n\log a-(1-1/k)\sum_{i=1}^{n}\log(1-kx_i/a) \text{ for } k\neq 0\\ &= -n\log a-\sum_{i=1}^{n}x_i/a \text{ for } k=0\end{aligned} \tag{31}$$

The range for a is $a>0$ for $k\le 0$ and $a>kx_{(n)}$ for $k>0$. Davison [14] pointed out that, by a change of parameters to $\theta=k/a$ and $k=k$, the problem is reduced to a unidimensional search. A local maximum of the profile log-likelihood (the log-likelihood maximized over k) is

$$L^*(\theta)=-n-\sum_{i=1}^{n}\log\ (1-\theta x_i)-n\log\left[-(n\theta)^{-1}\sum_{i=1}^{n}\log\ (1-\theta x_i)\right] \tag{32}$$

for $\theta<1/x_{(n)}$.

A local maximum $\hat{\theta}$ of (32) can be found [5] as

$$\hat{k} = -\left(n^{-1}\right) \sum_{i=1}^{n} \log\left(1 - \hat{\theta}\, x_i\right) \tag{33}$$

$$\hat{a} = \hat{k}/\hat{\theta} \tag{34}$$

By equation (33) and (34), one gets

$$\hat{k} = -\left(n^{-1}\right) \sum_{i=1}^{n} \log\left(1 - \frac{\hat{k}}{\hat{a}} x_i\right) \tag{35}$$

The test procedure is as follows:

1. Parameters are estimated by equation (35).
2. The transformation $Z_i = F(x_i)$, where F(x) is given in (29).
3. EDF statistics W^2 and A^2 are calculated from equation (A).
4. The decision is made by comparing the calculated value with the tabulated value [5] whether the hypothesis would be accepted or rejected.

3 Internet Traffic Data Collection

In our research we studied the http traffic data which are publicly available in [15]. A special care was taken to check the sanity of the data, as any irregularity or mistakes could set back the entire analysis for an extended period of time. The original data file was divided into smaller file and then the size (byte) of the file was arrayed into different bins for convenience of the compactness of the data file. It is very congenial to array the file size into a bin as millions of samples can be analyzed in an expected range.

4 Results and Discussion

4.1 Case Study with EPA Traffic

Figure 1 compares the cumulative probability distribution (CDF) of the generated data points using Pareto model (described in Subsection 2.1) with those from the original traces. The Pareto model data do not match with the trace data especially in the middle range of web file size. There is consistent discrepancy between the actual distribution (web file data) and the Pareto (simple) model. Figure 2 compares CDF of the actual trace data, the extreme value, Weibull, exponential, logistic and Pareto (generalized) distributions. It is clearly seen from Fig. 2 that the generalized Pareto shows quite smoother fitting than the simple Pareto (Subsection 2.1) distribution. There are close match observed between experimental and models data, particularly generalized Pareto and Weibull model show the best fitness of the data points.

Empirical Distribution Function (EDF) test has been performed to identify if the data fit with those distributions. The Anderson–Darling test (A^2), the most powerful test, was performed for normality and Cramer von Mises (W^2) and Watson (U^2) tests were used as well. A^2, W^2 and U^2 tests did reject the hypothesis in favour of normal, extreme value, Weibull, logistic and generalized distributions, i.e. the trace data do not follow these distributions. It is

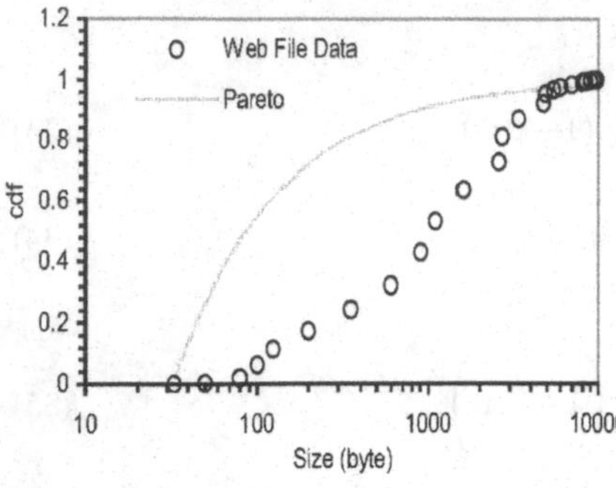

Fig. 1: Web file size Cumulative distribution for EPA traffic

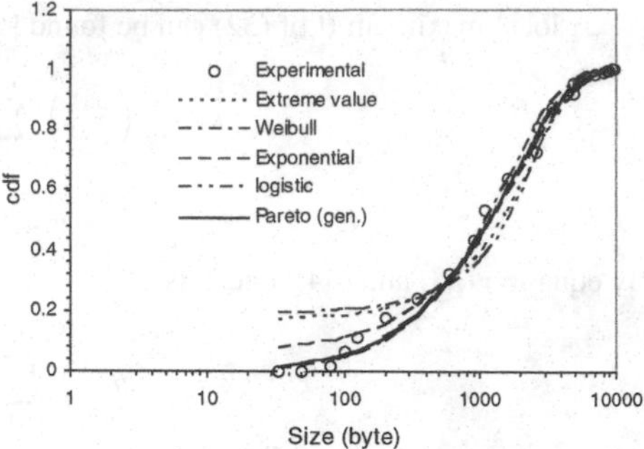

Fig. 2: Web file size cumulative distribution of different analytic models for EPA traffic

Table 1: Parameters used in the distribution for EPA

Data	Distribution	Parameters used in the distribution (α = location parameter, β = scale parameter, for Simple Pareto β = shape parameter, a = scale parameter, k = shape parameter, m = shape parameter)		
EPA	Extreme value	$\alpha = 899.455$	$\beta = 1514.583$	
	Weibull	$\alpha = 15$	$\beta = 1553.941$	$m = 1.05$
	Exponential	$\alpha = -122.88$	$\beta = 1935.13$	
	Logistic	$\alpha = 1594.9$	$\beta = 1093.909$	
	Pareto (generalized)	$a = 1692$	$k = 0.0086$	
	Pareto (simple)	$\alpha = 32$	$\beta = 0.695$	

noted that the lowest level of significance considered accepting the hypothesis in the analysis is 0.025. The test statistics are consistent with the graphical results as shown in Fig. 2 except Weibull (three parameters) and generalized Pareto distributions. In case of these two distributions, the hypothesis was rejected as the test statistics of A^2, W^2, and U^2 are greater than the values of corresponding specified level. However, there is a close matching of trace data and model data as shown in Fig. 2. Such discrepancy might be due to the effect of location, scale and shape parameters as these three parameters play a significant role in fitting the curves as close as possible.

4.2 Case Study with NASA (http-95-August) Traffic

Figure 3 compares the CDF of the trace data with the data points generated using Pareto (simple) model. The model data do not fit the trace data well though the actual (web file data) distribution seems to be heavy-tailed. Figure 4 compares CDF of the actual trace data, the exponential, Pareto (generalized), Weibull (three parameters), logistic and extreme value distributions. Like the previous case, the generalized Pareto and Weibull distributions show the best matching of the data points as shown in Fig. 4. Both the experimental and models data track each other. This is the similar case (i.e. parameters influence) observed when plotting the graphs by iteration.

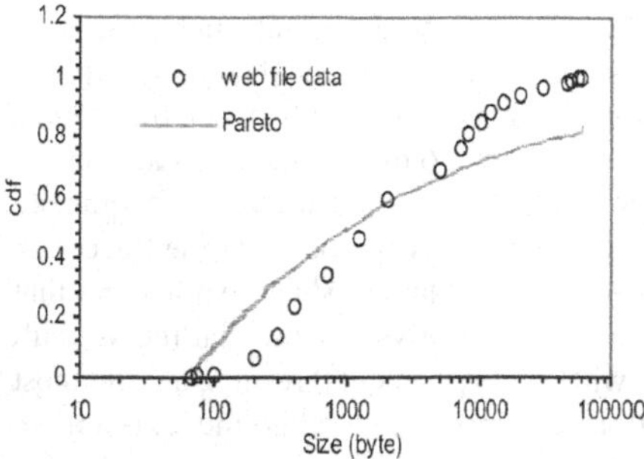

Fig. 3: Web file size cumulative distribution for NASA (http-95-August) traffic

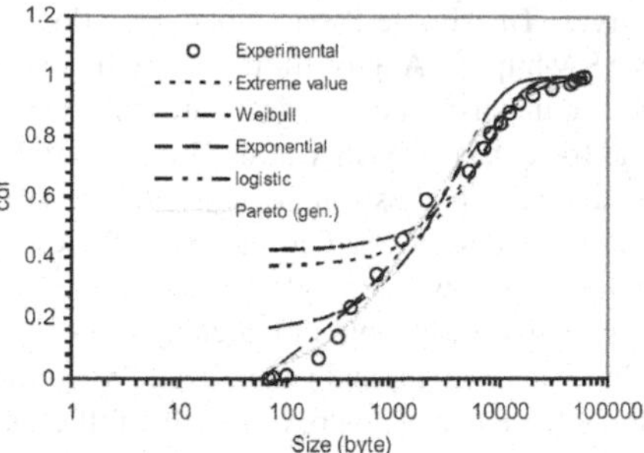

Fig. 4: Web file size cumulative distribution of different analytic models for NASA (http-95-August) traffic

Table 2: Parameters used in the distribution for NASA-http-95-August

Data	Distribution	Parameters used in the distribution (α = location parameter, β = scale parameter, for Simple Pareto β = shape parameter, a = scale parameter, k = shape parameter, m = shape parameter)		
NASA-http-95-August	Extreme value	$\alpha = 0.5$	$\beta = 5{,}800.201$	
	Weibull	$\alpha = 67$	$\beta = 3{,}050$	$m = 0.59$
	Exponential	$\alpha = -650.607$	$\beta = 3{,}850.75$	
	Logistic	$\alpha = 1{,}519.43$	$\beta = 4{,}902.155$	
	Pareto (generalized)	$a = 1{,}758$	$k = -0.8$	
	Pareto (simple)	$\alpha = 66$	$\beta = 0.255$	

The iterative methods described (in Subsections 2.4–2.8) are used to determine the parameters of the distributions. The parameters used in the distributions for different traces are represented in Tables 1 and 2. Since three parameter Weibull model and generalised Pareto model (GPD) show the best matching with different web traffic traces, we have provided several graphical results (i.e. matching) from these two models in appendices (Figs. 5, 6). The length (N) of number of samples (i.e. document sizes) used in these experiments is 10,000.

5 Conclusions

The analyses of web traffic have been described here using the traces collected from different working groups. We analysed several traces of Web traffic. Because of space limitations we cannot provide all the results here. Statistical testing was used to evaluate how well a particular distribution describes the experimental data. The Empirical Distribution Function (EDF) test statistics (A^2, W^2 and U^2) was performed in this case. The Pareto (simple), exponential, Pareto (generalized), Weibull (three parameters), logistic and extreme value distributions were employed in the research.

In all traffic cases, the EDF test statistics confirm that the trace data do not follow normal distributions. In EPA traffic, the test statistics support the exponential distribution to fit the trace data, though other distributions (model data) match with the trace data pretty well with Weibull (three parameters) and generalized Pareto models showing the best fittings of curves. The simple Pareto model data do not match with the trace data in all traffic cases. In NASA-http-95-August traffic, the EDF test statistics confirm that the trace data do not follow any of the distributions. There are satisfactory fittings of curves observed between the models and trace data in NASA-http-95-August traffic as shown in Fig. 6 though the test statistics rejected the hypothesis in favour of the models. The Weibull (three parameters) and generalized Pareto models contradict the test statistics in all traffic as these two models fit the trace data quite closely which might be due to the effect of location, scale and shape parameters that play a significant role in matching the curves smoothly. The analyses show that the Weibull (three parameters) and generalized Pareto models, with the experimental results, are the most suitable models to approximate the traffic. In addition, the generalized Pareto model is a more suitable model for analysing traffic behaviour than the simple Pareto model in terms of heavy-tailedness. Hence, as an efficient analytical tool, thc generalised Pareto model can be used for identifying a heavy-tail nature based on samples from the Web traffic.

References

1. Khayari REA, Ramin S, Haverkort BR (2003) Fitting world-wide web request traces with the EM-algorithm. Internet Performance and Control of Network Systems, April 2003 52(2–3):175–191
2. Paxson V (1994) Empirically derived analytic models of wide-area TCP connections. IEEE/ACM Transactions on Networking, August 1994 2(4):316–336
3. Frost V, Melamed B (1994) Traffic modeling for telecommunications networks. IEEE Communications Magazine 33:70–80
4. Gusella R (1990) A measurement study of diskless workstation traffic on an Ethernet. IEEE Transactions on Communications 38:1557–1568
5. Fowler HL, Leland WE (1991) Local area network traffic characteristics. IEEE Journal on Selected Areas in Communications 9:1139–1149
6. Danzig P, Jamin S, Caceres R, Mitzel D, Estrin D (1992) Journal of Internetworking Research and Experience 3:1–26
7. Paxson V, Floyd S (1995) Wide area traffic: The failure of Poisson Modeling. IEEE/ACM Transactions on Networking 3(3):226–244
8. D'Agostino RB, Stephens MA (1986) Tests based on EDF statistics, in Goodness-of-fit Techniques. Ch 4, Dekker, New York
9. Johnson NL, Kotz S (1970) Distributions in Statistics: Continuous Univariate Distribution. Vol 1, Houghton Mifflin, Boston, MA
10. Lockhart RA, Stephens MA (1994) Estimation and tests of fit for the three-parameter Weibull distribution. Journal of the Royal Statistical Society Series B 56:491–500
11. Stephens MA (1979) Tests of fit for the logistic distribution based on the empirical distribution function. Biometrika 66:591–595
12. Choulakian V, Stephens MA (2001) Goodness-of-Fit for the Generalized Pareto Distribution. Tech-nometrics 43:478–484
13. Pickands J (1975) Statistical interference using extreme order statistics. The Annals of Statistics 3:119–131
14. Davison AC (1984) Modelling Excesses over High Thresholds, with an Application. In: de Oliveira JT (ed) Statistical Extremes and Applications, pp 461–482. D. Reidel, Dordrecht
15. The Internet Traffic archive http://ita.ee.lbl.gov/html/traces.html

Appendices

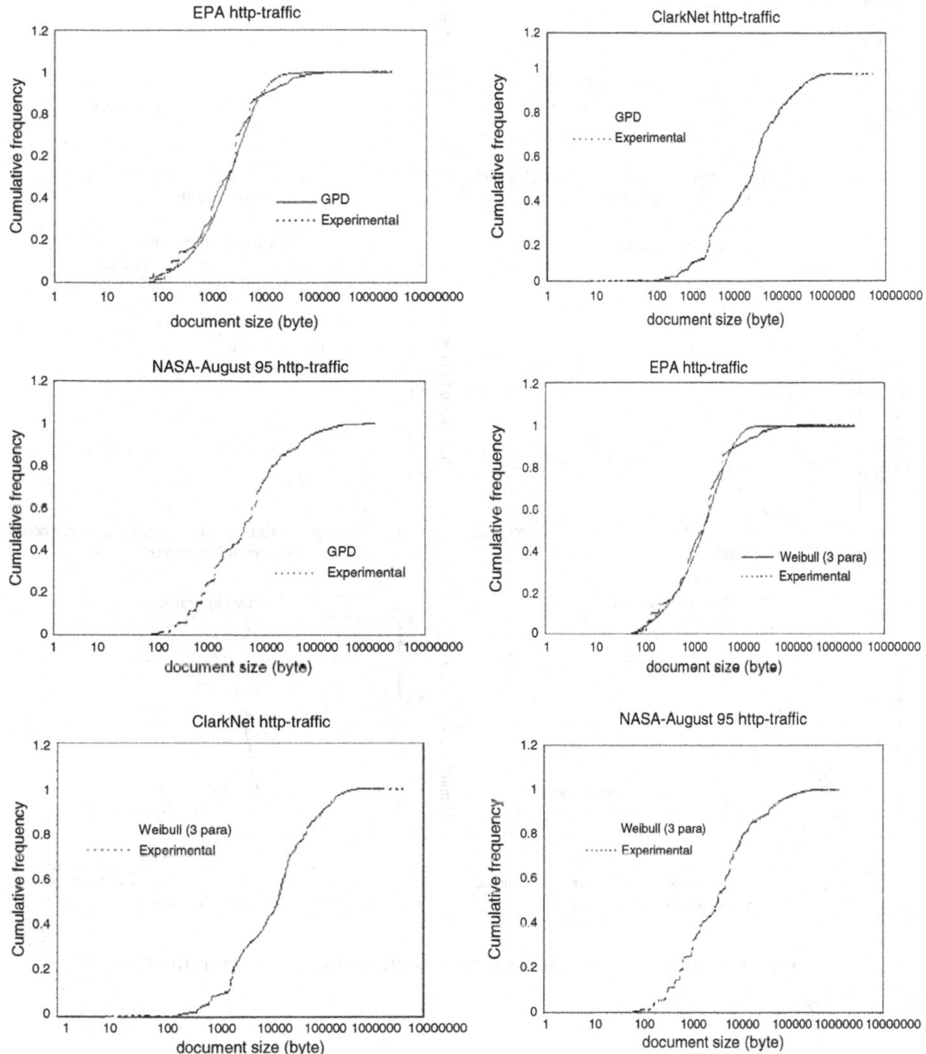

Fig. 5: Cumulative frequency for web document size of http-traffic (1)

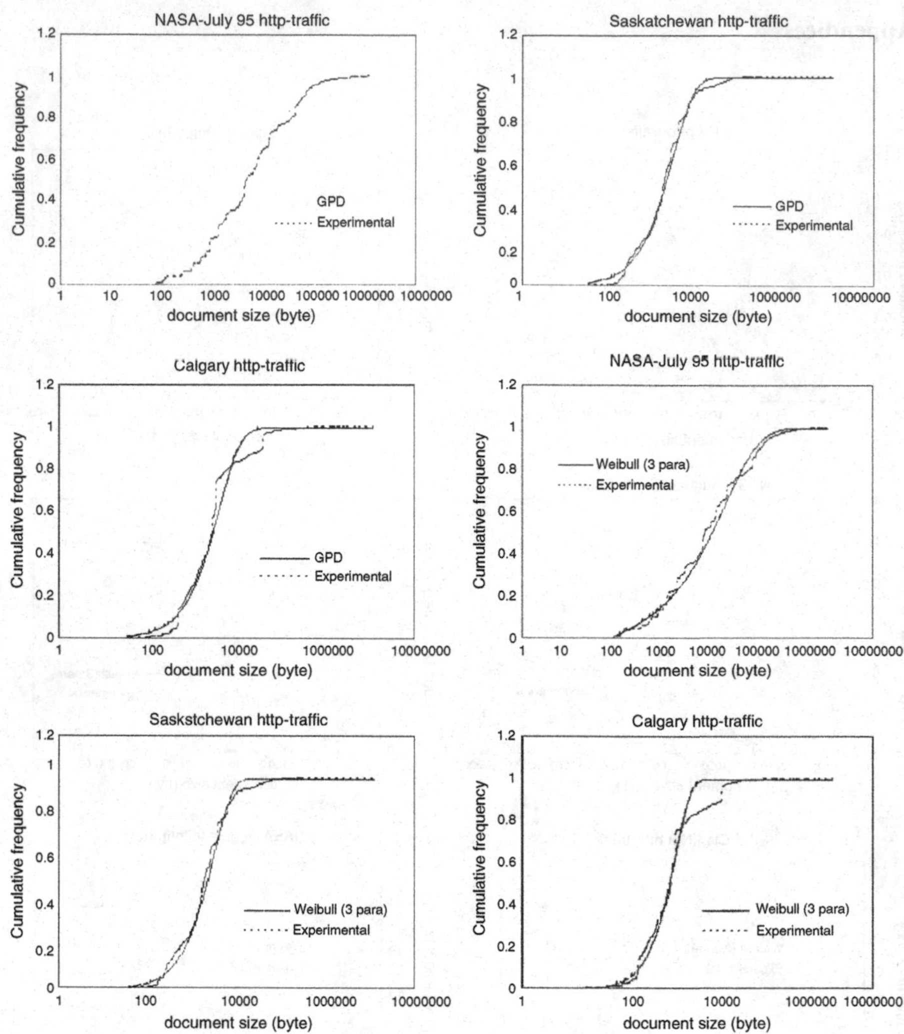

Fig. 6: Cumulative frequency for web document size of http-traffic (2)

Capturing the Meaning of Internet Search Queries by Taxonomy Mapping

Domonkos Tikk, Zsolt T. Kardkovács, and Zoltán Bánsághi

Department of Telecommunications and Media Informatics
Budapest University of Technology and Economics
H-1117 Budapest, Magyar Tudósok krt 2, Hungary
{tikk,kardkovacs}@tmit.bme.hu

Summary. Capturing the meaning of internet search queries can significantly improve the effectiveness of search retrieval. Users often have problem to find relevant answer to their queries, particularly, when the posted query is ambiguous. The orientation of the user can be greatly facilitated, if answers are grouped into topics of a fixed subject taxonomy. In this manner, the original problem can be transformed to the labelling of queries – and consequently, the answers – with the topic names. Thus the original problem is transformed into a classification set-up. This paper introduces our Ferrety algorithm that performs topic assignment, which also works when there is no directly available training data that describes the semantics of the subject taxonomy. The approach is presented via the example of ACM KDD Cup 2005 problem, where Ferrety was awarded for precision and creativity.

1 Introduction

Most web search queries contain only a very few terms (mostly just one). This provides very limited information about the user's information need to the search engine (SE). Utilizing this information effectively is a key factor when constructing internet SEs. A possible solution of this problem is to assign the intended meaning of a query by topic(s) of a given subject taxonomy. Query "jaguar", e.g., could be equally assigned to the topic of automobiles or zoology. Topics can be used as the basis of organization when displaying search results. It can significantly improve the search facilities of the user, if the engine presents search results organized into topics. Hence, the problem of capturing the semantic of a query is reduced to the problem of assigning the query to a set of topics of an arbitrary, but fixed subject taxonomy.

In this paper we present our solution for assigning queries to topics of a given taxonomy. This is a supervised learning task: an automated text categorization problem. However, there are some peculiarities of the particular problem: (1) the method should work without available training data as well; (2) queries are very short, compared to usual text classification tasks.

Since the early phase of internet SE's evolution (mid 1990s), there has been made much effort to determine the meaning of search queries, mostly by means of word sense disambiguation. When queries are sufficiently long, the entire set of query words can provide enough information to determine the context of the query. Despite the discouraging early results surveyed in [1], it has been shown that this technique can slightly improve precision of IS and, in

J.A.T. Machado et al. (eds.), *Intelligent Engineering Systems and Computational Cybernetics,*

more general context, information retrieval. Schütze and Pederson [2] could improve retrieval effectiveness by 14%, though, on a relatively small corpus, Gonzalo et al. [3] used WordNet synonyms to enrich the representation of words and thus achieved a few percent increase in precision, Krovetz [4] found that the connection of morphological variants of words can also improve slightly retrieval effectiveness.

Query expansion or query reformulation is another direction of research in this subject (see e.g. [5–7]). These techniques work in an interactive manner, which we considered to be a weak point, since users are not very likely to use such search services that require significantly more attention than in regular.

Automatic clustering and categorization of search result is another possible way to determine the meaning of a search query. In this direction we refer to the following rccent works [8–10].

Our solution is based on a taxonomic mapping between one or more internet directories and the subject taxonomy. The problem of assigning documents from one taxonomy to another has been investigated by several authors. This problem often occurs at the work of patent offices when assigning patent categorized into one of the main patent classification into another one (e.g. when making IPC–USPC concordance) [11]. In work [12], authors developed an enhanced naive Bayes algorithm that is based on the intuition that if two documents are assigned the same category in one taxonomy, they are likely to be also in the same category in another taxonomy. Similar approach was applied in [13] but in combination with Support Vector Machines (SVM).

Instead of assigning the documents of one taxonomies to another one Doan et al. [14] learn a direct mapping between taxonomy nodes using the joint distribution of concepts. These approaches are common that they require fair amount of of training examples in both source and target taxonomies. However, it is often the case that at least one of the above training set is missing. Ferrety is able to deal also with this case as it is shown on the data set of ACM KDD cup 2005. Here we extend the results of our previous paper [15], and give a brief overview of the other two awarded solutions [16,17].[1]

This paper is organized as follows. Section 2 presents the KDD CUP 2005 problem and place it in the fields of machine learning and advanced IS. Section 3 describes Ferrety algorithm, while Section 4 evaluates it on the test data set of the contest. Section 5 presents two other solutions from KDD Cup 2005.

2 KDD Cup 2005 Problem

The contest of KDD Cup 2005[2] set focus on internet query categorization. A two-level *target taxonomy* is given with 67 leaf-level categories, where 800,000 internet user search queries have to be classified into in a multi-label way: a query can be assigned to at most five categories. As illustrative example, another set of only 111 labelled queries is given. Beside that no other training documents were delivered together with the original problem.

Usually, document categorization requires supervised learning methods, however, in the very case the lack of sufficiently large number of training examples excludes the simple use of an arbitrary categorizer. Next to this, the peculiarity of internet user search queries as documents makes the problem even harder, because the queries are very short (up to five words for

[1] Papers are available at www.acm.org/sigs/sigkdd/explorations/, 2005 December issue.
[2] kdd05.lac.uic.edu/kddcup.html

more than 90% of the corpus), and the the corpus is very noisy (more than 30% of the corpus contains incorrectly represented non-English documents or trash queries).

3 Ferrety Algorithm

The main idea of Ferrety algorithm is the following. It executes query mapping in several steps. First, source taxonomies are sought, and a mapping between categories of target and source taxonomies are created based on an extension of original category names. Then, it attempts to fetch training examples from internet directories by posting the slightly processed original queries to their search service. The results can serve both as preliminary results of query categorization (here the taxonomy mapping is exploited again), and as training set for such queries where the above technique specifies an empty result. Let us now have a look on the algorithm step-by-step (see also Fig. 1):

1. Find source: Determine the proper semantics for target categories by creating a basic dictionary with an extension. Create valid mappings between target taxonomy and source taxonomies of existing internet directories of SEs by means of category semantics just determined.
2. Stem queries: Preprocess all queries by performing stemming on them. The result is the set of stemmed queries.
3. Query the Web: Exploiting the connections between target and source taxonomies gather relevant training documents to (stemmed) queries by posting them to SEs.
4. Parse results: Parse the results of Step 3.
5. Train a classifier: Feed the parsed results to a text categorizer as training set for target taxonomy. (In our experiments we used HITEC text categorizer[3] [18, 19].)
6. Classify queries: Run the trained text categorizer on all queries. Rank the results obtained from SEs and categorization, then determine the top five categories for each query.

3.1 Internet as Knowledge Base

Ferrety uses the Internet as a knowledge base: it asks selected SEs and explore what they may know about a query. Consequently, Ferrety can be considered as a meta-search engine. In this experiment only two engine are asked: LookSmart and Zeal,[4] partly because their source taxonomy showed some similarity with the target taxonomy, and partly, because they have a clear semantic description on each category. This latter property could be exploited when making creating taxonomy mapping.

The result pages of L&Z have two main parts: *local categories* to which the query may belong (if any) and *relevant contexts* of the words contained in the query. Each category has a *semantic description* that describes the content of the category with a few words. The entirety of these description constitutes the Basic Corpus ($\mathscr{BC}$), with elements referring to the $\mathscr{BC}$-categories.

[3] categorizer.tmit.bme.hu

[4] search.looksmart.com and www.zeal.com, together we refer to them as L&Z.

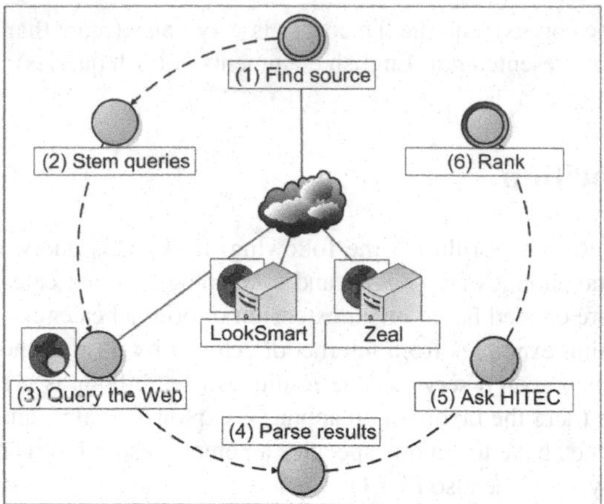

Fig. 1: The steps of the Ferrety Algorithm

3.2 Building and Mapping Taxonomies

The taxonomy building procedure was designed to determine the maximum relevant set of words which may correspond to a target category. It assumes that the name of target category describes sufficiently well the semantic of the category. This description was extended with WordNet synonyms, that are also supposed to be valid terms to that category. We call this step a semantic closure. Let

$$W(0) = \bigcup_i w_i(0)$$

stand for this initial set of partitions where $w_i(0)$ denotes the semantic closure of the ith target category.

We applied $W(0)$ to find relevant $\mathscr{BC}$-categories, i.e. we searched $\mathscr{BC}$-categories with semantic description containing the given set of words. Let $\mathscr{C}_i^0 \subseteq \mathscr{BC}$ be the subset of the Basic Corpus determined by $w_i(0)$. Ferrety calculates the well-known tf-idf measure (see e.g. [20]) for the words of the descriptions in $\mathscr{C}_i^0$, which have frequency greater then ω in at least one $\mathscr{C}_i^0$, and occur no more than in α $\mathscr{C}_i^0$. Let A^0 denote the set of words that holds this property. Then we can apply the recursive formula:

$$w_i(n+1) = w_i(n) \cup \{a | a \in A^n \cap \mathscr{C}_i^n\} \quad (n = 0, 1, \ldots) \tag{1}$$

We also use this step for the top-level categories: we create the union of sub-levels, and apply the above mentioned procedure. This step is important, because the limitation in idf factor might eliminate words that determine the valid context of a top level category. When classifying queries, those ones that are assigned only a top-level categories, are consider to be in "other" (for example, Computer → Computer/Other).

One can easily see that by the finiteness of words occurring in finite many semantic descriptions the algorithm described by Equation (1) always terminates. Categories of

$$\mathscr{C}^- = \mathscr{BC} \setminus \bigcup_i \mathscr{C}_i^N$$

are classified using the L&Z default hierarchy in a way, that we propagate mapping results downward in the hierarchy until an element in $\mathscr{C}^+ = \mathscr{BC} \setminus \mathscr{C}^-$ (a *flagged category*[5]) is found, or the tree ends. Note that, at this point we obtained both the basic dictionary (it is the union of w_i^Ns), and the mapping between target and source taxonomies, since $\mathscr{C}_i^N$s are target categories while their elements belong to sources (L&Z) categories. The flowchart of the algorithm is illustrated in Fig. 2, while its pseudo-code can be found in [15].

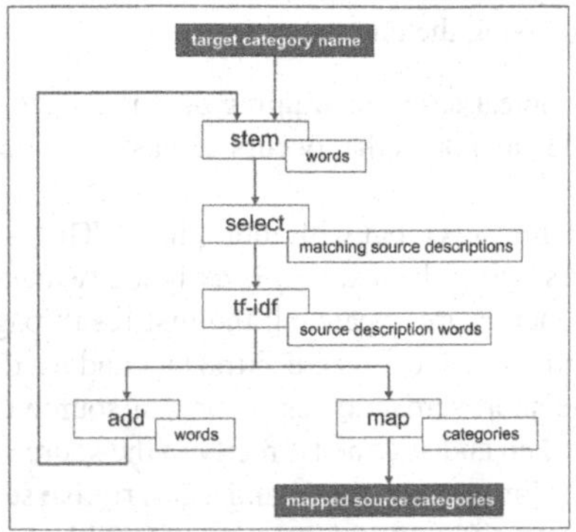

Fig. 2: The steps of taxonomy mapping

Note that the algorithm described in Fig. 2 can be applied to connect a *source* taxonomy and a *target* taxonomy, if the source taxonomy has semantic description for categories. This condition can easily be fulfilled if there are available training documents filed in the source taxonomy. The semantic description of a category can be obtained from the most frequent terms of assigned training documents. A profile or a prototype vector of a category can also be considered as semantic description.

3.3 Training and Categorization

After having the taxonomy mapping we posted all queries to the L&Z search interface and determined the target taxonomy. Before executing this step, we applied the Porter stemmer [21] to all queries. Because of the noisy queries, the lack of complete semantic description of target taxonomy, and the existing difference of target and source taxonomies, the technique described produces results only for about the half of the posted queries. Nevertheless, these queries can be employed as training set of a classifier.

For this purpose, i.e. to determine the target categories of the remaining queries, we applied HITEC [18, 19] categorizer. HITEC is a hierarchical online categorizer that implements a neural network based learning algorithm. It creates a prototype for each category based on the training samples. When categorizing unknown documents, HITEC compares them to prototype of categories, and selects recursively downward in the taxonomy the most similar ones. The depth and the width of the searching path can be controlled by various parameters both at training and at categorization.

[5] A category is flagged in the source taxonomy if it has a corresponding target category.

On the Training Sets

Because the queries themselves contains very few words, we applied query expansion and created four different training sets. The first three alternatives assign training data to source categories. This is done by posting the (stemmed) queries to SEs. The training data is assigned to such source categories that occurs at least once among the local categories on the SEs result page. The link between L&Z and query categories is established by the taxonomy mapping described in Subsection 3.2. The last alternative of training set assigns training data directly to source categories, without using the queries.

Q QUERY: This is the simplest case, the training data is the stemmed query itself. This is assigned to such source categories that occurs at least once on the result page generated by the query.

WQ WEIGHTED QUERY: Same as **Q**, but with multiplicity. That is, if on the result page of q query category c occurs twice, the text of q is assigned twice to c.

T TEXT: The stemmed queries are expanded: the first result page generated by the query is parsed and meaningful text excerpts are extracted, and used as query expansion.

C CATEGORY INFO: The short semantic description of source categories provided by the internet directories is taken and assigned directly to the source category as training document. The short description contains a title and a descriptive sentence about the category; the two field can be given different weights (see also Table 1). This procedure is applied only for flagged categories.

Feature Selection

At feature selection one has to find the optimal balance between discarding rare terms and keeping discriminative ones in the dictionary. HITEC has two simple parameters to control the dictionary size ($|\mathcal{D}|$): a lower limit for the minimal occurrence (d_1) of a term in the entire (training) corpus, and a maximum allowed value (d_2) for the overall distribution of a term over the corpus. These parameters are related to the tf-idf frequency scheme: d_1 and d_2 can be considered as lower threshold for tf, and upper threshold for df, resp.

For the given problem, we applied very low d_1 (2–5), and very high d_2 (over 0.5) because the entire corpus contains relatively few terms, and terms occur very rarely. We also found that d_1 parameter should be kept low for training set C, because a lot of discriminative words occurred only one or two times in the entire corpus. We applied two weight factors when training set C was concerned: w_t is for the title and w_d for the description of the category, respectively. We tabulated some significant feature selection runs in Table 1.

Table 1: The size of dictionary in terms of the parameters of feature selection

Run	d_1	d_2	w_t	w_d	$\lvert\mathcal{D}\rvert$	Training set
R1	2	0.7	5	3	888565	Q+T+C
R2	5	1.0	10	3	149792	Q+C
R3	5	1.0	10	3	809165	Q+T+C
R4	0	1.0	10	3	1088171	WQ+T+C
R5	1	1.0	10	3	763872	T+C

Parameters of Training

At training we fixed the number of iteration for 5, which had been found quasi-optimal for other large corpora (e.g. Reuters Corpus Volume 1). We used all training documents for the training.

HITEC has two important parameters to control the size of the result set of a categorization. By max-variance (v_{max}) one can specify the minimal deviation (ratio) of a node to be considered for further search w.r.t. the maximal confidence value at a given hierarchy level. Setting this value low (around 0.5), one can have many firing categories at each level of the taxonomy. By threshold (θ), one can set the minimal confidence value of a node. Setting this value low ($0.05 \sim 0.15$) results in better recall values and more result categories. With the help of these two parameters, we were able to make a trade-off between recall and precision. For the current task, our goal was to obtain large result set for the categorization, therefore we set $v_{max} = 0.5$ and $\theta = 0.1$.

In HITEC, one can use different training corpora for dictionary creation and for training. We exploited this feature and in most run by disregarding Q, since the usage of Q at training decreased the effectiveness. This observation can be explained by the argumentation that query texts are too short for effective categorization. On the other hand, Q is useful at dictionary creation since it could raise the importance of certain occurring terms in the entire Q+T+C training corpus. The learning capability of various settings of Table 1 – i.e. how HITEC could learn the training data – is indicated on Fig. 3.

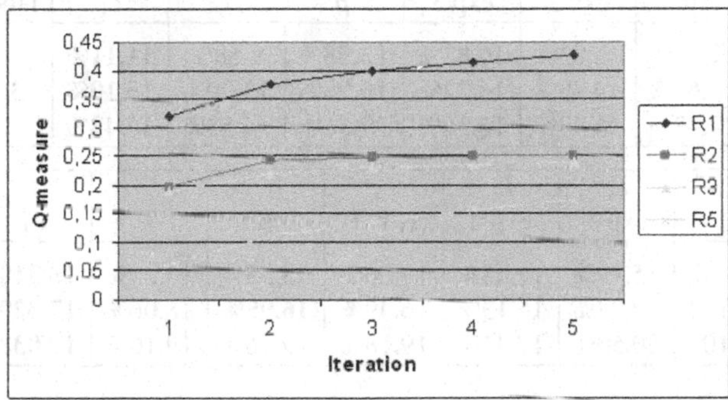

Fig. 3: The effect of learning during iterations in terms of HITEC's inner quality measure

4 Evaluation

We evaluated query categorization results achieved by solely taxonomy mapping, and by taxonomy mapping and categorization by means of the 111 sample queries. We found that results categorized by HITEC were considerably worse than those ones that were obtained by taxonomy mapping. This is not at all surprising: HITEC's result cumulates two errors: the one of inherited from the training set and the one of HITEC's training and categorization. We remark that, there are also terms in the 111 sample queries that do not occur any more in the entire corpus.

As for HITEC, the efficiency of training was checked in terms of two factors. First, on the 111 samples queries, and second, on the queries of the training set used as validation (F_1 measure). We found that the training set (T + C) gave the most promising result with $d_1 = 2$, $\theta = 0.1$ and $v_{max} = 0.5$.

We divided queries into two sets: Set A contained queries that receive target category via searching internet directories and category mapping (cca. 400 k), other queries were put into Set B. About 80% of Set B queries (cca. 320 k) was assigned category by HITEC (setting R2). The remaining 80 k queries had no category.

The effectiveness of category assignment was evaluated on 800 manually categorized queries have been made public after the contest. Each query was classified by three human experts. On this test set our submitted solution had 0.340883 and 0.34009, for precision and F_1-measure, resp.

Let us examine how the components of Ferrety contributed to this results. Six hundred and sixty-five queries from the test set is in Set A, and the others in Set B. There are Set B queries for which HITEC was unable determine a category. See Table 2 for detailed analysis.

Table 2: Effectiveness of HITEC runs on (a) the 135 Set B queries (b) Set B queries labelled (exp #n – labeller expert n; $\overline{p}$ – average precision, $\overline{r}$ – average recall, $\overline{F_1}$ – average F_1; $|N|$ – number of Set B queries labelled by HITEC

Run	F_1 exp 1	F_1 exp 2	F_1 exp 3	$\overline{p}$	$\overline{r}$	$\overline{F_1}$	No label
R1	12.47%	10.03%	10.82%	**13.55%**	9.58%	**11.11%**	39
R2	16.61%	14.48%	14.22%	**16.95%**	13.90%	**15.10%**	31
R3	17.43%	15.43%	16.39%	**19.36%**	14.53%	**16.42%**	32

Run	$\|N\|$	F_1 exp 1	F_1 exp 2	F_1 exp 3	$\overline{p}$	$\overline{r}$	$\overline{F_1}$
R1	96	15.19%	11.68%	13.06%	**13.55%**	13.35%	**13.31%**
R2	104	19.33%	16.13%	16.49%	**16.95%**	18.06%	**17.32%**
R3	103	20.50%	17.41%	19.18%	**19.36%**	19.10%	**19.03%**

The tables give unexpected results: their effectiveness are in reverse order compared to validation results (Fig. 3). This can be argued by two reasons. First, better results on the training data means the ability of better learning Set A queries, and those were not taken into consideration in the above evaluation. Second, it can be attributed to overfitting: better validation result means lower generalization ability for the classifier. We note that the low figures in the tables are misleading: these are the most difficult queries, without available training data, hence, for those much better results cannot be expected.

We can state that the good performance of our approach is based on two factors: first, the effectiveness of taxonomy mapping, and the careful selection of source taxonomies (internet directories); second, on the generalization and learning ability of the applied classifier.

5 Alternative Solutions

5.1 Committee of Classifiers

The first prize was taken in all the three category by the team on HKUST university of Hong Kong [16]. They created a multi-component classifier. Their approach has similar structure as Ferrety. The first step of query categorization is the query expansion by means of internet directories and SEs. For this purpose three SEs were employed: Looksmart, Google, and their own engine created by Lemur[6] SE constructor and equipped with ODP taxonomy.[7] About 40 millions of web pages were collected generating 50 GBs of data. Analogously to Ferrety, the training set has been prepared by posting the queries to SEs and then processing the result pages. Here, obviously, a taxonomy mapping between the target and the three source taxonomies was also a necessary component.

The taxonomy mapping has been established in two steps. First, a keyword matching technique was applied, where also the WordNet synonyms were utilized to expand the names of target categories. This guaranteed a mapping with high precision, but with low recall, since there were relatively few source categories that can be mapped via (partial) name matching to target categories. In order to increase recall, an SVM based learning model was applied with large training data set (15 million web pages) obtained in the first step (search result download). This procedure extended the vector space of the model sufficiently.

The main difference compared to Ferrety is in the next step. The various mappings created so far yield different classification functions and results. By combining them one can often achieve better results. This idea was implemented by means of a classifier committee. The goodness of each classifier was evaluated by means of the 111 sample queries, this affected the weight of the classifier in the committee. In order to balance the the bias of overfitting, they also used uniformly weighted committee of classifiers, and the final results were aggregated from the two committees.

5.2 Regression Based Classification

The second place of F_1 contest was achieved by the team of Florida IT companies A.I. Insight and MEDai, and Humboldt University of Berlin [17]. In the first step of their approach they employed Google on the ODP taxonomy, with somewhat modified queries. They exploited the "did you mean" feature. The mapping between ODP and the target taxonomies were created manually, in general, the nodes of the top two levels of ODP taxonomy were mapped to target categories, but where refinement was necessary, this was extended until the forth level. This mapping was supplemented with deeper categories automatically based on a recommendation system. At the final mapping one source category was assigned to at most three target categories.

The classification was performed by logical regression based categorizer of A.I. Insight. This software has three types of parameter (category weights, category rank and margin between best two categories) which are combined to a single conditional probability. The target categories was determined based on this value. The advantage of the method is that it allows optimization between recall and precision.

[6] www.lemurproject.org/

[7] Open Directory Project; dmoz.com

6 Conclusions

In this paper we presented the Ferrety algorithm, that is able to categorize internet search queries into a subject taxonomy even without training data. We combined a web search based taxonomy mapper and a hierarchical categorizer to perform this task. The approach was presented via the example of ACM KDD Cup 2005, were it had been awarded for precision and creativity.

Acknowledgements

Domonkos Tikk was supported by the János Bolyai Research Scholarship of the Hungarian Academy of Science.

References

1. Sanderson M (2000) Information Retrieval 2:49–69
2. Schütze H, Pedersen JO (1995) Information retrieval based on word senses. In: Proceedings of the 4th Symposium on Document Analysis and Information Retrieval, Las Vegas, NV
3. Gonzalo J, Verdejo F, Chugur I, Cigarran J (1998) Indexing with WordNet synsets can improve text retrieval. In: Proceedings of the COLING/ACL '98 Workshop on Usage of WordNet for NLP, Montréal, Canada
4. Krovetz R (1997) Homonomy and polysemy in information retrieval. In: Proceedings of the ACL/EACL-97, Madrid, Spain
5. Allan J, Raghavan H (2002) Using part-of-speech patterns to reduce query ambiguity. In: Proceedings of the 25th Annual International ACM SIGIR Conference on Research and Development in Information Retrieval, Tampere, Finland
6. Dennis S, Bruza P, McArthur R (2002) American Society for Information Science and Technology 53:120–133
7. Glance NS (2001) Community search assistant. In: Proceedings of the 6th International Conference on Intelligent User Interfaces, Santa Fe, NM
8. Beeferman D, Berger A (2000) Agglomerative clustering of a search engine query log. In: Proceedings of the 6th ACM SIGKDD International Conference on Knowledge Discovery and Data Mining, Boston, MA
9. Chen H, Dumais S (2000) Bringing order to the Web: Automatically categorizing search results. In: Proceedings of the ACM SIGCHI Conference on Human Factors in Computing Systems (CHI), The Hague, The Netherlands
10. Kang IH, Kim G (2003) Query type classification for web document retrieval. In: Proceedings of the 26th Annual International ACM SIGIR Conference on Research and Development in Information Retrieval, Toronto, Canada
11. Adams S (2001) World Patent Information 23:15–23
12. Agrawal R, Srikant R (2001) On integrating catalogs. In: Proceedings of the Conference on the World Wide Web
13. Zhang D, Lee W (2004) Web Semantics 2:131–151
14. Doan A, Madhavan J, Domingos P, Halevy A (2004) Ontology matching: A machine learning approach. In: Staab S, Studer R (eds) Handbook on Ontologies in Information Systems, Springer, Berlin-Heidelberg, pp 397–416
15. Kardkovács ZT, Tikk D, Bánsághi Z (2005) ACM SIGKDD Explorations Newsletter 7:111–116

16. Shen D, Pan R, Sun JT, Pan JJ, Wu K, Yin J, Yang Q (2005) ACM SIGKDD Explorations Newsletter 7:100–110
17. Vogel D, Bickel S, Haider P, Schimpfky R, Siemen P, Bridges S, Scheffer T (2005) ACM SIGKDD Explorations Newsletter 7:117–122
18. Tikk D, Biró G, Yang JD (2004) Australian Journal of Intelligent Information Proceedings Systems 8:123–131
19. Tikk D, Biró G, Yang JD (2005) Experiments with a hierarchical text categorization method on WIPO patent collections. In: Attok-Okine NO, Ayyub BM (eds) Applied Research in Uncertainty Modelling and Analysis, Springer, New York, NY, USA, pp 283–302
20. Salton G, McGill MJ (1983) An Introduction to Modern Information Retrieval, McGraw-Hill, New York, NY, USA,
21. Porter MF Program (1980) 14:130–147

Scheduling Jobs with Genetic Algorithms

António Ferrolho[1] and Manuel Crisóstomo[2]

[1] Department of Electrotechnical Engineering, Superior School of Technology, Polytechnic Institute of Viseu, 3504-510 Viseu, Portugal
antferrolho@elect.estv.ipv.pt

[2] Institute of Systems and Robotics, University of Coimbra, Polo II, 3030-290 Coimbra, Portugal
mcris@isr.uc.pt

Summary. Most scheduling problems are *NP*-hard, the time required to solve the problem optimally increases exponentially with the size of the problem. Scheduling problems have important applications, and a number of heuristic algorithms have been proposed to determine relatively good solutions in polynomial time. Recently, genetic algorithms (GA) are successfully used to solve scheduling problems, as shown by the growing numbers of papers. GA are known as one of the most efficient algorithms for solving scheduling problems. But, when a GA is applied to scheduling problems various crossovers and mutations operators can be applicable. This paper presents and examines a new concept of genetic operators for scheduling problems. A software tool called hybrid and flexible genetic algorithm (HybFlexGA) was developed to examine the performance of various crossover and mutation operators by computing simulations of job scheduling problems.

1 Introduction

The strong performance of the genetic algorithm (GA) depends on the choice of good genetic operators. Then, the selection of appropriate genetic operators is very important for constructing a high performance GA. Various crossover and mutation operators have been examined for sequencing problems in the literature (for example, see [1,4–7,11–13]).

This paper presents and examines a new concept of genetic operators for scheduling problems. Each one of these genetic operators was evaluated with the objective of selecting the best performance crossover and mutation operators. When the performance of a crossover operator is evaluated, a GA without mutation is employed and the evaluation of a mutation operator is carried out by a GA without crossover. In the literature, various crossover operators and mutation operators were examined in this manner (for example, see [7,13]).

2 Genetic Operators

This section presents a new concept of genetic operators for scheduling problems.

2.1 Crossover Operators

Crossover is an operation to generate a new sequence (child chromosome) from two sequences (parent chromosomes). Figure 1(a) presents a one-point crossover: 1 child (OPC1C). In OPC1C

J.A.T. Machado et al. (eds.), *Intelligent Engineering Systems and Computational Cybernetics,*

we randomly chose one parent, and then one point is also randomly selected in this parent. The jobs on one side are inherited from the parent to the child, and the other jobs are placed in the order of they appeared in the other parent.

Figure 1(b) presents a two-point crossover: 1 child (Version I) – TPC1CV1. In TPC1CV1 we chose one parent, and then two points are also randomly selected in this parent. The jobs outside the two selected points are always passed down from one parent to the child, and the other jobs are placed in the order of they appeared in the other parent. Figure 1(c) presents a two-point crossover: 1 child (Version II) – TPC1CV2. In TPC1CV2 we chose one parent, and after two points are also randomly selected in this parent. The jobs between the two selected points are always passed down from one parent to the child, and the other jobs are placed in the order of they appeared in the other parent.

Figure 1(d) presents a one-point crossover: 2 children (OPC2C). OPC2C is similar to OPC1C but in this crossover operator we always obtain two children. Fig. 1 e) presents a two-point crossover: 2 children (Version I) – TPC2CV1. TPC2CV1 is similar to TPC1CV1 but in this crossover operator we always obtain two children.

Figure 1(f) presents a two-point crossover: 2 children (Version II) – TPC2CV2. TPC2CV2 is similar to TPC1CV2 but in this crossover operator we always obtain two children.

Crossover operators with three and four children were also developed. The crossover operators with three children are called two-point crossover: 3 children (Version I) – TPC3CV1 and two-point crossover: 3 children (Version II) – TPC3CV2. TPC3CV1 is a mix of TPC1CV1 plus TPC2CV1 and TPC3CV2 is a mix of TPC1CV2 plus TPC2CV2. The crossover operator with four children is called two-point crossover: 4 children (TPC4C). This operator is a mix of TPC2CV1 plus TPC2CV2.

In our computational tests in Section 4 we also used the following crossover operators: order crossover (OX) in Goldberg [1], cycle crossover (CX) in Oliver [11], and position based crossover (PBX) in Syswerda [4].

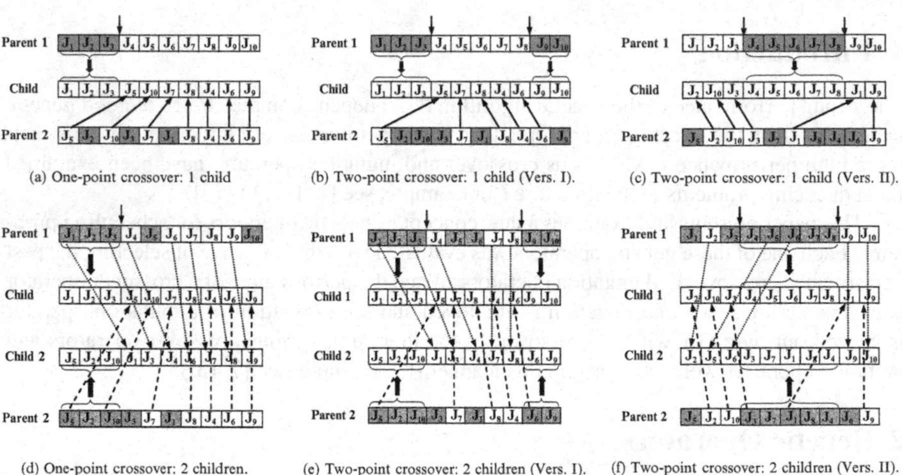

Fig. 1: Illustration of crossover operators

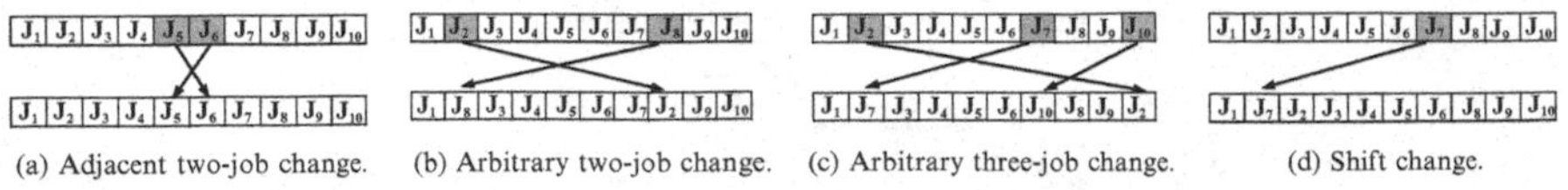

(a) Adjacent two-job change. (b) Arbitrary two-job change. (c) Arbitrary three-job change. (d) Shift change.

Fig. 2: Illustration of mutation operators

2.2 Mutation Operators

Mutation is an operation to change the order of n jobs in the generated child. The mutation operators are used to prevent the loss of genetic diversity. We examined the following four mutations used by Murata in [12, 13]: adjacent two-job change (Adj2JC), arbitrary two-job change (Arb2JC), arbitrary three-job change (Arb3JC) and shift change (SC).

As we can see in Fig. 2 a), in Adj2JC the two adjacent jobs to be changed are randomly selected. In Arb2JC (see Fig. 2(b)) the two jobs to be changed are arbitrarily and randomly selected. In Arb3JC (see Fig. 2(c)) the three jobs to be changed are arbitrarily and randomly selected. In SC a job at one position is removed and placed at another position, as shown in the Fig. 2(d).

A new mutation operator called the arbitrary 20%-jobchange (Arb20%JC) was developed, as we can see in Fig. 3. This mutation selects 20% of the jobs in the child chromosome. The 20% of the jobs to be changed are arbitrarily and randomly selected, and the order of the selected jobs after the mutation is randomly specified. The percentage in this mutation operator gives the operator some flexibility, i.e., the number of jobs to be changed depends on the size of the chromosome. For example, if we have a chromosome with 40 jobs and other with 100 jobs, the number of the jobs to be changed is 8 and 20 respectively.

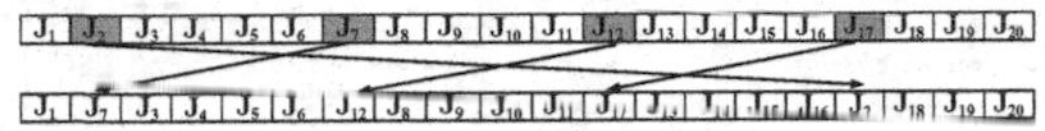

Fig. 3: Arbitrary 20%-job change

3 Developed Software

We developed a software tool, called Hybrid and Flexible Genetic Algorithm (HybFlexGA), to examine the performance of various crossover and mutation operators by computing simulations on scheduling problems. The HybFlexGA was coded in C++ language, and Fig. 4 shows its architecture. Its architecture is composed of three modules: interface, preprocessing, and scheduling module.

The interface module with the user is very important for the scheduling system's success. Thus, this interface should be user friend and dynamic so as to allow easy manipulation of the scheduling plan, jobs, and so forth. This interface allows the connection between the user and the scheduling module, facilitating data entry (for example, parameter definition and problem definition) and the visualization of the solutions for the scheduling module. Figure 5 shows the interface window.

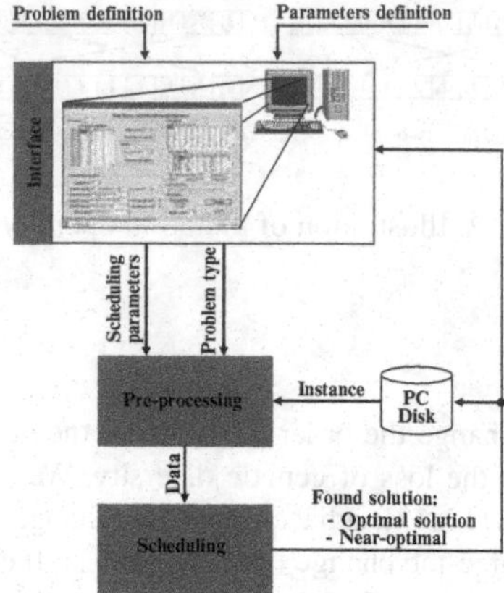

Fig. 4: Architecture of the HybFlexGA

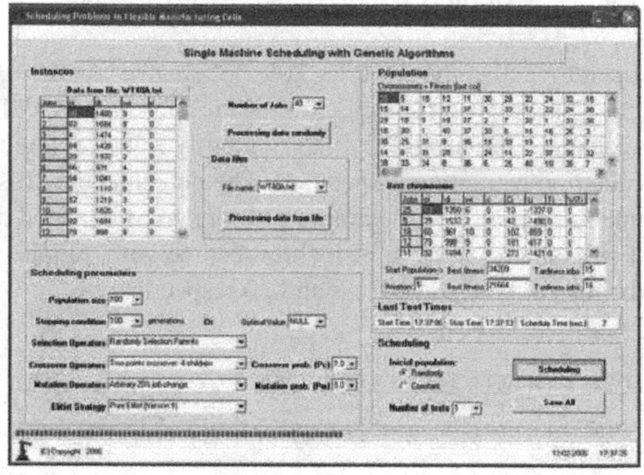

Fig. 5: Interface window

The inputs of the pre-processing module are the problem type and the scheduling parameters. The instance of the scheduling problem can be randomly generated or generated by PC file, as shown in Fig. 4. This module, pre-processes the inputs information and then sends the data to the next module – the scheduling module.

The objective of the scheduling module is to give the optimal solution of any scheduling problem. If the optimal solution is not found, the GA gives the best solution found (near-optimal solution). Figure 6 shows the GA implemented in scheduling module.

4 Computer Simulations

This section presents the results of the computational tests done to examine the crossover and mutation operators presentedin Section 2. As a test problem, we randomly generated a single

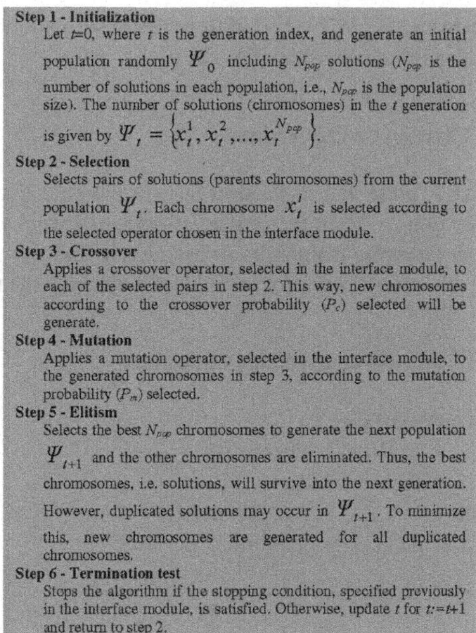

Step 1 - Initialization
Let t=0, where t is the generation index, and generate an initial population randomly Ψ_0 including N_{pop} solutions (N_{pop} is the number of solutions in each population, i.e., N_{pop} is the population size). The number of solutions (chromosomes) in the t generation is given by $\Psi_t = \{x_t^1, x_t^2, ..., x_t^{N_{pop}}\}$.

Step 2 - Selection
Selects pairs of solutions (parents chromosomes) from the current population Ψ_t. Each chromosome x_t^i is selected according to the selected operator chosen in the interface module.

Step 3 - Crossover
Applies a crossover operator, selected in the interface module, to each of the selected pairs in step 2. This way, new chromosomes according to the crossover probability (P_c) selected will be generate.

Step 4 - Mutation
Applies a mutation operator, selected in the interface module, to the generated chromosomes in step 3, according to the mutation probability (P_m) selected.

Step 5 - Elitism
Selects the best N_{pop} chromosomes to generate the next population Ψ_{t+1} and the other chromosomes are eliminated. Thus, the best chromosomes, i.e. solutions, will survive into the next generation. However, duplicated solutions may occur in Ψ_{t+1}. To minimize this, new chromosomes are generated for all duplicated chromosomes.

Step 6 - Termination test
Stops the algorithm if the stopping condition, specified previously in the interface module, is satisfied. Otherwise, update t for t:=t+1 and return to step 2.

Fig. 6: GA implemented in the scheduling module

machine total weighted tardiness (SMTWT) problem with 40 jobs. In the SMTWT problem n jobs have to be sequentially processed on a single machine. Each job i has a processing time p_i, a weight w_i, and a due date d_i associated, and the job becomes available for processing at time zero. The tardiness of a job i is defined as $T_i = max\{0, C_i - d_i\}$, where C_i is the completion time of job i in the current job sequence. The objective is to find a job sequence which minimizes the sum of the weighted tardiness given by:

$$\sum_{i=1}^{n} w_i T_i \qquad (1)$$

Because the SMTWT problem is *NP*-hard, optimal algorithms for this problem would require a computational time that increases exponentially with the problem size [2, 3, 8–10, 14]. Several methods have been proposed to solve this, for example: Earliest Due Date, Simulated Annealing, Tabu Search, Genetic Algorithms and Ant Colony [2, 14]. The computational tests were carried out on a PC Pentium 4, 3 GHz and 1 GB of RAM.

4.1 Test Conditions

For each job i (i = 1, ..., n) the processing time p_i is an integer randomly generated from the closed interval [1, 100] and the weight w_i is also an integer randomly generated from the closed interval [1, 10]. The due date d_i for each job is an integer randomly generated from the interval [P(1 - TF - RDD/2), P(1 - TF + RDD/2)], where RDD is the range of due dates, TF is the tardiness factor and $P = \sum_{i=1}^{n} p_i$. RDD and TF can assume the following values: 0.2, 0.4, 0.6, 0.8 and 1.0. By varying RDD and TF we can generate instance classes of different hardness.

We randomly generated a SMTWT scheduling problem with 40 jobs (SMTWT40) to use in computational tests. This instance is shown in Table 1.

4.2 Examination of Crossover Operators

We applied the HybFlexGA presented in Section 3, with the objective of examining the 12 crossover operators in Section 2. When the crossover operators were examined the mutation operator was not used, in the same manner as in Manderick and Spiessens [7], Murata and Ishibuchi [12, 13]. Each crossover operator was examined by using the following conditions:

- Number of tests: 20
- Initial population Ψ_t : constant and randomly generated
- Number of jobs: 40
- Instance used: constant (see Table 1)
- Population size: N_{pop} = 20, 40, 60, 80 and 100
- Stopping condition: 1,000 generations
- Crossover probabilities P_c: 0.2, 0.4, 0.6, 0.8 and 1.0
- Mutation operators and mutation probabilities: not used

Table 1: Instance used in computational tests

	Jobs	1	2	3	4	5	6	7	8	9	10
	p_i	71	58	89	62	8	31	74	52	71	85
	d_i	1,419	1,682	1,683	1,617	1,703	1,549	1,741	1,634	1,580	1,588
	w_i	6	3	9	3	8	8	3	8	10	4
	r_i	0	0	0	0	0	0	0	0	0	0
	Jobs	11	12	13	14	15	16	17	18	19	20
	p_i	1	77	35	30	96	12	4	29	64	34
	d_i	1,694	1,574	1,548	1,730	1,535	1,438	1,501	1,504	1,587	1,687
	w_i	7	8	10	5	5	6	5	10	3	6
SMTWT40	r_i	0	0	0	0	0	0	0	0	0	0
	Jobs	21	22	23	24	25	26	27	28	29	30
	p_i	8	98	86	22	6	6	24	61	86	76
	d_i	1,472	1,507	1,389	1,454	1,404	1,522	1,526	1,681	1,506	1,584
	w_i	9	4	3	2	7	10	6	10	9	3
	r_i	0	0	0	0	0	0	0	0	0	0
	Jobs	31	32	33	34	35	36	37	38	39	40
	p_i	17	36	63	83	81	37	80	56	11	57
	d_i	1,720	1,767	1,621	1,677	1,487	1,513	1,591	1,620	1,771	1,712
	w_i	2	6	6	9	10	9	7	5	5	6
	r_i	0	0	0	0	0	0	0	0	0	0

With the objective of evaluating each genetic operator we used the following performance measure:

$$Performance = f(\overline{x}_{initial}) - f(\overline{x}_{end}), \quad (2)$$

where $x_{initial}$ is the best chromosome in the initial population and x_{end} is the best chromosome in the last population. That is, $f(\overline{x}_{initial})$ is the fitness average (of the 20 computational tests) of the best chromosomes in the initial population and $f(\overline{x}_{end})$ is the fitness average of the best chromosomes in the end of the 1,000 generations. The performance measure in (2) gives the total improvement in fitness during the execution of the genetic algorithm.

We used 20 computational tests to examine each crossover operator. The average value of the performance measure in (2) was calculated for each crossover operator with each crossover probability and each population size.

Table 2 shows the best average value of the performance measure obtained by each crossover operator with its best crossover probability and its best population size. This table shows the crossover operator by classification order. From Table 2 we can see that:

- The best average performance is obtained for the crossover operator two-point crossover: 4 children with $P_c = 1.0$ and $N_{pop} = 100$. In this case the average performance is 3,834.1.
- The worst average performance is obtained for the crossover operator one-point crossover: 1 child with $P_c = 0.6$ and $N_{pop} = 100$. In this case the average performance is 3,570.5.

Table 2: Classification of the crossover operators

Position	Crossover	P_c	N_{pop}	Performance
1st	TPC4C	1.0	100	3,834.1
2nd	TPC3CV2	1.0	100	3,822.9
3rd	TPC2CV2	1.0	100	3,821.8
4th	PBX	1.0	80	3,805.8
5th	TPC1CV2	0.8	100	3,789.3
6th	CX	0.8	80	3,788.7
7th	TPC3CV1	0.8	80	3,680.2
8th	TPC2CV1	1.0	80	3,662.1
9th	OPC2C	0.6	100	3,647.8
10th	OX	1.0	100	3,635.4
11th	TPC1CV1	1.0	100	3,624.7
12th	OPC1C	0.6	100	3,570.5

We used 20 computational tests to examine each crossover operator. Figure 7 presents the results obtained in the 20 computational tests for the six best crossover operators obtained in this work. In all of the graphs, the bars represent the results obtained for every 100 generations (along the 1,000 generations) in the computational tests. The three firsts crossover operators (TPC4C, TPC3CV2 and TPC2CV2) present a very fast evolution in the first 100 generations (see Fig. 7(a)). But, as we can see in Fig. 7(b), the three firsts crossover operators require more CPU time than the other crossover operators. On the other hand, Fig. 7(c) shows these crossovers operators present a lower standard deviation average along the 1,000 generations. This means that the dispersion of the values (in the 20 computational tests) relative to the fitness average is smaller in these crossover operators. PBX, TPC1CV2 and CX require less CPU time relative to the first three crossover operators (see Fig. 7(b)). But, as we can see in Fig. 7(a) they present a slower evolution relative to the first three crossover operators, and Fig. 7(c) shows these crossovers operators present a higher standard deviation average along the 1,000 generations.

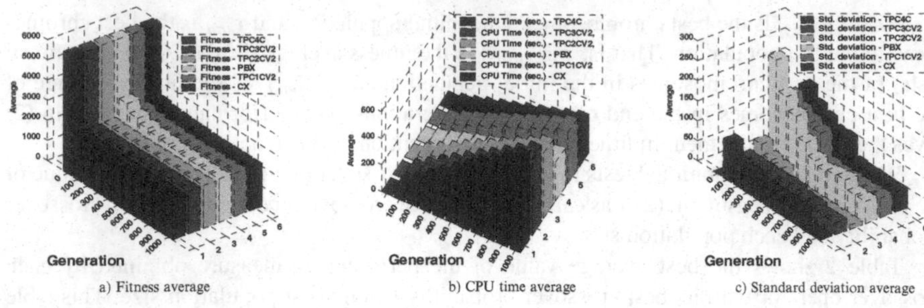

a) Fitness average b) CPU time average c) Standard deviation average

Fig. 7: The six best crossover operators

4.3 Examination of Mutation Operators

We applied the HybFlexGA presented in Section 3, with the objective of examining the five mutation operators in Section 2. When the mutation operators were examined the crossover operator was not used, in the same manner as in Manderick and Spiessens [7], Murata and Ishibuchi [12, 13]. Each mutation operator was examined by using the same conditions used in the examination of crossover operators.

The average value of the performance measure in (2) was calculated for each mutation operator with each mutation probability and each population size. Table 3 shows the best average value of the performance measure obtained by each mutation operator with its best mutation probability and its best population size. This table shows the mutation operator by classification order. From Table 3 we can see that:

- The best average performance is obtained by the arbitrary 20%-job change with $P_m = 1.0$ and $N_{pop} = 100$. In this case the average performance is 3,833.9.
- The worst average performance is obtained by the adjacent two-job change with $P_m = 0.4$ and $N_{pop} = 100$. In this case the average performance is 3,250.4.

Table 3: Classification of the mutation operators

Position	Mutation	P_m	N_{pop}	Performance
1st	Arb20%JC	1.0	100	3,833.9
2nd	Arb2JC	0.8	100	3,826.4
3rd	Arb3JC	1.0	60	3,814.9
4th	SC	0.8	60	3,673.5
5th	Adj2JC	0.4	100	3,250.4

Figure 8 presents the results obtained in the 20 computational tests for the three best mutation operators obtained in this work. In all of the graphs, the bars represent the results obtained for every 100 generations (along the 1,000 generations) in the computational tests. Fig. 8(a) shows the Arb20%JC presents a good fitness average along the 1,000 generations. Figure 8(b) shows that Arb20%JC requires a little more CPU time than the Arb2JC. On the other hand, the Arb20%JC presents a lower standard deviation average along the 1,000 generations (see Fig. 8(c)).

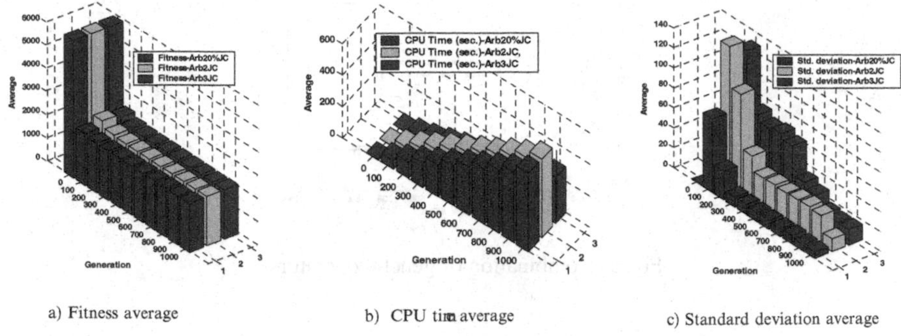

a) Fitness average

b) CPU tim average

c) Standard deviation average

Fig. 8: The three best mutation operators

We examined the Arb20%JC for other percentage values, with the objective of finding the best percentage for this operator. Thus, we examined this operator for 10% and 30%. Table 4 shows the performance of the computational test results. From the Table 4, we can conclude that this operator has a better performance for 20%.

Table 4: Arb20%JC for different percentage

Mutation	P_m	N_{pop}	Performance
Arb10%JC	1.0	100	3,829.2
Arb20%JC	1.0	100	3,833.9
Arb30%JC	1.0	100	3,830.6

4.4 Combination of Genetic Operators

Based on the simulation results in Tables 2 and 3, we applied in the HybFlexGA the best genetic operators obtained from our computational tests (TPC4C with $P_c = 1.0$ and Arb20%JC with $P_m = 1.0$). The objective is to examine the performance of the genetic algorithm with both genetic operators. We examined this combination of genetic operators in the same manner as in the last two subsections.

This means that we used the same instance (see Table 1), the same starting population N_{pop} (100 chromosomes), the same stopping condition (1,000 generations) and we also carried out 20 computational tests. From this computer simulation, we had the following average performance: 3,836. This result and the result obtained in Fig. 9 suggest the existence of a positive combination crossover and mutation effect, because the average performance was improved by the combination of genetic operators. As we can see in Fig. 9, starting from the generation 250 the improvement of the average performance is small because, in our opinion, the HybFlexGA found the optimal solution for this instance. In other computational tests (with other instances) we obtained a larger improvement.

Murata [13] demonstrated that, with other combinations of genetic operators (other crossovers and other mutations), we can obtain a negative combination crossover and mutation effect.

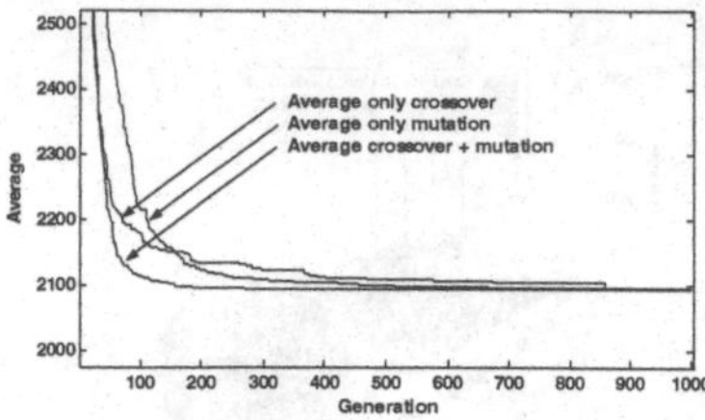

Fig. 9: Combination of genetic operators

5 Conclusions

When a GA is applied to scheduling problems, various crossovers and mutations can be applicable. This paper proposed and examined a new concept of genetic operators for scheduling problems. A software tool, called HybFlexGA, was developed to examine the performance of various crossover and mutation operators by computing simulations on job scheduling problems. Computational results show that the TPC4C and Arb20%JC are the best genetic operators and a positive combination with both operators (TPC4C + Arb20%JC) was obtained. With the combination TPC4C + Arb20%JC, the HybFlexGA presents the following advantages: founds more optimal solutions, requires fewer generations to find the optimal solutions, and requires less CPU time to find the optimal solutions.

References

1. Goldberg DE (1989) Genetic Algorithms in Search, Optimization and Machine Learning. Addison-Wesley, Reading, MA
2. Thomas ME, Pentico DW (1993) Heuristic Scheduling Systems. John Wiley & Sons, ISBN: 0471578193, New York
3. Baker KR (1974) Introduction to Sequencing and Scheduling. Wiley, New York
4. Syswerda G (1991) Scheduling optimization using genetic algorithms. In: Davis L (ed) Handbook of Genetic Algorithms, pp 332–349, Van Nostrand Reimhold, New York
5. Ferrolho A, Crisóstomo M (2005) Genetic Algorithms: Concepts, Techniques and Applications. WSEAS Transactions on Advances in Engineering Education 2:12–19, January, ISSN:1790-1979
6. Ferrolho A, Crisóstomo M (2005) Scheduling and Control of Flexible Manufacturing Cells Using Genetic Algorithms. WSEAS Transactions on Computers 4:502–510, June, ISSN:1109-2750
7. Manderick B, Spiessens P (1994) How to select genetic operators for combinatorial optimization problems by analyzing their fitness landscape. In: Zurada JM, Marks II RJ, Robinson CJ (eds) Computational Intelligence Imitating Life, pp 170–181, IEEE Press
8. Lawer EL (1977) A Pseudopolinomial Algorithm for Sequencing Jobs to Minimize Total Tardiness. Annals of Discrete Mathematics 1977:331–342
9. Abdul-Razaq TS, Potts CN, Van Wassenhove LN (1990) A Survey for the Single-Machine Scheduling Total WT Scheduling Problem. Discrete Applied Mathematics 26:235–253
10. Potts CN, Van Wassenhove LN (1991) Single Machine Tardiness Sequencing Heuristics. IIE Transactions 23(4)346–354

11. Oliver J, Smith D, Holland J (1987) A study of permutation crossover operators on the traveling salesman problem. In: Proceedings of the Second ICGA, pp 224–230, ISBN:960-8457-29-7, Lawrence Erlbaum Associates, Mahwah, NJ, USA
12. Murata T, Ishibuchi H (1994) Performance evaluation of genetic algorithms for flowshop scheduling problems. In: Proceedings of the 1st IEEE International Conference on Evolutionary Computation, pp 812–817
13. Murata T, Ishibuchi H (1996) Positive and negative combination effects of crossover and mutation operators in sequencing problems. In: Proceedings of the IEEE International Conference on Evolutionary Computation, pp 170–175, ISBN: 0-7803-2902-3, Nagoya, Japan
14. Madureira A, Ramos C, Silva S (2001) A GA based scheduling system for dynamic single machine problem. In: Proceedings of the 4th IEEE International Symposium on Assembly and Task Planning Soft Research Park, pp 262–267, ISBN: 0-7803-7004-x, Fukuoka, Japan

Self-Referential Reasoning in the Light of Extended Truth Qualification Principle

Mohammad Reza Rajati, Hamid Khaloozadeh, and Alireza Fatehi

Department of Electrical Engineering, K. N. Toosi University of Technology, Tehran, Iran
rajati@ee.kntu.ac.ir,
{h_khaloozadeh, fatehi}@kntu.ac.ir

Summary. The purpose of this paper is to formulate truth-value assignment to self-referential sentences via Zadeh's truth qualification principle and to present new methods to assign truth-values to them. Therefore, based on the truth qualification process, a new interpretation of possibilities and truth-values is suggested by means of type-2 fuzzy sets and then, the qualification process is modified such that it results in type-2 fuzzy sets. Finally, an idea of a comprehensive theory of type-2 fuzzy possibility is proposed. This approach may be unified with Zadeh's Generalized Theory of Uncertainty (GTU) in the future.

1 Introduction

In the era of information and communications, truth and reliability play a significant role and intelligent machinery are demanding new tools of reliable data interpretation and processing. Now, human beings tend to develop new means to handle the great bulk of available information. Unquestionably, mimicry of human's ability to process the incoming data intelligently and deduce conclusions about its reliability as well as "meaning" is of equal importance with obtaining information, if not more important.

After the development of fuzzy computation theory, the great contribution of L. A. Zadeh [1] revealed the fact that there might be rigorous tools via which the "meaning" of the analyzed information could be manipulated besides its statistical nature. He investigated the "meaning-oriented" approach to information processing via the well-founded theory of fuzzy computation and established a new theory of possibility as a counterpart of the probability theory.

One of the novel applications of possibility theory is the resolution of the liar paradox (and henceforth, self-referential sentences bearing a paradox). Zadeh proposed a method to assign a truth-value to the liar sentence [2, 3].

In this paper, we investigate the application of Zadeh's method to resolve the paradox borne by self-referential sentences and propose an extension to Zadeh's method, such that it can be useful in a broader sense. This extension is based on the concept of type-2 fuzzy sets.

We first introduce type-2 fuzzy sets, possibility theory, and the concept of self-reference, and then based on Zadeh's method we try to assign truth-values to self-referential sentences. In the next step, we extend Zadeh's truth qualification principle to handle the existing uncertainties in fuzzy possibilities. Finally, a comprehensive theory of type-2 fuzzy possibility is touched.

J.A.T. Machado et al. (eds.), *Intelligent Engineering Systems and Computational Cybernetics,*

2 Preliminaries

Here, we introduce some basic concepts which are necessary for our paper to be understood, including type-2 fuzzy sets, the concept of fixed points, and possibility theory.

2.1 Type-2 Fuzzy Sets

The basic notion of type-2 fuzzy sets was suggested by Zadeh [4] to illustrate uncertainty in the membership values of a fuzzy set.

Mendel and John distinguish four types of uncertainty in type-1 (conventional) fuzzy systems [5]: (1) The meanings of the words which are used in the premises and conclusions of the rules can be uncertain (words mean different things to different individuals). (2) Conclusions may have a histogram of values assigned to them, especially if the rules are obtained from a group of people who do not agree (and it is almost always the case). (3) Measurements that activate a type-1 fuzzy system may be noisy and hence uncertain. (4) The data that are used to tune the parameters of a type-1 fuzzy system may also be noisy.

Mendel and John then conclude that all of these kinds of uncertainty could be interpreted as uncertainties about the membership functions and the best way to model them is making use of type-2 fuzzy sets.

Although these types of uncertainty are directly related to rule-based fuzzy systems, other applications of fuzzy computation may also encounter either such uncertainties or other types of uncertainty. We will see this fact in the sequel.

Below, some definitions and concepts related to type-2 fuzzy sets are presented. The expressions and concepts are somewhat similar to those used by Mendel and John [5]. It is worth noting that there are several papers on basic set theoretic operations of type-2 fuzzy sets [6], centroid of a type-2 fuzzy set [7], and composition of type-2 relations [8, 9]. Also, [10–13] developed the use of interval-valued sets in fuzzy logic. The approach of [14] is somewhat related to our work in this paper, because it discusses "interval-valued degrees of belief".

Definition (1): A type-2 fuzzy set $\tilde{A}$ is associated with a type-2 membership function $\mu_{\tilde{A}}(x,y) \in [0,1]$ where $x \in U$ (universe of discourse) and $y \in V_x \subseteq [0,1]$:

$$\tilde{A} = \{((x,y), \mu_{\tilde{A}}(x,y)) | \forall x \in U, \forall y \in V_x \subseteq [0,1]\} \quad (1)$$

$\tilde{A}$ can also be expressed as:

$$\int_{x\in U} \int_{y\in V_x} \mu_{\tilde{A}}(x,y)/(x,y), V_x \subseteq [0,1] \quad (2)$$

Here, $\int\int$ denotes union over all x, y in U, V. See [5] to understand this notion of union in which, every element of a type-2 (or correspondingly, type-1) fuzzy set is viewed as a subset, distinctly. In case of a discrete universe of discourse, the $\int$ is replaced by $\sum$. (Thus, each of U, V or both could be discrete.)

Definition (2): A *secondary membership function* is the basic tool to show the nature of uncertainty in the membership functions of type-1 and is defined as:

$$\mu_{\tilde{A}}(x=b,y) = \mu_{\tilde{A}}(b) = \int_{y\in V_b} f_b(y)/y \; V_b \subseteq [0,1] \quad (3)$$

It should be noted that a secondary membership function is obtained by retaining $x = b$ and evaluating the membership value of every member of V_b . (V_b defines all membership values which could be assigned to b) . Now, it is obvious why we call $\mu_{\tilde{A}}(b)$ a *secondary membership function.*

Definition (3): The uncertainty in the primary memberships of a type-2 fuzzy set $\tilde{A}$ builds a region which is called the *footprint of uncertainty* (FOU). It is defined as:

$$FOU(\tilde{A}) = \{(x,y)|\mu_{\tilde{A}}(x,y) > 0\} \quad (4)$$

Speaking more precisely, FOU could be defined as the closure of the abovementioned set, i.e. the smallest closed set which contains it. FOU is essentially the result of aggregation of all of the intervals associated with the members of the universe of discourse. FOU can contain some points on its boundary or not, depending on V's.

2.2 Fixed Points of Real-Valued Mappings

Consider a mapping $f: W \to W$ where $W \subseteq \Re^n$. A point $x \in W$ is called a fixed point if $f(x) = x$.

2.3 Possibility Theory

Possibility theory is initiated with the concept of a fuzzy restriction [15]. If x is a variable which takes values from the universe of discourse U, and A is a fuzzy set in U, the sentence "x is A" is considered as putting a fuzzy restriction on x and this "elastic" restriction is characterized by the membership function of A. This leads us to consider μ_A as the degree of possibility of $x = u$.

Given a fuzzy set A in U, the sentence "x is A" provides a possibility distribution assigned to x, denoted by π_x:

$$\forall x \in U \;\; \pi_x(u) = \mu_A(u) \quad (5)$$

For example, "John is young" is equivalent to "Age (John) is A", in which Age is an attribute of "John" and A is a fuzzy set denoting the concept of young. Hence, we can define: $\pi_{Age}(u) = \mu_A(u)$. The universe of discourse could be taken as $U = [0, 130]$.

Zadeh establishes rather a simple, but extremely useful mathematical basis to extend the theory of possibility. The development of this theory is well-known in the fuzzy computation community, therefore we avoid repeating the material and suggest the interested readers to study Zadeh's original work deeply, as a contribution which is still highly readable.

Also, there are other approaches to the theory of possibility, which are axiomatic and based on Dempster–Shafer theory of evidence [16]. Despite the fact that they are very helpful, the approach of Zadeh is very capable in the "meaning-oriented" possibility, and as we denote in this paper, it is very well-behaved for interpretation of self-referential sentences. Thereby we adhere to his approach and try to extend it.

In his paper, Zadeh proposes three principles for analyzing the information content of sentences, namely truth qualification principle, probability qualification principle and possibility qualification principle. As our paper concerns these principles, we reconsider them briefly in the sequel.

Truth Qualification Principle

This principle tries to judge the truth of sentences which speak about the truth of other sentences.

If x is $A \rightarrow \pi_x(u) = \mu_A(u)$
then
x is A is $\tau \rightarrow \pi_x(u) = \mu_\tau(\mu_A(u))$
in which τ is a linguistic truth-value such as: true, very true, more or less true, false, etc.

Probability Qualification Principle

If we need to assign a possibility to a sentence which asserts some facts about the probability of a sentence we have:
If x is $A \rightarrow \pi_x(u) = \mu_A(u)$
then
x is A is $\lambda \rightarrow \pi_x(\int_U p(u)\mu_A(u)du) = \mu_\lambda(\int_U p(u)\mu_A(u)du)$
where λ is a linguistic probability value, e.g. likely, very likely, etc. $p(u)\mu_A(u)du$ is the probability that the value of x falls in the interval $[u, u+du]$ and so $\int_U p(u)\mu_A(u)du$ is actually the probability of fuzzy event A [17]. The principle assigns a possibility to a probability density.

Possibility Qualification Principle

It is obviously realized that this principle qualifies sentences speaking about the possibilities of other sentences and events:
If x is $A \rightarrow \pi_x(u) = \mu_A(u)$
then
x is A is $\alpha - possible \rightarrow \tilde{\pi}_x = \int_{u \in U} \int_{v \in V_u} \mu_{\tilde{\pi}}(u,v)/(u,v)$
$V_u = [min(\alpha, \mu_A(u)), min(1, 1-\mu_A(u))]$
Here $\alpha - possible$ is a linguistic possibility value, and $\tilde{\pi}$ is a type-2 fuzzy set. This type-2 fuzzy set is constructed to show the uncertainty caused by weakening the proposition by a linguistic possibility value. Zadeh, proposes that $\mu_{\tilde{\pi}} = a$, $a \in [0,1]$, i.e. an interval-valued fuzzy set. Replacing $\alpha = 1$ for "completely possible" makes V_u reduce to $[\mu_A(u), 1]$. Zadeh suggests some other versions of the possibility qualification principle, but for the sake of brevity we don't mention them. The basic concept of assigning a type-2 fuzzy set as the possibility of a proposition suffices us.

3 Self-Referential Sentences and Paradox

Self-referential sentences have been a source of debate in logic and mathematics, mainly because some of self-referential sentences make logical paradoxes. Since the time of ancient Greek, the paradoxical nature of self-referential sentences was known [18–21].

Maybe the best known self-referential paradox is the liar, which could be formally expressed as the following statement, known as the modern form of the liar:
$p :=$ Sentence p is false
where := stands for "is defined to be"

"This sentence is false" is also a form of the liar. The paradoxical nature of this sentence is obvious: its truth leads to its wrongness and its wrongness leads to its truth and this circular debate will continue ad infinitum. So it could not be determined whether this sentence is true or false.

It is believed that the works of Eubulides of Miletus, are the most ancient resources on the liar. He included it among a list of seven puzzles: "A man says that he is lying. Is what he says true or false?" Epimenides, in his works, reportedly uttered the following sentence: "All Cretans are liars" although he was a Cretan.

It can be shown that there are other self-referential sentences which bring about paradox. A collection of sentences containing reference to each other or themselves can generate paradox. The paradox is due to an *indirect* self-reference, so we call them self-referential sentences as well. An example is the inconsistent dualist:

$p_1 :=$ Sentence p_2 is true
$p_2 :=$ Sentence p_1 is false

It is worth noting that self-referential sentences may be consistent, i.e. without any paradoxical nature. Good examples are the truth-teller:

$p_1 :=$ This sentence is true

and the consistent dualist:

$p_1 :=$ Sentence p_2 is true
$p_2 :=$ Sentence p_1 is true

These sentences generate no paradox, but the problem faces when we realize that we can consider them either false or true!

Other kinds of self-referential sentences are those which assign numerical truth-values to each other, in the framework of fuzzy logic. See the following example:

$p_1 :=$ The truth-value of p_2 is 0.9 and the truth-value of p_3 is 0.2
$p_2 :=$ The truth-value of p_1 is 0.8 and the truth-value of p_3 is 0.3
$p_3 :=$ The truth-value of p_1 is 0.1

Another example is the following set of sentences containing biconditionals:

$p_1 :=$ The truth-value of p_2 is 0.9 if and only if the truth-value of p_1 is 0.6
$p_2 :=$ The truth-value of p_2 is 0.7 if and only if the truth-value of p_1 is 0.2

It will be confusing to imagine how we can assign truth-values to the abovementioned set of sentences.

Considering the vagueness in the natural language expressions, we can also investigate self-referential sentences containing linguistic hedges or uncertain expressions. For example see the following sentences:

Modest liar 1 [22]:

$p := p$ is at least a little false

Modest liar 2 [23]:

$p := p$ is more or less false

Modest and emphatic truth-teller [23]:

$p_1 := p_1$ is less and more true
$p_2 := p_2$ is very true

Fuzzy logistic [23]:

$p :=$ It is very false that p is true if it is false

Two fuzzy equivalents of the inconsistent dualist are [23]:

$p_1 := p_1$ is true if and only if p_2 is true
$p_2 := p_2$ is true if and only if p_1 is very false

and:

$p_1 :=$ It is very false that $p_1 \leftrightarrow p_2$
$p_2 :=$ It is less and more false that $p_1 \leftrightarrow \neg p_2$

The so-called triplist:

$p_1 :=$ It is very false that $p_1 \leftrightarrow \neg(p_1 \leftrightarrow p_2)$
$p_2 :=$ It is very false that $p_1 \leftrightarrow \neg p_3$
$p_3 :=$ It is very false that $p_3 \leftrightarrow \neg(p_1 \leftrightarrow p_3)$

3.1 Zadeh's Method for Resolution of the Liar

Zadeh in [2] proposed a method based on truth qualification principle to resolve the paradox of the liar. Almost all of the work done in the fuzzy community to cope with paradoxes is inspired by the work of Zadeh [22–24].

To use the landscape of possibility theory, Zadeh interprets the modern form of the liar as:
$p := p \text{ is } \tau$
The truth qualification principle asserts that for the liar:

$$\pi_p(u) = \pi_\tau(\pi_p(u)) \; u \in U \tag{6}$$

Regarding the definition of possibility, we should assume that $U = [0,1]$ or at least $U \subseteq [0,1]$. From the definition of a fixed point, we can realize that $\pi_p(u)$ is a fixed point of $\pi_\tau : [0,1] \to [0,1]$.

If τ is replaced by "true", Zadeh's method is trying to evaluate the "truth-teller". So $\pi_\tau(v) = v$ and then the resultant mapping becomes $\pi_p(u) = \pi_p(u)$, which says that p is true for any $u \in U$. It could be supposed that $\pi_p(u)$ is a possibility distribution which can take any value in $[0,1]$.

Evaluating the liar in the same way we have:

$$\pi_p(u) = 1 - \pi_p(u) \tag{7}$$

This induces a possibility distribution on U defined by:

$$\pi_p(u) = 1/2 \; \forall u \in U \tag{8}$$

This result determines that the possibility for the truth-value of p to be u is $1/2$ for all $u \in U$. This interpretation is somewhat different from saying that "the liar is half-true", which assigns a truth-value to the liar. The same discussion applies to other self-referential sentences as well. However, the result is heuristically interpreted as a truth-value assignment to such sentences.

4 Truth-Value Assignment to Self-Referential Sentences

Zadeh's method is a proper way for truth-value assignment to self-referential sentences. First, we try to use his interpretation to assign truth-values (or possibility distributions) to self-referential sentences. Actually, Zadeh accepts the fixed point of the mapping $\pi_\tau : [0,1] \to [0,1]$ as the solution of the liar paradox. It is a natural extension to use the fixed points of the mappings generated by self-referential sentences as their solutions. However, as we show, the mappings could be assumed in different ways. First, observe the following theorem about fixed points:

Theorem 1(Brouwer fixed point theorem): Any continuous mapping $f : W \to W$ has at least one fixed point provided that W is a nonempty, compact and convex subset of $\Re^n$ [25].

So, provided that our interpretation functions are continuous, Zadeh's method always yields a solution.

4.1 An Extension to Zadeh's Method

We use standard fuzzy logic to translate logical operators, hedges and etc. to create mappings related to self-referential sentences. For example, for the inconsistent dualist:

$$\begin{aligned} \pi_{p1}(u) &= \pi_{p2}(u) \\ \pi_{p2}(u) &= 1 - \pi_{p1}(u) \end{aligned} \tag{9}$$

(Note that for "false" we could use many other functions, but the standard function is $f(v) = 1 - v$. Selection of specific logic and logical functions does not affect our discussion, provided that the functions are continuous.)

Solving (9), we obtain $\pi_{p1}(u) = \pi_{p2}(u) = 0.5$. Similar discussion is applicable to the consistent dualist:

$$\begin{aligned} \pi_{p1}(u) &= \pi_{p2}(u) \\ \pi_{p2}(u) &= \pi_{p1}(u) \end{aligned} \tag{10}$$

So the solution is:

$$\pi_{p1}(u) = \pi_{p2}(u) = a \in [0,1] \tag{11}$$

For self-referential sentences which assign truth-values to each other in the framework of multiple-valued logic, we can still utilize Zadeh's method. Consider the sentence p: = The truth-value of q is 0.9.

Naturally, we can interpret it as:

$$\pi_p(u) = \pi_{0.9}(\pi_q(u)) \tag{12}$$

$\pi_{0.9}$ is a fuzzy number which reflects the uncertainty about the meaning of "is 0.9". It could be taken for example by the approach of Vezerides and Kehagias [26]. They assume that for "p : = The truth-value of q is a" we can assert that:

$$\pi_p(u) = 1 - abs(\pi_q(u) - a) \tag{13}$$

Thereby equation (12) results in $\pi_p(u) = 0.95$.

Accordingly, the fixed point for the liar is obtained:

$$\pi_p(u) = 1 - abs(\pi_p(u) - 0) \Rightarrow \pi_p(u) = 1/2$$

Applying this rule for the truth-teller:

$$\pi_p(u) = 1 - abs(\pi_p(u) - 1) \Rightarrow \pi_p(u) = a \in [0,1]$$

In the same way, Zadeh's method could be directly applied to self-referential sentences with fuzzy hedges and vague expressions.

Consider the set of sentences which we called triplist in Section 3. We use $f(v) = \sqrt{1-v}$ for "more or less" and $g(v) = v^2$ for "very". However, there are many other choices (e.g. $(f(v) = \sqrt[3]{1-v}$) and $g(v) = v^3$. The corresponding mapping is:

$$\begin{aligned} \pi_{p1}(u) &= abs(\pi_{p1}(u) - (1 - abs(\pi_{p1}(u) - \pi_{p2}(u)))^2 \\ \pi_{p2}(u) &= abs(\pi_{p1}(u) - (1 - \pi_{p3}(u)))^2 \\ \pi_{p3}(u) &= abs(\pi_{p3}(u) - (1 - abs(\pi_{p1}(u) - \pi_{p3}(u)))^2 \end{aligned}$$

We found some fixed points for the mapping numerically:

$$(0,0,1),(0.29969,0.19997,0.35284),$$
$$(1,0,0),(0.29295,0.12715,.35047)$$

Correspondingly, we could assign a mapping to any set of self-referential statements containing linguistic and numerical truth-values.

5 An Extension to Zadeh's Truth Qualification Principle

Previously, we studied the usefulness of the truth qualification principle in resolution of self-referential sentences and proposed a new method for interpretation of them. However, this approach leaves some questions unanswered. One may be on the actual "meaning" of a truth-value less than one for a sentence, say 0.5. Obviously, without the "meaning-oriented" theory of possibility and fuzzy constraints, it seems to be of very little meaning and usage. However, justifications for this interpretation of the liar need to be probed. We provided a justification after reviewing Zadeh's resolution of the liar.

Another problem is that different maps are obtained using different logics, i.e. different t-norms, s-norms, hedge interpreters and etc. In addition, a mapping $\pi_\tau : [0,1] \to [0,1]$ may have several fixed points. So, it is difficult to select a suitable *solution* to the set of self-referential sentences.

It might be firstly imagined to ask an expert to choose a *solution* or define a proper criterion for selecting it. But how could one accomplish this task when the interpretation of the possibilities seems to be controversial? It is an open problem for further investigation.

Another way is using proper tools for modeling uncertainty in the possibilities which are membership functions themselves. The uncertainty of having multiple fixed points somewhat resembles the second kind of uncertainty in membership functions presented in Section 2. Therefore, it is quite natural to model the "output" of the whole process of truth qualification as a type-2 fuzzy set. This idea is supported by Zadeh's possibility qualification principle very well: When he tends to model the uncertainty (or relaxation) induced by the $\alpha - possibility$ constraint in this sentence: "A is τ is α-possible" he makes use of interval-valued fuzzy sets which are obviously type-2 fuzzy sets with constant $\mu_{\tilde{\pi}}(x,y)$.

Therefore, in the first step we suggest that the possibility distribution function π_p be interpreted as a type-2 fuzzy set:

$$\tilde{\pi}_p = \int_{x\in U} \sum_{y\in F} \mu_{\tilde{\pi}}(x,y)/(x,y) \tag{14}$$

where F is the set of known fixed points of π_τ. F could be defined in a wider sense as a set of fixed points of all available candidates for π_τ.

It is an accepted convention (which is quite reasonable [5]) to select $\mu_{\tilde{\pi}}(x,y) = 1$ or another constant in the interval $[0,1]$. This reduces $\tilde{\pi}_p$ to an interval-valued fuzzy set.

In his paper [27], Mendel proposes new methods for modeling uncertainties via type-2 fuzzy sets. He believes that asking persons to assign anything other than a uniform weighting to their entire FOU would be very difficult. So, he is inclined to interpret the uncertainties about a word's membership function as an interval-valued fuzzy set. However, someone may disagree him and assert that a person may be inclined to weight some membership functions for a linguistic variable more than the others.

He also suggests to model uncertainties which may exist between a group of people by means of aggregation of each person's equally weighted FOU. He also mentions that we may trust some people more, so we can weight their suggestions more than the others. Thereby, instead of assigning a predetermined $\mu_{\tilde{\pi}}(x,y)$ to the set of solutions, we could refer to a group of experts in the area of fuzzy logic and build a type-2 fuzzy set according to Mendel's suggestions. For an application of uncertainty bounds in designing fuzzy systems see [28].

Inspired by the solution of the truth-teller which has an infinite number of fixed points, we try to propose a type-2 fuzzy set, of which the universes of discourse are continuous. We characterize this set as follows:

$$\tilde{\pi}_p = \int_{x \in U} \int_{y \in \hat{F}} \mu_{\tilde{\pi}}(x,y)/(x,y) \quad \hat{F} = [\inf(F), \sup(F)] \tag{15}$$

Here F is the set of available fixed points, as before. It is obvious that if $\inf(F) = \sup(F)$ then the type-2 fuzzy set will reduce to the so-called "singleton-valued type-2 fuzzy set"[1] which is actually analogous to a conventional fuzzy set. For a similar approach see [29, 30].

6 Toward a Theory of Type-2 Fuzzy Possibility

We believe that the approach of Zadeh to the theory of possibility should be investigated more such that it is fertilized to handle type-2 fuzzy sets. Many types of uncertainty could be present in the interpretation of important linguistic variables such as truth, probability, and possibility. Furthermore, as we showed, the procedure of truth qualification of self-referential statements introduces some uncertainty to the possibility distributions. Therefore, a thorough study and consistent reformulation of Zadeh's approach is needed for type-2 fuzzy possibility. We believe that it is yet another step in completing the recently proposed Generalized Theory of Uncertainty [31]. An example of proper tools for type-2 possibility formulation is the concept of a fuzzy-valued measure [32].

7 Conclusions

In this paper, we tried to formulate truth-value assignment to self-referential sentences following Zadeh's approach in interpreting the truth-values as possibility distributions of the truth of sentences. We also established a method for truth-value assignment to sentences which assign numerical truth-values to each other. Then, we broadened the idea of truth-values to type-2 fuzzy sets such that we can handle the uncertainty induced by existence of multiple fixed points for possibility mappings. Immediately, we modified Zadeh's truth qualification principle taking uncertainties in membership functions into account. At last, we touched a comprehensive theory of type-2 fuzzy possibility to formulate type-2 fuzzy constraints and possibilities.

References

1. Zadeh L. A. (1978) Fuzzy sets as a basis for a theory of possibility. Fuzzy Sets and Sysems 1(1): 3–28.
2. Zadeh L. A. (1979) Liar's Paradox and Truth Qualification Principle. ERL Memorandum M79/34. University of California, Berkeley, CA, USA.

[1] This expression is due to the authors.

3. Klir G. J., Yuan B. (eds.) (1996) Fuzzy Sets, Fuzzy Logic, and Fuzzy Systems: Selected Papers by Lotfi A. Zadeh. World Scientific Publishing Company, Inc. RiverEdge, NJ, USA.
4. Zadeh L. A. (1975) The concept of a linguistic variable and its applications to approximate reasoning I. Information Sciences 8: 199–249.
5. Mendel J. M., John R. I. (2002) Type-2 fuzzy sets made simple. IEEE Transactions on Fuzzy Systems 10(2): 117–127.
6. Mizumoto M., Tanaka K. (1976) Some properties of fuzzy sets of type-2. Information and Control 31: 312–340.
7. Karnik N. N., Mendel J. M. (2001) Centroid of a type-2 fuzzy set. Information Sciences 132: 195–220.
8. Dubois D., Prade H. (1978) Operations on fuzzy numbers. International Journal of System Sciences 9: 613–626.
9. Dubois D., Prade H. (1979) Operations in a fuzzy-valued logic. Information and Control 43: 224–240.
10. Bustince H., Burillo P. (2000) Mathematical analysis of interval-valued fuzzy relations: Applications to approximate reasoning. Fuzzy Sets and Systems 113: 205–219.
11. Mabuchi S.(1979) An interpretation of membership functions and the properties of general probabilistic operators as fuzzy set operators II: Extension to three-valued and interval-valued fuzzy sets. Fuzzy Sets and Systems 92: 31–50.
12. Schwartz D. G. (1985) The case for an interval-based representations of linguistic truth. Fuzzy Sets and Systems 17: 153–165.
13. Turksen I. B. (1986) Interval-valued fuzzy sets based on normal forms. Fuzzy Sets and Systems 20: 191–210.
14. Nguyen H. T., Kreinovich V., Zuo Q. (1997) Interval-valued degrees of belief: Applications of interval computations to expert systems and intelligent control. International Journal of Uncertainty, Fuzziness, and Knowledge-Based Sysems 5: 317–385.
15. Zadeh L. A. (1975) Calculus of fuzzy restrictions. In: Zadeh L. A., Fu K. S., Tanaka K. and Shimura M. (eds.) Fuzzy Sets and Their Applications to Cognitive and Decision Processes. Academic, New York, USA.
16. Klir G. J., Yuan B. (1995) Fuzzy Sets and Fuzzy Logic: Theory and Applications. Prentice-Hall, Englewood Cliffs, NJ, USA.
17. Zadeh L. A. (1968) Probability measures of fuzzy events. Journal of Mathematical Analysis and Applications 23: 421–427.
18. Barwise J., Etchemendy J. (1987) The Liar. Oxford University Press, New York, USA.
19. Martin R. L. (1978) The Paradox of the Liar. Ridgeview Press, USA.
20. Martin R. L. (1984) Recent Essays on Truth and the Liar Paradox. Oxford University Press.
21. McGee V. (1991) Truth, Vagueness and Paradox. Hackett, Indianapolis, IN, USA.
22. Hajek P., Paris J., Shepherdson J. (2000) The liar paradox and fuzzy logic. The Journal of Symbolic Logic 65(1): 339–346.
23. Grim P. (1993) Self-reference and chaos in fuzzy logic. IEEE Transactions on Fuzzy Systems 1(4): 237–253.
24. Chen Y. H. (1999) A revisit to the liar. Journal of the Franklin Institute 336: 1023–1033.
25. Binmore K. (1992) Fun and Games. Heath and Company, Lexington, MA, USA.
26. Vezerides K., Kehagias A. (2003) The liar and related paradoxes: Fuzzy truth-value assignment for collections of self-referential sentences. Technical Report.
27. Mendel J. M. (2003) Fuzzy sets for words: A new beginning. Proceedings of 2003 IEEE FUZZ. pp. 37–42.

28. Wu H., Mendel J. M. (2002) Uncertainty bounds and their use in the design of type-2 fuzzy logic systems. IEEE Transactions on Fuzzy Systems 10(5): 622–639.
29. Dubois D., Prade H. (2005) Interval-valued fuzzy sets, possibility theory and imprecise probability. Proceedings of International Conference in Fuzzy Logic and Technology (EUSFLAT'05), Barcelona, Spain.
30. Dubois D., Prade H. (1987) Two-fold fuzzy sets and rough sets—some issues in knowledge representation. Fuzzy Sets and Systems 23: 3–18.
31. Zadeh L. A. (2005) Toward a generalized theory of uncertainty (GTU)—an outline. Information Sciences 172: 1–40.
32. Lucas C., Nadjar Araabi B. (1999) Generalization of the Dempster-Shafer theory: A fuzzy-valued measure. IEEE Transactions on Fuzzy Systems 7(3): 255–270.

28. Wu H, Mendel JM (2002) Uncertainty bounds and their use in the design of interval type-2 fuzzy logic systems. IEEE Transactions on Fuzzy Systems 10(5):622–639
29. Dubois D, Prade H (2005) Interval-valued fuzzy sets, possibility theory and imprecise probability. Proceedings of International Conference in Fuzzy Logic and Technology (EUSFLAT'05), Barcelona, Spain
30. Dubois D, Prade H (1987) Two-fold fuzzy sets and rough sets – some issues in knowledge representation. Fuzzy Sets and Systems 23:3–18
31. Zadeh LA (2005) Toward a generalized theory of uncertainty (GTU)—an outline. Information Sciences 172:1–40
32. Lucas C, Nadjar Araabi B (1999) Generalization of the Dempster-Shafer theory: A fuzzy-valued measure. IEEE Transactions on Fuzzy Systems 7(3):255–270

Part IV

Intelligent Mechatronics

Control of Differential Mode Harmonic Drive Systems

László Lemmer[1] and Bálint Kiss[2]

[1] Budapest University of Technology and Economics,
Department of Control Engineering and Information Technology
lemmer@seeger.iit.bme.hu

[2] Budapest University of Technology and Economics,
Department of Control Engineering and Information Technology
bkiss@iit.bme.hu

Summary. This paper reports the modeling and control of harmonic drives in differential gearing configuration where none of its shafts is fixed. This configuration makes it possible to solve a control problem where one applies two actuators in order to carry out simultaneous position and torque control on two different axes.

1 Introduction

Currently, harmonic drives in most applications are applied to provide high efficiency gearing without complex mechanisms and structures. Harmonic drives have high speed reduction and torque multiplication ratios using single stage and coaxial configuration of shafts. Other benefits include nearly zero backlash, small size, lightweight and high torque transmission capacity due to the high number of teethes in contact. The concept of the harmonic drive was patented by C.W. Musser in 1955 [1]. This new gear concept was applied in aerospace and other specific applications. Nowadays, applications and industry examples include but are not limited to robotics, machine tools, medical equipment and automotive industry.

The harmonic drive is made up of three basic components: the wave generator, the flexspline and the circular spline as depicted in Fig. 1. The wave generator is an elliptical cam enclosed in an antifriction ballbearing assembly. It is inserted into the bore of the externally toothed flexspline. The number of teethes on the flexspline is less than the number of teethes on the internally toothed circular spline but two. The flexspline is deformable and takes on the elliptical shape of the wave generator causing its external teeth to engage with the internal teeth of the circular spline at two opposite points hence the greater number of contacted teethes compared to traditional gears.

Rotation of the wave generator causes rotation of the flexspline relative to the circular spline in an opposite direction due to the difference in the number of teethes. Every harmonic drive is assigned to a transmission ratio that describes how many revolution of the wave generator causes one revolution in the flexspline relative to the circular spline. Usually, the circular spline is fixed and input is through the wave generator while output is via the flexspline that rotates N times slower than the wave generator in the opposite direction relative to the circular spline where the notation of the above mentioned transmission ratio is N. In a configuration

J.A.T. Machado et al. (eds.), *Intelligent Engineering Systems and Computational Cybernetics,*

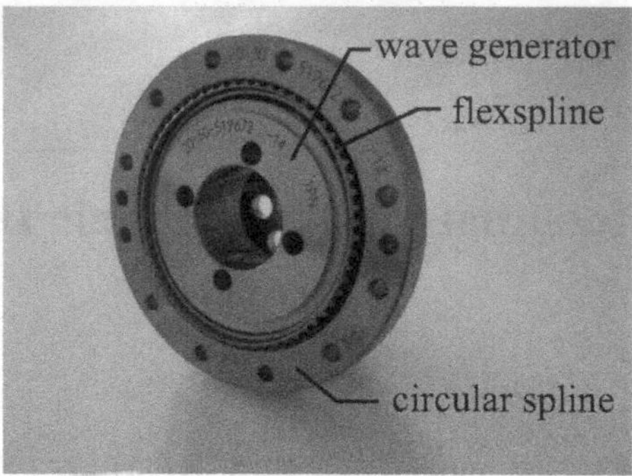

Fig. 1: Harmonic drive

where the flex spline is held stationary and the transmission of motion is from the wave generator to the circular spline, the circular spline rotates $N+1$ times slower as the wave generator in the same direction relative to the flexspline. Both of these driving configurations are so-called reduction gearing configurations. The modeling and control of harmonic drives in this configuration has a rich literature such as [2–4] including the consideration of the nonlinear compliance and torque transmission due to the nonrigid behavior of the flexspline. The position control of the harmonic drive is also addressed in several papers, e.g. in [5]. The authors of [6] report a robust controller synthesis approach whereas adaptive control is suggested in [7].

This paper discusses an alternative configuration of harmonic drives where none of the shafts is fixed i.e. a differential gearing configuration. In this case, the harmonic drive does not serve as a simple gear train device with two rotating shafts where one input shaft is actuated by a motor and the velocity or position of the output shaft subject to variable load is controlled. In our setup, there are two motors on two different axes and we control the angular position on one of the three axes and the torque on another axis. The motivation of this setup is its potential use in automotive applications as multifunctional actuators. The fact that none of the three axes are fixed requires a more general model which differs from the classic case. This paper derives this model and gives the control algorithm. Nonlinear torque transmission and compliance issues are not considered here.

The remaining part of the paper is organized as follows. The next section presents two approaches to modeling of a harmonic drive system. The first is based on the application of the constrained torque and the second is formulated using the generalized coordinates. The compensator design is detailed in Section 3. Simulation results and closed-loop measurements are presented in Section 4.

2 Dynamical Model

To give the mathematical model of the harmonic drive we consider the kinematical condition introduced by the geometry

$$\varphi_{\mathrm{wg}} - (N+1)\varphi_{\mathrm{cs}} + N\varphi_{\mathrm{fs}} + \varphi_0 = 0 \tag{1}$$

where φ_{wg}, φ_{cs}, φ_{fs} stand for the angular positions of the wave generator, circular spline and flexspline axes, respectively. The integer number N is the transmission ratio assigned to the harmonic drive. The constant φ_0 can introduce offset to the angular positions, e.g. in the case of relative angle measurement devices such as optical encoders. We will set it to zero without loss of generality in the sequel.

Let us consider the configuration space of the mechanical system that consists of the three shafts connected by the harmonic drive. The motion of the shafts is subject to the kinematical constraint (1) restricting its freedom of motion. As we have only this constraint the degrees of freedom of the system is two. Consequently, the configuration of the system with the three shafts subject to the kinematical condition (1) can be described uniquely by two independent configuration variables (q_1, q_2) that are referred to as generalized coordinates. Because the system is subject to geometrical constraint it is holonomic and it also is scleronomic since (1) does not contain the time explicitly. Notice that constraint (1) may be also interpreted as restricting the motion to a subspace of the vector space spanned by the shaft angles $(\varphi_{\mathrm{wg}}, \varphi_{\mathrm{cs}}, \varphi_{\mathrm{fs}})$, namely the plain orthogonal to the vector $\mathbf{n} = (1, -(N+1), N)^{\mathrm{T}}$.

We present two ways to assure the satisfaction of constraint (1). The first method introduces the vector of the so-called constraint torques that supplements the vector of external torques in such a manner that the dynamics evolves on the plane determined by (1). The second method eliminates φ_{fs}, the shaft angle of the flexspline and gives the dynamics in the space of the generalized coordinates chosen as the remaining two shaft angles.

2.1 Constraint Torques

Consider the dynamical equation (2) of three rotating shafts. The vector of their angular positions is denoted by φ.

$$H\ddot{\varphi} + f(\dot{\varphi}) = T. \tag{2}$$

H is the three-by-three nonsingular and constant diagonal matrix and it contains the equivalent shaft inertias. Term f includes the frictional forces. The variables $\ddot{\varphi}$, $\dot{\varphi}$ and T are vectors of the shaft angular accelerations, velocities and torques, respectively. Let us express $\ddot{\varphi}$ from (2)

$$\ddot{\varphi} = H^{-1}\left[T - f(\dot{\varphi})\right]. \tag{3}$$

Equation (1) implies that the shaft velocities and accelerations are tangent to the surface of normal $\mathbf{n}$. This means that the right hand side of (3) must satisfy

$$\left\langle H^{-1}\left[T - f(\dot{\varphi})\right], \mathbf{n}\right\rangle = 0 \tag{4}$$

where $\langle .,. \rangle$ denotes the standard scalar product. The vector of torques T may be decomposed as

$$T = T_{\mathrm{x}} + T_{\mathrm{c}}, \tag{5}$$

where T_{x} is the vector of the external torques and T_{c} is the vector of the constraint torques, which ensures that condition (1) holds true. From (4) and (5) we have

$$\left\langle H^{-1}T_{\mathrm{c}}, \mathbf{n}\right\rangle = -\left\langle H^{-1}\left[T_{\mathrm{x}} - f(\dot{\varphi})\right], \mathbf{n}\right\rangle. \tag{6}$$

Let us define T_{c} such that it is orthogonal to the surface determined by constraint (1) and does not introduce power to the system. So T_{c} is scalar multiple of $\mathbf{n}$

$$T_{\mathrm{c}} = \lambda\,\mathbf{n}. \tag{7}$$

Substituting (7) into (6) we obtain

$$\lambda = -\frac{\langle H^{-1}[T_x - f(\dot{\varphi})], \mathbf{n}\rangle}{\langle H^{-1}\mathbf{n}, \mathbf{n}\rangle} \tag{8}$$

and accordingly, T_c reads

$$T_c = -\frac{\langle H^{-1}[T_x - f(\dot{\varphi})], \mathbf{n}\rangle}{\langle H^{-1}\mathbf{n}, \mathbf{n}\rangle}\mathbf{n}. \tag{9}$$

The knowledge of the explicit value of the constraint torques is of practical interest if we want to consider the harmonic drive shafts separated and we want to determine the torques that affect these separated axes. Nevertheless, in order to describe the motion of the axes, it seems to be simpler to use generalized coordinates, which is done in the next subsection.

2.2 Model with Generalized Coordinates

We introduce the vector of the generalized coordinates $q = (q_1, q_2)^{\mathrm{T}}$. The relationship between the generalized coordinates and the shaft angles is given by the linear expression:

$$\varphi = Rq \tag{10}$$

Note that R involves the linear constraint (1). Moreover, since R is constant, one has $\dot{\varphi} = R\dot{q}$ and $\ddot{\varphi} = R\ddot{q}$.

To calculate the kinetic energy of the system, let us use (2) and (10) to obtain

$$K = \frac{1}{2}\langle H\dot{\varphi}, \dot{\varphi}\rangle = \frac{1}{2}\langle HR\dot{q}, R\dot{q}\rangle = \frac{1}{2}\left\langle R^{\mathrm{T}}HR\dot{q}, \dot{q}\right\rangle. \tag{11}$$

We can read out from (11) that the inertia matrix in the generalized coordinates reads

$$\tilde{H} = R^{\mathrm{T}}HR. \tag{12}$$

To see how the torques have to be transformed, we consider the expression of power, which is applied to the system

$$P = \langle T, \dot{\varphi}\rangle = \langle T, R\dot{q}\rangle = \left\langle R^{\mathrm{T}}T, \dot{q}\right\rangle. \tag{13}$$

From (13) we can read that the transformed vector of the torques is

$$\tilde{T} = R^{\mathrm{T}}T = R^{\mathrm{T}}T_x. \tag{14}$$

The second equation is implied by the fact that the constraint component T_c in (5) does not apply any power to the system because it is orthogonal to the motion. Now we can give the dynamical equation in generalized coordinates:

$$R^{\mathrm{T}}HR\ddot{q} + R^{\mathrm{T}}f(R\dot{q}) = R^{\mathrm{T}}T_x. \tag{15}$$

Based on the above considerations, let us give the dynamical equation of the harmonic drive system. In order to apply linear controller design techniques, we suppose only viscous friction. The abbreviations *wg*, *cs* and *fs* stand for wave generator, circular spline and flexspline, respectively. One may rewrite (2) as

$$H\begin{pmatrix}\ddot{\varphi}_{wg}\\ \ddot{\varphi}_{cs}\\ \ddot{\varphi}_{fs}\end{pmatrix} + D\begin{pmatrix}\dot{\varphi}_{wg}\\ \dot{\varphi}_{cs}\\ \dot{\varphi}_{fs}\end{pmatrix} = \begin{pmatrix}T_{wg}\\ T_{cs}\\ T_{fs}\end{pmatrix}, \tag{16}$$

where $H = \mathrm{diag}\left(J_{\mathrm{wg}}, J_{\mathrm{cs}}, J_{\mathrm{fs}}\right)$ and $D = \mathrm{diag}\left(d_{\mathrm{wg}}, d_{\mathrm{cs}}, d_{\mathrm{fs}}\right)$. The diagonal elements J_{wg}, J_{cs}, J_{fs} in H and d_{wg}, d_{cs}, d_{fs} in D are defined as the equivalent inertias and equivalent viscous damping coefficients reduced to the wave generator, the circular spline and the flexspline shafts, respectively. We introduce the generalized coordinates as $q = (\varphi_{\mathrm{wg}}, \varphi_{\mathrm{cs}})^{\mathrm{T}}$. Then (10) reads

$$\begin{pmatrix} \varphi_{\mathrm{wg}} \\ \varphi_{\mathrm{cs}} \\ \varphi_{\mathrm{fs}} \end{pmatrix} = \begin{bmatrix} 1 & 0 \\ 0 & 1 \\ -\frac{1}{N} & \frac{N+1}{N} \end{bmatrix} \begin{pmatrix} \varphi_{\mathrm{wg}} \\ \varphi_{\mathrm{cs}} \end{pmatrix} = Rq. \tag{17}$$

Accordingly, the dynamical equations of motion in the generalized coordinates read

$$\tilde{H}\ddot{q} + \tilde{D}\dot{q} = R^{\mathrm{T}} T_{\mathrm{x}} \tag{18}$$

where

$$\tilde{H} = R^{\mathrm{T}} H R = \begin{bmatrix} J_{\mathrm{wg}} + \frac{1}{N^2} J_{\mathrm{fs}} & -\frac{N+1}{N^2} J_{\mathrm{fs}} \\ -\frac{N+1}{N^2} J_{\mathrm{fs}} & J_{\mathrm{cs}} + \left(\frac{N+1}{N}\right)^2 J_{\mathrm{fs}} \end{bmatrix}, \tag{19}$$

$$\tilde{D} = R^{\mathrm{T}} D R = \begin{bmatrix} d_{\mathrm{wg}} + \frac{1}{N^2} d_{\mathrm{fs}} & -\frac{N+1}{N^2} d_{\mathrm{fs}} \\ -\frac{N+1}{N^2} d_{\mathrm{fs}} & d_{\mathrm{cs}} + \left(\frac{N+1}{N}\right)^2 d_{\mathrm{fs}} \end{bmatrix}, \tag{20}$$

and we used the fact that $R^{\mathrm{T}} f(R\dot{q}) = R^{\mathrm{T}} D R \dot{q}$. Note that one may obtain similar results using the Lagrangian subject to (1) and the Rayleigh's dissipation function to cope with the viscous friction terms.

3 Compensator Design

Figure 2 depicts the schematic view of the harmonic drive system with the inertia and friction elements on all three axes with most of the notations introduced so far. We actuate the wave generator and the circular spline with independent torque inputs provided by motors mounted to both axes. Angle sensors measure q.

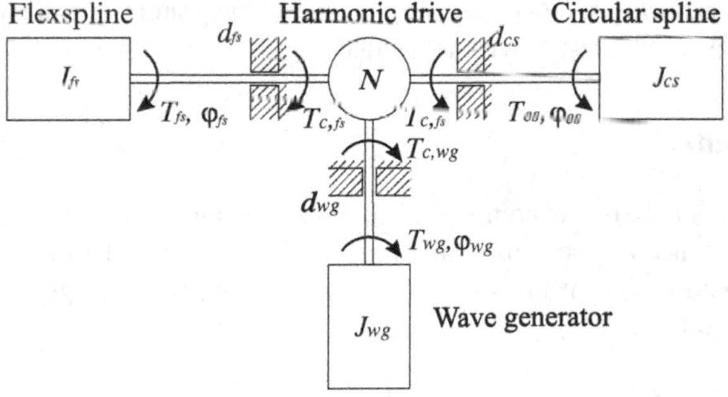

Fig. 2: Schematic view of the harmonic drive system

We assume external torque disturbances that will be denoted by subscript d. Subscript a stands for the actuator torque. The torque T_{fs} is considered as disturbance. The input vector then reads $T_{\mathrm{a}} = (T_{\mathrm{a,wg}}, T_{\mathrm{a,cs}})^{\mathrm{T}}$ and the vector of torque disturbances is $T_{\mathrm{d}} = (T_{\mathrm{d,wg}}, T_{\mathrm{d,cs}}, T_{\mathrm{fs}})^{\mathrm{T}}$

hence the vector of external torques in (2) reads $T_x = (T_{a,wg} + T_{d,cs}, T_{a,cs} + T_{d,cs}, T_{fs})^T$. Recall from Subsection 2.1 that the constraint torque $T_c = (T_{c,wg}, T_{c,cs}, T_{c,fs})^T$ is expressed by (9) corresponding to the geometric constraint on the shafts.

The controller has a twofold objective. First, the angular position $\varphi_{cs} = q_2$ of the circular spline shaft has to track a reference denoted by φ_{cs}^{ref}. Second, the constraint torque $T_{c,fs}$ has to follow a reference torque, which is denoted by $T_{c,fs}^{ref}$.

The disturbance torques are estimated with the help of a load estimator and are fed forward as well as the friction terms. The feed-forward terms and the linear feedback for the position control are described first. Then the tracking of the torque reference is presented in details. At last, the load observer is discussed.

3.1 Position Control

We suppose that $\dot{q}$ and T_d are observed that will be described in more details in Subsect. 3.3. The right side of (18) can be written as

$$R^T T_x = R^T \begin{pmatrix} T_{a,wg} + T_{d,wg} \\ T_{a,cs} + T_{d,cs} \\ T_{fs} \end{pmatrix} = T_a + R^T T_d. \tag{21}$$

Let us cancel the frictional effect and the disturbances:

$$T_a = T_a' + T_{a,f}, \tag{22}$$

where $T_a' = (T_{wg}', T_{a,cs}')^T$ and $T_{a,f} = \tilde{D}\dot{q} - R^T T_d$. We obtain

$$\tilde{H}\ddot{q} = T_a'. \tag{23}$$

Position control is required for the variable $q_2 = \varphi_{cs}$ hence we consider the second equation of (23) after pre-multiplication by the inverse of $\tilde{H}$:

$$\ddot{q}_2 = (0 \quad 1)\tilde{H}^{-1}T_a' = k_{11}T_{wg}' + k_{12}T_{a,cs}' = u, \tag{24}$$

for some real constants k_{11} and k_{12}. To control a double integrator, several methods are available (e.g. PID control or state-feedback techniques).

3.2 Torque Control

Note that (24) gives only one condition for the linear combination of the input signals $T_{a,wg}$ and $T_{a,cs}$. The reference for the constraint torque on the flexspline shaft is another condition. This torque, denoted by $T_{c,fs}$, is the last element of T_c. From (9), which gives T_c we obtain the following expression

$$\begin{pmatrix} T_{c,wg} \\ T_{c,cs} \\ T_{c,fs} \end{pmatrix} = -\frac{\langle H^{-1}[T_x - f(R\dot{q})], \mathbf{n}\rangle}{\langle H^{-1}\mathbf{n}, \mathbf{n}\rangle} \begin{pmatrix} 1 \\ -(N+1) \\ N \end{pmatrix}. \tag{25}$$

Let us take the last row of (25) and replace the frictional term with its expression

$$T_{c,fs} = -\frac{\langle H^{-1}[T_x - DR\dot{q}], \mathbf{n}\rangle}{\langle H^{-1}\mathbf{n}, \mathbf{n}\rangle} N. \tag{26}$$

Let us substitute $T_{\mathrm{c,fs}}^{\mathrm{ref}}$ for $T_{\mathrm{c,fs}}$ and, after some lengthy but elementary calculations involving expression (22), we obtain

$$\frac{1}{J_{\mathrm{wg}}}T'_{\mathrm{wg}} - \frac{N+1}{J_{\mathrm{cs}}}T'_{\mathrm{a,cs}} = \left\langle H^{-1}\left[DR\dot{q} - T_{\mathrm{d}} - \begin{pmatrix} T_{\mathrm{a,f}} \\ 0 \end{pmatrix} - \frac{T_{\mathrm{c,fs}}^{\mathrm{ref}}}{N}\mathbf{n}\right], \mathbf{n}\right\rangle. \tag{27}$$

Equations (24) and (27) uniquely define T'_{a}:

$$\begin{bmatrix} k_{11} & k_{12} \\ k_{21} & k_{22} \end{bmatrix} T'_{\mathrm{a}} = z, \tag{28}$$

where z is a vector of the right hand sides of (24) and (27) and $k_{21} = 1/J_{\mathrm{wg}}$, $k_{22} = -(N+1)/J_{\mathrm{cs}}$. The coefficient matrix in (28) is nonsingular so there exists the solution for T'_{a}. Since the elements of the coefficient matrix are constant its inverse can be computed offline. The solution of (28) has to be applied to (22) and we obtain the actuator torques in dependence of signal u from (24).

3.3 Observer Design

In the feed-forward terms we supposed the knowledge of $\dot{q}$ and T_{d} but normally, we only measure q and know the system input T_{a}. Consequently, $\dot{q}$ and T_{d} have to be observed. Nevertheless, for a two-degree-of-freedom system only two equivalent disturbing torques can be observed, namely two linear combinations of the three disturbing torques. Let us introduce

$$\tau_{\mathrm{d}} = R^{\mathrm{T}} T_{\mathrm{d}} \tag{29}$$

In order to obtain the values of the disturbing torques, one of them will be measured. Then from (29) we obtain

$$T_{\mathrm{d}} = \begin{bmatrix} R^{\mathrm{T}} \\ 0\ 0\ 1 \end{bmatrix}^{-1} \begin{pmatrix} \tau_{\mathrm{d}} \\ T_{\mathrm{fs}} \end{pmatrix} = \begin{bmatrix} 1 & 0 & \frac{1}{N} \\ 0 & 1 & -\frac{N+1}{N} \\ 0 & 0 & 1 \end{bmatrix} \begin{pmatrix} \tau_{\mathrm{d}} \\ T_{\mathrm{fs}} \end{pmatrix} \tag{30}$$

Equation (18) can be written as

$$\tilde{H}\ddot{q} + \tilde{D}\dot{q} = T_{\mathrm{a}} + \tau_{\mathrm{d}}. \tag{31}$$

From (31) the state equation of the extended system (with load change) can be given as

$$\dot{x}_{\mathrm{o}} = \begin{bmatrix} -\tilde{H}^{-1}\tilde{D} & 0 & \tilde{H}^{-1} \\ I & 0 & 0 \\ 0 & 0 & 0 \end{bmatrix} x_{\mathrm{o}} + \begin{bmatrix} \tilde{H}^{-1} \\ 0 \\ 0 \end{bmatrix} T_{\mathrm{a}}, \tag{32}$$

where $x_{\mathrm{o}} = (\dot{q}^{\mathrm{T}}, q^{\mathrm{T}}, \tau_{\mathrm{d}}^{\mathrm{T}})^{\mathrm{T}}$. The output equation reads

$$y = [0 \quad I \quad 0]x_{\mathrm{o}}. \tag{33}$$

The observability of this extended system can be easily proven checking the Kalman rank condition. For the model (32) and (33) a state observer can be designed with standard pole placement methods.

4 Closed-Loop Behavior

We analyze the closed-loop behavior with simulation and measurement data, as well. The measurements were performed on the rapid prototyping test bench setup of ThyssenKrupp R&D Institute Budapest, which will be presented in Subsection 4.2. The model parameters for controller design were identified with a simple LS technique, which is described in [8]. The identified parameters are given in Table 1 where d_{ij} stands for the element in the ith row and jth column of matrix $\tilde{D}$. The control input u of the position loop in (24) was produced with a pole placement method. We placed the poles of the double integrator to -20 and the pairs of poles of the state-observer designed for the extended system (32), to -50.0, -50.1 and -50.2, respectively.

Table 1: Parameter values

Parameter	Value	Unit
J_{wg}	130.0000	kg mm^2
J_{cs}	10,250.3400	kg mm^2
J_{fs}	10,297.4100	kg mm^2
d_{11}	0.0005	Nm s/rad
$d_{12} = d_{21}$	-0.0004	Nm s/rad
d_{22}	0.0081	Nm s/rad
N	50	1

4.1 Simulation

Figure 3 presents the simulation of the step response for the closed-loop system. The references change at 1 s from 0 to 1. We achieved the decoupling of the harmonic drive system and the results of the simulation proves that the derived control algorithm allows us to control the angular position of the circular spline axis and the constraint torque on the flexspline axis independently. Notice that the flexspline constraint torque can follow the reference without dynamics.

4.2 Measurement

We performed measurements on the rapid prototyping test bench setup of ThyssenKrupp R&D Institute Budapest (Fig. 4). This setup allows us to actuate the harmonic drive at any of the three shaft. The motor mounted to the wave generator shaft can apply a maximum torque of 1.3 Nm in both directions and the motors mounted to the other two shafts are limited to ± 8 Nm. The resolution of the flexspline and circular spline shaft angle encoder is 8,192 per turn and the resolution of the wave generator shaft angle encoder is 3,520 per turn. The sampling time during the measurement was 1 ms.

Figure 5 shows the measurements of the closed-loop step responses. The output φ_{cs} behaves in a similar way as in the simulation results. For the output $T_{c,fs}$, the steady-state values are as respected but the transients are high, especially the transient in the step response from φ_{cs}^{ref} to $T_{c,fs}$. Notice that there is no differential filter between the inputs and the torque output that makes this output extremely sensitive to disturbances and parameter deviations.

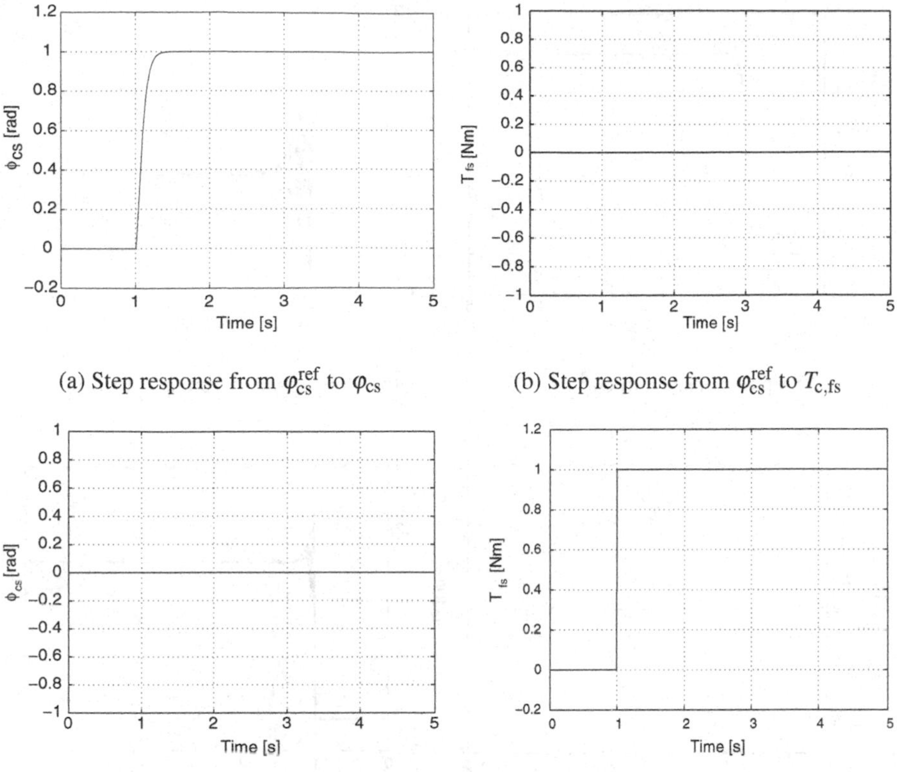

(a) Step response from φ_{cs}^{ref} to φ_{cs}

(b) Step response from φ_{cs}^{ref} to $T_{c,fs}$

(c) Step response from $T_{c,fs}^{ref}$ to φ_{cs}

(d) Step response from $T_{c,fs}^{ref}$ to $T_{c,fs}$

Fig. 3: Simulation of the step response of the closed-loop system

Fig. 4: Rapid prototyping test bench at ThyssenKrupp R&D Institute Budapest

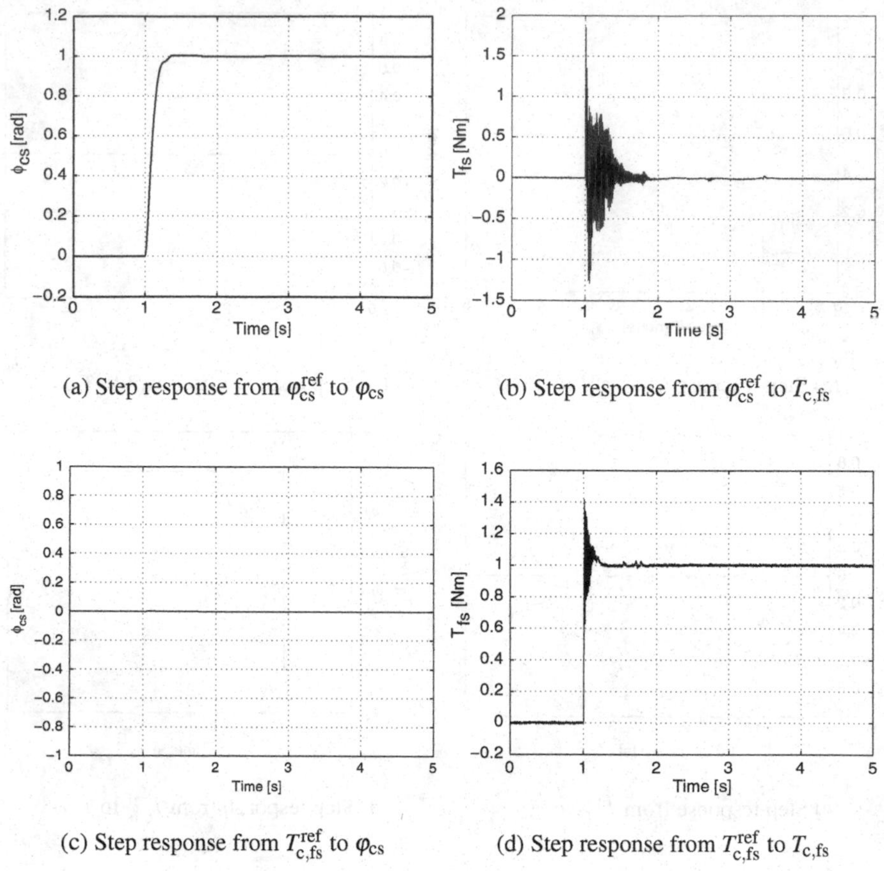

(a) Step response from $\varphi_{\text{cs}}^{\text{ref}}$ to φ_{cs}

(b) Step response from $\varphi_{\text{cs}}^{\text{ref}}$ to $T_{\text{c,fs}}$

(c) Step response from $T_{\text{c,fs}}^{\text{ref}}$ to φ_{cs}

(d) Step response from $T_{\text{c,fs}}^{\text{ref}}$ to $T_{\text{c,fs}}$

Fig. 5: Measurement of the closed-loop step response

The difference between the simulation results and the measurement data suggests that further improvement could be achieved if we considered compliance issues as well. On the other hand, a robustness analysis would also be reasonable.

5 Conclusions

The paper reported the three port modeling of harmonic drives. A control strategy were proposed where the simultaneous tracking control for the angular position of the circular spline shaft and the constraint torque on the flexspline axis are assured using two actuators. This setup may have applications in the automotive industry and may contribute to the development of new intelligent mechatronic actuators featuring harmonic drives.

Acknowledgments

This research was partially founded by the Advanced Vehicles and Vehicle Control Knowledge Center under grant RET 04/2004.

References

1. Musser C W (1995) Strain Wave Gearing. US Patent 2 906 143
2. Tuttle T D (1992) Understanding and Modeling the Behavior of a Harmonic Drive Gear Transmission. M.Sc. thesis, Massachusetts Institute of Technology, Cambridge, MA
3. Tuttle T D, Seering W P (1996) A Nonlinear Model of a Harmonic Drive Gear Transmission. IEEE Transactions on Robotics Automation 12:368–374
4. Taghirad H D, Bélanger P R (1998) Modeling and Parameter Identification of Harmonic Drive Systems. ASME Journal of Dynamic Systems, Measurement and Control 120:439–444
5. Gandhi P S, Ghorbel F H (2002) Closed-Loop Compensation of Kinematic Error in Harmonic Drives for Precision Control Applications. IEEE Transactions on Control Systems Technology 10:759–768
6. Taghirad H D, Bélanger P R (2001) H_∞-Based Robust Torque Control of Harmonic Drive Systems. ASME Journal of Dynamic Systems, Measurement and Control 123:338–345
7. Zhu W H, Doyon M (2004) Adaptive Control of Harmonic Drives. 43rd IEEE Conference on Decision and Control 3:2604–2608, Bahamas
8. Lemmer L, Kiss B (2006) Modeling, Identification, and Control of Harmonic Drives for Mobile Vehicles. IEEE 3rd International Conference on Mechatronics 369–374, Budapest

Intelligent Control of an Inverted Pendulum

Webjorn Rekdalsbakken

Aalesund University College, Institute of Technology and Nautical Science
N-6025 Aalesund, Norway
wr@hials.no

Summary. An inverted pendulum represents an unstable system which is excellent for demonstrating the use of feedback control with different kinds of control strategies. In this work state feedback of the inverted pendulum is examined. First a pole placement algorithm is explored. After that artificial intelligence (AI) methods are investigated to better cope with the nonlinearities of the physical model. The technique used is based on a hybrid system combining a neural network (NN) with a genetic algorithm (GA). The NN controller is trained by the GA against the behaviour of the physical model. The results of the training process show that the chromosome population tends to station at a suboptimal level, and that changes in the environmental parameters have to take place to reach a new optimal level. By systematically changing these parameters the NN controller will gradually adapt to the pendulum behaviour.

1 Introduction

At Aalesund University College (AUC) the control of an inverted pendulum has been a standard laboratory exercise for many years. Several physical models have been built, and many control strategies have been explored. In recent time the control focus has been on AI methods, like fuzzy logic (FL), neural networks (NN) and genetic algorithms (GA). Hybrid combinations of these methods are also very relevant. All physical models of the inverted pendulum will have inherent nonlinearities like friction and dead bands. These characteristics represent obstacles to the control system, which are not always simple to overcome. In such cases self-learning, adaptive control methods represent a way to deal with the control problem. In this work a hybrid control system combining a NN controller with a GA is investigated in an attempt to manage these challenges. These problems are common to a large group of control systems, and thus the solutions may be of comparative value to other systems.

2 Physical System

The physical system consists of a rod placed on a cart driven by a rubber belt. An AC servo drive with a motor of 400 W is controlling the motion of the belt. The servo drive is programmable to either control the shaft speed or torque. The system state variables are measured with a camera in front of the pendulum. The relevant state variables are the position and

J.A.T. Machado et al. (eds.), *Intelligent Engineering Systems and Computational Cybernetics,*

speed of the cart, and the angle and angular velocity of the pendulum. All these states can be measured in real-time with the camera through a frame grabber card. Thus full-order state feedback of the system can be performed. The integrals of the cart position and the pendulum angle can be estimated for use in the feedback control to obtain zero stationary error of the cart position. The software of the control system is implemented with C++. Descriptions of the system components and software are found in [1, 2]. The physical model including the camera is shown in Fig. 1.

3 Mathematical Model

A mathematical model of the system is derived on basis of classic Newtonian mechanics. A sketch of the system is shown in Fig. 2. With reference to Fig. 2, the acceleration of the mass centre, C of the pendulum can be expressed by the radial and tangential component vectors:

$$\mathbf{a}_C(t) = \mathbf{a}_P + \mathbf{a}_t + \mathbf{a}_r \tag{1}$$

In (1) $\mathbf{a}_P$ is the acceleration of the pendulum connection point, P and $\mathbf{a}_t$ and $\mathbf{a}_r$ are the tangential and radial acceleration of the mass centre, C of the pendulum. When expressed by the position vector, $\mathbf{r}_{PC}$ and the angular velocity, ω the acceleration will be given as follows:

$$\mathbf{a}_C(t) = \mathbf{a}_P + \varepsilon \times \mathbf{r}_{PC} + \omega \times (\omega \times \mathbf{r}_{PC}) \tag{2}$$

With reference to Fig. 1, the vectors are expressed on component form:

$$r_{PC} = \begin{bmatrix} -l\sin\theta & l\cos\theta & 0 \end{bmatrix}, \quad \omega = \begin{bmatrix} 0 & 0 & \dot{\theta} \end{bmatrix}, \quad \varepsilon = \begin{bmatrix} 0 & 0 & \ddot{\theta} \end{bmatrix}$$

The acceleration of the mass centre of the pendulum on component form, now becomes:

$$\mathbf{a}_C = \begin{bmatrix} \ddot{x} - \ddot{\theta} l\cos\theta + \dot{\theta}^2 l\sin\theta & -\ddot{\theta} l\sin\theta - \dot{\theta}^2 l\cos\theta & 0 \end{bmatrix} \tag{3}$$

The torque balance of the pendulum about the connection point P is expressed as follows:

$$\sum \mathbf{M}_P = \mathbf{r}_{PC} \times \mathbf{G} = I_P\varepsilon + (\mathbf{r}_{PC} \times m\mathbf{a}_P) \tag{4}$$

Here I_P is the moment of inertia of the pendulum about the connection point, and the force of gravitation is given by $\mathbf{G} = [0 \;\; -mg \;\; 0]$. On component form 4 is expressed like this:

$$[\,0 \;\; 0 \;\; mgl\sin\theta\,] = [\,0 \;\; 0 \;\; I_P\ddot{\theta}\,] + [\,0 \;\; 0 \;\; -ml\ddot{x}\cos\theta\,] \tag{5}$$

The z-component of 5 in combination with Steiner's formula, $I_P = I_C + ml^2$ results in the following equation:

$$\left(I_C + ml^2\right)\ddot{\theta} - mgl\sin\theta - ml\ddot{x}\cos\theta = 0 \tag{6}$$

The forces on the coupled system in the x-direction is expressed as:

$$F_x - b\dot{x} - ma_x = M\ddot{x} \tag{7}$$

Here b is the coefficient of friction of the rubber belt against the supporting rails. In combination with 3 this gives the equation:

$$(M+m)\ddot{x} + b\dot{x} - ml\ddot{\theta}\cos(\theta) + ml\dot{\theta}^2\sin(\theta) = F_x \tag{8}$$

Equations 6 and 8 now represent the dynamic description of the system. For mathematical modelling and modal control of the pendulum, see [3].

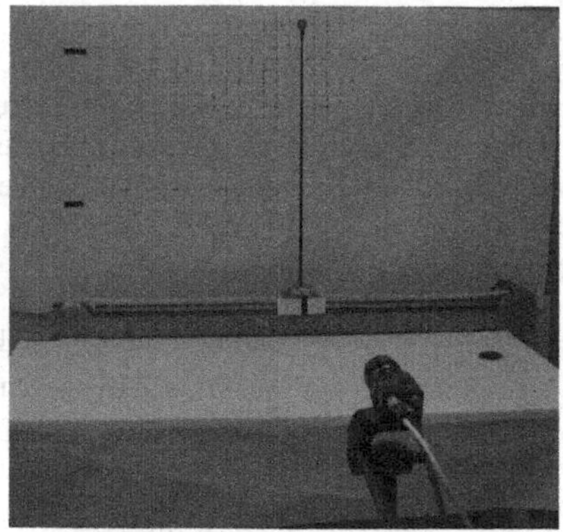

Fig. 1: The physical inverted pendulum

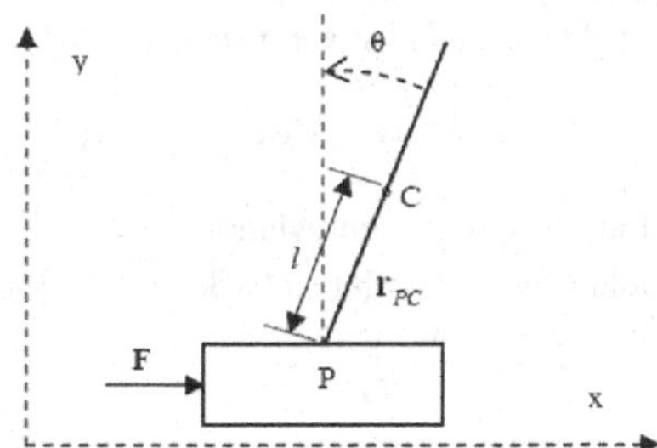

Fig. 2: A sketch of the pendulum

4 Feedback Control of the Pendulum

4.1 State Space Model

Linear approximations about the vertical position of the pendulum are derived for 6 and 8. By using the moment of inertia for a uniform rod, $I_C = \frac{ml^2}{3}$ this gives the following equations:

$$\frac{4l}{3}\ddot{\theta} - g\theta - \ddot{x} = 0 \tag{9}$$

$$(M+m)\ddot{x} + b\dot{x} - ml\ddot{\theta} = F_x \tag{10}$$

To decompose the equations into a state space model it is necessary to separate the highest derivatives. This is done by linear combinations of 9 and 10, which gives:

$$\frac{4M+m}{4}\ddot{x} + b\dot{x} - \frac{3mg}{4}\theta = F_x \tag{11}$$

$$\frac{(4M+m)\,l}{3}\ddot{\theta} + b\dot{x} - (M+m)\,g\theta = F_x \tag{12}$$

Equations (11) and (12) are reduced to a system of first order equations in the state variables $z_1 = x$, $z_2 = \dot{x}$, $z_3 = \theta$, $z_4 = \dot{\theta}$. The force F_x on the cart is the input signal to the system. For torque control of the motor one has to use the relation $F_x = \frac{T_m}{R_m}$, where R_m is the radius of the motor shaft and $u = T_m$ is the control signal to the system. The state space model of the fourth order system will be as follows:

$$\dot{z} = \begin{bmatrix} 0 & 1 & 0 & 0 \\ 0 & -\frac{4b}{4M+m} & \frac{3mg}{4M+m} & 0 \\ 0 & 0 & 0 & 1 \\ 0 & -\frac{3b}{(4M+m)\,l} & \frac{3\,(M+m)\,g}{(4M+m)\,l} & 0 \end{bmatrix} \cdot z + \begin{bmatrix} 0 \\ \frac{4}{(4M+m)\,R_m} \\ 0 \\ \frac{3}{(4M+m)\,lR_m} \end{bmatrix} \cdot u \tag{13}$$

4.2 Modal Control

In the feedback control of the system a fifth state variable, z_5 is defined to obtain the integration of the cart position. It is expressed as $\dot{z}_5 = r - z_1$, where r is the reference position of the cart. For a pole placement strategy there has to be defined a control law from the five state variables:

$$u = -G \cdot z + g_5 z_5 = -\begin{bmatrix} g_1 & g_2 & g_3 & g_4 & -g_5 \end{bmatrix} \cdot \begin{bmatrix} z_1 & z_2 & z_3 & z_4 & z_5 \end{bmatrix}^T \quad (14)$$

Introducing the abbreviations, $A = \frac{4}{4M+m}$, $B = \frac{3}{(4M+m)l}$, $C = \frac{3mg}{4}$, and $D = (M+m)g$, the pendulum system with feedback according to the control law in (14) is expressed by the state space model below:

$$\dot{z} = \begin{bmatrix} 0 & 1 & 0 & 0 & 0 \\ -\frac{A}{R_m}g_1 & -Ab - \frac{A}{R_m}g_2 & AC - \frac{A}{R_m}g_3 & -\frac{A}{R_m}g_4 & \frac{A}{R_m}g_5 \\ 0 & 0 & 0 & 1 & 0 \\ -\frac{B}{R_m}g_1 & -Bb - \frac{B}{R_m}g_2 & BD - \frac{B}{R_m}g_3 & -\frac{B}{R_m}g_4 & \frac{B}{R_m}g_5 \\ -1 & 0 & 0 & 0 & 0 \end{bmatrix} \cdot z + \begin{bmatrix} 0 \\ 0 \\ 0 \\ 0 \\ 1 \end{bmatrix} \cdot r \quad (15)$$

The cart reference position is the input signal, and the motor torque is generated by the demand of zero errors in cart position and pendulum angle. The control problem is to decide where to place the five poles of the controlled system, and to derive the corresponding **G**-matrix. Starting with a Butterworth polynomial and an estimated system bandwidth the poles of the system were calculated to be:

$$s_1 = -1 + j\sqrt{2}, \; s_2 = -1 - j\sqrt{2}, \; s_3 = s_4 = s_5 = -3 \quad (16)$$

The **G**-matrix was derived with Ackermann's formula to be:

$$\mathbf{G} = \begin{bmatrix} -0.2584 & -0.3860 & 1.4643 & 0.3213 & -0.1551 \end{bmatrix} \quad (17)$$

The dynamics of the controlled system according to 15 is shown by the diagram in Fig. 3. The response of the simulated system to a step in the position of the cart is shown in Fig. 4, where the motor torque and all the state variables are shown. Position is given in *metres* and angle in *radians*.

4.3 Measurement of the State Variables

A camera was used to measure the state variables of the system. The camera configuration is described in [2, 4]. The camera acquires up to 30 monochromatic pictures each second with a resolution of 8 bits. The matrix size is 768 x 576 pixels. The picture matrix is scaled in metres by using a linear gauge on the pendulum background. The colour of the cart and the pendulum is dark while the background is white. In the picture matrix three horizontal lines are selected. The lowest line is used to localize the centre of the cart in the x-direction. The two uppermost lines are used to calculate the angle of the pendulum from the vertical line. The remaining state variables are derived from succeeding measurements of the cart position and the pendulum angle by numeric differentiation and integration.

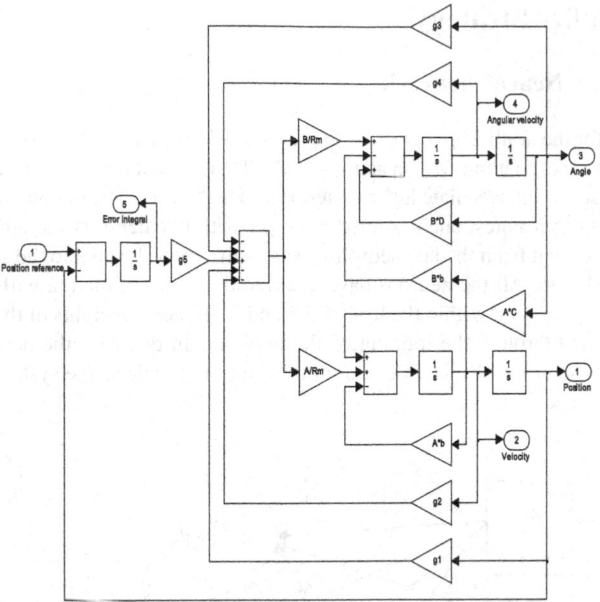

Fig. 3: Control system with state feedback

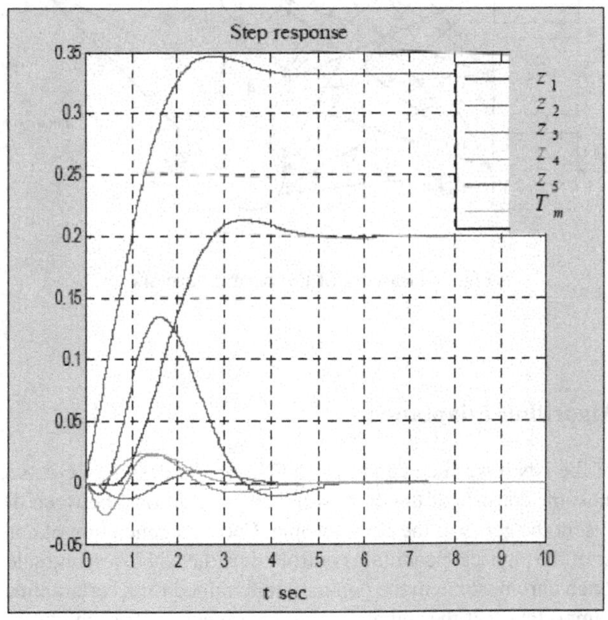

Fig. 4: System step response

5 AI Controller Strategy

5.1 Modelling a Neural Network

To compensate for the nonlinear friction of the rubber belt, it was decided to try an adaptive AI method. The choice of method was an adaptive NN. To represent a nonlinear function, the NN must have at least one intermediate layer of neurons. The NN consists of one input layer representing the five system states, one intermediate layer with five neurons and an output layer of one neuron. The output from the last neuron serves as the control signal to the servo drive. The NN is shown in Fig. 5. All the neurons have a sigmoid response function with a bias of -0.5, restricting the output signal within the limits -0.5 and 0.5. The 30 weights of the neuron inputs must be determined through the training of the network. In this case the network is trained against the response of the physical pendulum to adapt to the role as the system controller; on the design of NN, see [5, 6].

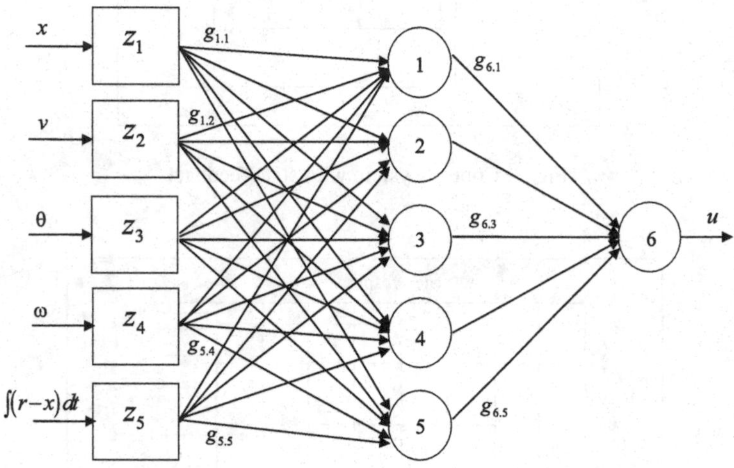

Fig. 5: Drawing of the neural network

5.2 Genetic Algorithm Adaptation

In the training of the NN it was decided to use a GA; on the design of GA, see [5, 7]. In this case each chromosome consists of the thirty weights (connections) between the neurons. The six neurons represent the genes of the chromosome. For each generation of chromosomes one uses the response of the physical pendulum controlled by the NN to test and select among these chromosomes. Each chromosome in the population is trained for a certain time (an epoch). In this testing it is important that the initial conditions are equal for each chromosome. This is obtained by starting the epoch with the pendulum in the vertical position with the cart at a given x-position. The pendulum is first controlled by the modal regulator to stand in the upright position at the reference point. The regulator is then turned off, and the NN is taking over the control of the pendulum. During the epoch the control software restricts the pendulum angle to be within 15 degrees of the vertical position and the x-position to be within 0.35 m from the reference point, for the control signal to be active. A crucial point in any adaptive technique is the optimization criterion. In this experiment it is chosen to minimize a weighted sum of the running errors in cart position and pendulum angle during one epoch. The fitness factors for the chromosomes are calculated by the formulae shown below:

$$Q = \frac{1}{\sum_{t=0}^{t=T_E/\Delta t} \left((\Delta x)^2 + 2\,(\Delta \theta)^2 \right) \Delta t} \tag{18}$$

The sum in denominator of (18) runs from the start to the end, of one training epoch, and T_E is the cycle time of the control loop. Different values for the weights of the quadric errors were tried, but this gave only small differences in the relative magnitude of the fitness values of the chromosomes. All of the chromosomes in the current population are successively tested in this way. The total fitness is accumulated for the whole population, and the relative fitness is calculated for each of the chromosomes. A new population is generated from the best fitted chromosomes of the current generation.

5.3 The Hybrid Control System

In this experiment the NN serves the role of the regulator in the control loop. The NN is trained by the GA to obtain the optimum weights on basis of the criterion in 18. In the initial population of chromosomes the gene values are allocated randomly within given limits. The selection process of the chromosomes is based on the "Roulette Wheel" technique using the relative fitness ratio obtained in the test as each chromosome's share of the wheel area. In the forming of the new population both crossover and mutation of the genes are taking place. A random selection of 70% of the new generation of chromosomes is chosen for crossover. The crossover of two chromosomes takes place between genes located from a randomly chosen site of the chromosomes. Each chromosome of the new generation has a 10% chance of being exposed to a mutation. In Table 1 is a summary of the different parameters used in the GA.

Table 1: Parameters used in the genetic algorithm

Name	Value
No. of chromosomes in population	10
No. of genes in chromosome	6
No. of variables in each gene	5
Initial values of gene variables	Random no between -5 and 5
Epoch time	15 sec
Sample time	30 msec
Probability for crossover	70%
Probability for small mutations	10%
Probability for large mutations	1%
Change in small mutation	Random no between -1 and 1
Change in large mutation	Random no between -10 and 10

6 Results and Discussion

The training of the chromosomes was done off-line. The training process is indeed a time consuming process. First the pendulum must be made to balance with the modal controller and then a chromosome can be tested. With an epoch time of 15 sec. the average fitness showed very little improvement from generation to generation. This is presented in Fig. 6. After 22 generations it was decided to lower the epoch time to 8 sec, which was observed to be in better accordance with the time the NN controller actually managed to balance the pendulum. With the intention to get a more steady progress of the chromosome fitness, the mutation probabilities were also reduced to one tenth of the original values, both for small and large mutations. Figure 7 shows the average fitness for generations 23 to 75. There is very little sign of a general improvement in the fitness factor. It looks like the chromosomes have been stationed in a suboptimal region. It was therefore decided to further reduce the epoch time to 4 sec. This gave, however, still no tendency to improvement in the fitness factor for the next 50 generations. The only apparent way to leave such a suboptimal region was to raise the mutation probability, so accordingly the probabilities for both small and large mutations were raised to 20%. Chromosome generations from 130 to 200 were now trained under these conditions. The results show that the average fitness has stabilised in a somewhat higher suboptimal region. It was noticed that in generation 192 there was a chromosome with an especially high relative fitness value. This chromosome was explored in the NN controller and showed promising tendencies in balancing the pendulum. In the further testing, all the chromosomes in the generation were set equal to this special chromosome, i. e. the population was a clone of this chromosome. At the same time the mutation probability for small mutations was set equal to 30%, while that of large mutations was set to 0%. Ten new generations were tested, and the average fitness was raised to a higher level. The average fitness of the generations from 76 to 211 is shown in Fig. 8.

7 Conclusions

The many parameters involved in the GA made it difficult to carry through systematic research in a practical experiment. During the testing many promising chromosomes showed up, but the average fitness value only seemed to be raised when some principal condition to the population was changed, e.g. the mutation rate. The best chromosomes of the last generation were tested in the NN controller. Some of the chromosomes were able to balance the pendulum, but none of them any better than the modal control algorithm. The main problem seems to be that the population reaches a local suboptimum level after some generations. This level seems to be inherently stable, and it is necessary to change some of the system environmental conditions, to leave this suboptimum. A complete search should therefore gradually change all test parameters, such as mutation probability and epoch time, to be sure to reach a global optimum point. This is, however, very difficult without an automated test procedure. One major difference between this kind of real-time laboratory experiment and the evolution processes in nature is that nature has a lot of machines to run its chromosomes in parallel. Those machines not optimally fitted to the demands, will just vanish. In the lab, however, there is only one machine. The challenge is to design an automatic adaptive system so that testing and optimization can continue while the system is up and running. This task is quite possible if one can find a starting chromosome that is able to balance the pendulum, and then design an algorithm that systematically tests each influential parameter to eventually obtain the optimal chromosome.

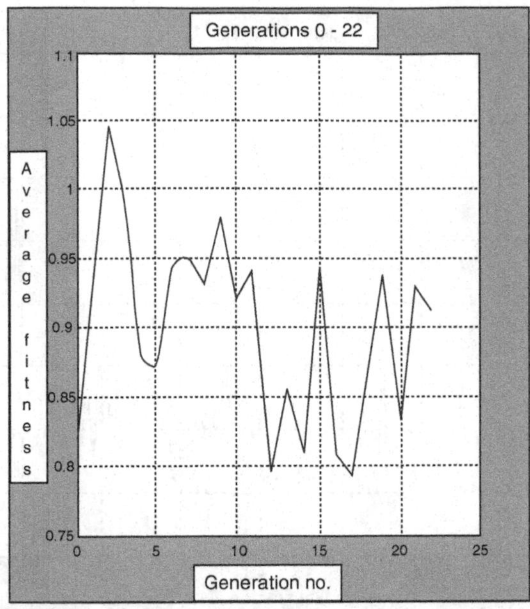

Fig. 6: The average fitness as a function of the generations from 0 to 22

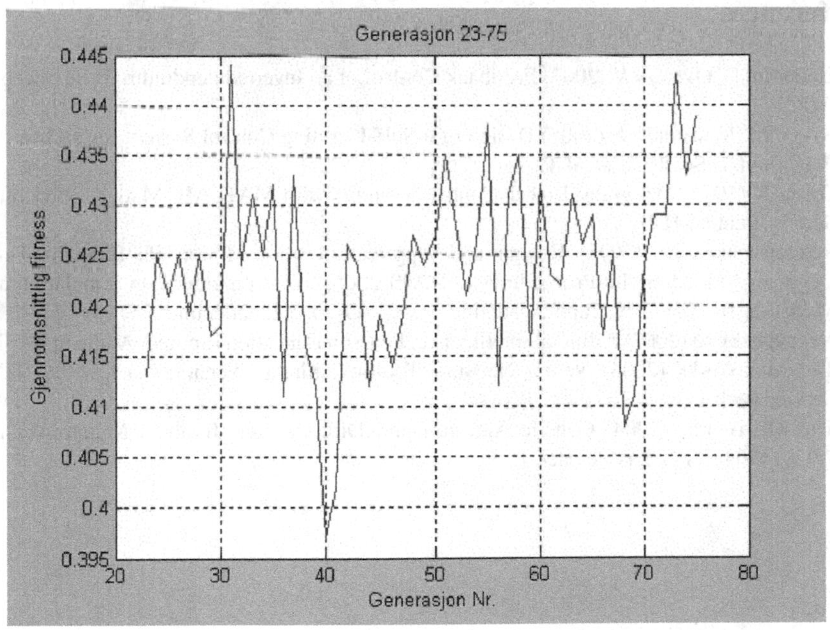

Fig. 7: The average fitness as a function of the generations from 23 to 75

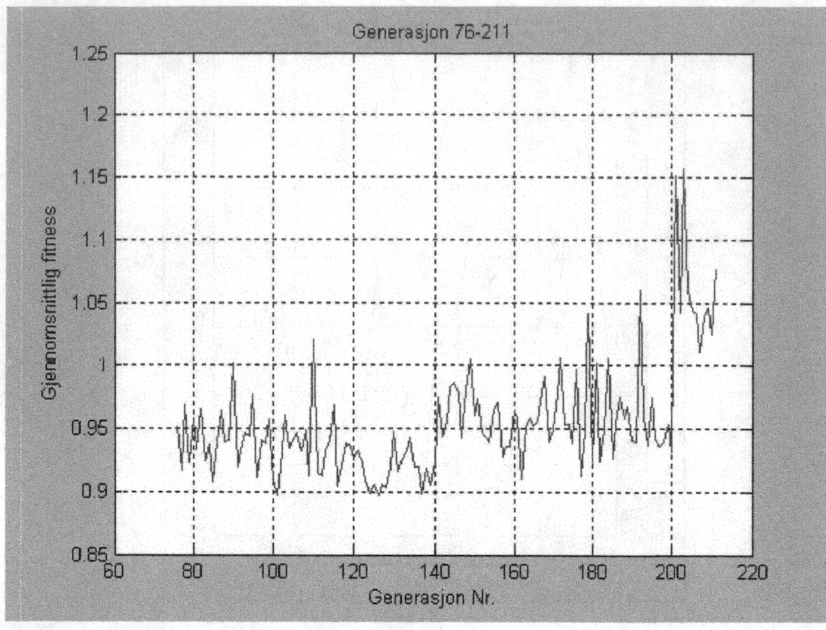

Fig. 8: The average fitness as a function of the generations from 76 to 211

References

1. Undertun O, Overaas V (2003) Feedback Control of an Inverted Pendulum. B.Sc. thesis at AUC
2. Salen IM, Voldsund TA (2005) Design of a Self-Learning Control System for an Inverted Pendulum. B.Sc. thesis at AUC.
3. Ogata K (1994) Designing Linear Control Systems with MATLAB. Matlab Curriculum Series. Prentice-Hall
4. Rekdalsbakken W (2005) Design and Application of a Motion Platform in Three Degrees of Freedom. In: Proceedings of SIMS 2005, 46th Conference on Simulation and Modelling, pp 269–279, Tapir Academic Press, NO-7005 Trondheim
5. Negnevitsky M (2002) Artificial Intelligence. Pearson Education Limited. Addison-Wesley
6. The MathWorks (2004) Neural Network Toolbox. Matlab Version 7.0.1.24704 (R14) Service pack 1
7. The MathWorks (2004) Genetic Algoritm and Direct search Toolbox. Matlab Version 7.0.1.24704 (R14) Service pack 1

Tuning and Application of Integer and Fractional Order PID Controllers

Ramiro S. Barbosa, Manuel F. Silva, and J. A. Tenreiro Machado

Department of Electrotechnical Engineering
Institute of Engineering of Porto
Rua Dr. António Bernardino de Almeida, 431
4200-072 Porto, Portugal
{rsb,mss,jtm}@isep.ipp.pt

Summary. Fractional calculus (FC) is widely used in most areas of science and engineering, being recognized its ability to yield a superior modeling and control in many dynamical systems. In this perspective, this article illustrates two applications of FC in the area of control systems. Firstly, is presented a methodology of tuning PID controllers that gives closed-loop systems robust to gain variations. After, a fractional-order PID controller is proposed for the control of an hexapod robot with three dof legs. In both cases, it is demonstrated the system's superior performance by using the FC concepts.

1 Introduction

The fractional calculus (FC) has been adopted in many areas of science and engineering [1, 2, 6], enabling the discovery of exciting new methodologies and the extension of several classical results. In what concerns the area of automatic control, the fractional-order algorithms are extensively investigated. Podlubny [3, 4] proposed a generalization of the PID scheme, the so-called PI$^{\lambda}$D$^{\mu}$ controller, involving an integrator of order $\lambda \in \Re^{+}$ and differentiator of order $\mu \in \Re^{+}$. The transfer function $G_c(s)$ of such a controller is:

$$G_c(s) = \frac{U(s)}{E(s)} = K_P + K_I s^{-\lambda} + K_D s^{\mu}, \quad (\lambda, \mu > 0) \tag{1}$$

where $E(s)$ is the error signal, and $U(s)$ is controller's output. The parameters (K_P, K_I, K_D) are the proportional, integral, and derivative gains of the controller, respectively.

The PI$^{\lambda}$D$^{\mu}$ algorithm is represented by a fractional integro-differential equation of type [4]:

$$u(t) = K_P e(t) + K_I D^{-\lambda} e(t) + K_D D^{\mu} e(t) \tag{2}$$

Clearly, depending on the values of the orders λ and μ, we get an infinite number of choices for controller's type (defined through the (λ, μ)-plane). For instance, taking $(\lambda, \mu) \equiv (1, 1)$ yields the classical PID controller. Moreover, $(\lambda, \mu) \equiv (1, 0)$ leads to the PI controller, $(\lambda, \mu) \equiv (0, 1)$ to the PD controller, and $(\lambda, \mu) \equiv (0, 0)$ to the P controller.

J.A.T. Machado et al. (eds.), *Intelligent Engineering Systems and Computational Cybernetics*,

All these classical types of PID controllers are particular cases of the $PI^{\lambda}D^{\mu}$ algorithm (1). However, the $PI^{\lambda}D^{\mu}$ controller is more flexible and gives the possibility of adjusting more carefully the dynamical properties of the closed-loop system.

For the definition of the generalized operator ${}_aD_t^{\alpha}$ $(\alpha \in \Re)$, where a and t are the limits and α the order of operation, we usually adopted the Riemann–Liouville (RL) and the Grünwald–Letnikov (GL) definitions. The RL definition is given by $(\alpha > 0)$:

$$ {}_aD_t^{\alpha}x(t) = \frac{1}{\Gamma(n-\alpha)}\frac{d^n}{dt^n}\int_a^t \frac{x(\tau)}{(t-\tau)^{\alpha-n+1}}d\tau, \quad (n-1<\alpha<n) \tag{3} $$

where $\Gamma(x)$ represents the Gamma function of x. The GL definition is $(\alpha \in \Re)$:

$$ {}_aD_t^{\alpha}x(t) = \lim_{h\to 0}\frac{1}{h^{\alpha}}\sum_{k=0}^{\left[\frac{t-a}{h}\right]}(-1)^k\binom{\alpha}{k}x(t-kh) \tag{4} $$

where h is the time increment and $[x]$ means the integer part of x. As indicated above, the definition (4) is valid for $\alpha > 0$ (fractional derivative) and for $\alpha < 0$ (fractional integral) and, commonly, these two notions are grouped into one single operator called *differintegral*. Moreover, expressions (3) and (4) show that the fractional-order operators are *global* operators having a memory of all past events, making them adequate for modeling hereditary and memory effects in most materials and systems.

The Laplace transform (L) of the fractional derivative defined by (3) has the form [3,4]:

$$ L\{{}_0D_t^{\alpha}x(t)\} = s^{\alpha}X(s) - \sum_{k=0}^{n-1}s^k\,{}_0D_t^{\alpha-k-1}x(t)\Big|_{t=0}, \quad (n-1<\alpha\le n) \tag{5} $$

where $X(s) = L\{x(t)\}$. For $\alpha < 0$ (*i.e.*, for the case of a fractional integral) the sum in the right-hand side must be omitted. The Laplace transform is a valuable tool for the analysis and synthesis of control systems. The expression (5) also suggests that the fractional operators are more easily handled in the s-domain, and that the classical techniques for the analysis of control systems can also be employed in the fractional-order case.

Bearing these ideas in mind, the article presents two applications of the FC concepts in the area of control systems. Section 2 introduces a methodology for the tuning of PID controllers based on basic concepts of FC. It is shown that the resulting compensation system, tuned by the proposed method, is robust against gain variations. In Section 3 we apply a fractional PID controller in the control of an hexapod robot with three degrees of freedom (dof) legs. It is shown the superior performance of the overall system in the presence of a fractional-order controller. Finally, in Section 4 we address the main conclusions.

2 Tuning of PID Controllers Using Basic Concepts of Fractional Calculus

In this section we present a methodology for tuning PID controllers such that the response of the closed-loop system has an almost constant overshoot defined by a prescribed value. The proposed method is based on the minimization of the integral of square error (ISE) between the step responses of a unit feedback control system, whose open-loop transfer function $L(s)$ is given by a fractional-order integrator, and that of the PID compensated system. The controller specifications consist in the gain crossover frequency and the slope at that frequency of the fractional-order integrator. In this way, we can ensure the nearly flatness of the phase curve around the gain crossover frequency of the compensated system. This fact implies that the system will be more robust to gain variations, exhibiting step responses with an almost constant overshoot, that is, with the iso-damping property.

2.1 Basics Concepts of Fractional-Order Control

Figure 1 illustrates the fractional-order control system with open-loop transfer function $L(s)$ given by a fractional-order integrator. This is the fractional system that will be used as reference model for the tuning of PID controllers.

The open-loop transfer function $L(s)$ has the form $\left(\alpha \in \Re^{+}\right)$:

$$L(s) = \left(\frac{\omega_c}{s}\right)^{\alpha} \tag{6}$$

where ω_c is the gain crossover frequency, that is, $|L(j\omega_c)| = 1$. The parameter α is the slope of the magnitude curve, on a log-log scale, and may assume integer as well noninteger values. In this study we consider $1 < \alpha < 2$, such that the output response may have a fractional oscillation (with some similarities to the response of an underdamped second-order system). This transfer function is also known as the Bode's ideal loop transfer function, since Bode studied the design of feedback amplifiers in the 1940s [5].

The Bode diagrams of magnitude and phase of $L(s)$ are illustrated in Fig. 2. The magnitude curve is a straight line of constant slope -20α dB/dec, and the phase curve is a horizontal line positioned at $-\alpha\pi/2$ rad. The Nyquist curve is simply the straight line through the origin, $\arg L(j\omega) = -\alpha\pi/2$ rad.

This choice of $L(s)$ gives a closed-loop system with the desirable property of being insensitive to gain changes. If the gain changes the crossover frequency ω_c will change but the phase margin (PM) of the system remains PM $= \pi(1 - \alpha/2)$ rad, independently of the value of the gain. This can be seen from the curves of magnitude and phase of Fig. 2.

The closed-loop transfer function of system of Fig. 1 is given by:

$$G(s) = \frac{L(s)}{1+L(s)} = \frac{1}{\left(\frac{s}{\omega_c}\right)^{\alpha} + 1}, \quad (1 < \alpha < 2) \tag{7}$$

The unit step response of $G(s)$ is given by the expression:

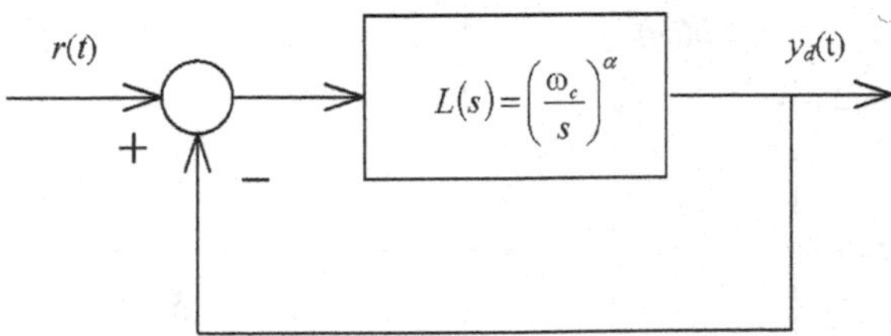

Fig. 1: Fractional-order control system with open-loop transfer function $L(s)$

$$y_d(t) = L^{-1}\left\{\frac{1}{s}G(s)\right\} = L^{-1}\left\{\frac{\omega_c^\alpha}{s(s^\alpha+\omega_c^\alpha)}\right\}$$
$$= 1 - \sum_{n=0}^{\infty}\frac{\left[-(\omega_c t)^\alpha\right]^n}{\Gamma(1+\alpha n)} = 1 - E_\alpha\left[-(\omega_c t)^\alpha\right] \quad (8)$$

where $E_\alpha(x)$ is the one-parameter Mittag-Leffler function [2,6]. This function is a generalization of the common exponential function, since for $\alpha = 1$ we have $E_1(x) = e^x$.

2.2 Tuning of PID Controllers

For the tuning of PID controllers we address the fractional-order control system of Fig. 1 as the reference system. After defining the order α and the crossover frequency ω_c we can establish the overshoot and the speed of the output response, respectively. For that purpose we consider the closed-loop system shown in Fig. 3, where $G_c(s)$ and $G_p(s)$ are the PID controller transfer function and the plant transfer function, respectively.

The transfer function $G_c(s)$ of the classical PID controller has the form ($\lambda = 1$ and $\mu = 1$ in (1)):

$$G_c(s) = \frac{U(s)}{E(s)} = K_P + \frac{K_I}{s} + K_D s \quad (9)$$

The design of the PID controller will consist on the determination of the optimum PID set gains (K_P, K_I, K_D) that minimize J, that is, the integral of the square error (ISE), which is defined as:

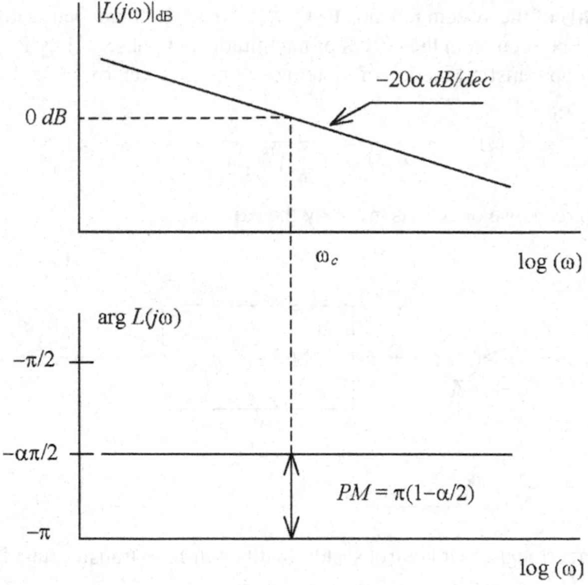

Fig. 2: Bode diagrams of magnitude and phase of $L(j\omega)$ for $1 < \alpha < 2$

$$J = \int_0^\infty [y(t) - y_d(t)]^2 dt \tag{10}$$

where $y(t)$ is the step response of the closed-loop system with the PID controller (Fig. 3) and $y_d(t)$ is the desired step response of the fractional-order control system of Fig. 1 given by expression (8).

To illustrate the effectiveness of the proposed methodology we consider the following four-order plant transfer function:

$$G_p(s) = \frac{K}{(s+1)^4} \tag{11}$$

with nominal gain $K = 1$. Figure 4 shows the step responses and the Bode diagrams of phase of the closed-loop system with the PID for the transfer function $G_p(s)$, and gain variations around the nominal gain ($K = 1$) corresponding to $K = \{0.6, 0.8, 1.0, 1.2, 1.4\}$, that is, for a variation up to $\pm 40\%$ of its nominal value. The system is tuned for $\alpha = 4/3$ (PM = 60°), $\omega_c = 0.5$ rad/s. We verify that we get the desired iso-damping property corresponding to the prescribed (α, ω_c)-values. In fact, we observe that the step responses have an almost constant overshoot, independently of the variation of the plant gain around the gain crossover

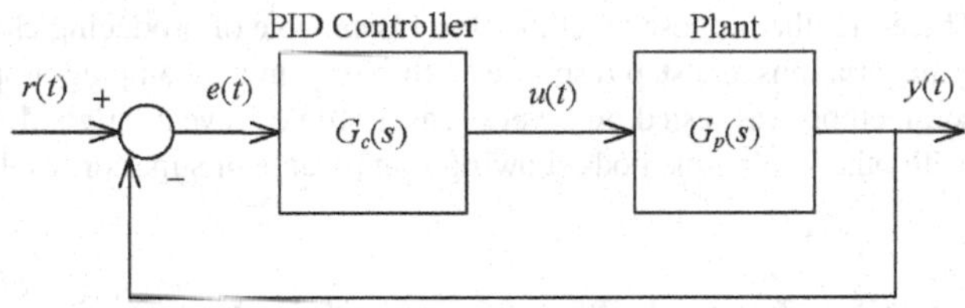

Fig. 3: Closed-loop control system with PID controller $G_c(s)$

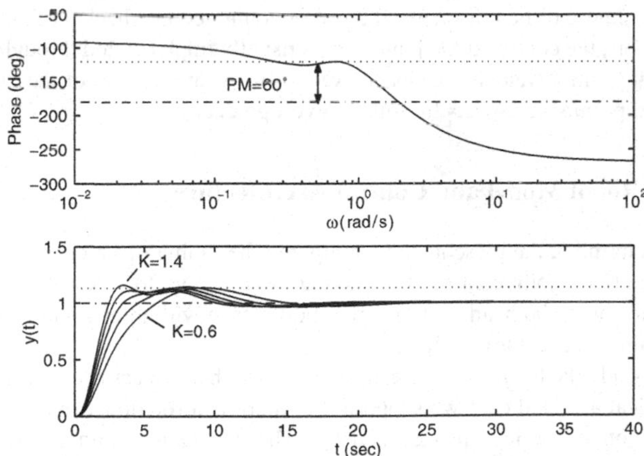

Fig. 4: Bode phase diagram and step responses for the closed-loop system with a PID controller (tuned by the proposed methodology) for $G_p(s)$. The PID parameters are $K_P = 1.3774$, $K_I = 0.8088$ and $K_D = 2.3673$

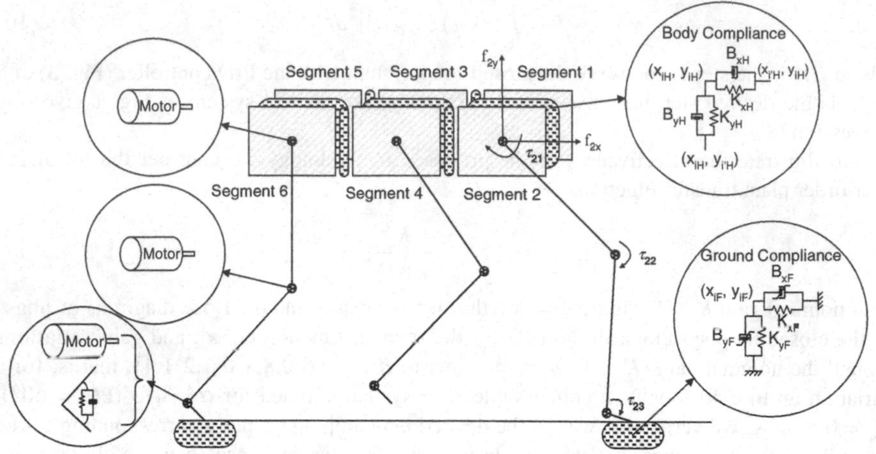

Fig. 5: Model of the robot body and foot-ground interaction

frequency ω_c. Therefore, the proposed methodology is capable of producing closed-loop systems robust to gain variations and step responses exhibiting an iso-damping property.

The proposed method was tested on several cases studies revealing good results. It was also compared with other tuning methods showing comparable or superior results [7,8].

3 Fractional PD$^{\mu}$ Control of an Hexapod Robot

Walking machines allow locomotion in terrain inaccessible to other type of vehicles, since they do not need a continuous support surface. However, the requirements for leg coordination and control impose difficulties beyond those encountered in wheeled robots. Usually, for multi-legged robots, the control at the joint level is usually implemented through a simple PID like scheme with position/velocity feedback. Recently, the application of the theory of FC to robotics revealed promising aspects for future developments [9].

3.1 Hexapod Robot Model and Control Architecture

With these facts in mind, the present study compares the tuning of Fractional Order (FO) algorithms, applied to the joint control of a walking system (Fig. 5). The robot has $n = 6$ legs, equally distributed along both sides of the robot body, each with three rotational joints (*i.e.*, $j = \{1, 2, 3\} \equiv \{\text{hip, knee, ankle}\}$) [10].

During this study the leg joint $j = 3$ can be either mechanical actuated, or motor actuated. For the mechanical actuated case we suppose that there is a rotational pre-tensioned spring-dashpot system connecting leg links L_{i2} and L_{i3}. This mechanical impedance maintains the angle between the two links while imposing a joint torque [10].

Figure 5 presents the dynamic model for the hexapod body and foot-ground interaction. It is considered the existence of robot body compliance because walking animals have a spine that allows supporting the locomotion with improved stability. The robot body is divided in n

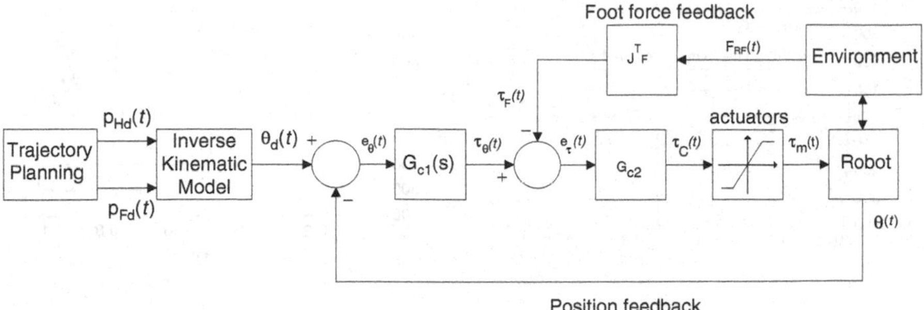

Fig. 6: Hexapod robot control architecture

identical segments (each with mass $M_b n^{-1}$) and a linear spring-damper system (with parameters defined so that the body behaviour is similar to the one expected to occur on an animal) is adopted to implement the intra-body compliance [10]. The contact of the ith robot feet with the ground is modelled through a non-linear system [12], being the values for the parameters based on the studies of soil mechanics [12].

The general control architecture of the hexapod robot is presented in Fig. 6. We evaluate the effect of different PD$^\mu$ controller implementations for $G_{c1}(s)$, while G_{c2} is a P controller. For the PD$^\mu$ algorithm, implemented through a discrete-time 4th-order Padé approximation $(a_{ij}, b_{ij} \in \Re)$, we have:

$$G_{c1j}(z) \approx Kp_j + K\mu_j \sum_{i=0}^{i=u} a_{ij} z^{-i} \Big/ \sum_{i=0}^{i=u} b_{ij} z^{-i}, \quad j = 1,2,3 \tag{12}$$

where Kp_j and $K\mu_j$ are the proportional and derivative gains, respectively, and μ_j is the fractional order, for joint j.

The performance analysis is based on the formulation of two indices measuring the mean absolute density of energy per travelled distance (E_{av}) and the hip trajectory errors (ε_{xyH}) during walking [11]. It is analyzed the system performance of the different PD$^\mu$ controller tuning, when adopting a periodic wave gait at a constant forward velocity V_F, for two distinct cases: the hip and knee joints are motor actuated while the ankle joint is mechanically (passively) actuated, and the three leg joints are fully motor actuated [10].

3.2 Simulation Results

To tune the different controller implementations we adopt a systematic method, testing and evaluating several possible combinations of parameters, for all controller implementations. Therefore, we adopt the $G_{c1}(s)$ parameters that establish a compromise in what concerns the simultaneous minimization of E_{av} and ε_{xyH}, and a proportional controller G_{c2} with gain $Kp_j = 0.9$ ($j = 1, 2, 3$). It is assumed high performance joint actuators (i.e., with almost negligible saturation), having a maximum actuator torque of $\tau_{ijMax} = 400$ Nm. The desired angle between the foot and the ground (assumed horizontal) is established as $\theta_{i3hd} = -15^o$. We start by considering that leg joints 1 and 2 are motor actuated and joint 3 has a passive spring-dashpot system. For this case we tune the PD$^\mu$ controllers for values of the fractional order in the interval $-0.9 < \mu_j < +0.9$ and $\mu_j \neq 0.0$,establishing $\mu_1 = \mu_2 = \mu_3$. Afterwards, we

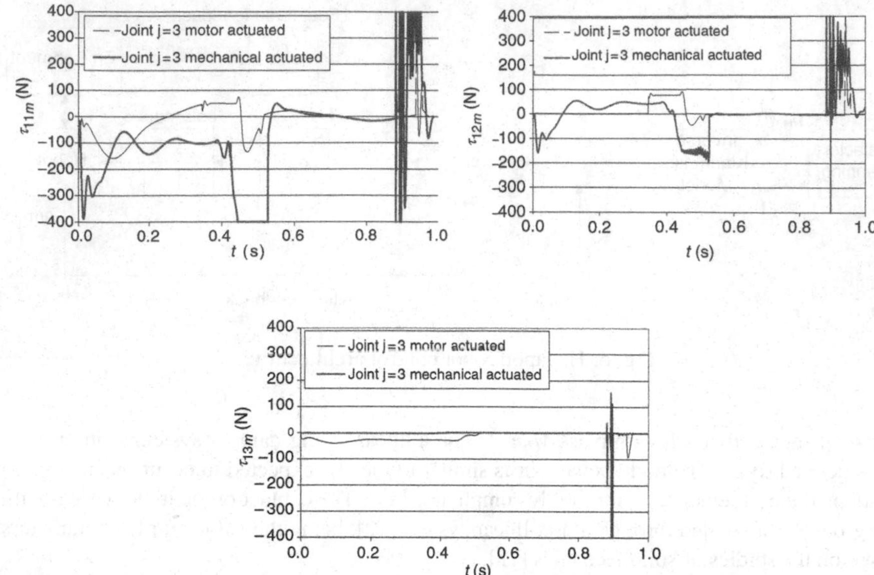

Fig. 7: Plots of τ_{1jm} *vs. t*, with joints 1 and 2 motor actuated and joint 3 mechanical actuated and all joints motor actuated, for $\mu_j = 0.5$

consider that joint 3 is also motor actuated, and we repeat the controller tuning procedure seeking for the best parameters.

When joint 3 is mechanically actuated, we observe that the value of $\mu_j = 0.6$ presents the best compromise situation in what concerns the simultaneous minimization of ε_{xyH} and E_{av}. When all joints are motor actuated, we find that $\mu_j = 0.5$ presents the best compromise between ε_{xyH} and E_{av}. Furthermore, we conclude that the best case corresponds to all leg joints being motor actuated.

In conclusion, the experiments reveal the superior performance of the FO controller for $\mu_j \approx 0.5$ and a robot with all joints motor actuated. The good performance can be verified in the joint actuation torques τ_{1jm} (Fig. 7) and the hip trajectory tracking errors Δ_{1xH} and Δ_{1yH} (Fig. 8).

Since the objective of the walking robots is to walk in natural terrains, in the sequel we test how the different controllers behave under distinct ground properties.

Considering the previously tuning controller parameters, and assuming that joint 3 is mechanically actuated, the values of the ground model parameters are varied simultaneously through a multiplying factor varied in the range $K_{mult} \in [0.1; 4.0]$. This variation for the ground model parameters allows the simulation of the ground behaviour for growing stiffness, from peat to gravel [12]. On a second phase, the experiments are repeated considering that joint 3 is also motor actuated. We conclude that the controller responses are quite similar, meaning that these algorithms are robust to variations of the ground characteristics.

The performance measures *versus* the multiplying factor of the ground parameters K_{mult} are presented on Figs. 9 and 10, for the cases of joint 3 being mechanically actuated and motor actuated, respectively.

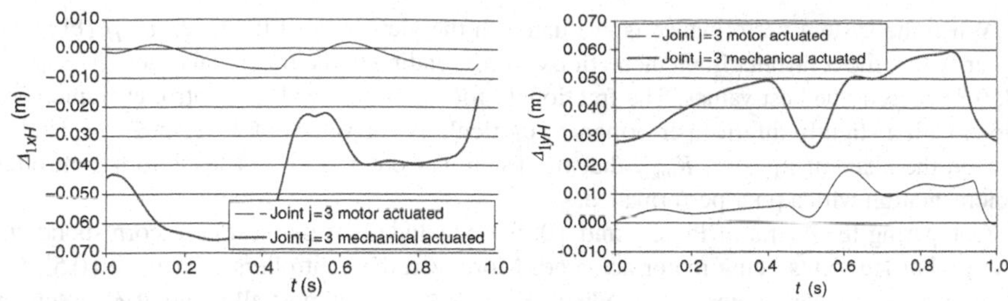

Fig. 8: Plots of Δ_{1xH} and Δ_{1yH} *vs.* t, with joints 1 and 2 motor actuated and joint 3 mechanical actuated and all joints motor actuated, for $\mu_j = 0.5$

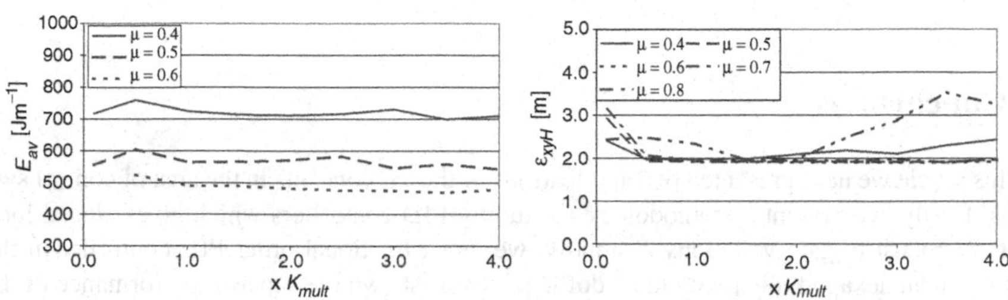

Fig. 9: Performance indices E_{av} and ε_{xyH} *vs.* K_{mult} for the different $G_{c1}(s)$ PD$^{\mu}$ controller tuning with joints 1 and 2 motor actuated and joint 3 mechanical actuated

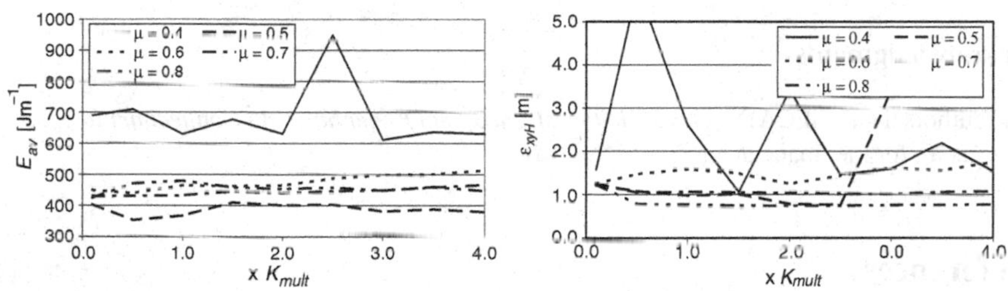

Fig. 10: Performance indices E_{av} and ε_{xyH} *vs.* K_{mult} for the different $G_{c1}(s)$ PD$^{\mu}$ controller tuning with all joints motor actuated

For the case of joint 3 being motor actuated, and analysing the system performance from the viewpoint of the index E_{av} (Fig. 10), it is possible to conclude that the best PD$^{\mu}$ implementation occurs for the fractional order $\mu_j = 0.5$. Moreover, it is clear that the performances of the different controller implementations are almost constant on all range of the ground parameters, with the exception of the fractional order $\mu_j = 0.4$, where E_{av} presents a significant variation.

When the system performance is evaluated in the viewpoint of the index ε_{xyH} (Fig. 10) we verify that the controller implementations corresponding to the fractional orders $\mu_j = \{0.6, 0.7, 0.8\}$ present the best values. The fractional order $\mu_j = 0.5$ leads to controller implementations with a slightly inferior performance, particularly for values of $K_{mult} > 2.5$. It is also clear on the chart of ε_{xyH} *vs.* K_{mult} that the fractional order $\mu_j = 0.4$ leads to a controller implementation with a poor performance.

Comparing the charts of Figs. 9 and 10, we conclude that the best case corresponds to all the robot leg joints being motor actuated. Moreover, the controllers with $\mu_j = \{0.5, 0.6, 0.7, 0.8\}$ present lower values of the indices E_{av} and ε_{xyH} on almost all range of K_{mult} under consideration. The only exception to this observation occurs for the PD^{μ} controller implementation when $\mu_j = 0.5$, that presents slightly higher values of the index ε_{xyH} for values of $K_{mult} > 2.5$, when all robot leg joints are motor actuated. We conclude that the controller responses are quite similar, meaning that these algorithms are robust to variations of the ground characteristics [11].

4 Conclusions

In this article we have presented two applications of the FC concepts in the area of control systems. Firstly, we present a methodology for tuning PID controllers which gives closed-loop systems robust to gain variations. Secondly, we use a fractional-order PID controller in the control of an hexapod robot with three dof legs. It was shown the superior performance of the overall system when adopting a fractional-order controller with $\mu_j \approx 0.5$, and a robot having all joints motor actuated. The superior performance of the PD^{μ} joint leg controller is kept for different ground properties. In both cases, we demonstrate the advantages of applying FC concepts, which encourages to pursue this line of investigation and to extend the FC concepts to other dynamical systems.

Acknowledgments

The authors thank GECAD – *Grupo de Investigação em Engenharia do Conhecimento e Apoio à Decisão*, for the financial support to this work.

References

1. Oldham, K. B. and Spanier, J. (1974) The Fractional Calculus. Academic, New York
2. Podlubny, I. (1999) Fractional Differential Equations. Academic, San Diego, CA
3. Podlubny, I., Dorcak, L. and Kostial, I. (1997) On Fractional Derivatives, Fractional-Order Dynamics Systems and $PI^{\lambda}D^{\mu}$-Controllers. In: 36th IEEE Conference on Decision and Control. San Diego, CA
4. Podlubny, I. (1999) Fractional-Order Systems and $PI^{\lambda}D^{\mu}$-Controllers. IEEE Transactions on Automatic Control 44(1):208–214
5. Bode, H. W. (1945) Network Analysis and Feedback Amplifier Design. Van Nostrand, New York
6. Miller, K. S. and Ross, B. (1993) An Introduction to the Fractional Calculus and Fractional Differential Equations. Wiley, New York

7. Barbosa, R. S., Machado, J. A. T., and Ferreira, I. M. (2004) PID Controller Tuning Using Fractional Calculus Concepts. Journal of Fractional Calculus and Applied Analysis 7(2):119–134
8. Barbosa, R. S., Machado, J. A. T., and Ferreira, I. M. (2004) Tuning of PID Controllers Based on Bode's Ideal Transfer Function. Nonlinear Dynamics 38(1–4):305–321
9. Silva, M. F., Machado, J. A. T. and Lopes, A. M. (2003) Comparison of Fractional and Integer Order Control of an Hexapod Robot. In: Proceedings of VIB 2003 – ASME International 19th Biennial Conference on Mechanical Vibration and Noise. USA
10. Silva, M. F., Machado, J. A. T. and Jesus, I. S. (2006) Modelling and Simulation of Walking Robots With 3 dof Legs. In: MIC 2006 – The 25th IASTED International Conference on Modelling, Identification and Control. Lanzarote, Spain
11. Silva, M. F., Machado, J. A. T. (2006) Fractional Order PD^{α} Joint Control of Legged Robots. Journal of Vibration and Control - Special Issue on Modeling and Control of Artificial Locomotion Systems 12(12):1483–1501
12. Silva, M. F., Machado, J. A. T. (2003) Position/Force Control of a Walking Robot. MIROC – Machine Intelligence and Robot Control 5:33–44

Fractional Describing Function of Systems with Nonlinear Friction

Fernando B. M. Duarte[1] and J. A. Tenreiro Machado[2]

[1] Department of Mathematics, School of Technology, Viseu, Portugal
fduarte@mat.estv.ipv.pt
[2] Department of Electrotechnical Engineering, Institute of Engineering, Porto, Portugal
jtm@isep.ipp.pt

Summary. This paper studies the describing function (DF) of systems consisting in a mass subjected to nonlinear friction. The friction force is composed in three components namely, the viscous, the Coulomb and the static forces. The system dynamics is analyzed in the DF perspective revealing a fractional-order behaviour. The reliability of the DF method is evaluated through the signal harmonic content and the limit cycle prediction.

1 Introduction

The phenomenon of vibration due to friction is verified in many branches of technology where it plays a very useful role. However, its occurrence is often undesirable, because it causes additional dynamic loads, as well as faulty operation of machines and devices. Despite the investigation that was carried out so far, this phenomenon is not yet fully understood, due to the diversity of reasons underlying the energy dissipation involving the dynamic effects [5,6,11,12]. In this paper we investigate the dynamics of systems that contain nonlinear friction, namely the Coulomb and the static forces, in addition to the linear viscous component. Bearing these ideas in mind, the article is organized as follows. Section 2 introduces the fundamental aspects of the describing function (DF) method. Section 3 studies the DF of mechanical systems with nonlinear friction. Section 4 analyzes the prediction of cycle limit in the friction system under the action of a PID controller. Finally, Section 5 draws the main conclusions and addresses perspectives towards future developments.

2 Fundamental Concepts

Let us consider the feedback system of Fig. 1 with one nonlinear element N and a linear system with transfer function $G(s)$.

Suppose that the input to a nonlinear element is sinusoidal $x(t) = X\sin(\omega t)$. In general the output of the nonlinear element $y(t)$ is not sinusoidal. Nevertheless, $y(t)$ is periodic, with the same period as the input, and contains higher harmonics in addition to the fundamental harmonic component.

J.A.T. Machado et al. (eds.), *Intelligent Engineering Systems and Computational Cybernetics,*

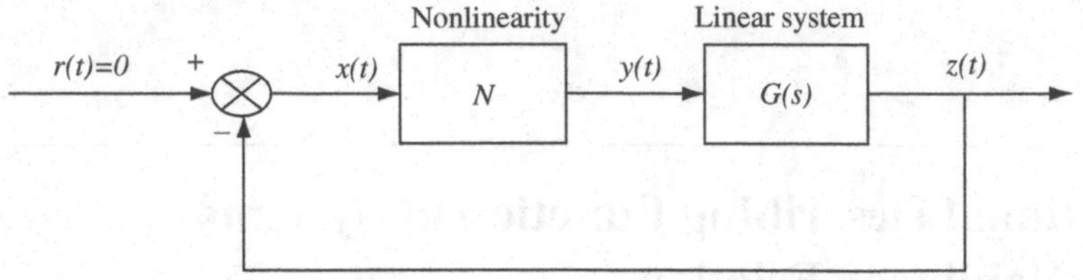

Fig. 1: Nonlinear control system

If we assume that the nonlinearity is symmetrical with respect to the variation around zero, the Fourier series becomes:

$$y(t) = \sum_{k=1}^{\infty} Y_k \cos(k\,\omega t + \phi_k) \tag{1}$$

where Y_k and ϕ_k are the amplitude and the phase shift of the kth harmonic component of the output $y(t)$, respectively.

In the DF analysis, we assume that only the fundamental harmonic component of the output is significant. Such assumption is often valid since the higher harmonics in the output of a nonlinear element are usually of smaller amplitude than the fundamental component [1]. Moreover, most control systems are "low-pass filters" with the result that the higher harmonics are further attenuated [9, 14, 15].

The DF, or sinusoidal DF, of a nonlinear element, $N(X, \omega)$, is defined as the complex ratio of the fundamental harmonic component of the output $y(t)$ and the input $x(t)$, that is:

$$N(X, \omega) = \frac{Y_1}{X} e^{j\phi_1} \tag{2}$$

where the symbol N represents the DF, X is the amplitude of the input sinusoid, Y_1 and ϕ_1 are the amplitude and the phase shift of the fundamental harmonic component of the output, respectively. Several DFs of standard nonlinear system elements can be found in the references [7, 8, 10].

For nonlinear systems not involving energy storage or dissipation the DF is merely amplitude-dependent, that is $N = N(X)$. However, when we have nonlinear elements that involve energy, the DF method is both amplitude and frequency dependent yielding $N(X, \omega)$. In this case, to determine the DF, we have to adopt a numerical approach because it is impossible to find a closed-form analytical solution. Once calculated, the DF can be used for the approximate stability analysis of a nonlinear control system.

Let us consider again the standard control system shown in Fig. 1 where the block N denotes the DF of the nonlinear element. If the higher harmonics are sufficiently attenuated, N can be treated as a variable gain and the closed-loop frequency response becomes:

$$\frac{C(j\omega)}{R(j\omega)} = \frac{N(X, \omega)G(j\omega)}{1 + N(X, \omega)G(j\omega)} \tag{3}$$

The characteristic equation yields:

$$1 + N(X, \omega)G(j\omega) = 0 \tag{4}$$

If (4) can be satisfied for some value of X and ω, then a limit cycle is *predicted* for the nonlinear system. Moreover, since (4) is valid only if the nonlinear system is in a steady-state limit cycle, the DF analysis predicts only the presence or the absence of a limit cycle and cannot be applied to the analysis of other types of time responses.

3 Systems with Nonlinear Friction

In this section we calculate the DF of a dynamical system with nonlinear friction and we study its properties. In Subsection 3.1 we start by a combination of the viscous and Coulomb components. In Subsection 3.2 we complement the study by including also the static friction.

3.1 Coulomb and Viscous Friction

Let us consider a system (Fig. 2) with a mass M, having displacement $x(t)$ under the action of an input force $f(t)$. The friction $F_f(t)$ force has two components: a non-linear Coulomb K part and a linear viscous $B\dot{x}$ part (so-called CV model). In the model of the nonlinear friction, for avoiding numerical problems, the discontinuity at the origin, due to the Coulomb component, is avoided by introducing a straight line in the interval $-\Delta \leq \dot{x} \leq \Delta$ that connects the two opposite points (Fig. 2b).

The equation of motion in this system is as follows:

$$M\,\ddot{x}(t) + F_f(t) = f(t) \tag{5}$$

For the simple system of Fig. 2 we can calculate, numerically, the polar plot of $N(F,\omega)$ considering as input a sinusoidal force $f(t) = F\cos(\omega t)$ applied to mass M and as output the position $x(t)$.

Figure 3, shows $N(F,\omega)$ for $M = 9$ kg, $B = 0.5$ Ns m^{-1}, $K = 5$ N, $\Delta = 0.01$ m s^{-1}. Alternatively Fig. 4 illustrates the log-log plots of $|Re\{N\}|$ and $|Im\{N\}|$ *vs* the exciting frequency ω, for different values of the input force $F = \{10, 50, 100\}$. We verify that the locus of $N(F,\omega)$, with the exception of low F, consists in nearly straight lines, both for the cases of F and ω constant. Moreover, the phases for the different values of F reveal, clearly, a fractional-order behaviour.

In Fig. 5 it is depicted the harmonic content of the output signal $x(t)$ for an input force of $F = 10$ N. From this chart we verify that the output signal has a half-wave symmetry because the harmonics of even order are negligible. Moreover, the fundamental component of the output signal is the most important one, while the amplitude of the high order harmonics decays significantly. Therefore, we can conclude that, for the friction CV model, the DF method may lead to a good approximation.

In order to gain further insight into the system nature, we repeat the experiments for different mass values $M = \{1, 2, 3, 5, 7, 9\}$ kg.

The results show that $Re\{N\}$ and $Im\{N\}$ vary with M. In order to study the relation between $Re\{N\}$ and $Im\{N\}$ *versus* F and M, we approximate the numerical results through power functions:

$$|Re\{N\}| = a\,\omega^b,\ |Im\{N\}| = c\,\omega^d,\ \{a,b,c,d\} \in \mathbb{R} \tag{6}$$

Figure 6 illustrates the variation of the parameters $\{a,b,c,d\}$ with F and M. Moreover, $Re\{N\}$ and $Im\{N\}$ reveal a distinct fractional order relationships with ω [2].

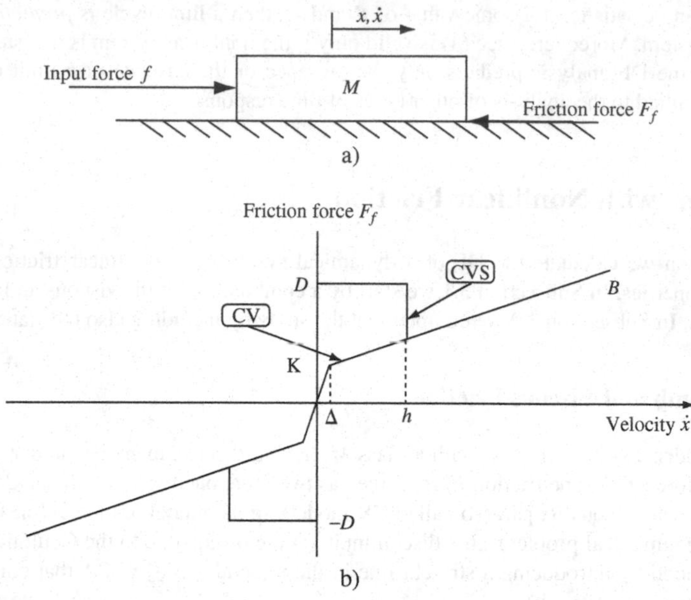

Fig. 2: (a) Elemental mass system subjected to nonlinear friction and (b) non-linear friction with Coulomb, Viscous (CV model) and Static components (CVS model)

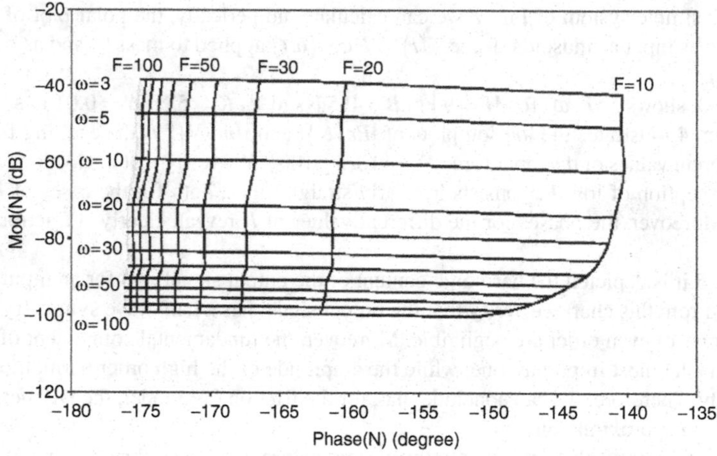

Fig. 3: Nichols plot of $N(F,\omega)$ for the system subjected to nonlinear friction (CV model), $10 \leq F \leq 100$ N, $3 \leq \omega \leq 100$ rad s^{-1} and $\{K,B\} = \{0.5\text{ N}, 0.5\text{ Ns m}^{-1}\}$

3.2 Coulomb, Viscous and Static Friction

In this sub-section we incorporate the static friction (D, h) in the CV model (see Fig. 2b) leading to the so-called CVS model. In this line of thought, we develop a study similar to the one adopted previously, with $M = 9$ kg, $B = 0.5$ Ns m^{-1}, $K = 5$ N, $D = 7$ N, $h = 0.5$ m s^{-1}.

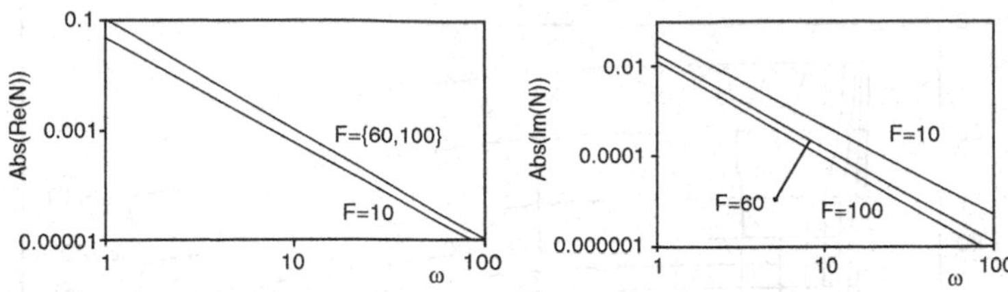

Fig. 4: Log-log plots of $|Re\{N\}|$ and $|Im\{N\}|$ *vs.* the exciting frequency $3 \leq \omega \leq 100$ N, with de CV model and and $\{K, B\}$ ={0.5 N, 0.5 Ns m^{-1}}

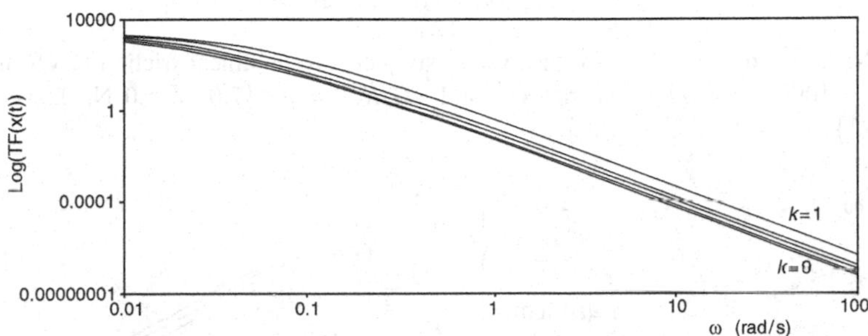

Fig. 5: Fourier transform of the output position $x(t)$, for the CV model, *vs.* the exciting frequency ω and the harmonic frequency index $k = \{1, 3, 5, 7, 9\}$ for an input force F = 10 N

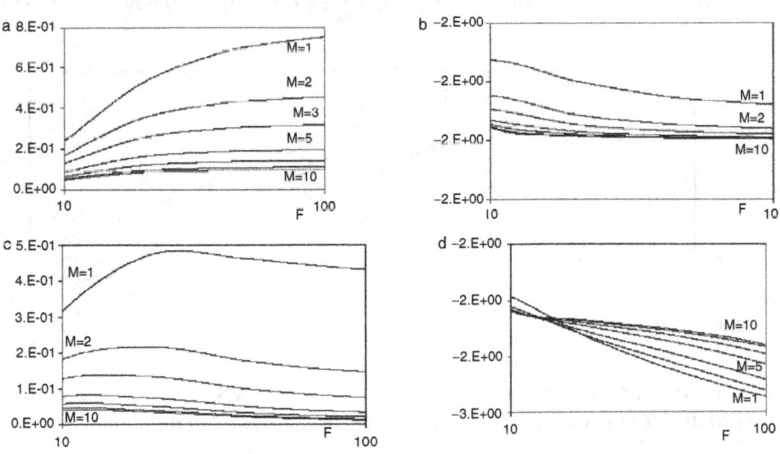

Fig. 6: The parameters $\{a, b, c, d\}$ *vs.* $10 \leq F \leq 100$ N for $1 \leq M \leq 10$ kg, in the CV model with $\{K, B\} = \{0.5$ N, 0.5 Ns m$^{-1}\}$

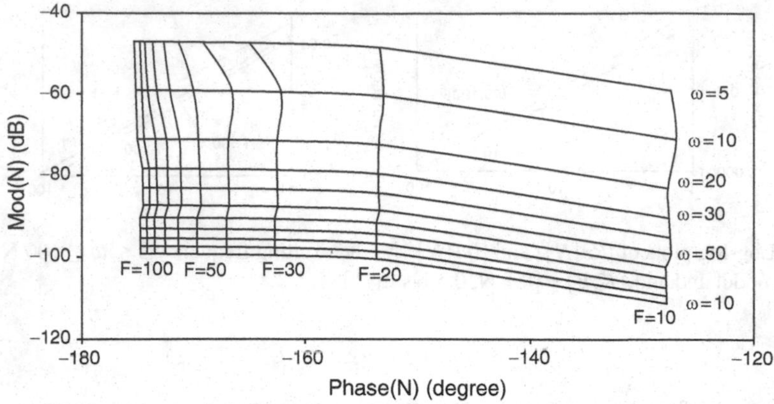

Fig. 7: Nichols plot of $N(F,\omega)$ for the system subjected to nonlinear friction (CVS model), $10 \leq F \leq 100$ N, $5 \leq \omega \leq 100$ rad s^{-1} and $\{D,K,B,h\} = \{7.0$ N, 5.0 N, 0.5 Ns m^{-1}, 0.5 m s$^{-1}\}$

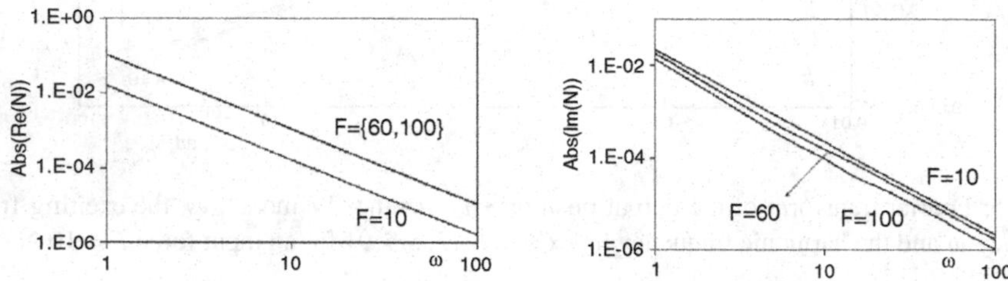

Fig. 8: Log-log plots of $|Re\{N\}|$ and $|Im\{N\}|$ *vs.* the exciting frequency ω for $F = \{10,50,100\}$ N, with de CVS model and $\{D,K,B,h\} = \{7.0\,\text{N}, 5.0\,\text{N}, 0.5\,\text{Ns m}^{-1}, 0.5\,\text{m s}^{-1}\}$

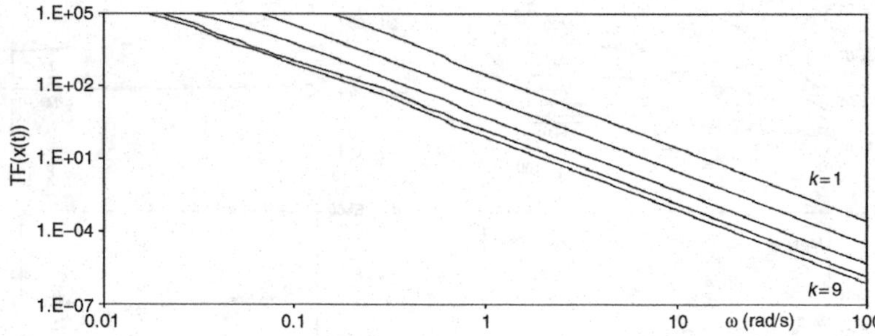

Fig. 9: Fourier transform of the output position $x(t)$, for the CVS model, *vs.* the exciting frequency ω and the harmonic frequency index $k = \{1,3,5,7,9\}$ for an input force $F = 10$ N

Figures 7–10 depict the corresponding charts. Comparing the results of the VC and CVS models we conclude that $Re\{N\}$ and $Im\{N\}$ are, in the two cases, of the same type, following a

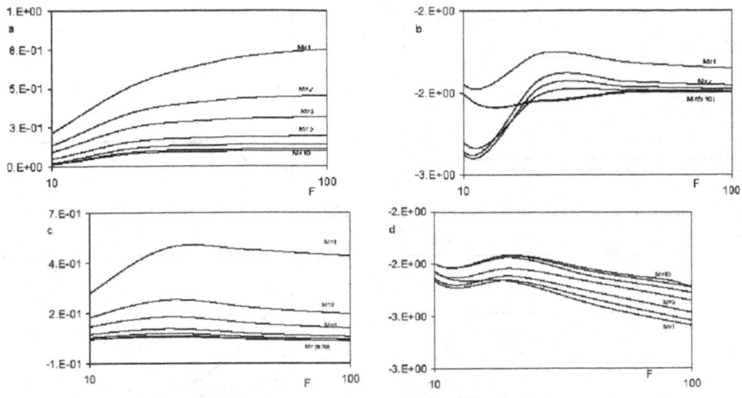

Fig. 10: The parameters $\{a,b,c,d\}$ *vs* $10 \leq F \leq 100$ N in the CVS model, for $1 \leq M \leq 10$ kg, and $\{D,K,B,h\} = \{7.0 \text{ N}, 5.0 \text{ N}, 0.5 \text{ Ns m}^{-1}, 0.5 \text{ m s}^{-1}\}$

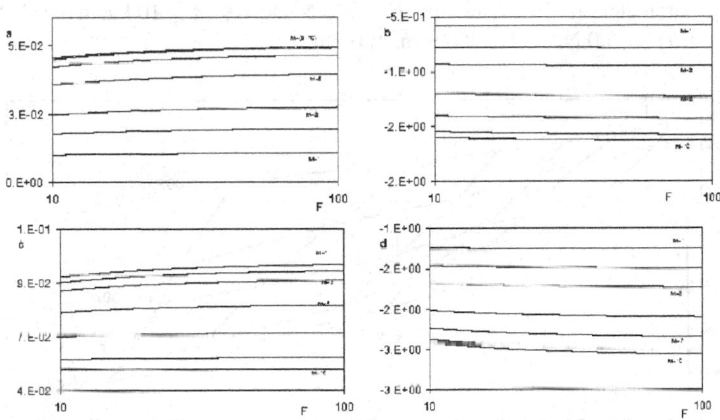

Fig. 11: The parameters $\{a,b,c,d\}$ *vs* $10 \leq F \leq 100$ N and $1 \leq M \leq 10$ kg, in the CVS model, with $\{D,K,B,h\} = \{1.0 \text{ N}, 0.5 \text{ N}, 10.0 \text{ Ns m}^{-1}, 0.05 \text{ m s}^{-1}\}$

power law according with expression (6). Furthermore, once again we obtain fractional-order dynamics as revealed clearly by the phase of the Nichols chart in Fig. 7. Nevertheless, the CVS model is very sensitive to small input forces F (stimulation mainly of the static component) leading to large values of N and to a higher harmonic content [3, 13].

To have a deeper insight into the effects of the different CVS components several experiences were performed varying $\{D,K,B,h\}$. For example, Figs. 11 and 12 present the values of the parameters $\{a,b,c,d\}$ when approximating $|Re\{N\}|$ and $|Im\{N\}|$, for $F = \{10,20,30,40,50,60,70,80,90,100\}$ N, and $M = \{1,2,3,5,7,9,10\}$ kg, with $\{D,K,B,h\} = \{1.0 \text{ N}, 0.5 \text{ N}, 10.0 \text{ Ns m}^{-1}, 0.05 \text{ m s}^{-1}\}$ and $\{D,K,B,h\} = \{5.0 \text{ N}, 1.0 \text{ N}, 1.0 \text{ Ns m}^{-1}, 0.1 \text{ m s}^{-1}\}$, respectively.

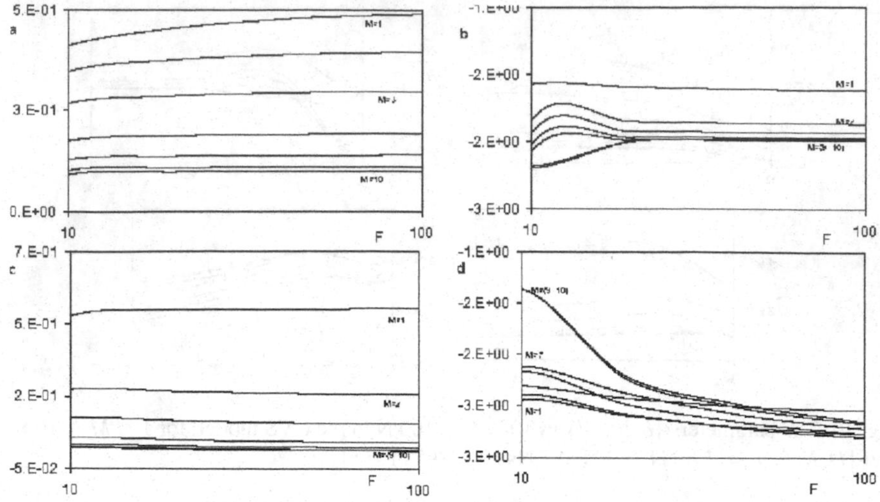

Fig. 12: The parameters $\{a,b,c,d\}$ *vs* $10 \leq F \leq 100$ N and $1 \leq M \leq 10$ kg, in the CVS model, with $\{D,K,B,h\}$ = $\{5.0$ N, 1.0 N, 1.0 Ns m^{-1}, 0.1 m s$^{-1}\}$

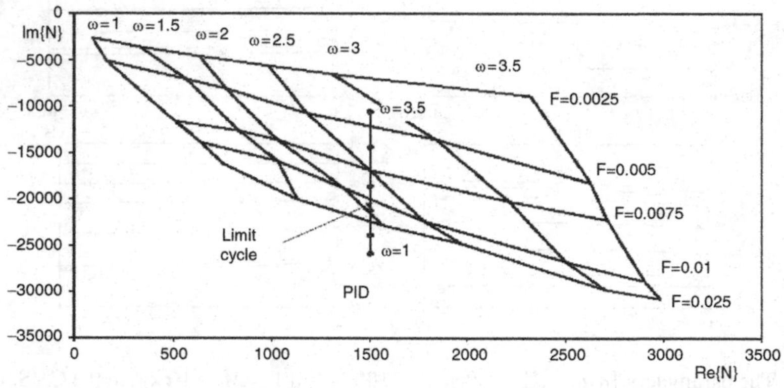

Fig. 13: Limit cycle for the CVS friction model with $M = 5$ kg, $B = 1$ Ns m^{-1}, $K = 0.5$ N, $D = 0.5$ N, $h = 0.01$ m s^{-1}

4 Limit Cycle Prediction

The characteristic equation (4) involves two nonlinear equations with the variables X and ω. It may be difficult to solve this equation by analytical methods and, therefore, a numerical and graphical approach is more adequate [4].

Adopting a classical PID controller $f(t) = K_p\ e(t) + K_d\ e(t) + K_i \int e(t)\ dt$, with gains $K_p = 1{,}500$, $K_d = 44{,}250$, and $K_i = 130$ in the closed-loop system, the FD *predicts* (Fig. 13) a limit cycle with $\omega = 2.2$ rad s^{-1} and $F = 0.01$ N which is very close to the real response $f(t)$ depicted in Fig. 14. This result confirms the reliability of the DF method.

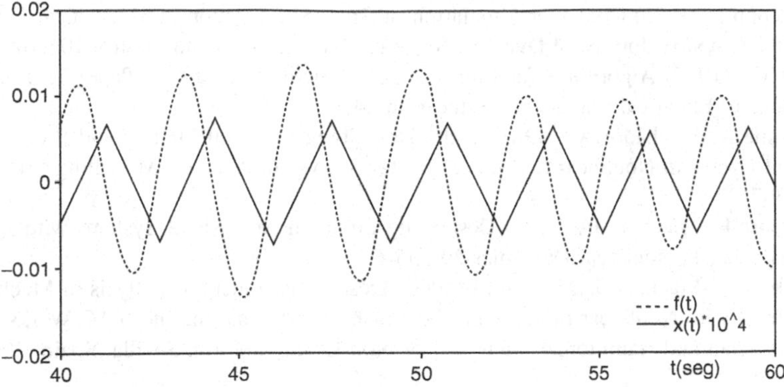

Fig. 14: Time response of system with CVS friction with $M = 5$ kg, $B = 1$ Ns m^{-1}, $K = 0.5$ N, $D = 0.5$ N, $h = 0.01$ m s^{-1}: Actuation force $f(t)$, Output displacement $x(t)$

5 Conclusions

This paper addressed the limit cycle prediction of systems with nonlinear friction. The dynamics of elemental mechanical system was analyzed through the describing function method and compared with standard models. The results encourage further studies of nonlinear systems in a similar perspective and the adoption of the tools of fractional calculus. The conclusions may lead to the development of compensation schemes capable of improving the control system performance.

Acknowledgments

The authors thank GECAD – *Grupo de Investigação em Engenharia do Conhecimento e Apoio à Decisão*, for the financial support to this work.

References

1. Slotine JE, Li W (1991) Applied Nonlinear Control. Prentice-Hall, NJ
2. Podlubny I (1999) Fractional Differential Equations. Academic, San Diego, CA
3. Duarte F, Machado JA (2006) Fractional Dynamics in the Describing Function Analysis of Nonlinear Friction. In: 2nd IFAC Workshop on Fractional Differentiation and Its Applications, Porto, Portugal
4. Azenha A, Machado JA (1996) Limit Cycles Prediction of Robot Systems with Nonlinear Phenomena in the Joints. In: 27th International Symposium on Industrial Robots, Milan, Italy, 1996, pp 1003–1008
5. Armstrong B, Dupont B, de Wit C (1994) A Survey of Models, Analysis Tools and Compensation Methods for the Machines with Friction. Automatica 30:1083–1183
6. Armstrong B, Amin B (1996) PID Control in the Presence of Static Friction: A Comparison of Algebraic and Describing Function Analysis. Automatica 32:679–692
7. Haessig DA, Friedland B (1991) On the Modelling and Simulation of Friction. ASME Journal of Dynamic Systems, Measurement and Control 113:354–362

8. Karnopp D (1985) Computer Simulation of Stick-Slip Friction in Mechanical Dynamic Systems. ASME Journal of Dynamic Systems, Measurement and Control 107:100–103
9. Cox CS (1987) Algorithms for Limit Cycle Prediction: A Tutorial Paper. International Journal of Electrical Engineering Education 24:165–182
10. Azenha A, Machado JA (1998) On the Describing Function Method and Prediction of Limit Cycles in Nonlinear Dynamical Systems. System Analysis-Modelling-Simulation 33:307–320
11. Barbosa R, Machado JA (2002) Describing Function Analysis of Systems with Impacts and Backlash. Nonlinear Dynamics 29:235–250
12. Barbosa R, Machado JA, Ferreira I (2003) Describing Function Analysis of Mechanical Systems with Nonlinear Friction and Backlash Phenomena. In: 2nd IFAC Workshop on Lagrangian and Hamiltonian Methods for Non Linear Control, Sevilla, Spain, 2003, pp 299–304
13. Duarte F, Machado JA (2005) Describing Function Method in Nonlinear Friction. In: IEEE, 1st International Conference on Electrical Engineering, Coimbra, Portugal, 2005
14. Atherton DP (1975) Nonlinear Control Engineering. IEEE, 1st International Conference on Electrical Engineering, Van Nostrand Reinhold Company, London
15. Dupont PE (1992) The Effect of Coulomb Friction on the Existence and Uniqueness of the Forward Dynamics Problem. In: Proceedings of the IEEE International Conference on Robotics and Automation, Nice, France, pp 1442–1447

Generalized Geometric Error Correction in Coordinate Measurement

Gyula Hermann

BrainWare Ltd., H-1021 Budapest, Völgy utca 13/A, Hungary
hermgy@iif.hu

Summary. Software compensation of geometric errors in coordinate measuring is hot subject because it results the decrease of manufacturing costs. The paper gives a summary of the results and achievements of earlier works on the subject. In order to improve these results a method is adapted to capture simultaneously the new coordinate frames in order use exact transformation values at discrete points of the measuring volume. The interpolation techniques published in the literature have the draw back that they could not maintain the orthogonality of the rotational part of the transformation matrices. The paper gives a technique, based on quaternions, which avoid this problem and leads to better results.

1 Introduction

Three dimensional coordinate metrology is a firmly established technique in industry. Its universal applicability and high degree of automation accounts for its success in the last 30 years. In order to fulfill its task, to verify the geometry of products on the basis of the measured results, CMM-s must be principally an order of magnitude more accurate than the machine tool used to produce the part.

Over the last 50 years one can observe enormous enhancement in positioning and measuring accuracy. The main portion of this enhancement is the result of improved knowledge about high precision machine design [14]. A fundamental principle was recognized by professor Abbe already in the 1890s about the alignment of the displacement measuring system with the distance to be measured. Another fundamental principle is the separation of the structural and measuring functions in a machine. Various novel constructions of high precision coordinate measuring machines are given in [8, 12, 17].

As mechanical accuracy is costly, whereas repeatability is not expensive, software techniques were used from the beginning to compensate for the systematic errors in order to keep manufacturing costs low [9].

One of the earliest papers on error compensation of coordinate measuring machines is by Zhang et al. [19]. They describe the compensation of a bridge type industrial three-coordinate measuring machine, which resulted in an accuracy improvement by approximately a factor 10. The correction vectors are determined at equally spaced points in the measuring volume. At intermediate point they are calculated by linear interpolation.

J.A.T. Machado et al. (eds.), *Intelligent Engineering Systems and Computational Cybernetics*,

An analytical quadratic model for the geometric error of a machine tool was developed by Ferreira and Liu [6] using rigid body kinematics. They introduced the notion of shape and joint transform. The former describes the transformation between the coordinate system on the same link and latter the transformation across a joint. To represent these transformations homogeneous transformation were introduced and a quadratic expression was developed for the individual error components. The global error description was obtained by concatenating the transformation matrices.

Duffie and Yang [2] invented a method to generate the kinematic error functions from volumetric error measurements. To represent the displacement error a vectorial approach was followed The error components were approximated by cubic polynomials. To find the polynomial's coefficients least square fit was applied.

Teeuwsen [16] described the error motion of the kinematic components of a coordinate measuring machine by using homogeneous transformations and concatenating these transformations to calculate the resulting global error. To obtain a continuous description of the correction vector, between these points, regression was used to establish a piecewise polynomial representation.

Vermeulen [17] has designed a novel high precision 3D-CMM, meant for small products, that have to be measured with a volumetric measuring uncertainty less than 0.1 μm. The structural loop of this CMM differs from the structural loop of a conventional construction. The horizontal slides have their own vertical support to avoid changing gravitational load caused by moving. Abbe errors are eliminated by introducing two intermediate bodies. For reasons of repeatability aerostatic bearings are applied for all slides. To attain the desired volumetric accuracy, the residual geometric errors are calibrated individually and compensated by software.

Ruijl [12] has build a high precision coordinate measuring machine with a measuring uncertainty of 50 nm in a 100 x 100 x 40 mm measuring volume. The machine has a novel construction, where the air bearing table performs the measuring motion in all the three principal directions. The measuring systems are aligned with the center of the probe tip hence it is possible to comply with the Abbe principle.

Recently in a paper Tan and his coauthors [15] describe the application of neural networks for the error compensation of a single-axis, a gantry and X-Y stage. Using this technique the authors could improve the positioning accuracy depending on the configuration investigated by a factor between 2 and 3.

2 Overview of the Errors and Their Sources

When considering the mechanical accuracy of coordinate measuring devices three primary sources of quasi-static errors can be identified:

- Geometric errors due to the limited accuracy of the individual machine components such as guideways and measuring systems
- Errors related to the final stiffness of those components, mainly by moving parts
- Thermal errors as expansion and bending of guideways due to uniform temperature changes and temperature gradients

Geometric errors are caused by out of straightness of the guideways, imperfect alignment of the axis and flatness errors.

Deformations in the metrology frame introduce measuring errors. During measurement the deformation of the metrology frame is caused by the probing force. Its effect can be predicted with relatively high precision if the probing force is known and therefore it easily incorporated into the model.

The static deformation of the table is caused be gravity forces. It manifests itself as a contribution to out of flatness error. That means it can be handled on a similar way.

The largest deformations of the metrology frame are thermally induced. The main sources of the thermal disturbance are:

- Heating and cooling source in the environment, like lighting, air conditioning, people around the machine etc.
- Heat generated by the machine itself
- Thermal memory: heat stored in the machine components from a previous thermal state

The compensation of thermally induced errors is rather cumbersome, because of the complexity of the problem [12]. Based on results from the literature a linear thermal compensation model can be used.

3 Geometric Error Model

A coordinate measuring machine consists of a chain of translational and/or rotational kinematic components. The geometric deviation of a CMM is originating from the deviations of its components. In order to discuss a general model the error model of the components are discussed first. The measuring loop of a coordinate measuring machine in general is given in

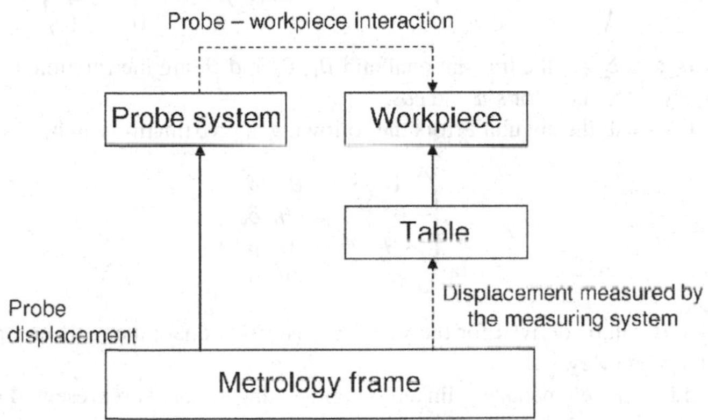

Fig. 1: Schema of the metrology loop

Fig. 1. Depending on the machine configuration all measuring motions are performed by the probe or by the table, or divided between the two.

A linear stage of precision machinery is expected to travel along a straight line and stop at a predefined position. However in practice the actual path deviate from the straight line due to the geometric errors of the guideways and it results also in angular errors as it is given in Fig. 2. For each linear axis a transformation matrix can be used to describe in homogeneous co-ordinates the deviations from the ideal motion. The general form of the transformation is given by:

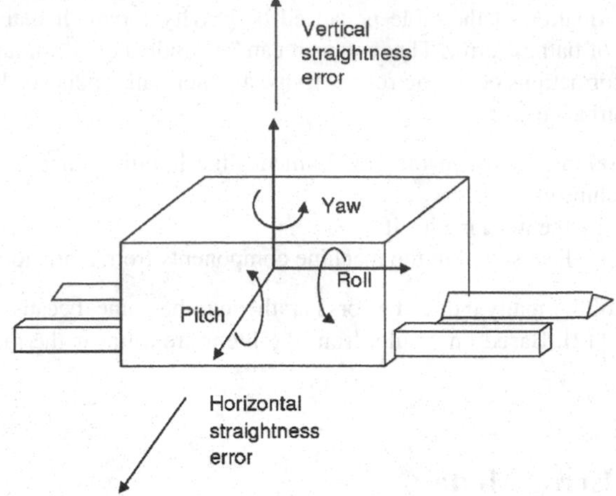

Fig. 2: Representation of the six deviations of a translational kinematic component

$$T_{err} = \begin{pmatrix} c\theta_y c\theta_z & -c\theta_y s\theta_z & s\theta_y & \delta_x \\ c\theta_x s\theta_z + s\theta_x s\theta_y s\theta_z & c\theta_x s\theta_z - s\theta_x s\theta_y s\theta_z & -s\theta_x c\theta_y & \delta_y \\ s\theta_x s\theta_z - c\theta_x s\theta_y c\theta_z & s\theta_x c\theta_z - c\theta_x s\theta_y s\theta_z & c\theta_x c\theta_y & \delta_z \\ 0 & 0 & 0 & 1 \end{pmatrix} \quad (1)$$

Where δ_x,δ_y and δ_z are the translational and θ_x, θ_y and θ_z are the rotational components and s respectively c are short for *sin* and *cos*.

In case of the small the angular errors the following approximation can be made:

$$T_{err} = \begin{pmatrix} 1 & -\theta_z & \theta_y & \delta_x \\ \theta_z & 1 & -\theta_x & \delta_y \\ -\theta_y & \theta_x & 1 & \delta_z \\ 0 & 0 & 0 & 1 \end{pmatrix} \quad (2)$$

Similar results can be derived for the y and z axis. Rotational kinematic components can presented on the same way.

The guided moving elements are linked by connecting elements represented by matrices with similar structure having only constant elements. The resulting error matrix can be obtained by multiplying the individual matrices in the sequence as they follow each other in the kinematic chain.

In case of a measuring probe, where the probe signal is proportional with the deflection of the probe tip, the error components can be handled on the similar way as it was done in case of a carriage and a rotational element.

4 Errors of a Coordinate Measuring Machine with Three Translational Components

In order to illustrate the application of the technique, let us consider a coordinate measuring machine with three translational components, given in Fig. 3. The carriage consists of two nested tables; each of them on vacuum preloaded air bearings. The reference plane surface is lapped to an accuracy of 0.5 μm. The axes are driven by piezomotors. The position are determined by an incremental two co-ordinate optoelectronic measuring system, having 50 nm resolution. This results in a lightweight construction, which in turn ensures fast (acceleration up to 20 m/s^2) and accurate positioning (less than 0.1 μm). The probe is attached to a pinole

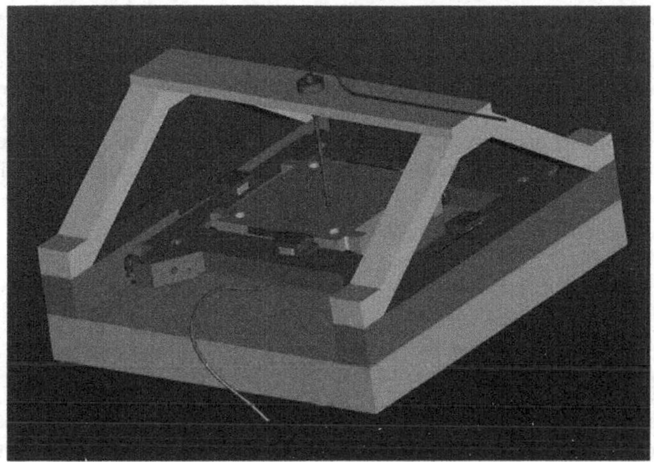

Fig. 3: The investigated coordinate measuring machine

running in air bushing driven by a piezomotor, with approximately 5 nm resolution. A counterweight minimizes the force needed for lifting the probe. The displacement is measured with a linear optoelectronic scale having a resolution of 50 nm. The position dependent transfor-

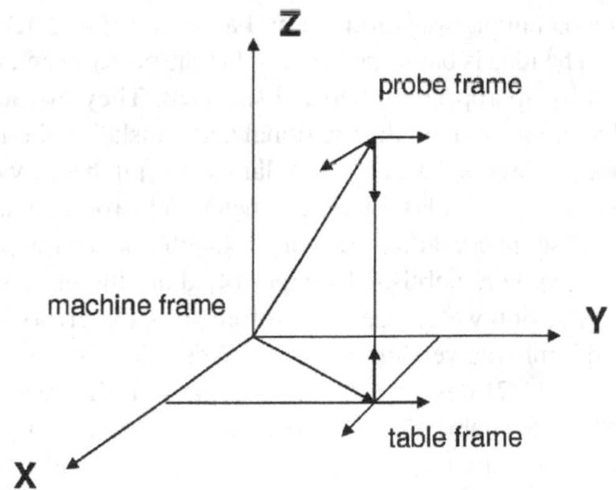

Fig. 4: The frames of the investigated CMM

mation matrices for the table T_t and the probe T_p coordinate frames (Fig. 4) are given below:

$$T_t = \begin{pmatrix} c\theta_y c\theta_z & c\theta_y s\theta_z & s\theta_y & \delta_x \\ c\theta_x s\theta_z + s\theta_x s\theta_y s\theta_z & c\theta_x s\theta_z - s\theta_x s\theta_y s\theta_z & -s\theta_x c\theta_y & \delta_y \\ s\theta_x s\theta_z - c\theta_x s\theta_y c\theta_z & s\theta_x c\theta_z - c\theta_x s\theta_y s\theta_z & c\theta_x c\theta_y & \delta_z \\ 0 & 0 & 0 & 1 \end{pmatrix} \begin{pmatrix} 1 & 0 & 0 & X_c \\ 0 & 1 & 0 & Y_c \\ 0 & 0 & 1 & 0 \\ 0 & 0 & 0 & 1 \end{pmatrix} \quad (3)$$

$$T_p = \begin{pmatrix} c\varphi_y c\varphi_z & c\varphi_y s\varphi_z & s\varphi_y & \varepsilon_x \\ c\varphi_x s\varphi_z + s\varphi_x s\varphi_y s\varphi_z & c\varphi_x s\varphi_z - s\varphi_x s\varphi_y s\varphi_z & -s\varphi_x c\varphi_y & \varepsilon_y \\ s\varphi_x s\varphi_z - c\varphi_x s\varphi_y c\varphi_z & s\varphi_x c\varphi_z - c\varphi_x s\varphi_y s\varphi_z & c\varphi_x c\varphi_y & \varepsilon_z \\ 0 & 0 & 0 & 1 \end{pmatrix} \begin{pmatrix} 1 & 0 & 0 & X_c \\ 0 & 1 & 0 & Y_c \\ 0 & 0 & 1 & Z_{ref} - Z_l \\ 0 & 0 & 0 & 1 \end{pmatrix} \quad (4)$$

Here again δ_x, δ_y, δ_z and ε_x, ε_y, ε_z stand for the translational θ_x, θ_y, θ_z and φ_x, φ_y, φ_z for the rotational errors of respectively the table and the probe. The constant values X_c, Y_c and Z_c are the offset coordinate distances between the machine coordinate system and the coordinate system attached to the workpiece to be measured. Z_{ref} is the probe reference point and Z_l is the probe length.

The actual coordinate values captured by the CMM are:

$$T_{actual} = \begin{bmatrix} x_{actual} \\ y_{actual} \\ z_{actual} \\ 1 \end{bmatrix} = T_{table} = \begin{bmatrix} x_{meas} \\ y_{meas} \\ 0 \\ 1 \end{bmatrix} + T_{probe} \begin{bmatrix} 0 \\ 0 \\ z_{meas} \\ 0 \end{bmatrix} \quad (5)$$

If for accuracy reason one may not substitute α for $\sin\alpha$ and 1 for $\cos\alpha$ than the components of each matrix describing the transformation should be captured simultaneously. The subsequent paragraph gives a suitable method.

5 Determination of the Geometric Errors by Measurement

For the calibration of coordinate measuring machines Zhang et al. [18] proposed to determine the angular errors by measuring the displacement errors along two parallel lines to the axis of motion but separated by a distance in the appropriate orthogonal direction.

A simple measuring technique was invented by Fan et al. [3] to determine the motion accuracy of a linear stage. The idea is based on the fact that the position and orientation of a rigid body can be determined by appropriately selected six point. They measure the displacement of these points and calculate from them the rotational and translational error components. The displacement in the motion direction and the angular errors (pitch and yaw) perpendicular to this directions are measured by three laser interferometers. The roll and the straightness errors are captured by an optical setup containing two quadrant photodetectors.

The above mentioned authors published a paper [4] about the measurement to determine the accuracy of a high precision wafer stage. Therefore six-DOF errors of its positioning accuracy is significant. An improved version of the above described system was used.

In their paper Gao et al. [7] describe the measurement straightness and rotational error motions of a commercially available linear airbearing stage actuated by a linear motor. The pitch and yaw errors were measured by an autocollimator. For the roll error measurement two capacitive displacement probes scan the flat surface in the X-Z plane The probes with their sensing axis in the Y direction were aligned with a certain spacing. The roll error is obtained by dividing the difference of the outputs of the two probes by the spacing between them. The horizontal and vertical straightness error were measured by using the straightness kit of a laser interferometer.

The setup to detect motion errors of the linear stage uses two laser interferometers [13] and three capacitive sensors [10] is given in Fig. 5. The stationary part consists of a single and a dualbeam laser interferometer and three capacitive sensors perpendicular to each other. Both translational and rotational errors can be derived out of the displacement values captured by the transducers (Fig. 6). Where d_1, d_2 and d_3 represent the distance between the laserbeams

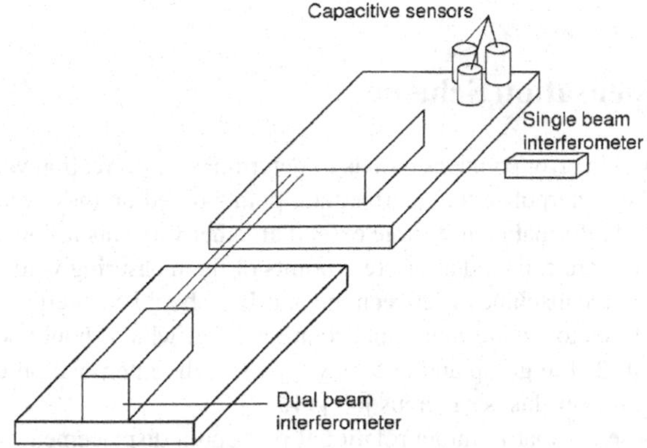

Fig. 5: Setup for determining the motion error of a linear stage

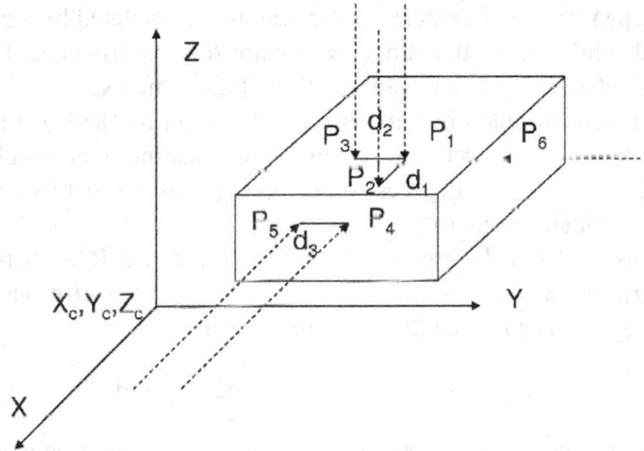

Fig. 6: The measuring points on the surface of the artifact and their relation to each other

respectively the capacitive sensors parallel to the coordinate axis.

By expanding the following determinants the equations of the table's boundary planes can be determined. These equations can be used to compute the origo of the new coordinate system and the orientation of its principal axis. Having these values one can directly draw up the transformation matrix.

$$\begin{vmatrix} x-x_1 & y-y_1 & z-z_1 \\ d_1 & 0 & d_1 tg\theta_y \\ 0 & d_2 & d_2 tg\theta_x \end{vmatrix} = 0, \begin{vmatrix} x-x_4 & y-y_4 & z-z_4 \\ d_3 tg\theta_z & d_3 & 0 \\ tg\theta_y & tg\theta_z & -1 \end{vmatrix} = 0 \tag{6}$$

$$\begin{vmatrix} x-x_6 & y-y_6 & z-z_6 \\ tg\theta_y & tg\theta_x & -1 \\ -1 & tg\theta_z & tg\theta_z tg\theta_x - tg\theta_y \end{vmatrix} = 0 \tag{7}$$

6 Error Compensation Scheme

The conventional way to error compensation is to determine the correction vectors at the measuring points and then interpolate the intermediate points based on these vectors. In order to maintain the captured information to a large extend an other way was followed.

The error matrices are captured at discrete points of the measuring volume. To be able to model the behavior of the machine in between the matrices should be interpolated. Componentwise interpolation leads to a nonorthonormal rotational part which should be avoided. These matrices form the SE(3) Lie group and one may try to do the interpolation using techniques from their theory. However this is a tedious job [11, 12].

Let us decompose the matrix into a rotational part and a displacement vector describing the motion of the machine components as follows:

$$T = \begin{pmatrix} r_{11} & r_{12} & r_{13} \\ r_{21} & r_{22} & r_{23} \\ r_{31} & r_{32} & r_{33} \end{pmatrix} \begin{pmatrix} x \\ y \\ z \end{pmatrix} + \begin{pmatrix} \delta_x \\ \delta_y \\ \delta_z \end{pmatrix} \tag{8}$$

Here the position dependent displacement vectors can be interpolated by a piecewise polynomial space curve [5], while the rotational matrix remains to be interpolated. However there are simple techniques available based on the application of quaternions.

Quaternions [1] were invented by Sir William Hamilton in 1843. He realized that four number are needed to describe a rotation followed by a scaling. One number describes the size of scaling, one the number of degrees to be rotated, and the first two numbers give the plane in which the vector should be rotated.

Quaternions consist of a scalar part $s \in \mathbf{R}$ and $\mathbf{v} = (x,y,z) \in \mathbf{R}^3$: where $\mathbf{i}^2 = \mathbf{j}^2 = \mathbf{k}^2 = \mathbf{ijk} = -1, \mathbf{ij} = kand\mathbf{ji} = -\mathbf{k}$ if q is a quaternion with $q = [cos\theta, sin\theta\mathbf{n}]$ and p is quaternion $p = [0,\mathbf{r}]$ then $p' = qpq^{-1}$ is p rotated 2θ about the axis $\mathbf{n}$

$$q \equiv [s,\mathbf{v}] \equiv [s,(x,y,z)] \equiv s + x\mathbf{i} + y\mathbf{j} + z\mathbf{k} \tag{9}$$

Given a transformation matrix M the corresponding unit quaternion can be calculated in two steps: first we must find s which is equal to:

$$s = \frac{1}{2}\sqrt{(1 + M_{11} + M_{22} + M_{33})} \tag{10}$$

Now the other values follow:

$$x = \frac{M_{32} - M_{23}}{4s}, y = \frac{M_{13} - M_{31}}{4s}, z = \frac{M_{21} - M_{12}}{4s} \tag{11}$$

Where the M-s are the determinants of the respective submatrices.

A so-called spherical linear quaternion interpolation (Slerp) can used to compute the intermediate quaternions. The quaternions generated by Slerp are unit quaternions, which means that they represent pure rotation matrices. The formula for for Slerp is:

$$\cos(\Omega) = q_0 \bullet q_1, \mathrm{Slerp}(q_0, q_1, h) = \frac{q_0 sin((1-h)\Omega) + q_1 \sinh\Omega}{\sin\Omega} \tag{12}$$

where $\bullet$ stands for the inner product defined as $q \bullet q' = ss' + xx' + yy' + zz'$ An even better (smoother) interpolation can be formulated which is the spherical cubic equivalent of a Beziér curve. Let $q_1, ..., q_n$ point on the unit sphere. Find the cubic spline which interpolate the points in the given sequence. This can be achieved by the following formula:

$$Squad(q_i, q_{i+1}, s_i, s_{i+1}, h) = Slerp(Slerp(q_i, q_{i+1}, h)Slerp(s_i, s_{i+1}, h), 2h(1-h)) \tag{13}$$

where s_i are:

$$s_i = q_i.exp(-\frac{\log(q_i^{-1}q_{i+1}) + \log(q_i^{-1}q_{i-1})}{4}) \tag{14}$$

and $\log q$ respectively $\exp q$ are defined as follows:

If $q = [\cos\theta, \mathbf{v}\sin\theta]$, then $\log q \equiv [0, \theta\mathbf{v}]$

If $q = [0, \mathbf{v}\theta]$, then $\exp q \equiv [\cos\theta, \mathbf{v}\sin\theta]$.

The suggested procedure to find the intermediate rotation matrices consists of the following steps:

- Convert the matrices captured using the procedure described in paragraph 4 in quaternions
- Find the Squad interpolation of these points
- Convert the quaterinon splines back into matrix form

On this way a spherical spline representing pure rotation of the object will be obtained. The translational error can be handled by finding a spatial interpolation spline function using for example least square fit. In the possession of these functions one can easily reconstruct the real coordinate values of the measured point.

In case of two-parameter motion (the table moving on a flat reference surface) the above described technique can be extended to surfaces by repeated application of bilinear interpolation on the quaternions at the four corner points.

7 Conclusions

The paper presents a new approach to the compensation of geometric errors in coordinate measuring machines. It consists of a measuring procedure which captures simultaneously the six error components of moving rigid body. The transformation matrices obtained on this way are interpolated by using quaternion representation. Hereby the orthonormality of the rotation matrices are maintained. Simulation values with randomly generated error components showed that the intermediate values lead to accuracy improvement.

Acknowledgments

The author gratefully acknowledges the support obtained within the frames of National Research Fund (OTKA) K 0429305.

References

1. Dam EB, Koch M, Lillholm M (1998) Quaternions, Interpolation and Animation, Technical Report DIKU-TR-98/5, Department of Computer Science, University of Copenhagen
2. Duffie NA, Yang SM (1985) Generation of Parametric Kinematic Error-Correction Function from Volumetric Error Measurement, Annals of the CIRP 34(1):435–438
3. Fan KC, Chen MJ, Huang WM (2010) International Journal of Machine Tools and Manufacturing 38(3):155–164
4. Fan KC, Chen MJ (2000) Precision Engineering 24:15–23
5. Farin G (1993) Curves and Surfaces for Computer Aided Geometric Design: A Practical Guide, Academic, Boston, MA 3rd ed.
6. Ferreira PM, Liu CR (2010) Journal of Manufacturing Systems 5(1):51–62
7. Gao W, Arai Y, Shibuya A, Kiyono S, Park CH (2006) Precision Engineering 30:96–103
8. Hermann Gy (2006) Design Consideration for a Modular High Precision Coordinate Measuring Machine. In: Proceedings of IEEE International Conference on Mechatronics, July 3–5, 2006 Budapest, Hungary, pp. 161–165
9. Huang PS, Ni J (1995) International Journal of Machine Tools and Manufacturing (3):725–738
10. Lion Precision: C6-E Capacitive Probe and Capacitive Compact Driver
11. Richardson RM (2003) Designing Smooth Motion of Rigid Objects, Harvey Mudd College, Clarement, USA
12. Ruijl TAM (2001) Ultra Precision Coordinate Measuring Machine, Ph. D. thesis TU Delft 2001
13. SIOS: Messtechnik GmbH: Miniatur interferometer mit Planspiegel-reflektor SP 2-12/50/2000
14. Slocum AH (1992) Precision Machine Design, Prentice-Hall, Englewood Cliffs, NJ
15. Tan KK, Huang SN, Lim SY, Leow YP, Liaw HC (2006) IEEE Transaction on Systems, Man and Cybernetics 36(6):797–809
16. Teeuwsen JWMC, Soons JA, Schellekens PHJ (1989) Annals of the CIRP 38(1):505–510
17. Vermeulen MMPA (1999) High-Precision 3D-Coordinate Measuring machine Ph. D. thesis TU Eindhoven 1999
18. Zhang G, Ouyang R, Lu B (1988) Annals of the CIRP 37(1):515–518
19. Zhang G, Veale R, Charlton T, Borchardt B, Hocken R (1985) Annals of the CIRP 34(1):445–448

Part V

Systems Engineering

Fixed Point Transformations Based Iterative Control of a Polymerization Reaction

József K. Tar and Imre J. Rudas

Budapest Tech Polytechnical Institution
H-1034 Budapest, Bécsi út 96/b, Hungary
tar.jozsef@nik.bmf.hu, rudas@bmf.hu

Summary. As a paradigm of strongly coupled non-linear multi-variable dynamic systems the mathematical model of the free-radical polymerization of *methyl-metachrylate* with *azobis (isobutyro-nitrile)* as an initiator and *toluene* as a solvent taking place in a jacketed Continuous Stirred Tank Reactor (CSTR) is considered. In the adaptive control of this system only a single input variable is used as the control signal (*the process input, i.e. dimensionless volumetric flow rate of the initiator*), and a single output variable is observed (*the process output, i.e. the number-average molecular weight of the polymer*). Simulation examples illustrate that on the basis of a very rough and primitive model consisting of two scalar variables various fixed-point transformations based convergent iterations result in a novel, sophisticated adaptive control.

1 Introduction

The mathematical foundation of the modern *Soft Computing (SC)* techniques goes back to the middle of the $20th$ century, namely to the first rebuttal of David Hilbert's $13th$ conjecture [1] that was delivered by Arnold [2] and Kolmogorov [3] in 1957. Hilbert supposed that there exist such continuous multi-variable functions that cannot be decomposed as the finite superposition of continuous functions of less variables. Kolmogorov provided a constructive proof stating that arbitrary continuous function on a compact domain can be approximated with arbitrary accuracy by the composition of single-variable continuous functions. Though the construction of Kolmogorov's functions that are used in this theorem is difficult, his theorem later was found to be the mathematical basis of the present *SC* techniques.

From the late eighties several authors proved that different types of neural networks possessed the universal approximation property [4–7]. Similar results have been published from the early nineties in fuzzy theory claiming that different fuzzy reasoning methods are related to universal approximators, too [8–10].

In spite of these theoretically inspiring and promising conclusions, from the point of view of the practical applicability of these methods various theoretical doubts emerged. The most significant problem was, and remained important problem even in our days, the *"curse of dimensionality"* that means that the approximating models have exponential complexity in terms of the number of components i.e. the number of components grows exponentially as the approximation error tends to zero. If the number of the components is bounded, the resulting

J.A.T. Machado et al. (eds.), *Intelligent Engineering Systems and Computational Cybernetics,*

set of models is nowhere dense in the space of the approximated functions. These observations frequently were formulated in a negatory style, as e.g. in [11] stating that *"Sugeno controllers with a bounded number of rules are nowhere dense"*, and initiated various investigations on the nowhere denseness of certain fuzzy controllers containing prerestricted number of rules e.g. in [12–14].

In contrast to these observations *SC* techniques obtained very wide range of real practical applications. As examples implementation of backward identification methods [15], the control of a furnace testing various features of plastic threads [16, 17], sensor data fusion [18] can be mentioned. The methodology of the *SC* techniques, partly concerning control applications, had fast theoretical development in recent years, too. Various operators concerning the operation of the fuzzy inference processes were investigated in [19, 20], minimum and maximum fuzziness generalized operators were invented [21], and new parametric operator families were introduced [22], etc.

To resolve the seemingly "antagonistic" contradiction between the successful practical applications and the theoretically proved "nowhere denseness properties" of *SC* methods one became apt to arrive at the conclusion that the problem roots in the fact that Kolmogorov's approximation theorem is valid for the very wide class of *continuous functions* that contains even very *"extreme"* elements at least from the point of view of the technical applications. The concepts of *continuity* and *smoothness* on the basis of which the "extremeness" of ceratin continuous functions was hypothetically supposed in the 19*th* century only slowly crystallized even amongst the mathematicians. The first example of a function that everywhere is continuous but nowhere is differentiable was given by Weierstraß in 1872. At that time mathematicians believed that such functions are only "rare extreme examples", but nowadays it became clear that the great majority of the continuous functions have "extreme" properties. Intuitively it can be expected that if we restrict our models to the far better behaving "everywhere differentiable" functions the problems with the dimensionality *ab ovo* could be evaded or at least reduced.

The first efforts in this direction were summarized in [23] in which the sizes of the necessary uniform structures used for developing *partial*, *temporal*, and *situation-dependent* models that needed continuous maintaining were definitely determined by the degree of freedom of the system to be controlled. These considerations were based on the modification of the *Renormalization Transformation*, and were valid only for *"increasing systems"* in which the "increase" in the necessary response could be achieved by also increasing the necessary excitation, and *vice versa*.

In the present paper, proceeding further on the path initiated in [24], this idea is systematically extended for *Single Input – Single Output (SISO) "increasing"* and *"decreasing" systems* by developing various *Parametric Fixed Point Transformations* more or less akin to the *Renormalization Transformation*. The applicability of the proposed method is illustrated by the use of a paradigm, namely the control of a polymerization process. In the sequel at first the main idea of the novel adaptive control is developed, then the mathematical model of the paradigm is discussed, and finally simulation results and conclusions are delivered.

2 The Idea of the Novel Adaptive Control

Each control task can be formulated by using the concepts of the appropriate *"excitation"* $\mathbf{Q}$ of the controlled system to which it is expected to respond by some prescribed or *"desired response"* $\mathbf{r}^d$. The appropriate excitation can be computed by the use of some *inverse dynamic model* $\mathbf{Q} = \varphi(\mathbf{r}^d)$. Since normally this inverse model is neither complete nor exact, the actual response determined by the system's dynamics, ψ, results in a *realized response*

$\mathbf{r}^r$ that differs from the desired one: $\mathbf{r}^r \equiv \psi(\varphi(\mathbf{r}^d)) \equiv f(\mathbf{r}^d) \neq \mathbf{r}^d$. It is worth noting that the functions $\varphi()$ and $\psi()$ may contain various hidden parameters that partly correspond to the dynamic model of the system, and partly pertain to unknown external dynamic forces acting on it. Due to phenomenological reasons the controller can manipulate or "deform" the input value from $\mathbf{r}^d$ so that $\mathbf{r}^r \equiv \psi(\mathbf{r}^d_*)$. Other possibility is the manipulation of the output of the rough model as $\mathbf{r}^r \equiv \psi(\varphi^*(\mathbf{r}^d))$. In the sequel it will be shown that for *SISO* systems the appropriate deformation can be defined as some *Parametric Fixed Point Transformation*.

2.1 The Parametric Fixed Point Transformations

Now restrict ourselves to *SISO* systems, and for obtaining the necessary deformation, consider the following functions inspired by simple schematic pictures of geometrically similar triangles for a *decreasing function* with the parameters $\Delta_- < x^d$, and $x > D_-$:

$$g(x|x^d, D_-, \Delta_-) := \frac{f(x) - \Delta_-}{x^d - \Delta_-}(x - D_-) + D_-, f'(x) < 0. \tag{1}$$

It is evident that if $f(x_\star) = x^d$ then $g(x_\star|x^d, D_-, \Delta_-) = x_\star$. That is the original control problem, i.e. finding a properly deformed input $x_\star$ is transformed to finding the solution of a fixed point problem (1). Fixed point problems in general have the advantageous feature that they can be solved via simple iteration provided that this iteration is convergent. Really, consider the sequence of points $\{x_0, x_1 = \Psi(x_0), ..., x_{n+1} = \Psi(x_n), ...\}$ obtained via iteration! Let us suppose that this series converges to some limit value: $x_n \to x_\star$. In order to apply iterations let us consider the set of the real numbers as a *linear normed space with the common addition and multiplication with real numbers, and with the absolute value* $|\bullet|$ *as a norm*! It is well known that this space is *complete*, i.e. it is a *Banach Space* in which the *Cauchy Sequences* are convergent. Due to that, using the norm inequality it is obtained that

$$|\Psi(x_\star) - x_\star| \leq |\Psi(x_\star) - x_n| + |x_n - x_\star| = |\Psi(x_\star) - \Psi(x_{n-1})| + |x_n - x_\star|. \tag{2}$$

It is evident from (2) that if Ψ is continuous then $\Psi(x_\star) = x_\star$, that is the desired fixed point is found by the iteration because the right hand side of (2) converges to 0 as $x_n \to x_\star$. The next question is giving the necessary or at least a *satisfactory condition of this convergence*. It also is evident that for this purpose contractivity of $\Psi(\bullet)$, i.e. the property that $|\Psi(a) - \Psi(b)| \leq K|a - b|$ with $0 \leq K < 1$ is satisfactory since it leads to a *Cauchy Sequence* ($|x_{n+L} - x_n| \to 0$ as $n \to \infty\ \forall L \in \mathbb{N}$):

$$|x_{n+L} - x_n| = |\Psi(x_{n+L-1}) - \Psi(x_{n-1})| \leq ... \leq K^n|x_L - x_0| \to 0. \tag{3}$$

The next question is whether a similar iteration obtained for the function g is convergent in the vicinity of the appropriate fixed point? For this purpose let us restrict ourselves to differentiable $\Psi(\bullet)$ functions with $|\Psi'| \leq K < 1$:

$$|\Psi(a) - \Psi(b)| = \left|\int_a^b \Psi'(x)dx\right| \leq \int_a^b |\Psi'(x)|dx \leq K|a - b| \tag{4}$$

that means that if Ψ is flat enough around the fixed point the iteration will converge to it. Now consider the derivative of g at $x_\star$ where $f(x_\star) = x^d$:

$$g'(x_\star) = f'(x_\star)\frac{x_\star - D_-}{x^d - \Delta_-} + \frac{f(x_\star) - \Delta_-}{x^d - \Delta_-} = f'(x_\star)\frac{x_\star - D_-}{x^d - \Delta_-} + 1 \tag{5}$$

If $x_\star > D_-$, $x^d > \Delta_-$, $f'(x_*) < 0$, and $|f'(x_*)|$ is small enough then the condition $-1 < g'(x_\star|x^d, D_-, \Delta_-) < 1$ can evidently be maintained. Furthermore, it can be also maintained in some vicinity of x_* that $-1 < g'(x|x^d, D_-, \Delta_-) < 1$. By manipulating the properties of the rough model function $\varphi()$ such a situation can evidently be achieved. (A formal possibility for estimating the upper and lower bound of the basin of attraction of $x_\star$ first order Taylor series expansion of $f(x)$ around $x_\star$ can be considered.) On the basis of quite similar considerations further fixed point transformations can be proposed for *decreasing functions* as

$$\begin{aligned} h(x|x^d, D_-, \Delta_+) &= \frac{x^d - \Delta_+}{f(x) - \Delta_+}(x - D_-) + D_-, f'(x) < 0 \\ h(x_\star|x^d, D_-, \Delta_+) &= x_\star \\ h'(x|x^d, D_-, \Delta_+) &= \frac{x^d - \Delta_+}{f(x) - \Delta_+} - \frac{x^d - \Delta_+}{(f(x) - \Delta_+)^2}(x - D_-)f'(x) \\ h'(x_\star|x^d, D_-, \Delta_+) &= 1 - \frac{x_\star - D_-}{x^d - \Delta_+}f'(x_\star) \end{aligned} \tag{6}$$

The condition $-1 < h'(x_\star|x^d, D_-, \Delta_+) < 1$ evidently can be maintained in (6) at $x_\star$ and in its vicinity if $x_\star > D_-$, $x^d < \Delta_+$, $f'(x_\star) < 0$, and $|f'(x_\star)|$ is small enough. In similar manner, for *increasing functions*

$$g(x|x^d, D_-, \Delta_+) := \frac{f(x) - \Delta_+}{x^d - \Delta_+}(x - D_-) + D_-, f'(x) > 0, \tag{7}$$

and

$$h(x|x^d, D_-, \Delta_-) := \frac{x^d - \Delta_-}{f(x) - \Delta_-}(x - D_-) + D_-, f'(x) > 0 \tag{8}$$

can be applied with the appropriate fixed point and convergence properties. These transformations to some extent are different to the two-parametric transformation proposed for decreasing systems in [24] and illustrated in the adaptive control of a cart-pendulum system:

$$w_{\zeta,D}(x) = \frac{f(x) + D + \zeta(f(x) - x^d)}{f(x) + D}x \tag{9}$$

$$w_{\zeta,D}(x_\star) = x_\star \tag{10}$$

$$\left(\frac{dw_{\zeta,D}(x)}{dx}\right)_{x=x_\star} = 1 + \frac{\zeta f'(x_\star)x_\star}{x_d + D} \tag{11}$$

that also may be convergent if $x_\star \approx x^d + D$, the small $\zeta > 0$, $f'(x_\star) < 0$, and $|f'(x_\star)|$ is small, too. However, the behavior of the newer constructions mathematically seems to be simpler than that of (9). In the sequel the potential applicability of this method will be illustrated in the paradigm chosen, that is in the control of the polymerization process.

3 The Model of the Polymerization Process

The chemical reaction considered is the free-radical polymerization of methyl-metachrylate with azobis(isobutyro-nitrile) as an initiator and toluene as a solvent taking place in a jacketed Continuous Stirred Tank Reactor (CSTR). The mathematical model of this process was taken from [25]. According to that the equations of state propagation of this system are given by

$$\dot{x}_1 = A(B - x_1) - Cx_1x_2^{1/2}, \quad \dot{x}_2 = Du - Ex_2, \quad \dot{x}_3 = Fx_1x_2^{1/2}, \quad \dot{x}_4 = Fx_1x_2^{1/2} - Jx_4, \\ y := x_4/x_3, \tag{12}$$

in which the state variables $x_1, \ldots, x_4$ denote *dimensionless* concentrations of various chemical components taking part in the reaction. For our purposes the really interesting variable is the output of the system, that is the *number-average molecular weight* denoted by y. The process input, that is the control signal, u is the *dimensionless volumetric flow rate of the initiator*. The constants in (12) have the following numerical values: $A = 10, B = 6, C = 2.4568, D = 80, E = 10.1022, F = 0.024121, G = 0.112191, H = 10, I = 245.978$, and $J = 10$. It is worth noting that though certain negative values for u may have physical meaning (i.e. a kind of subtraction of the initiator from the system), its practically realizable values are positive numbers or zero. It is easy to see that for a constant process input u (12) yields a stationary solution in which the time-derivatives of the state variables are equal to zero. The stability of these stationary states can easily be shown either by perturbation calculus or by more sophisticated numerical calculations. To substantiate that computations were done for a quite drastic step-function ($5 \times 10^{-3} \to 15 \times 10^{-3}$) for u. As it is revealed by Fig. 1 the state variables move from one stationary state to another one through approximately exponential behavior that shows a quasi-linear part. For control purposes it is expedient to take it into account that we have special functions

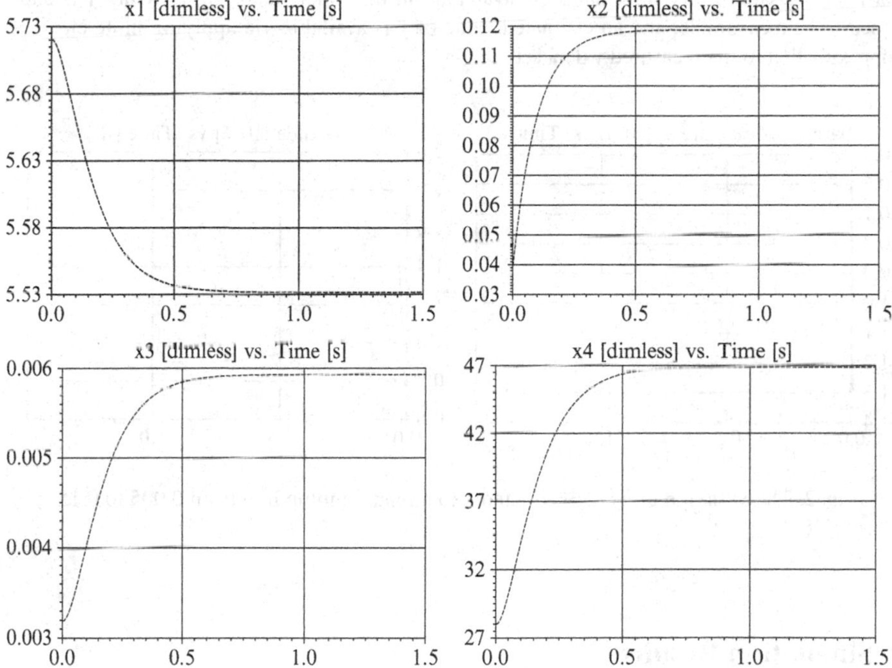

Fig. 1: The reaction of the state variables to a random jump in u from 0.005 to 0.15

as $y(x)$, $\dot{y}(x,\dot{x})$, $\dot{x}(u,x)$, therefore $\dot{y}(u,x)$, and $\ddot{y}(u,\dot{u},x)$. More specifically it can be obtained that

$$\dot{y} = \dot{x}_4 x_3^{-1} - x_4 x_3^{-2} \dot{x}_3, \qquad \ddot{y} = \ddot{x}_4 x_3^{-1} - 2\dot{x}_4 x_3^{-2} \dot{x}_3 - 2x_4 x_3^{-3} \dot{x}_3^2 - x_4 x_3^{-2} \ddot{x}_3 \tag{13}$$

Again, according to (12):

$$\ddot{x}_4 = I\dot{x}_1 x_2^{1/2} + 0.5 I x_1 x_2^{-1/2}\dot{x}_2 - A\dot{x}_4, \qquad \ddot{x}_3 = F\dot{x}_1 x_2^{1/2} + 0.5 F x_1 x_2^{-1/2}\dot{x}_2 + D\dot{x}_2 - H\dot{x}_3 \quad (14)$$

Taking into account that in the whole $\{\dot{x}_i\}$ set only $\dot{x}_2$ depends directly on u, it is obtained that

$$\frac{\partial \ddot{x}_4}{\partial u} = 0.5 I x_1 x_2^{-1/2} D, \qquad \frac{\partial \ddot{x}_3}{\partial u} = 0.5 F x_1 x_2^{-1/2} D + GD, \quad (15)$$

therefore the relatively simple result is obtained that

$$\frac{\partial \ddot{y}}{\partial u} = \frac{0.5 I D x_1}{x_3 x_2^{1/2}} - x_4 x_3^{-2}\left(0.5 F x_1 x_2^{-1/2} + G\right) D. \quad (16)$$

Equation (16) evidently suggests similarity with the state propagation of *Classical Mechanical Systems* in the sense that $\ddot{y}$ is proportional to u as the second derivatives of the generalized co-ordinates are proportional to the generalized force components: $\ddot{y} = \tilde{a}(\mathbf{x})u + \tilde{b}(\mathbf{x})$. The "inertia" of this system is not constant but depends on $\mathbf{x}$. The only essential difference is, that according to Fig. 2, this "inertia" is stably *negative* value during the whole relaxation process. On this basis the terminology of *Classical Mechanics* will be used in the sequel, and a very simple dynamic and adaptive controller will be designed to control it. The adaptive nature of the controller is important since even in the possession of the precise model of the process normally no satisfactory information is available on the exact state $\{x_i, \dot{x}_i\}$. Only y is supposed to be measurable, and an estimated value of $\ddot{y}$ is available via applying finite element approximation to the measured y data set.

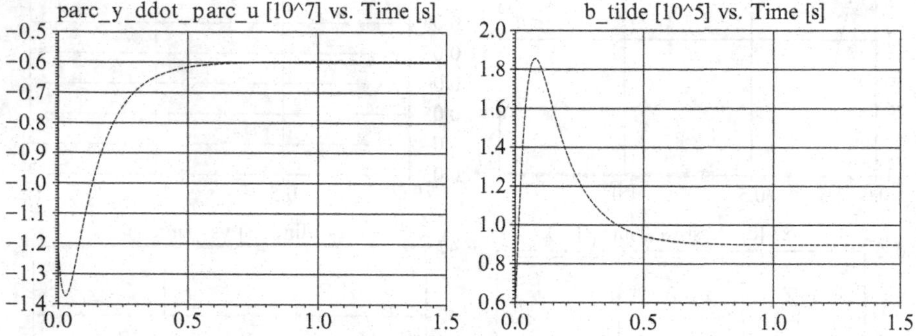

Fig. 2: The variation of $\tilde{a} := \partial\ddot{y}/\partial u$ and $\tilde{b}$ to a random jump in u from 0.005 to 0.15

4 Simulation Results

In the forthcoming simulations a very rough "constant" model of $\tilde{a}_{Mod} = -6 \times 10^6$ and $\tilde{b}_{Mod} = 10^5$ was used in harmony with the results of the calculation shown in Fig. 2. The necessary trajectory tracking was designed by using only "kinematic needs" in a PD style as $\ddot{y}^d = \ddot{y}^{Nom} + d(\dot{y}^{Nom} - \dot{y}) + p(y^{Nom} - y)$. The appropriate excitation was calculated according to the "desired acceleration" and the model as $u = \left(\ddot{y}^d - \tilde{b}_{Mod}\right)/\tilde{a}_{Mod}$ resulting in the actual acceleration $\ddot{y} = \frac{\tilde{a}\cdot\ddot{y}^d}{\tilde{a}_{Mod}} - \frac{\tilde{a}\cdot\tilde{b}_{Mod}}{\tilde{a}_{Mod}} + \tilde{b}$ that is an *increasing function*of $\ddot{y}^d$ if $\tilde{a} < 0$ and

$\tilde{a}_{Mod} < 0$, therefore for the purposes of the adaptive control the functions $g(x|x^d, D_-, \Delta_+)$ and $h(x|x^d, D_-, \Delta_-)$ can be used. During one control cycle only one step of iteration was executed that is $\ddot{y}(t_n) = g(\ddot{y}(t_{n-1})|\ddot{y}^d(t_n), D_-, \Delta_+)$ or $\ddot{y}(t_n) = h(\ddot{y}(t_{n-1})|\ddot{y}^d(t_n), D_-, \Delta_-)$ with $D_- = -7 \times 10^4$, $\Delta_- = -2 \times 10^6$, and $\Delta_+ = 2 \times 10^6$ with a time resolution $t_{n+1} - t_n = 0.067$ s. The adaptive factor was $s_{rel} = \frac{\ddot{y}^d(t_n)-\Delta_-}{\ddot{y}(t_n)-\Delta_-}$ for the function h, and $s_{rel} = \frac{\ddot{y}(t_n)-\Delta_+}{\ddot{y}^d(t_n)-\Delta_+}$ for the function g, respectively. If the adaptation is faster than the dynamics of the system to be controlled appropriate result can be expected. Figure 3 well represents the inadequacy of the rough model and the successful operation of the adaptive controller designed for this system for a considerable nominal "velocity" and acceleration range occurring in the harmonic nominal trajectory.

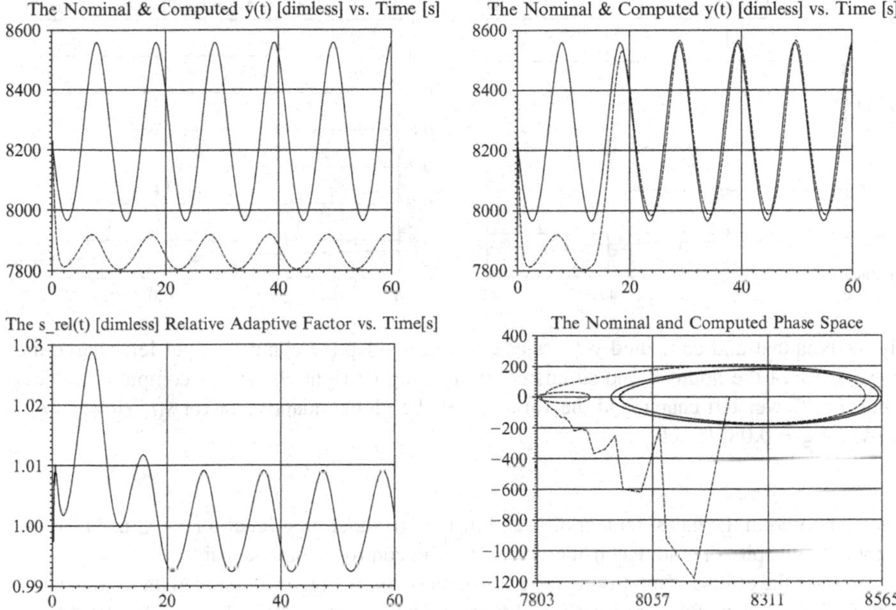

Fig. 3: Nominal and computed $y(t)$ trajectory for the rough model based control (upper left chart), for the adaptive control using function g (upper right chart), the variation of the adaptive factor $s_{rel}(t)$ (lower left chart), and the phase trajectories of the nominal and simulated motion in the adaptive control (lower right chart)

It is very interesting that similar results can be achieved by applying the fixed-point transformation defined in (9) with $\zeta = 0.08$, $D = 1$ (all the other parameters of the controls applied were the same) (Fig. 4).

In this latter case $s_{rel}(t_n) = \frac{f(x(t_{n-1})+D+\zeta\times(f(x(t_{n-1})))-x^d(t_n)}{f(x(t_{n-1})+D}$, and the adaptive factor is defined as $s(t_n) := \prod_{j=0}^{n} s_{rel}(t_j)$. (In this case this cumulative factor is more informative since (9) does not have additive correction after multiplying x with it.)

Regarding robustness issues either the effects of the measurement noises and that of the modification of the control parameters in tracking the same nominal motion with the same model can be investigated. Figure 5 represents the noisycounterpart of Fig. 3 when noises

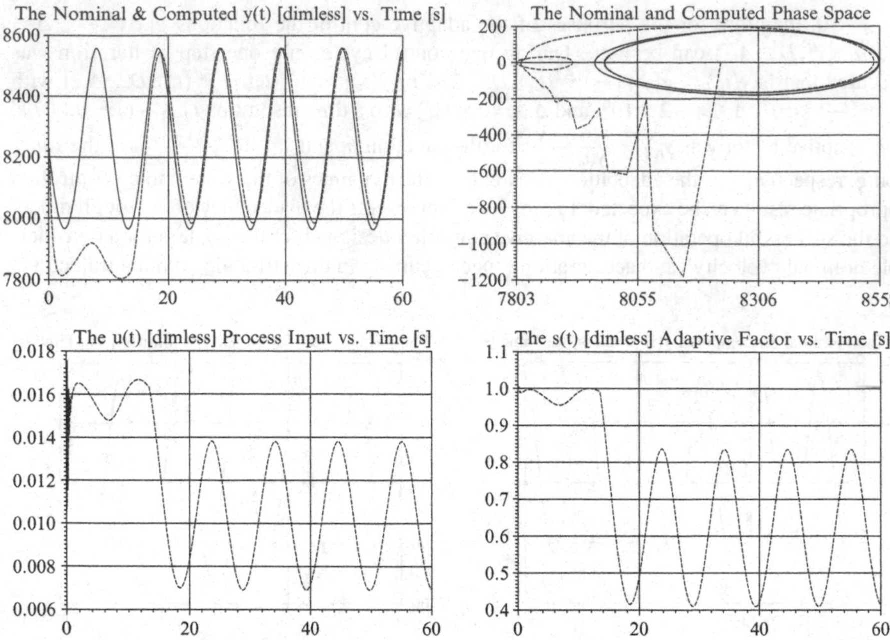

Fig. 4: Nominal and computed $y(t)$ trajectory of the adaptive control (upper left chart), the phase space of the nominal and simulated motion (upper right chart), the computed process input $u(t)$ (lower left chart), and the variation of the global adaptive factor $s(t)$ (lower right chart) for $\zeta = 0.08, D = 1$

were supposed in the measurement of $\ddot{y}$, and instead of function g function h was used for the adaptation. In spite of a quite considerable noise the control remained stable.

To trace the effects of modifying the control parameters Fig. 6 was created that is the counterpart of Fig. 3 using function g with modified parameters ($D_- = -7\times10^4 \to -8\times10^4, \Delta_+ = 2\times10^6 \to \Delta_+ = 10^6$). It can well be seen that in spite of the quite significant modification of the control parameters the control remained stable though the tracking accuracy worsened.

5 Conclusions

To sum up, in this paper various parametric Fixed Point Transformations were introduced to implement iterative adaptive control for *SISO* systems. Satisfactory conditions were provided for the convergence of the method that was also exemplified by the control of a considerably nonlinear plant, a Continuous Stirring Tank Reactor controlling a polymerization process. Robustness and noise–tolerance of the method was demonstrated by drastic modification of certain control parameters that are responsible for the convergence and adaptivity of the method.

By the use of the concepts of “acute” and “obtuse” angles the method expectedly can be extended to MIMO systems.

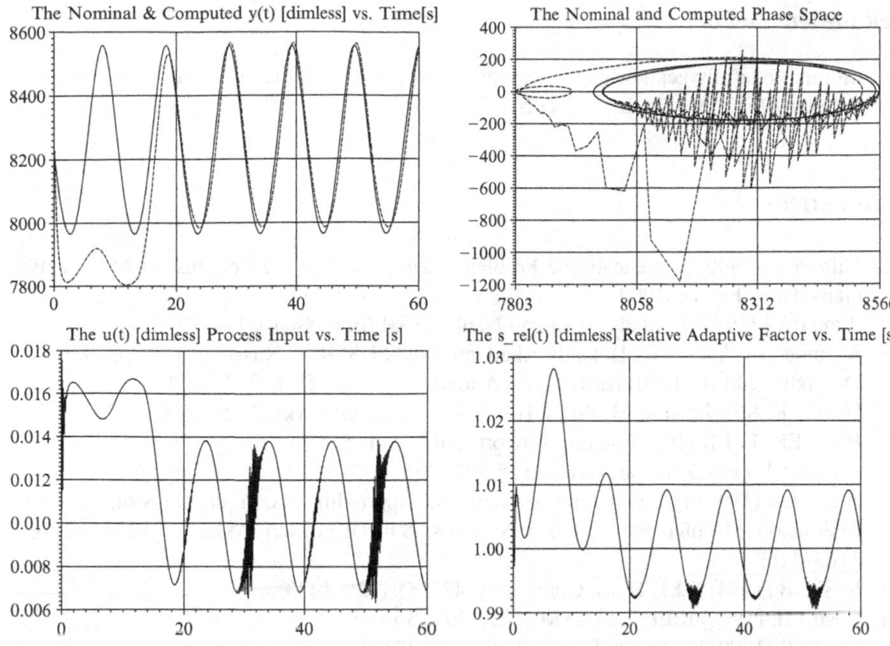

Fig. 5: Results for noisy measurement of $\ddot{y}$ in the adaptive control using function h: Nominal and computed trajectories (upper left chart), the phase space of the nominal and simulated motion (upper right chart), the computed process input $u(t)$ (lower left chart), and the relative adaptive factor s_{rel} vs time (lower right chart)

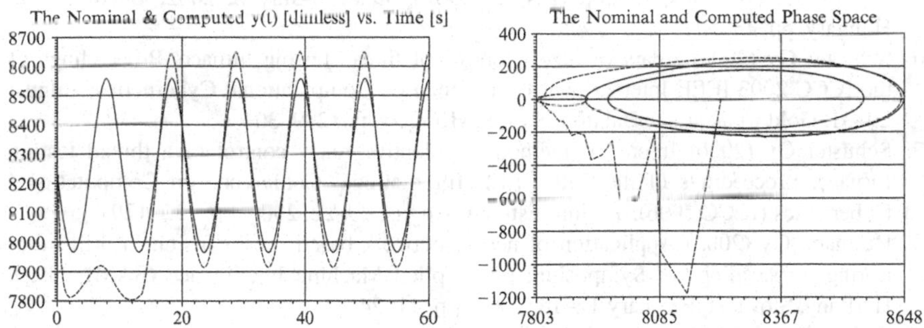

Fig. 6: Results for modified control parameters of function g as $D_- = -8 \times 10^4$, $\Delta_+ = 10^6$: Nominal and computed $y(t)$ trajectory of the adaptive control (upper left chart), the phase space of the nominal and simulated motion (upper right chart) (counterpart of the case described in Fig. 3)

Acknowledgments

The authors gratefully acknowledge the support obtained from the *Hungarian National Research Fund (OTKA)* within the project No. *K063405*.

References

1. Hilbert D (1900) Mathematische Probleme. 2*nd* International Congress of Mathematicians, Paris, France, 1900
2. Arnold VI (1957) Doklady Akademii Nauk USSR (in Russian), 114:679–668
3. Kolmogorov AN (1957) Doklady Akademii Nauk USSR (in Russian), 114:953–956
4. De Figueiredo JP (1980) IEEE Trans. Automat. Contr., 25(6):1227–1231
5. Hornik K, Stinchcombe M, White H (1989) Neural Networks, 2:359–366
6. Blum EK, Li LK (1991) Neural Networks, 4(4):511–515
7. Kurková V (1992) Neural Networks, 5:501–506
8. Wang LX (1992) Fuzzy systems are universal approximators. Proceedings of the IEEE International Conference on Fuzzy Systems, San Diego, CA, March 8–12, 1992, pp. 1163–1169
9. Kosko B (1994) IEEE Trans. Computers, 43(11):1329–1333
10. Castro JL (1995) IEEE Trans. SMC, 25:629–635
11. Moser B (1999) Fuzzy Set. Syst., 104(2):269–277
12. Tikk D (1999) Tatra Mountains Math. Publ., 16:369–377
13. Klement EP, Kóczy LT, Moser B (1999) Int. J. Gen. Syst., 28(2):259–282
14. Tikk D, Baranyi P, Patton J (2002) Polytopic and TS model are nowhere dense in the approximation model space. Proceedings of IEEE International Conference on Systems, Man and Cybernetics SMC 2002, Hammamet, Tunisia, October 6–9, pp. 150–153
15. Schuster Gy (2002) Fuzzy approach of backward identification of quasi-linear and quasi-time-invariant systems. Proceedings of the 11th international Workshop on Robotics in Alpe-Adria-Danube Region (RAAD 2002), June 30–July 2, 2002, Balatonfüred, Hungary, pp. 43–50
16. Schuster Gy (2003) Adaptive fuzzy control of thread testing furnace. Proceedings of the ICCC 2003 IEEE International Conference on Computational Cybernetics, August 29–31, Gold Coast, Lake Balaton, Siófok, Hungary, pp. 299–304
17. Schuster Gy (2006) Improved method of adaptive fuzzy control of a thread testing furnace. Proceedings of the 2006 IEEE International Conference on Computational Cybernetics (ICCC 2006), Tallinn, Estonia, August 20–22, 2006, pp. 125–129
18. Hermann Gy (2003) Application of neural network based sensor fusion in drill monitoring. Proceedings of Symposium on Applied Machine Intelligence (SAMI 2003), Herl'any, Slovakia, February 12–14, 2003, pp. 11–24
19. Tick J, Fodor J (2005) Some classes of binary operations in approximate reasoning. Proceedings of the 2005 IEEE International Conference on Intelligent Engineering Systems, (INES'05), September 16–19, 2005, pp. 123–128
20. Tick J, Fodor J (2005) Fuzzy Implications and Inference Processes. CAI 24(6):591–602
21. Rudas IJ, Kaynak MO (1998) Fuzzy Set. Syst., 98(1):83–94
22. Rudas IJ (2000) Int. Fuzzy Syst., 2(4):236–243
    Industrial Technology (ICIT 2002), vol. II, pp. 1290–1295, Bangkok, Thailand, December 11–14, 2002

23. Tar JK, Rudas IJ, Szeghegyi Á, Kozłowski K (2006) Novel Adaptive Control of Partially Modeled Dynamic Systems. In: Lecture Notes in Control and Information Sciences, Springer, Berlin/Heidelberg, Robot Motion and Control: Recent Development, Part II - Control and Mechanical Systems, (Ed. Krzysztof Kozłowski) 335/2006:99–111
24. Tar JK (2005) Extension of the modified renormalization transformation for the adaptive control of negative definite SISO systems. In the Proceedings of the *2nd* Romanian-Hungarian Joint Symposium on Applied Computational Intelligence (SACI 2005), May 12–14, 2005, Timişoara, Romania, pp. 447–457
25. Doyle FJ, Ogunnaike BK, Pearson RK (1995) Automatica 31:697

Reasoning in Semantic Web Services

Igor Toujilov and Sylvia Nagl

Department of Oncology, Royal Free and University College Medical School,
Rowland Hill Street, London NW3 2PF, UK
i.toujilov@medsch.ucl.ac.uk; s.nagl@medsch.ucl.ac.uk

Summary. This article investigates what kind of web services is the most effective for intelligent systems. We present a new method of web services development that allows the execution of client-defined scripts on the server side. Then we demonstrate the new system architecture and features implemented as an ontology server for the CancerGrid project. The authors believe that there will be broad applicability of the proposed method for future web services.

1 Introduction

A major problem of existing computer systems is data integration between heterogeneous systems. Interfaces between different research systems are urgently needed because data mapping between multidisciplinary databases and further computer-aided analysis of the merged data are key to the discovery of new knowledge.

The problem is especially obvious in the context of rapid and numerous changes in the full lifecycle of systems. Change request management mechanisms of most computer systems are not flexible enough for preserving sufficient interoperability in modern heterogeneous environments like the Internet.

It is clear that we need a more flexible system architecture, a model driven architecture (MDA) [1]. New data and knowledge languages and standards have emerged recently for this purpose: XML, XMLS, RDF, RDFS, and ontology web language (OWL) (listed in the order of increasing level of abstraction). These W3C languages serve the main methodology of MDA, i.e. a model transformation approach: semantics and knowledge models can be described in OWL, RDF, and RDFS and then transformed to data and forms models in XML and XMLS; data can be transformed to forms and vice versa using XML, XSL, and XMLS.

Another big problem of existing computer systems is separation of data and semantics. The meaning of data is not coded explicitly in the vast majority of systems. That is why the data can be interpreted only in the context of a native system, not that of an external system. Usually even in a native system, semantics is not well defined, so change requests cause many errors in data interpretation. That is why future systems need knowledge / semantics models first of all. Data models can then be derived from semantic ones, using MDA principles. Having explicit representation of knowledge, reasoners can automatically map, merge and analyze data from various heterogeneous systems, solving the disambiguation problem.

J.A.T. Machado et al. (eds.), *Intelligent Engineering Systems and Computational Cybernetics*,

Web services technology [2] developed rapidly during the last few years is glue for numerous components and systems of the modern computing environment. This technology had created a new type of system architecture called a service-oriented architecture (SOA) [3].

We investigate what kind of web services is the most effective for intelligent systems. Then we present a new method of web services development that allows the execution of client-defined scripts on the server side. At last we demonstrate how the proposed system architecture works in the case of its example implementation as an ontology server for the CancerGrid project [4].

The purpose of the CancerGrid project is to create knowledge and data representation standards for multidisciplinary research in clinical oncology and biomedical informatics in the UK. The main activities of the project include ontology development, knowledge representation and metadata modeling. Our approach is based on MDA and SOA principles and on application of OWL to biomedical ontology development. Some clinical trial protocols are used as domain models for the research. Ensuring the reusability of established biomedical ontologies, for example, the NCI Thesaurus [5], is also among the primary goals of the research.

2 The Idea of a Client Script Execution

Consider an example of complex web services, e.g. the US National Cancer Institute Centre for Bioinformatics (NCICB) system [6]. The system supports the conventional web services that are based on Java API. Let us suppose a client extracts value domains for data elements. Table 1 illustrates a possible way of doing that, and the approximate quantity of data elements in this example is given in parentheses. Also suppose the approximate average traffic of a server request-response cycle is 1 Kbyte and the response time is 1 s. The time of client side analysis is neglected.

Table 1: The use case of NCICB web services

Pseudo code	Traffic (Kbyte)	Time (s)
For all data elements (30000)		
Get a data element	1	1
Analyze the data element	0	Neglected
Get the value domain	1	1
Analyze the value domain	0	Neglected
Total	60,000	60,000

This simple analysis shows that we need approximately 60 Mbyte of traffic and 16 h for executing the task using the conventional web services.

To overcome the associated waste of computational resources, and also to achieve the additional advantages described below, we propose a new architecture for web services that execute client-defined scripts or client scripts, for brevity. (For simplicity we here also do not make a difference between a client-defined script and a middleware-defined script, as it is not significant for the purposes of this paper.) Table 2 shows how the new web services might process the example use case.

Table 2: The use case of new web services that execute client scripts

Pseudo code	Traffic (Kbyte)	Time (s)
Submit a script to the server	Few	1
Execute the script on the server	0	Few
Get a result from the server	Final results only, no intermediate traffic	Duration of transferring the final results
Total	Final results only plus few Kbytes	Duration of transferring the final results plus few seconds

Although the idea of client custom script execution on the server side is simple, it is, at present, uncommon for standard web services. Indeed, there are several problems regarding implementation of this idea:

- It is not obvious which runtime environment, script language and execution mechanism may be the best ones for different use cases.
- Security problems: an attacker may submit a malignant script to the server or, accidentally, a wrong script may harm the system.

These problems are closely connected to each other, as it is shown below.

2.1 The Runtime Environment, Script Language and Execution Mechanism

Currently there are two widely used types of development frameworks for web services, Java based and .NET based ones. None of them has the sufficient support for the safe server-side execution of client-defined scripts. It is not easy to programmatically analyze for security purposes the structure of a client-defined script in Java, JavaScript, JScript, C#, Visual Basic or VBScript.

It is possible to use two different programming languages for web services implementation and for client scripts. This approach implies an implementation of a runtime script compiler or interpreter and requires significant efforts. An additional mechanism for getting the reusable functionality of the implementation system from client scripts has to be implemented which requires additional effort.

Another possibility is the usage of a meta-programming language like LISP or Prolog for both the system implementation and the client scripts. We define a language as a meta-programming language when it is well suited for both interpreting a program as data and data as a program. This ability makes a meta-programming language the ideal tool for implementing web services that execute client scripts. That is why we decided to use this approach for the development of our system.

More specifically, SWI-Prolog [7] was selected as the implementation and client-defined scripting language due to the following reasons:

- Prolog (Programming in Logics) has the logics-theoretical semantics connected to the description logics that is the base of the OWL.
- Prolog is a computer language for knowledge representation that is also a purpose of the OWL.
- Prolog is a fundamental, powerful programming language of very high level, and is well established.

- Prolog is a concise and rich query language, ideal for scripting.
- SWI-Prolog is one of the best open-source Prolog implementations, encompassing the Semantic Web library and the Triple20 package for OWL processing.

2.2 The Solution of the Security Problems

Each client script is being automatically analyzed before its execution. If the script contains an unacceptable predicate, the execution will be refused.

Then each acceptable script is being automatically filled with monitoring codes that watch for a processing time limit and safely interrupt the script during its execution when the processing time goes out of the limit.

2.3 The Advantages and Disadvantages of the New Web Services

The web services that execute client scripts have the following advantages:

- The performance has been dramatically increased; it is clear when comparing Table 1 and Table 2.
- As a client always gets final results only, not traffic for intermediate results, the network traffic is minimal; it is also obvious from Tables 1, 2.
- As a client always gets all data currently needed by one query only, there is no need to program transactions from the client side (each query is a transaction).
- It is possible to program very complex business logics in only one query (similar to the batch processing or submitting a job in the grid computing).
- Using the standard and library predicates in the scripts, the client has access to most functionality of the underlying Prolog system; there is no need to reinvent the wheel, because all the basic functionality, e.g. mathematical expressions, is reusable.
- Message structures for the web services are inherently very simple: in a request we need a data item to contain the script and probably we need a few general parameters; in a response we need a resulting data item whose structure may be defined by the script (and again we may need a few general data items); as long as the message structures are so simple they are going to be very stable; that means we will have stable and reliable interfaces leading to high interoperability of web services.

To be honest, the new method also has at least one known disadvantage: basic Prolog knowledge is required for a middleware and client software developer.

Comparing the advantages and disadvantages of the newly proposed method for developing web services, it can be concluded that the method may be applied successfully.

3 Software Engineering Ecology

For conceptual analysis of an innovation in software systems it is useful to consider how the innovation fits in the modern philosophy of software architecture and to which degree it follows (or does not do) the current trends in software engineering. For that purpose we analyze the proposed approach in the context of software engineering philosophy.

We introduce software engineering ecology based on an analogy between the general and industrial ecology and modern software engineering, see Table 3.

Table 3: Analogy between ecology and software engineering

General and industrial ecology	Software engineering
Energy	Human mind
To use energy sparingly	To lower consumption of human mind
Pollution	Software bugs
Replaceable parts	Software libraries and web service components

Human mind for software engineering is an equivalent of energy for ecology. The more energy we use, the more pollution (heat and contaminants) we emit and the greater our expenses are. In software engineering the more human mind we consume, the more software bugs we produce (it is not a surprise that software bugs are generated by humans, not machines) and the more expensive the software lifecycle is. All the history of software engineering shows a tendency to software lifecycle automation, i.e. lowering consumption of human mind. However nowadays software bugs might be considered as a global disaster.

One of the main approaches to overcoming the bug problems is reusability of well established software libraries and web service components those are equivalents of replaceable parts in industry. Starting from 18th century, it was the invention of replaceable parts, which boosted industry and economy. Today we still have a lack of reliable software libraries and web service components that could boost software engineering in the future.

In this context the idea of the web services based on the server side execution of client's scripts is very promising because of its basic reusability feature for an underlying server system. The proposed approach also fits in well with the modern architecture of grid computing where the client submits jobs to the grid for execution.

Another principal concept of replaceable parts is that their contact surfaces are very simple (cylinder, plane) and standardized. In the software engineering analogy, that means that web service interfaces should be very simple in order to provide runtime replace-ability potentially leading to very reliable web services based on multiple component reservation. And again that complies well with the proposed approach having inherently very simple structures of web service messages.

Thus we can see the web services based on the server side execution of client's scripts fit very well in the modern philosophy and ecology of software engineering. We believe that future web services can apply the proposed method to advantage.

The remaining part of the article describes details of the ontology server based on the proposed approach.

4 CancerGrid Ontology Server

The ontology server is an environment for integration of different ontologies that come from different sources. Each ontology is considered as a module. This modular approach implies the following rules:

- Third party ontologies are not editable, they are read-only, in order to simplify their upgrading to new versions.
- Newly constructed ontologies should be of reasonable size, i.e. their OWL files should not be overpopulated, in order to simplify exporting them to third parties and avoid unnecessary overload on memory.

- Like clusters, ontologies should be logically coupled more inside than outside as much as possible.

As we represent common data elements (CDEs) in the form of an ontology, general approaches for editing of ontologies and CDEs are the same.

The CancerGrid ontology server has the following features:

- Integrated vocabulary ontologies and CDEs in OWL as primary medium.
- Two interfaces (support the same functionality): an end-user can communicate directly via a web browser; software can communicate using SOAP over HTTP.
- The query language is Prolog.
- The implementation environment is SWI-Prolog with SemWeb and Triple20 packages [7].

After directing the web browser to the server, one gets a web form for querying the server. This form contains also a table of query examples, as shown in Fig. 1.

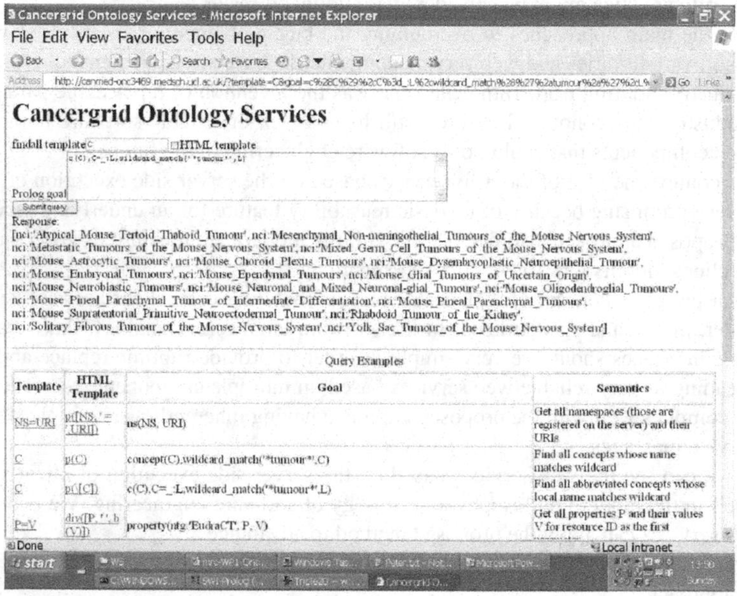

Fig. 1: The web browser interface of the CancerGrid ontology server

There are two main fields of the query form: the Prolog template and Prolog goal. Given a template `T` and a goal `G`, the server executes the standard predicate `findall(T, G, L)`, and then sorts list `L` in order to avoid duplicated elements and replies with the result list.

An optional feature, the HTML template, which is available in the HTML interface only, allows interpreting the template as a Prolog-encoded HTML text in order to customize a representation of the server response. The HTML encoding rules are specified in the `html_write` library of SWI-Prolog.

The CancerGrid server at present supports the following basic ontologies:

- Imported NCI Thesaurus, a large ontology containing general cancer-specific knowledge.

- Imported extended metadata registry (XMDR), containing extended ISO 11179 standard for CDE representation [8].
- CancerGrid CDE ontology that imports NCI Thesaurus and XMDR ontology (Fig. 2).

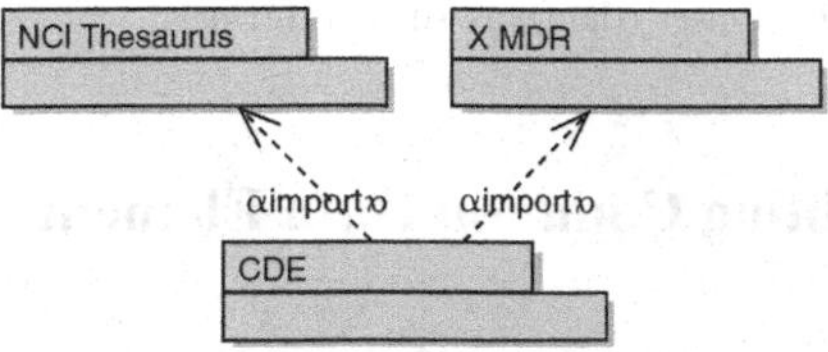

Fig. 2: Basic ontologies in UML 2.0

To get a list of all currently loaded ontologies, one needs to use a request like:

```
Template: OntURI
Goal: p(OntURI, rdf:type, owl:'Ontology')
```

5 Common Data Elements

CDEs are standardized by the XMDR framework which is a further development of the ISO 11179 standard. Practice of working with the NCICB core infrastructure (caCORE) [6] shows that a CDE repository conceptually has all features of ontologies. Moreover, each CDE must be linked with a well known concept from a vocabulary (in our case, ontology). That is why, for uniformity, our CDE repository is implemented as an ontology.

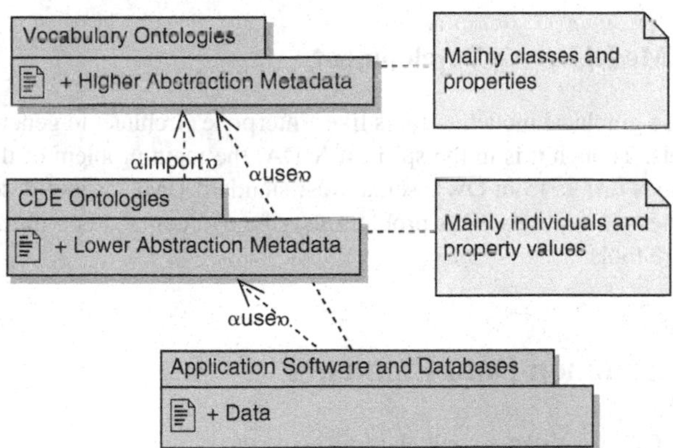

Fig. 3: CDE metadata provide a bridge between vocabularies and applications (UML 2.0)

The difference between a CDE ontology and a vocabulary ontology (like NCI Thesaurus or XMDR) is the following. A vocabulary ontology contains mainly specification of classes and properties. A CDE ontology contains mainly individuals of XMDR classes and property values for these individuals. Thus the abstraction level of vocabulary ontologies is higher than that of CDE ontologies, and CDE metadata provide a bridge between vocabularies (metadata) and software applications/databases (data), as shown in Fig. 3.

6 Creating and Editing Common Data Elements

6.1 OWL Coding

Direct OWL coding is always available, but it requires knowledge of OWL. While it provides the highest degree of freedom and flexibility, it has the highest exposure to human errors.

6.2 Using Protege

The intelligent software package Protege can be used for OWL code generation for CDEs. Unfortunately the following problems exist:

- Windows version of Protege requires a machine with at least 1 Gbyte of physical memory in order to import the large OWL file of NCI Thesaurus.
- Protege works very slowly with large OWL files like NCI Thesaurus.
- Protege reformats an entire OWL file, even if only a small piece of code should be edited; as the reformatting is unpredictable, it complicates usage of file compare features of a version control system for OWL files.

To overcome these problems, additional software tools are needed and may be developed in the future.

6.3 Graphical Model Driven Development

It is possible to use graphical modeling tools like Enterprise Architect to generate OWL code from UML models. Though it is in the spirit of MDA, the main problem of this approach is that UML semantics differs from OWL semantics: standard UML cannot describe all OWL features, e.g. property transitivity. This problem may be overcome later with development of additional software tools.

7 Querying Common Data Elements

To get a list of all CDEs, we can use e.g. the following query:

```
Template: CDE
Goal: rdfs_individual_of(CDE, xmdr:'DataElement')
```

Another way of doing so:

```
Template: CDE
Goal: p(CDE, rdf:type, xmdr:'DataElement')
```

Given a CDE name, Breast Tumour Size, the next query returns all properties and their values for the data element:

```
Template: P, V
Goal: p(cde:indDE_Breast_Tumour_Size, P, V)
Response: [ (rdf:type, xmdr:'DataElement'),
(xmdr:administrationRecord, cde:indARDE_Breast_Tumour_Size),
(xmdr:domain, cde:indVD_Size),
(xmdr:example, literal('Breast tumour size is 20 mm.')),
(xmdr:meaning, cde:indDEC_Breast_Tumour_Size),
(xmdr:submitter, cde:indSubmit_IgorToujilov),
(xmdr:terminologicalEntry, cde:indTEDE_Breast_Tumour_Size)]
```

8 Reasoner Features of Cancergrid Ontology Server

8.1 Sub-properties

Consider the following request:

```
Template: V
Goal: p(cde:indARDE_Breast_Tumour_Size,
xmdr:administrativeStatus, V)
Response: [literal('Draft new')]
```

The value `'Draft new'` was not directly specified for property
`xmdr:administrativeStatus`
of CDE administrative record
`indARDE_Breast_Tumour_Size`.
Instead it was specified for property `cde:administrativeStatusEnum`. Based on the fact that `cde:administrativeStatusEnum` is a subproperty of
`xmdr:administrativeStatus`,
it was deduced that `xmdr:administrativeStatus` has that value.

This simple example shows a property value inference based on a value of a sub-property of the first underlying hierarchical level. However this reasoning feature works also in the general case of a multilayered property hierarchy.

8.2 Inverse Properties

Let us explore a CDE conceptual domain representation in an OWL file:

```
<xmdr:NonEnumeratedConceptualDomain rdf:ID="indCD_Size">
  <rdfs:isDefinedBy rdf:resource="&nci;#Size"/>
  <xmdr:representation rdf:resource="#indVD_Size"/>
  <xmdr:administrationRecord rdf:resource="#indARCDSize"/>
  <xmdr:terminologicalEntry rdf:resource="#indTECDSize"/>
  <xmdr:submitter rdf:resource="#indSubmit_IgorToujilov"/>
</xmdr:NonEnumeratedConceptualDomain>
```

This part of the code specifies the conceptual domain individual `indCD_Size` and its four XMDR property values. The following request displays all XMDR property values of the conceptual domain:

```
Template: P, Obj
Goal: p(cde:indCD_Size, xmdr:P, Obj)
Response: [(administrationRecord, cde:indARCDSize),
(application, cde:indDEC_Breast_Tumour_Size),
(representation, cde:indVD_Size),
(submitter, cde:indSubmit_IgorToujilov),
(terminologicalEntry, cde:indTECDSize)]
```

As can be seen, all the four property values are displayed. However the additional value for the fifth property application is displayed also. The reason is that the application property value is inferred from its inverse domain property value specification which occurred elsewhere in the OWL file.

8.3 Immediate Subclasses and Superclasses

Predicate `sc/2` implements the immediate subclasssuperclass relationships including inferred ones. For example in order to get immediate superclasses of concept Mercaptopurine, we would use the following query:

```
Template: C
Goal: sc(nci:'Mercaptopurine', C)
Response: ['__Description116', '__Description117',
'__Description118', '__Description119',
'__Description120', '__Description121',
nci:'Immunosuppressant', nci:'Purine_Antagonist']
```

That means that Immunosuppressant and Purine Antagonist are immediate superclasses of Mercaptopurine. However if one looks at the OWL description of Mercaptopurine in NCI Thesaurus, one will find this fact is not directly specified:

```
<owl:Class rdf:ID="Mercaptopurine">
  <rdfs:label>Mercaptopurine</rdfs:label>
  <code>C195</code>
  <owl:equivalentClass>
    <owl:Class>
      <owl:intersectionOf rdf:parseType="Collection">
        <owl:Class rdf:about="#Immunosuppressant"/>
        <owl:Class rdf:about="#Purine_Antagonist"/>
        <owl:Restriction>
          ...
        </owl:Restriction>
        ...
      </owl:intersectionOf>
    </owl:Class>
  </owl:equivalentClass>
  ...
</owl:Class>
```

Immunosuppressant and Purine Antagonist are inferred to be immediate superclasses of Mercaptopurine, based on the class expression in the description of Mercaptopurine where Immunosuppressant and Purine Antagonist are members of an intersection. In the request response above, one can see also other members of the intersection which appear as anonymous descriptions.

References

1. Atkinson C, Kühne T (2003) IEEE Trans Software 20:36–41
2. Chappell D, Jewell T (2002) Java web services. O'Reilly, Beijing Cambridge Farnham Koln Paris Sebastopol Taipei Tokyo
3. Erl T (2004) Service-oriented architecture: a field guide to integrating XML and web services. Prentice-Hall, NJ
4. CancerGrid project website, http://www.cancergrid.org
5. NCI Terminology Browser, http://nciterms.nci.nih.gov/NCIBrowser
6. NCI Centre for Bioinformatics, http://ncicb.nci.nih.gov
7. SWI-Prolog website, http://www.swi-prolog.org
8. Extended Metadata Registry (XMDR) project, http://xmdr.org

Defining Requirements and Applying Information Modeling for Protecting Enterprise Assets

Stephen C. Fortier[1] and Jennifer H. Volk[2]

[1] George Washington University, School of Engineering and Applied Science, Washington, DC, USA
sfortier@gwu.edu

[2] George Mason University, The Volgenau School of Information Technology and Engineering Fairfax, Virginia, USA
jvolk@gmu.edu

Summary. The advent of terrorist threats has heightened local, regional, and national governments' interest in emergency response and disaster preparedness. The threat of natural disasters also challenges emergency responders to act swiftly and in a coordinated fashion. When a disaster occurs, an ad hoc coalition of pre-planned groups usually forms to respond to the incident. History has shown that these "system of systems" do not interoperate very well. Communications between fire, police and rescue components either do not work or are inefficient. Government agencies, non-governmental organizations (NGOs), and private industry use a wide array of software platforms for managing data about emergency conditions, resources and response activities. Most of these are stand-alone systems with very limited capability for data sharing with other agencies or other levels of government. Information technology advances have facilitated the movement towards an integrated and coordinated approach to emergency management. Other communication mechanisms, such as video teleconferencing, digital television and radio broadcasting, are being utilized to combat the challenges of emergency information exchange. Recent disasters, such as Hurricane Katrina and the tsunami in Indonesia, have illuminated the weaknesses in emergency response. This paper will discuss the need for defining requirements for components of ad hoc coalitions which are formed to respond to disasters. A goal of our effort was to develop a proof of concept that applying information modeling to the business processes used to protect and mitigate potential loss of an enterprise was feasible. These activities would be modeled both pre- and post-incident.

1 Introduction

The threat of terrorism has expanded the definition of critical infrastructure protection. The government, as well as private enterprises, must be aware of the threats and prepare themselves to protect their critical assets. The definition of "what is a critical asset" is also being reevaluated. Society has seen new threat modes, such as 9/11, improvised explosive devices and suicide bombings, that require much better planning to prevent these types of incidents.

J.A.T. Machado et al. (eds.), *Intelligent Engineering Systems and Computational Cybernetics*,

Emergency and disaster preparedness planning is an evolving science. Mechanisms such as continuity of operations planning (COOP), indication and warning systems, and vulnerability assessments are used to plan for emergencies or disasters and model the potential weaknesses in enterprise assets (such as policies, physical assets, personnel, procedures and methods). Threat models are used by the government and private industry to analyze and prepare for possible threats to critical infrastructure, and to mitigate the effects of a potential loss.

All of the above disciplines are usually created or engineered as a stove-pipe component. Within a design space, a piece of hardware, set of software, or process are not holistically engineered within the context or application to an emergency or disaster situation. Traditional system engineering approaches are not adequate to define and design solutions for ad hoc coalition systems during disasters. The application of "system of systems" engineering does provide hope for solving this problem. Additionally, the use of a system level design language will provide the mechanism to achieve effective and efficient interaction between components of ad hoc coalitions.

An information model is a formal description of an area of interest, a domain. It specifies the objects within the domain, the relationships between the objects, the basic attributes of the objects and the constraints upon the objects and their relationships.

Initially one uses the unambiguous properties of information models to document a common understanding of domains. In this case, we are looking at safety and security aspects of critical enterprise assets. There are two aspects to writing good information models. First, producing a model that describes the required information, and second, writing it in a good style. The second normally requires a modeling expert to talk to domain experts and write the information model from the information gathered.

The first aspect requires domain experts to review the model written by the modeling expert. This need for review means that mathematical based formalisms, such as VDM and Z, are inappropriate modeling formalisms as domain experts have found them difficult to comprehend. Experience has shown that programming language-like formalisms are comprehensible to domain experts. Hence, we use the EXPRESS language (ISO Standard 10303-11) [17] for our information modeling related work. Graphical formalisms, such as UML, are not used as these do not contain sufficient rigor, particularly in the description of constrains.[1] The domain of interest of this paper is a defense contractor as an enterprise.

The execution of continuity of operations planning (COOP), vulnerability assessments, business continuity planning (BCP), and threat analysis are usually done apart from one another and they are not associated with each other. The government and private industry would be better served if these analyses were interlinked and information was shared between the mechanisms.

We have researched this area and identified a new way to leverage the above mechanisms to achieve information reuse and improved information awareness. We used a systems integration approach to support risk mitigation and vulnerability reduction.

2 Background

A defense contractor is usually a for-profit business organization that provides the federal government products or services. The traditional customer is the Department of Department (DoD). The products that are typically provided include military aircraft, vehicles, ships, arms, and electronic systems. Services can include logistics, information technology, training (e.g.,

[1] Discussions with Professor Hilary Kahn, The University of Manchester.

warfare simulation), computer program development and communications support. Defense contractors generally do not provide direct support to military operations. But, defense contractors are an integral part of the combat solution. Typically, defense contractors are on or near the front lines during combat, and provide technical expertise as weapon systems become increasing sophisticated.

Defense contractors usually have their offices and factories in the United States, but the larger businesses tend to be more global and could potentially have facilities in foreign countries.

The main stakeholders are the customers, stockholders, senior management, and program managers. The customer, usually DoD, can make extraordinary adjurations on a defense contractor to deliver their contracted products or services. This is especially true during a time of war, or when a mission capability is incapacitated, or has the potential of being incapacitated.

Since the customer has such urgent demands, a defense contractor must be able to keep the production and manufacturing lines up and running, in the case of producing a product. In terms of services, a defense contractor must have hot backups or quickly get a cold site up and running so that the mission requirements are not interrupted. Since 9/11 the federal government has been emphasizing their continuity of government (COG) and continuity of operations planning (COOP). The following tables list the potential events or occurrences that could lead to a disaster. Each event has a suggested probability and relative impact. This type of analysis highlights to the stakeholders where the enterprise resources should be allocated. Table 1 lists natural or environmental disasters.

Table 1: Environmental disasters

Potential disaster	Probability rating	Impact rating
Tornado	3	2
Flood	5	4
Electrical storms	3	4
Fire	4	2
Contamination and environmental hazards	5	4
Earthquake	3	3

Table 2 reviews the organized and/or deliberate disruptions that could potentially occur. The legend that explains the ranking for the probability rating and the impact rating are contained in Table 3.

Table 2: Man-made disasters

Potential disaster	Probability rating	Impact rating
Act of terrorism	3	3
Act of sabotage	3	3
Act of war	2	3
Theft	3	5
Arson	3	5
Insider malfeasance	2	3

Table 3: Legend for Tables 1 and 2

Probability rating		Impact rating (financial)	
Score	Level	Score	Level
1	VERY HIGH	1	TERMINAL – 90%
2	HIGH	2	DEVASTATING – 50%
3	MEDIUM	3	CRITICAL – 25%
4	LOW	4	CONTROLLABLE – 10%

In the event of a disaster, information will need to be exchanged between the enterprise, emergency responders, public safety through related applications and systems. The following are examples of data-sharing activities:

- Warnings and indications, and alerting for first responders and the public
- Resource identification, including tasking and tracking
- Incident reporting and tracking
- Planning and monitoring of enterprise operations
- Staff, personnel and organizational management
- Geospatial characterizations and tracking of resources and potential hazards
- Sensor monitoring and data acquisition systems
- Emergency response activities

This is a complex problem with a lot of data available, but not much of it is coordinated in a near-real time fashion. The Fairfax County Government[2] estimates that they have 600 different information systems that could potentially provide information during a disaster.

2.1 Standards Activities

There are a number of emergency data standards activities on-going in the USA. They include:

- The National Information Exchange Model (NIEM)
- IEEE 1512
- The Common Alerting Protocol (CAP)
- The Emergency Data Exchange Language (EDXL)
- The Global Justice XML Data Model (GJXDM)
- Risk Analysis and Management for Critical Asset Protection (RAMCAP)

The NIEM is a joint venture between the U.S. Government, led by the Department of Homeland Security (HLS) and the Department of Justice with coordination of other government agencies. The program was established to facilitate growth and harmonization of the GJXDM data model while investigating new data components [20].

Recognizing the need for intelligent transportation systems (ITS), the U.S. Department of Transportation (USDOT) and the IEEE are developing a family of standards for ensuring safety, protecting the environment, and attempting to relieve traffic congestion [21]. The family of standards, known as IEEE 1512, provides incident management message sets for public safety, common traffic management, and hazardous material incident response activities.

[2] Fairfax County is a municipality of the state of Virginia in the USA. It abuts the District of Columbia (D.C.) to the south and has approximately 750,000 residents.

The CAP was developed to coordinate emergency warning systems. The CAP is a digital format for expressing content of warning messages, regardless of the technology of the delivery system. A CAP message is used to activate Emergency Alert Systems, trigger sirens for first responders, telephone notification systems and weather radios [22].

The EDXL is a product of the Emergency Interoperability Consortium (EIC), which is comprised of a partnership between the U.S. government and industry members [23]. The EDXL is a standard for emergency data message types including message routing instructions, situation status and forecasts, and resource queries and requests.

The GJXDM is an SML-based standard developed specifically for information exchange for the criminal justice system. It provides public safety agencies, law enforcement, public defenders, prosecutors, and the judicial branch with a tool to improve information exchange in a timely manner [24].

Developed by ASME Innovative Technologies Institute, RAMCAP is a framework for analyzing and managing risks associated with terrorist attacks against critical infrastructure [25]. It provides a methodology to identify, analyze, quantify and communicate characteristics and impacts that terrorists may use to target a specific asset. The work described in this paper is consistent with the work of other activities.

In order to understand "system of systems" and its application to my problem set, you need to understand its origins. System engineering can be thought of as a framework for the entire design process from which derives the need for analysis to reduce risk against each requirement. Thomas [1] suggests that "system engineering is an element in the product development process with three basic development paradigms: waterfall model, V model and the spiral model."

The V model was created principally by the software industry [2]. The requirements for any task are defined first. A corollary requirement definition is created to go with the definition of how the required task will be validated as complete. The task is then conducted with the engineer aware of all the related requirements and the acceptance criteria. Upon task completion, acceptance is carried out against the acceptance criteria which were already defined. System engineering methods are good, but they fail when applied to system of systems,[3] especially the V Model.

2.2 System of Systems

What is required to make system of systems more efficient when faced with risk, hazard and loss? Boehm [3] suggests that system of systems should be agile and applied to system of systems engineering. He also suggested that one must address the socio- and political aspects of the systems. One should look at:

- Operational Concepts
  - Elaboration of system objectives and scope by increment
  - Elaboration of operational concept by increment
- System Prototypes
  - Field test usage scenarios
  - Resolve major outstanding risks
- System Requirements
  - Definition of functions, interfaces, quality attributes, and prototypes by increment
  - Stakeholders' concurrence on their priority concerns

[3] Conversation with Dr. Bret Michael, Naval Post Graduate School, March 2005.

  - Stakeholders' concurrence on their priority concerns
- System and Software Architecture
  - Choice of architecture and elaboration by increment
    - Physical and logical components, connectors, configurations and constraints
    - Commercial off-the-shelf (COTS) and reuse
    - Domain architecture and architectural style choices. Enterprise architecture is required by the U.S. Government
  - Evolving architecture parameters
- Life Cycle Plan
  - Elicitation of who, what, where, when, why, how and how much
  - Partial elaboration, identification of key to be determined for later increments
- Feasibility Rationale
  - Assurance of consistency among elements above
  - All major risks resolved to be covered by a comprehensive risk management plan

The Government and the private sector have the need to protect its critical infrastructure. There are critical operations that need to operate uninterrupted, such as banking, telecommunications, the power grid or military operations. In order to be successful, we will apply the principles of system of systems engineering concepts. The risk of having these systems taken off line requires an enterprise to make a substantial investment to protect itself from various threats. In a number of industries, these threats are not clearly known at this time. There is a resource allocation issue in determining whether or not to commit resources to a problem that may never happen.

Threats to an enterprise come in various ways; they could be from terrorists, a natural disaster, insider threat, corporate espionage, or cyber attack. An enterprise can combat the threat by conducting analyses or assessment of potential threats. This should all occur, hopefully, before an incident happens.

As technologies make advances, problems of increasing complexity are facing decision-makers within government and industry. The key characteristic of these problems is that they are of system of systems type [12]. Most environments contain multiple, heterogeneous, distributed systems which makes effective analysis for decision support unmanageable within the stovepipe context.

The use of traditional system engineering methods is not enough to solve current problems [13]. The notion of System of Systems (SoS) is gaining popularity, especially in the DoD and other government agencies. "In its classic form as usually taught and practiced, a set of system requirements drives the design. But there is generally no questioning about where these requirements themselves came from. No one has first thought at a somewhat higher level about how to optimize the solution to the problem. System engineering is also predicated on the assumption that if you give the requirements you can build the system. Yet, how do we determine requirements? What if there is a problem with determining what the requirements are."

Interoperable systems can be summarized as "a combination of tight and loose coupling between various systems of systems components [14]." Tight coupling usually occurs between systems that perform closely related functions, while looser coupling occurs where opportunities for interoperation arise among systems not originally developed to interoperate.

Clark and Jones [15] proposed the Organizational Interoperability Maturity Model which is comprised of five levels of organization and maturity, describing the ability to interoperate. They include:

- Level 0: Independent
- Level 1: Ad hoc

- Level 2: Collaborative
- Level 3: Integrated (also called Combined)
- Level 4: Unified

"During low risk times an enterprise operates as if its communications and information systems were unified, yet during a crisis the communications and information elements tend to operate ad hoc or independently. In order to be effective; communications and information systems need to be unified, for example, they need to have seamless interoperability."

2.3 System Level Design Language

True system level design offers hope to integrate communication and information systems.

System-level design is characterized by the requirement to bring together different types of information from multiple domains to envision the impacts of design decisions. To sustain system-level design, a language must allow heterogeneous specification while at the same time provide mechanisms to construct information across domains.

The Rosetta language [6] was originally developed to facilitate the "system on a chip" design problem. The developers found that the language could also be extended to handle complex systems [4].

The goal of the Rosetta system-level design language [5] is to compose heterogeneous specifications in a single environment. Rosetta provides modeling support for different design domains employing semantics and syntax appropriate for each. Hence, individual specifications are written employing semantics and vocabulary appropriate for their domains. Information is composed across specification domains by defining interactions between them. "Rosetta provides a collection of domains for describing system models called facets. Interactions provide a mechanism for defining constraints between domains."

Facets construe system models from one engineering perspective. Facets are composed by extending a domain that provides vocabulary and semantics for the model. Using the design abstractions provided by its domain, a facet describes requirements, behavior, constraints, or function of a system [7]. Domains provide the vocabulary and semantics for defining facets. Each domain provides mechanisms for describing data, computation and communication models appropriate for one area of systems design.

Interactions define how the information is reflected from one engineering domain to another. Domains do not share a common semantic, but instead share information, when required, using interactions. Thus, each design facet is defined using appropriate design abstractions from its domain rather than forcing a common design model across all facets. This allows system developers to design solutions using their native languages and tools. Figure 1 illustrates the relationships between the components of the Rosetta System Level Design Language.

Facet algebra expressions use facets, domains and interactions to compose models into robust system descriptions. Local design decisions can be evaluated from a systems perspective by using interactions to understand impacts in other system domains.

3 Proposed Solution

The ultimate goal would be to model the information systems and communication systems of all components of emergency responders. This would allow swifter and efficient response to disaster situations.

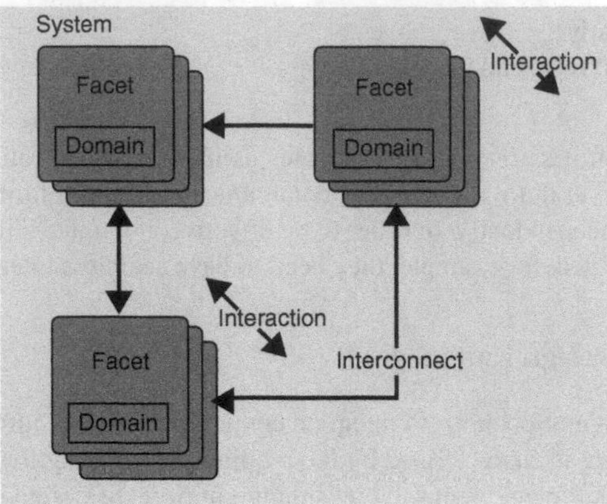

Fig. 1: Rosetta System Level Design Architecture

The question becomes, "How do you model the emergency response elements when you do not know which elements will come into play, given any random emergency or disaster?" To get a better understanding, we developed an abstract view of the disciplines associated with this problem. Figure 2 highlights the relationship among the disciplines.

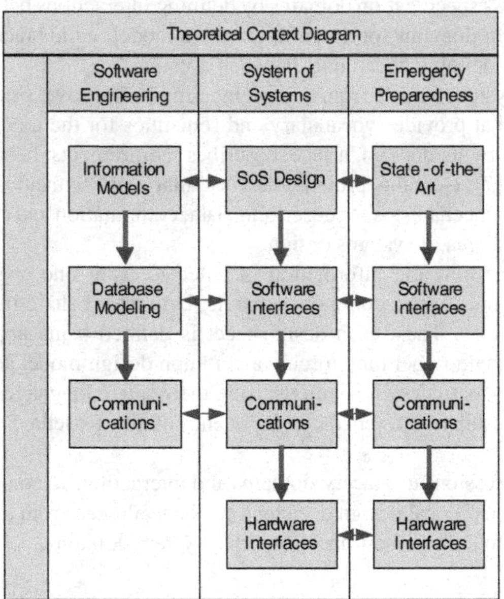

Fig. 2: Theoretical Context Diagram of domain of discourse

There are some assertions proposed for applying system of systems principles mitigating and protecting enterprise assets:

1. An enterprise is always in one of three states: risk/hazard/loss. The state will depend on the internal and external forces levied against the enterprise. Even in the best condition of preparedness, the enterprise is in a risk state.
2. During the risk state, the enterprise has the best chance to influence positive outcomes, thereby the best chance to prevent a loss and optimize the resources available.
3. During the hazard state, the enterprise encounters a first level of system of systems problems.
4. During the loss state, the enterprise experiences system of systems interactions, usually with unplanned interfaces and undefined requirements. Therefore, it is predicted that the "system of systems" operates in a chaotic state. These chaotic states tend to correct themselves over a period of time.
5. During a harmful contact incident the control of the health and well being of an enterprise is usually shifted to the emergency response mechanism; while the goal of the enterprise is to establish COOP/Continuity of Government (COG) and/or rapidly return operations to an on-line status. The objective of the enterprise during this state is to minimize the loss experienced during and post harmful contact incident while maintaining operation of the enterprise, even in a degraded state.
6. Situational awareness is critical during all three states. This has been proven in all emergency situations.
7. Information needs to be effectively shared during all three states. Again, this has been proven in all emergency situations.
8. An enterprise's understanding of its system of systems defense mechanism makes it better prepared as the enterprise shifts from a risk state à hazard state à loss state.
9. System of systems will morph from formalized to ad hoc to chaotic as it progresses through the three states.
10. Developing, practicing, and training for ad hoc system of systems allows an Enterprise to:
    (a) Minimize the chaos
    (b) Require less resources
    (c) Ensure less loss
11. There is a need to develop a mechanism to allow for system assets to enter and exit from protective mechanisms. This would be analogous to an application protocol interface (API) in software products.

The application of system of systems engineering requirements in understanding the interaction, quantify measurement methods and develop meaningful frameworks for analysis was the focus of this research.

A typical ad hoc coalition made up of pre-planned groups would look something like the picture in Fig. 3. Depending on the location and severity of the disaster, these types of elements would respond. Consider that each element, Fire and Rescue for example, has a unique set of information and communication systems. The initial requirement would be to define those systems.

Private individuals have responded to disasters by providing support to one of the other elements, or by providing physical asset, such as heavy equipment or vehicles. Non government organizations, such as the Red Cross or UNICEF, provide valuable resources during disasters. Yet, coordination and communication is not very efficient as experienced recently during Hurricane Katrina [8].

3.1 Application of System Level Design

The first step would be to examine technology specific model within our domain. One could develop specific reference models for the technology involved in potential ad hoc coalitions. Figure 4 illustrates the identification and abstractions of the technology models.

The goal is to write specifications with the semantics and vocabulary appropriate for each specific domain. Rosetta provides modeling support for different design domains employing semantics and syntax appropriate for each. The original intent of the semantics and syntax was to target at the complex system level, so we attempted to write code on an example of system communication constraints and system cost constraint for illustrative purposes.

Figure 5, below illustrates how we attempted to limit our example, where the domain is both communication and cost, and the facets are system communication constraint and system cost constraint.

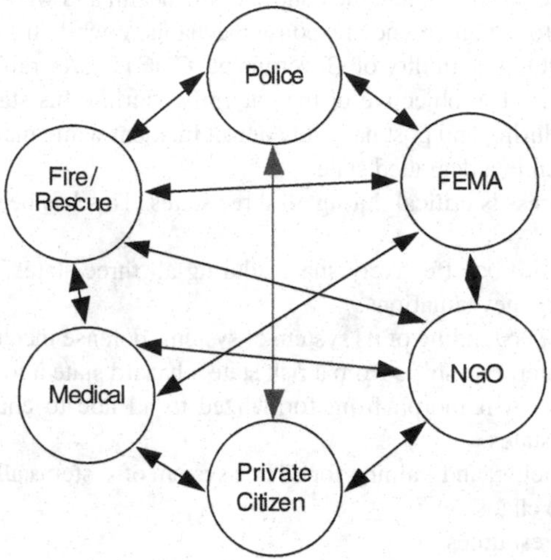

Fig. 3: Typical emergency response elements

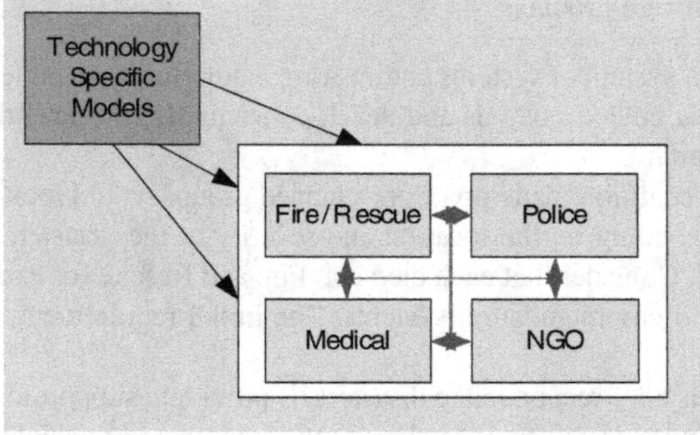

Fig. 4: Technology specific models are identified

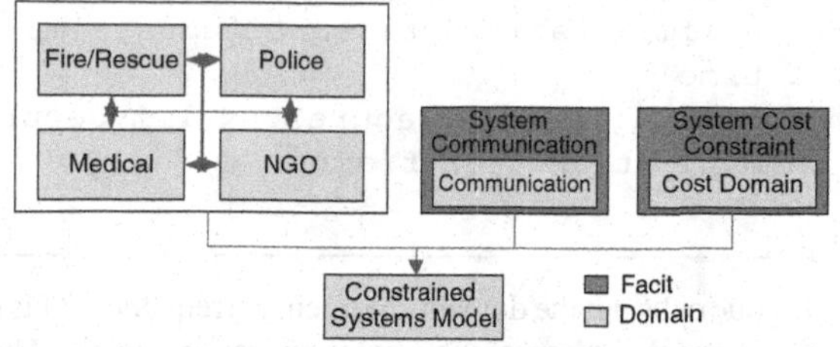

Fig. 5: Constrained system model using communication and cost domain

Provided below is example code for cost constraints. The initial cost creates the cost constraint facet and the latter functor allows the constraint set to be utilized among all of the domains [10].

```
facet cost_constraint(cost, limit::input real;
alert::output real)::continuous_time is cost::real;
begin
  ct: cost' = cost +
       if (cost >= limit) then alert else 0.0
  end if;
//this functor makes it possible for this constraint
  info to move among domains

limit_cost(lim::static;
ct::continuous_time)::continuous_time is
    add_term(forall(t::time | cost@t < lim.cost), ct);
```

The following example sets up a communication constraint with a security domain within the facet [9]. This also calls out the domain, emergency_frequency.

```
facet communications_constraint(freq, range, spectrum::real;
access, security::literal)::continuous_time is
  communication::literal;
begin
emergency_frequency
// terms
//example of security domain within communications facet:
domain security::state-based is riskType::posreal;
p,nominal::riskType;
activityType::real = sel(x::real | x >=0.0 and x=<1.0);
activity::activityType
begin
  p = activity*nominal+latent;
end domain security;
...
```

```
limit_security(lim::static; ct::continuous_time)
::continuous_time is
  add_term(forall(t::time | secure@t < lim.secure), ct);
end facet communications_constraint;
```

The above sample code calls out the domain emergency_frequency. This domain describes the approved frequency ranges for emergency communications in the United States. This would be used when an emergency communications asset enters a coalition communications link. Based on the emergency frequency asset specification, the entering asset could sync with the appropriate communications type. The sample code is highlighted below.

```
domain emergency_frequency (freq :: in real;
MHz :: out real) is

//declarations

begin frequency_specifications

MIL1: MHz =limit(30 < freq < 46);
MIL2: MHz = limit(30 < f req< 88);
COM1: MHz = limit(87.5 < freq< 108);
FAA1: MHz = limit(108 < freq <  118);
MIL3: MHz = limit(225 < freq < 400);
EM1: MHz = limit(450 < freq < 470);
EM2: MHz = limit(470 < freq< 572);
EM3: MHz = limit(849 < freq < 869);
EM4: MHz = limit(1850 < freq < 1910);

end emergency_frequency;
```

3.2 Information Modeling Methodology

We used a methodology, taken from business process reengineering, to rationalize the potential threats to an enterprise and the possible actions one could take to mitigate potential losses. The Government and the private sector have the need to protect its critical infrastructure. There are critical operations that need to operate uninterrupted, such as banking, telecommunications, the power grid or military operations. In this paper, we applied this initial model to a defense contractor.

The risk of having critical systems taken off line requires an enterprise to make a substantial investment to protect itself from various threats. In many industries, these threats are not clearly known at this time. Many enterprises struggle with committing resources to a problem that may never happen. An example of the dynamics that an enterprise faces, when protecting its assets, is highlighted in Fig. 6.

Threats to an enterprise come in various ways; they could be from terrorists, a natural disaster, insider threat, corporate espionage, or cyber attack. An enterprise can combat the threat

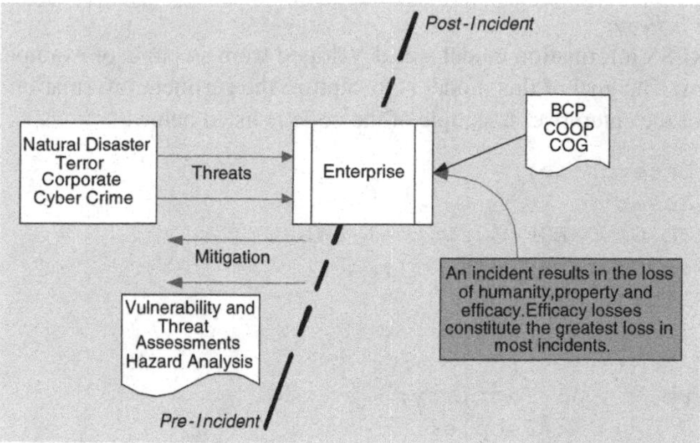

Fig. 6: Dynamics that an enterprise faces when attempting to protect its critical assets. This represents the domain of discourse for this paper

by conducting analyses or assessment of potential threats. This should all occur, hopefully, before an incident happens.

There is a potential for a loss of humanity, property and efficacy during any accident, natural disaster or terrorist event [18]. The loss of efficacy, as a result of an accident or disaster event, is usually the greater of the costs [19].

Many of the managers of critical infrastructure assets use the "acceptable risk" method for protecting their assets and they do not consult with people in the existing communities. For instance, the chemical industry views risk in three ways: chemical inventory, worst-case assessment and population at risk. "Since 9/11, in the absence of federal legislation, ACC members have led the way, investing nearly $3 billion on facility security enhancements such as intrusion prevention/detection and perimeter protection, screening employees and improving cyber-security" [26].

To understand our domain, we developed an IDEF0 [16] model to bind the problem space. Once bounded, review of IDEF0 models allowed us to decompose the activities of the enterprise. The IDEF0 model identified the information flow between the activities. These information flows are integral to understanding critical elements for the safety and security of the enterprise.

We then selected a few activities to develop information models. Information modeling allows one to have an unambiguous understanding of the domain of discourse. A meta-model (schema) was developed and the disparate information elements were turned into individual models and mapped back to the schema. Our goal was to understand the semantic contents of the information space.

The major benefit of information models is that they allow rules and constraints to be applied to a domain. The result of this modeling activity would support information sharing and consistency between the mechanisms of the BCP, vulnerability assessment and threat models. This activity could potentially improve emergency preparedness planning. Since we could not model the entire world, we selected three exemplars to demonstrate the EXPRESS information models. Each model is related to an activity or document highlighted in our enterprise diagram in Fig. 1.

Enterprise Schema

The EXPRESS information model was developed from a couple of examples of business continuity plans. The goal of this model is to capture the pertinent information related to the critical assets of the enterprise. A sample of the code is listed below:

```
SCHEMA Enterprise;
ENTITY Asset:
SUPERTYPE OF(ONEOF (Critical_Asset,
      Non_Critical_Asset));
END_ENTITY;
ENTITY Critical_Assets;
SUBTYPE OF (Asset);
    Where        : Facility;
    Who          : People;
    With_what1   : Equipment;
    With_what2   : Systems;
    With_what3   : Products;
    Info         : Information;
END_ENTITY;
ENTITY Facility;
  Name           : STRING;
  Address        : STRING;
  Geophysical_data    : REAL;
END_ENTITY;
ENTITY People;
  Name           : STRING;
  Title          : STRING;
  Location       : STRING;
  Telephone_no   : INTEGER;
END_ENTITY;
ENTITY Equipment;
LIST OF UNIQUE EQUIPMENT;
END_ENTITY;
ENTITY Systems;
  Critical_function     : STRING;
  Crit_infra         : Critical_Infrastructure;
END_ENTITY;
ENTITY Critical_Infrastructure;
  Security_system : STRING;
  Power           : STRING;
  Water           : STRING;
  Fuel            : STRING;
  Telecommunication  : STRING;
  Transportation     : STRING;
  Emergency-services : STRING;
END_ENTITY;
ENTITY Products;
  LIST OF UNIQUE PRODUCTS;
END_ENTITY;
```

```
ENTITY Information;
  IT_system :STRING
END_ENTITY;
END_SCHEMA;
```

Hazard Schema

The hazard schema was derived from a potential threat that is a natural disaster. The information related to an earthquake was specifically modeled. The code for the Hazard Schema is as follows:

Alert Schema

The following code was derived from a couple of examples of alert messages. This would be typical of emergency alert message traffic. Multiple warnings and indication systems produce this type of information. The information in this schema could potentially be used by an organization such as the Department of Homeland Security to share with the public.

```
SCHEMA Alert;
ENTITY HLS_Alert;
  Sent_from    : Sender;
  Stats        : Status;
  Info         : Information;
  Area         : Area_of_Interest;
END_ENTITY;
ENTITY Sender;
  Name         : STRING;
  Location     : STRING;
END_ENTITY;
ENTITY Status;
  Actual       : BOOLEAN;
  Drill        : STRING;
  Training     : STRING;
END_ENTITY;
ENTITY Information;
  Category     : STRING;
  Extent       : STRING;
  Urgency      : STRING;
  Severity     : LMH;
  Certainty    : STRING;
  Description : STRING;
  Instruction : STRING;
  Parameters   : GYOR;
END_ENTITY;
ENTITY LMH;
  Low          : BOOLEAN;
  Medium       : BOOLEAN;
  High         : BOOLEAN;
END_ENTITY;
ENTITY GYOR;
  Green        : BOOLEAN;
  Yellow       : BOOLEAN;
```

```
Orange          : BOOLEAN;
  Red           : BOOLEAN;
END_ENTITY;
ENTITY Area_of_Interest;
  Longitude     : REAL;
  Latitude      : REAL;
  Description : STRING;
END_ENTITY;
END_SCHEMA;
```

The next step in the process is to provide a mapping model from each schema to the overall schema. In our example we did this by hand, using EXPRESS I. Simulating a real earthquake, we proved that information provided active alerts and paper documents could be linked and interrelated. If implemented in an on-information system, the enterprise could react more swiftly to the report of the earthquake, more quickly response to the incident, and potentially mitigate cost associated with the event.

4 Conclusions

When a disaster occurs, an ad hoc coalition usually forms to respond to the incident. History has shown that these "system of systems" do not interoperate very well. Our research has demonstrated that the use of system level design language, combined with a system of systems engineering approach can potentially improve communications between fire, police and rescue components.

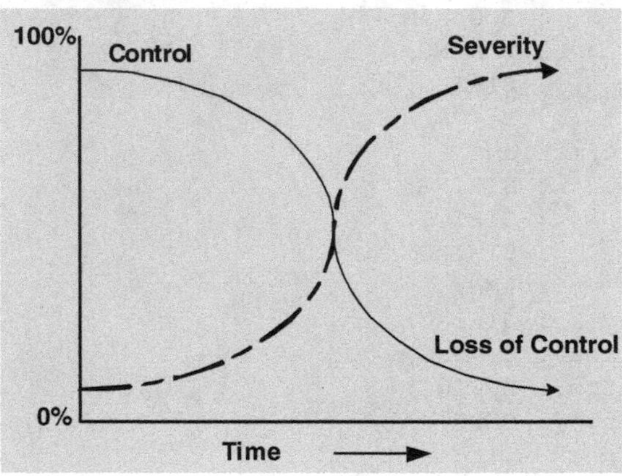

Fig. 7: Control versus severity of emergency or disaster

This paper demonstrated that there is practical application for defining requirements for components of ad hoc coalitions which are formed when responding to disasters. Further research will be conducted to expand the development of Rosetta to this problem space.

Requirements modeling will reduce the cycle time needed to integrate large and complex ad hoc coalitions by making the systems' specifications easier to integrate. It will allow the ad hoc coalitions to upgrade and reengineer their response systems by providing a clear and coherent picture of everything the systems are suppose to provide. It will allow system verification, define interfaces within itself and with other systems, and constraints and boundaries that it is expected to operate within.

Our goal is to provide control of an emergency or disaster situation. As control is not exercised on a disaster situation, the consequences of the event escalate rapidly. Figure 7 shows that as you lose control, the severity of the situation becomes great [11]. We have demonstrated that it is possible to create EXPRESS information models of elements and mechanisms that are involved with the protection of enterprise assets. The work is in line with other information and data modeling activities in the emergency planning and disaster preparedness. Our future work will involve modeling an actual enterprise. We will instantiate the EXPRESS code in XML, using an existing tool available from NIST. We could then simulate actual data flow and measure organizational responses.

Acknowledgments

The authors gratefully acknowledge the contributions of Perry Alexander at the University of Kansas, Bret Michael at the Naval Postgraduate School.

References

1. Carsten T (2003) Systems Engineering Development Process. Airbus/Universitat Oldenburg, System Engineering Lecture 2 -v1, May 2003
2. http://www.weirtechna.com/weir/techna/home.nsf/page/Case_Study_System_Engineering
3. Boehm B (2005) Teaching the Elephant to Dance: Agility Meets Systems of Systems Engineering and Acquisition. Keynote, GSAW, 3 March 2005
4. Barton DL, Fortier SC (1997) Using a Systems Description Language for Complete Avionics Systems. In: Proceedings of AUTOTESTCON, IEEE Computer Society Press, Anaheim, CA
5. http://www.ittc.ku.edu/Projects/rosetta/
6. http://www.ittc.ke.edu/Projects/SLDG/projects/project-rosetta.htm
7. Alexander P, Kamath R, Barton DL, Fortier SC (1999) Facets and Domains in Rosetta. Published in the Proceedings of the Forum on Design Languages, September 1999, Lyon, France
8. http://www.whitehouse.gov/reports/katrina-lessons-learned/
9. Alexander P, System Specification Using Rosetta. PowerPoint Slides Provided by the author
10. Kong C, Kimmell G, Streb J, Alexander P (2005) Semantic Support for Model Composition. Submitted to Integrated Formal Methods
11. Harrald J (2006) Class Notes from EMSE 332 Course at George Washington University. 13 April 2006
12. DeLaurentis D, Callaway R, A System-of-Systems Perspective for Public Policy Decisions. Review for Policy Research, 4th Quarter Edition
13. Popper S et al. (2004) System of Systems Symposium: Report on a Summer Conversation. Potomac Institute for Policy Studies, Arlington, VA, July 2004

14. Morris E et al. (2004) System of Systems Interoperability (SOSI). Final Report, Technical Report, CMU/SEI-2004-TR-004, April 2004
15. Clark T, Jones R (1999) Organisational Interoperability Maturity Model for C2. In: Command & Control Research & Technology Symposium, U.S. Naval War College, Newport, RI, 22 June – 1 July, 1999
16. FIPS Publication 183 (1993) IDEFO as a Standard for Function Modeling. The National Institute of Standards and Technology (NIST), December 1993
17. ISO Standard 10303-1 (1994) Industrial Automation Systems and Integration – Product Data Representation and Exchange – Part 1: Overview and Fundamental Principles, http://www.iso.org/iso/en/CatalogueDetailPage.CatalogueDetail? CSNUMBER=20579\&ICS1=25\&ICS2=40\&ICS3=40
18. Fortier SC, Michael JB (1993) A Risk-Based Approach to Cost-Benefit Analysis of Software Safety Activities. COMPASS '93, Proceedings of the Eight Annual Conference on Computer Assurance, NIST, June 1993
19. Bird FE Jr, Loftus RG (1976) LO$$ Control Management. Institute Press, Canada
20. http://niem.gov/aboutniem.htm
21. http://grouper.ieee.org/groups/scc32/imwg/
22. Effective Disaster Warnings. http://www.sdr.gov/reports.html
23. http://www.eic.org/
24. http://it.ojp.gov/topic.jsp?topic_id=43
25. Risk Analysis and Management for Critical Asset Protection (RAMCAP). ASME Innovative Technologies Institute, Washington, DC, Version 1.1e, October 23, 2005
26. http://www.americanchemistry.com/s_acc/sec_mediakits.asp?CID=258&DID=632

Investigating the Relationship Between Complex Systematic Concepts

Mohamed H. Al-Kuwaiti[1] and Nicholas Kyriakopoulos[2]

[1] School of Engineering and Applied Science
The George Washington University 801 22nd St. NW, Washington, DC, 20052 USA
kuwaiti@gwu.edu

[2] School of Engineering and Applied Science
The George Washington University 801 22nd St. NW, Washington, DC, 20052 USA
kyriak@gwu.edu

Summary. When a user or a designer tries to describe a system or a service provided by that system, then he/she faces some confusion in choosing the right qualitative and/or quantitative performance concepts. In fact, many concepts are used by many research and study groups in different ways; yet these same concepts are inconsistently used. Therefore, these qualitative/quantitative characteristic keep growing without limits and need to be clarified. This paper provides clear insights to the general problem of understanding these concepts and their relationship by integrating them into a one structure. It lays the foundation for achieving a unified design model for these characteristics, which will formulate the basic issues to develop unified quantitative approaches, tools, or evaluation methods. The expected outcomes would benefit many different entities in enhancing the design or optimizing it for achieving better systems and networks.

1 Introduction

Nowadays, describing complex computers and communications systems become a big challenge with all of these available different properties. In fact, "survivability", "dependability", "fault-tolerance", "reliability" and "security" properties have been the subject of many extensive studies, especially in the light of alarms generated by the President's Commission on Critical Infrastructure Protection (PCCIP) reports in the 1990s. These five concepts (and many other properties in general) become critical issues in the design of today's complex networks and systems. It is hard to find a single specific model that combines all of these features in one system. It is even harder to implement and maintain them in that system, if found [1, 14].

This paper is a part of a dissertation research that aims towards critically and comparatively analyzes the relationship between five most popular system's qualitative and/or quantitative concepts. These concepts are: survivability, dependability, fault-tolerance, reliability and security. In fact, the dissertation research focuses more on the definitions, requirements,

J.A.T. Machado et al. (eds.), *Intelligent Engineering Systems and Computational Cybernetics,*

attributes, similarities, differences, performance evaluations and measures of these concepts. However, due to space limitation; this paper briefly present these issues for complex computers and communications infrastructure, and it illustrates how these concepts along with possible measures can enhance and optimize the traditional system design.

Today's global infrastructure are growing increasingly dependent upon large-scale and highly distributed computers and communications systems that operate in unbounded environments; which like the Internet, have no central administrative control, no unified security policy, and new types of threats. This new infrastructure is based on the ability for diverse networks and system to be interconnected and provide global coverage for the transmission of data. However, it has also created the need for new concepts and tools for analyzing its behavior and impact on the services based on it [7, 8].

The critical infrastructures of today's modern society with the advanced technologies are built over traditional networks and systems that are relatively inadequate [7, 10]. These systems require some degree of dependability and survivability that were not clearly foreseen when they were first designed and built.

Traditionally, the main objective of any network design is to achieve some specified requirements under normal conditions, often in a link-to-link or an end-to-end basis, without explicit consideration of network or system features such as dependability, survivability, and fault-tolerance. However, as described in [17], *"While the distributed systems and services that sustain our day-to-day IT and communication infrastructures become increasingly complex; traditional solutions to manage and control them seem to have reached their limits. Researchers are thus testing alternate paradigms to organize and structure them."* Therefore, in order to enhance the design of computers and communications systems; we propose a conceptual framework that clarifies some concepts used in describing such complex systems. In fact, many of these concepts are described differently and sometimes incorrectly by different research programs and entities. Some entities are using and employing relatively inaccurate meanings of these terms. This on-going research would lay the foundation for achieving standardized, adequate and consistent meanings for some concepts that can replace the inarticulate, incomplete and incorrect descriptions of them.

In this paper, we explore the issues and challenges involved in clarifying some ambiguous concepts when used to describe complex systems. In Section 2, we begin by describing the traditional system design and the requirements for enhancing this design. In Section 3, we provide a brief survey on literature regarding works done to date in the field of the studied concepts as well as some drawbacks of them. In Section 4, the qualitative and quantitative approaches are described for such systematic concepts. Section 5, the concept analysis is illustrated here. Section 6, some discussion about the analysis results is shown. Finally, we conclude in Section 7 with the research and some directions for further research on the subject.

2 Traditional vs. Enhanced Designs

In our context, traditional networks and systems are the ones that have not being able to cover some of the needs for modern technology. These systems are not intended to fully address some of the requirements for the critical complex systems that have several of the following desirable features:

- Large-Scale Distribution: Networks and system are evolving towards large-scale and open distributed infrastructures, with a wide geographical span and inclusion of high variety of legacy components [7].

- Integration of Systems: Interconnection between communications infrastructure and other large-scale systems adds to the complexity of the interactions between components (concept of "system of systems") [1,7].
- Load Sharing (Graceful Degradation): Network or system performances are degrading gradually instead of rapidly to compensate for hardware or software faults [12].

With the rapid development and the advent of information-age services as well as the increased dependence of users on computers and communications systems, specifications of the five characteristics addressed here are preferred in many systems and networks. Thus, the traditional systems and networks need to be enhanced with some of these significant concepts. Figure 1 illustrates this idea.

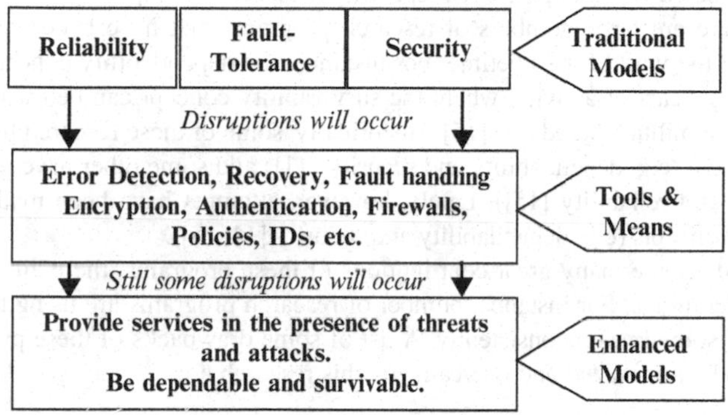

Fig. 1: Traditional and enhanced system properties

In fact, setting these concepts as performance objectives will help address the critical complex systems as well as will benefit the design to ensure that under given failure scenarios, network or system performance will not degrade blow predetermined levels. Beside that, systems will stand a variety of threats and keep providing essential services despite the presence of any disruptions.

However, in order to design or enhance the design of any network or system, these five concepts need to be clarified first. In fact, recently and based on this research, some interesting findings were discovered. The five important concepts are in fact described differently and sometimes incorrectly by different research programs and entities. Some entities are using and employing relatively inaccurate meanings of these words. Consequently, the system users or even designers may wonder about some of the definitions in these characteristics when they compare them to the dictionary meanings. Besides having different meanings, there is a significant gap between most entities that study these concepts. There is a lack of integration between all varieties of approaches for the same fundamental research area.

Therefore, this proposed research would lay the foundation for achieving standardized, adequate and consistent meanings for these five expressions that can replace the inarticulate, incomplete and incorrect descriptions of them. This research is intended, as well, to critically investigate, and analyze the relationship between these five most essential system's qualitative and quantitative concepts. Although there are many other synonymous terms that are used to describe comparable networks and systems, this research will look briefly at some of these terms but it will focus more on the following systematic concepts: survivable, dependable, fault-tolerant, reliable and secure systems.

3 Research Background and Examples of System Applications

3.1 Background

Numerous efforts and attempts have been made towards making computers and communications systems and/or their resulted services dependable, fault-tolerance, reliable, secure, or survivable. Based on various reviewed literature of this research, this section presents a brief overview of some current and past work regarding these five concepts.

In fact, there are large numbers of research programs that have been focusing on this field all around the world for sometime. For instance, the dependability concept is the result of nearly twenty years of activity, where the survivability concept can be traced back to the late sixties in the military standards [15]. Historically, some of these research fields have progressed separately (e.g. dependability and security [11]) and some other have grown together (dependability and reliability [15]). Lately, however, attempts have been made to integrate some of these concepts (e.g. dependability and security [11, 15]).

Although there are many great contributions of these programs, there are still some unfortunate shortcomings. For instance, number of research programs are using these concepts differently and sometimes inconsistently. A list of some drawbacks of these programs which are intended to be minimized and overcame by this research are:

- Little is said to clarify the ambiguity that resides in these concepts.
- No clear investigation for identifying the relationship and the overlaps between these concepts.
- Few illustrations about the attributes of these concepts as well as their performance evaluations and measures.
- Weak foundation for achieving standardized, adequate and consistent meanings for these five concepts that can replace the inarticulate, incomplete and incorrect descriptions of them.

3.2 Examples of Critical Computers and Communications Complex System Applications

Before presenting the detailed analysis of these concepts, some examples and background material about several critical systems are helpful in emphasizing the importance of having such systems possess some of these characteristics in their services.

With the rapid development and wide application of Information Technology (IT), these five characteristics become important challenges to everybody that includes system users, designers, developers, and architects. In fact, IT penetrates into many aspects of life for an increased number of people, enriching them but also producing systems of such complexity that create new dependencies and risks to the society [7]. In addition to that and with the sophistication of today's attacks and threats, attacker can easily penetrate a network and become a dangerous threat to the critical information systems.

Although there are challenges, several systems and services of systems are desired and/or required to have these characteristics so that they can achieve their required design goals. Some applications and examples of the systems are: defense systems, flight systems, air traffic control systems, financial services and banking systems, airline seat reservations, telephone systems, energy systems, transportation systems, and medical services. More detailed examples of these systems are presented in [1, 3, 7]. From all of these, it is evident that not only governments, militaries, businesses and emergency organizations are requiring enhanced levels of these concepts in their systems but also the general public.

4 Qualitative and Quantitative Evaluation Measures

Ways to quantify qualitative system's properties are urgently needed in these important areas of survivability, dependability, fault-tolerance, reliability, and security. Although it might be difficult and challenging to quantify the results of these aspects, quantification is absolutely essential to attain a mature capability for creating systems of any kind. There is in doubt that the old adage saying of Tom DeMarco: *"You can't control what you can't measure"* is relevant in designing these systems and setting the basic performance measures [9].

Although unified quantitative concepts are urgently required in today's systems for these addressed properties, there are no easy answers in achieving such standardized concepts; that is because of the multidimensionality of this problem. In fact, these systems with the addressed concepts are considered to be complex because of the unbounded and emergent properties of them with the dynamic changes of their behavior. In addition to that, they are lacking of central monitoring and controls, as well as they have an increased number of different disruptions [2,4].

It is crucial to address the evaluation methods when dealing with such systematic concepts. Thus, we briefly explain the differences between the qualitative (subjective) and quantitative (objective) evaluation approaches. These two distinct approaches are important in understanding the techniques for assessing and evaluating the performance characteristics as well as verifying if the intended performance has been achieved.

4.1 Qualitative Evaluation Methods

In general, a concept is basically a composite of many characteristics. It means mainly how good or bad the system is in relation to some characteristics. The qualitative approach usually involves more words than numbers and there are many qualitative methodologies that are used for this type of analysis. Some of these analysis methods as described in several reviewed literatures [5, 9] are:

- Brainstorming, Documentations, and Professional Experience
- Questionnaires, Surveys, Checklists, and Interviewing
- Policies and Standards of due cares

4.2 Quantitative Evaluation Methods

The quantitative methods have the advantage of producing figures and results that can be utilized to control or compare various systems or networks [5]. Some examples of quantitative methodologies that are used in evaluating the addressed concepts as discussed in [10, 12, 13, 15, 16] are:

- Deterministic and Probabilistic Analysis (e.g. Risk Analysis)
- Weight or Impact Analysis as well as Algorithms Analysis
- Analytical, Experimental, and Simulation Methods and Models

4.3 Qualitative or Quantitative Approach?

Both of these evaluation approaches are, in fact, applied in combinations, especially when dealing with the dependability, security, and survivability characteristics. It is a challenge to conduct a purely quantitative evaluation process because these general concepts have various attributes where some of them are hard to assess. Both of these approaches are very important and needed in conjunction to better design and control any system [5].

5 The Concept Analysis

Briefly, this section illustrates few important points of the analysis for the studied concepts.

5.1 Dependability

There are different meanings for this concept, but all have the same goal and idea. According to the International Federation for Information Processing (IFIP) Working Group (WG10.4) on Dependable Computing and fault Tolerance, dependability is used as an umbrella term and reliability is used as a well defined mathematical function [16].

Dependability is the ability of a system to deliver service that can justifiably be trusted [15]. Figure 2 shows the attributes, means and threats that affect this concept. The dependability concept is described by various non-functional properties. Although dependability performance can be evaluated using qualitative and quantitative approaches, it is complicated to make absolute measurements of dependability. Thus, it is not mathematically well-defined and it can be evaluated by using its attributes. Many dependability evaluation tools and models were developed by the computer engineering community.

5.2 Survivability

This concept describes a network or a system that has a single critical goal, which is to fulfill its mission in a timely manner, in the presence of attacks, failures, or accidents [2]. The CERT Coordination Center (CERT/CC) is developing many technologies and methods for analyzing and designing survivable network systems.

Similar to the dependability term, the survivability has minimum levels of quality attributes as shown in Fig. 3 which illustrates this concept. Various papers and articles try to define and describe approaches to achieve survivability [2, 3]. In fact, survivability evaluation is similar to dependability in a way that it is not mathematically well-defined. It can be evaluated by using qualitative and quantitative approaches as well as by using its attributes.

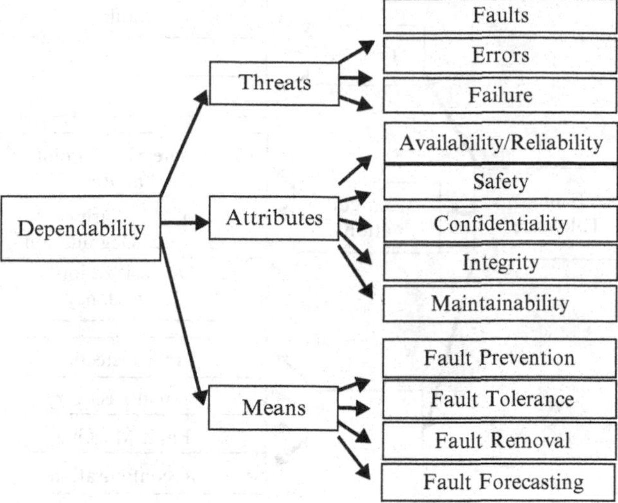

Fig. 2: Dependability taxonomy

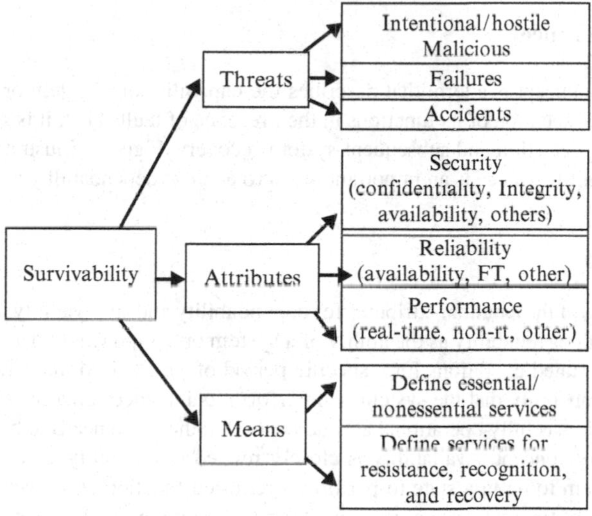

Fig. 3: Survivability taxonomy

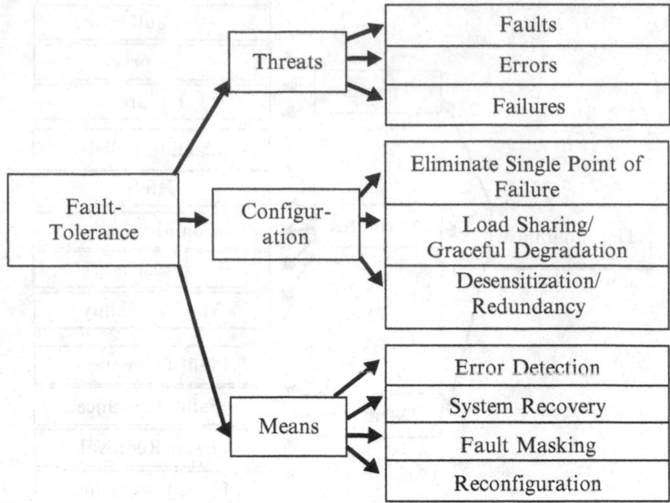

Fig. 4: Fault-Tolerance taxonomy

5.3 Fault-Tolerance

A fault-tolerant concept is a term that describes the capability of a system or service to continue the correct execution of its functions in the presence of faults [12]. It is generally implemented by error detection and subsequent system recovery. Figure 4 illustrates taxonomy of this concept. Fault-tolerance is an important mean to achieve dependability and reliability.

5.4 Reliability

Reliability is one of the essential attributes for dependability and survivability. Basically, IEEE 610.12-1990 defines reliability as the ability of a system or component to perform its required functions under stated conditions for a specific period of time [6]. Moreover, it is defined as a conditional probability that the system will perform its intended function without failure at time t provided it was fully operational at time $t = 0$. Another term needs to be addressed here is the availability concept. Availability is closely related to reliability, and is defined as the ability of a system to be in a state to perform a required function at a given instant of time or at any instant of time within a given time interval, assuming that the external resources, if required, are provided.

Reliability is a very important concept in designing a system and it is the most extensively studied concept. Figure 5 shows a simple taxonomy of this concept. A common measure of reliability is Mean Time To Failure (MTTF). To compute the reliability of a network, R(t), many methods and models are using the following quantified indexes [6]: Steady-State Availability of the system A(t), Mean Time Between Failures (MTBF), and Mean Time To Repair (MTTR).

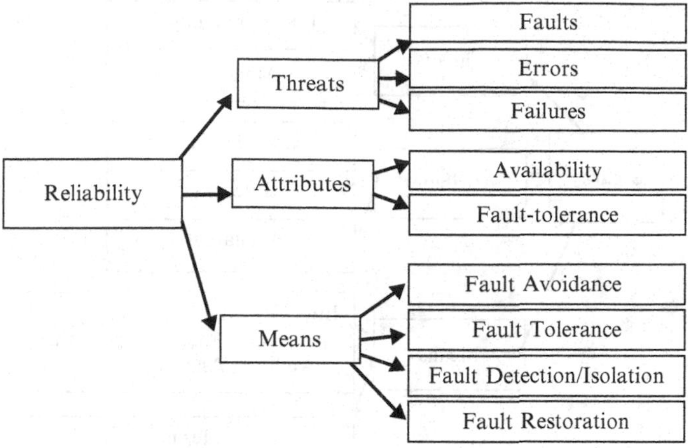

Fig. 5: Reliability taxonomy

5.5 Security

In general, security is protection against undesirable events. In fact, it means different things to different people. Security also considered as another attribute of the essential attributes for dependability and survivability. There are tremendous efforts to improve system and network security. These efforts have produced an amazing development of techniques and tools in hardware and software [5]. Figure 6 shows simple security taxonomy. The security of a system is not mathematically well-defined. It can be evaluated by using qualitative and quantitative approaches. Also, it can be evaluated by using different attributes or metrics of quality information assurance, i.e.. integrity, confidentiality, performance, and reliability. Many methods and models have been developed to measure system security [4].

6 Results Analysis

Briefly elaborating on this analysis, it revealed that some of these concepts are new ones and came out to be as an umbrella terms, and some other have been there since the beginning of the system design. The idea of decomposition of the general concepts (e.g. dependability and survivability) into several attributes (e.g. reliability and availability) is very helpful in provide a set of parameters that are measurable and quantifiable.

Further, the analysis of the dependability and survivability concepts has led into the conclusion that these two concepts converges, and are essentially equivalent in their goals, attributes, and address similar threats. Further, reliability and availability concepts are most common attributes to all characteristics and they have the most quantifiable parameters. It is also worth mentioning that accidental faults as well as intentional threats should be considering for such complex concepts. The cause of failure is different here and it is a challenge to accurately model intentional threats [16].

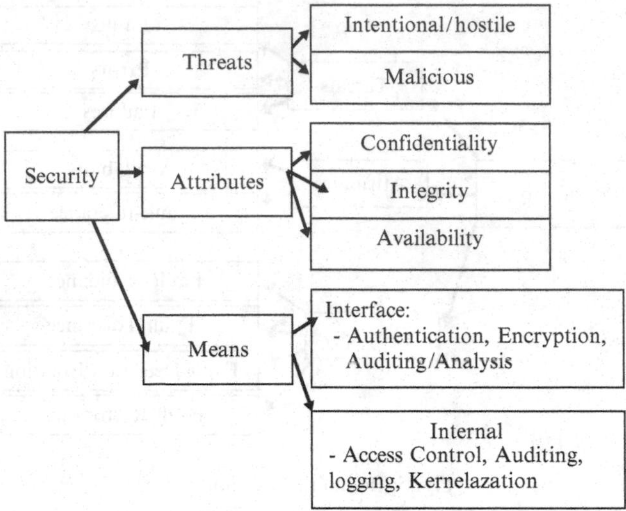

Fig. 6: Security taxonomy

7 Conclusions

In this paper, we critically investigate and analyze the relationship between five significant systematic qualitative/quantitative concepts that are commonly used in describing complex systems or network. This analysis is very useful in the future for designing better enhanced and optimized systems or networks. The traditional systems and networks need to be enhanced with some of these significant concepts so that they are able to express complex systems. Quantifying qualitative concepts or system properties are very essential and urgently needed in these important areas of survivability, dependability, fault-tolerance, reliability, and security.

Our future research points towards relating the classical network design approaches used at the physical level (i.e., of the OSI model) with the performance requirements as stated by the user at the application level on the IP-based telecommunications networks. We will look specifically at the Quality of Service (QoS) framework and taxonomy of the IP-based networks in order to enhance and optimize the current network.

References

1. Kyriakopoulos N, Wilikens M (2001) Dependability and complexity: exploring ideas for studying open systems. Report EUR 19797 EN, JRS, Brussels
2. Ellison R, Fischer D, Linger R, Lipson H, Longstaff T, Mead N (2001) Survivable network systems: an emerging discipline. Report CMU/SEI-2001-TN-001, Software Engineering Institute, Carnegie Mellon University, Pittsburgh
3. Zolfaghari A, Kaudel F (1994) IEEE Selected Areas in Communication 12:46–51
4. Reznik L (2003) Which models should be applied to measure computer security and information assurance?. In: 12th IEEE International Conference on Fuzzy Systems St. Louis, MO, 2:1243–1248

5. Orlandi E (1990) Computer security: a consequence of information technology quality. In: IEEE International Carnahan Conference on Crime Countermeasures, Security Technology, IEEE, New York NY, 109–112
6. Trivedi K (2002) Performance and availability of computers and networks. Course materials, Duke University, Durham, NC
7. Kyamakya K, Jobmann K, Meincke M (2000) Security and survivability of distributed systems: an overview. In: 21st Century Military Commissions Conference, Los Angeles, CA, 1:449–454
8. Kyriakopoulos N, Wilikens M (2000) Dependability of complex open systems: a unifying concept for understanding internet-related issues. In: 3rd Information Survivability Workshop, Boston, MA
9. Perkins T, Peterson R (2003) CrossTalk Jour of Defense Software Engineering 16:9–12
10. McCabe J (1998) Practical Computer Network Analysis and Design. Morgan Kaufmann, San Francisco, CA
11. Jonsson E (1998) An integrated framework for security and dependability. In: 1998 New Security Paradigm Workshop, Charlottesville, VA
12. Pradhan D (1996) Fault-Tolerant Computer System Design. Prentice-Hall,Upper Saddle River, NJ
13. Oliveto F (1997) The Four steps to achieve a reliable design. In: National Aerospace and Electronics Conference, Dayton, OH, 1:446–453
14. Iyer R (2004) Dependable and Secure Computing 1:2–3
15. Avizienis A, Laprie J-C, Randell B, Landwehr C (2004) Dependable and Secure Computing 1:11–33
16. Nicol D, Sanders W, Trivedi K (2004) Dependable and Secure Computing 1:48–65
17. Martin-Flatin J-P, Sventek J, Geihs K (2005) Topics in the IFIP/IEEE International Workshop on Self-Managed Systems and Services, Nice, France

Particle Swarm Design of Digital Circuits

Cecília Reis and J. A. Tenreiro Machado

Institute Engineering of Porto, Polytechnic Institute of Porto,
Engineering Department, Rua Dr. António Bernardino de Almeida, 4200-072 Porto, Portugal
{cmr, jtm}@isep.ipp.pt

Summary. Swarm Intelligence (SI) is the property of a system whereby the collective behaviors of (unsophisticated) agents interacting locally with their environment cause coherent functional global patterns to emerge. Particle swarm optimization (PSO) is a form of SI, and a population-based search algorithm that is initialized with a population of random solutions, called particles. These particles are flying through hyperspace and have two essential reasoning capabilities: their memory of their own best position and knowledge of the swarm's best position. In a PSO scheme each particle flies through the search space with a velocity that is adjusted dynamically according with its historical behavior. Therefore, the particles have a tendency to fly towards the best search area along the search process. This work proposes a PSO based algorithm for logic circuit synthesis. The results show the statistical characteristics of this algorithm with respect to number of generations required to achieve the solutions. It is also presented a comparison with other two Evolutionary Algorithms, namely Genetic and Memetic Algorithms.

1 Introduction

In recent decades Evolutionary Computation techniques have been applied to the design of electronic circuits and systems, leading to a novel area of research called Evolutionary Electronics (EE) or Evolvable Hardware [3]. EE considers the concept for automatic design of electronic systems. Instead of using human conceived models, abstractions and techniques, EE employs search algorithms to develop implementations not achievable with the traditional design schemes, such as the Karnaugh or the Quine–McCluskey Boolean methods.

Several papers proposed designing combinational logic circuits using evolutionary algorithms and, in particular, genetic algorithms (GAs) [1, 2, 4, 8] and hybrid schemes such as the memetic algorithms (MAs) [11]

Particle swarm optimization (PSO) constitutes an alternative evolutionary computation technique, and this paper studies its application to combinational logic circuit synthesis. Bearing these ideas in mind, the organization of this article is as follows. Section 2 presents a brief overview of the PSO. Section 3 describes the PSO based circuit design, while Section 4 exhibits the simulation results. Finally, Section 5 outlines the main conclusions and addresses perspectives towards future developments.

J.A.T. Machado et al. (eds.), *Intelligent Engineering Systems and Computational Cybernetics,*

2 Particle Swarm Optimization

2.1 Introduction

In the literature about PSO the term 'swarm intelligence' appears rather often and, therefore, we begin by explaining why this is so.

Non-computer scientists (ornithologists, biologists and psychologists) did early research, which led into the theory of particle swarms. In these areas, the term 'swarm intelligence' is well known and characterizes the case when a large number of individuals are able of accomplish complex tasks. Motivated by these facts, some basic simulations of swarms were abstracted into the mathematical field. The usage of swarms for solving simple tasks in nature became an intriguing idea in algorithmic and function optimization.

Eberhart and Kennedy were the first to introduce the PSO algorithm [5], which is an optimization method inspired in the collective intelligence of swarms of biological populations, and was discovered through simplified social model simulation of bird flocking, fishing schooling and swarm theory.

2.2 Parameters

In the PSO, instead of using genetic operators, as in the case of GAs, each particle (individual) adjusts its flying according with its own and its companions experiences. Each particle is treated as a point in a *D*-dimensional space and is manipulated as described below in the original PSO algorithm:

$$v_{id} = v_{id} + c_1\, rand()(p_{id} - x_{id}) + c_2\, rand()(p_{gd} - x_{id}) \quad (1)$$

$$x_{id} = x_{id} + v_{id} \quad (2)$$

where c_1 and c_2 are positive constants and *rand*() is a random function in the range [0,1], $X_i = (x_{i1}, x_{i2}, \ldots, x_{iD})$ represents the *i*th particle, $P_i = (p_{i1}, p_{i2}, \ldots, p_{iD})$ is the best previous position (the position giving the best fitness value) of the particle, the symbol g represents the index of the best particle among all particles in the population, and $V_i = (v_{i1}, v_{i2}, \ldots, v_{iD})$ is the rate of the position change (velocity) for particle i.

Expression (1) represents the flying trajectory of a population of particles. Equation (1) describes how the velocity is dynamically updated and equation (2) the position update of the "flying" particles. Equation (1) is divided in three parts, namely the momentum, the cognitive and the social parts. In the first part the velocity cannot be changed abruptly: it is adjusted based on the current velocity. The second part represents the learning from its own flying experience. The third part consists on the learning group flying experience [6,9].

The first new parameter added into the original PSO algorithm is the inertia weight. The dynamic equation of PSO with inertia weight is modified to be:

$$v_{id} = wv_{id} + c_1\, rand()(p_{id} - x_{id}) + c_2\, rand()(p_{gd} - x_{id}) \quad (3)$$

$$x_{id} = x_{id} + v_{id} \quad (4)$$

where w constitutes the inertia weight that introduces a balance between the global and the local search abilities. A large inertia facilitates a global search while a small inertia weight facilitates the local search.

Another parameter, called constriction coefficient k, is introduced with the hope that it can insure a PSO to converge. A simplified method of incorporating it appears in equation (3), where k is function of c_1 and c_2 as presented in equation (7).

$$v_{id} = k\left[v_{id} + c_1 rand\,()\,(p_{id} - x_{id}) + c_2 Rand\,()\,(p_{gd} - x_{id})\right] \tag{5}$$

$$x_{id} = x_{id} + v_{id} \tag{6}$$

$$k = 2\left(2 - \phi - \sqrt{\phi^2 - 4\phi}\right)^{-1}, \Phi = c_1 + c_2, \Phi > 4 \tag{7}$$

2.3 Topologies

There are two different PSO topologies, namely the global version and the local version. In the global version of PSO, each particle flies through the search space with a velocity that is dynamically adjusted according to the particle's personal best performance achieved so far and the best performance achieved so far by all particles. On the other hand, in the local version of PSO, each particle's velocity is adjusted according to its personal best and the best performance achieved so far within its neighborhood. The neighborhood of each particle is generally defined as topologically nearest particles to the particle at each side.

2.4 Algorithm

PSO is an evolutionary algorithm simple in concept, easy to implement and computationally efficient. Figures 1–3 present a generic EC algorithm, a hybrid algorithm, more precisely a MA and the original procedure for implementing the PSO algorithm, respectively.

1. Initialize the population
2. Calculate the fitness of each individual in the population
3. Reproduce selected individuals to form a new population
4. Perform evolutionary operations such as crossover and mutation on the population
5. Loop to step 2 until some condition is met

Fig. 1: Evolutionary computation algorithm

The different versions of the PSO algorithms are: the real-valued PSO, which is the original version of PSO and is well suited for solving real-value problems; the binary version of PSO, which is designed to solve binary problems; and the discrete version of PSO, which is good for solving the event-based problems. To extend the real-valued version of PSO to binary/discrete space, the most critical part is to understand the meaning of concepts such as trajectory and velocity in the binary/discrete space.

Kennedy and Eberhart [4] use velocity as a probability to determine whether x_{id} (a bit) will be in one state or another (zero or one). The particle swarm formula of equation (1) remains unchanged, except that now p_{id} and x_{id} are integers in [0.0,1.0] and a logistic transformation

1. Initialize the population
2. Calculate the fitness of each individual in the population
3. Reproduce selected individuals to form a new population
4. Perform evolutionary operations such as crossover and mutation on the population
5. Apply a local search algorithm
5. Loop to step 2 until some condition is met

Fig. 2: Memetic algorithm

1. Initialize population in hyperspace
2. Evaluate fitness of individual particles
3. Modify velocities based on previous best and global (or neighborhood) best
4. Terminate on some condition
5. Go to step 2

Fig. 3: Particle swarm optimization process

$S(v_{id})$ is used to accomplish this modification. The resulting change in position is defined by the following rule:

$$if\ [rand() < S(v_{id})] \qquad then\ x_{id} = 1; \quad else\ x_{id} = 0 \tag{8}$$

where the function $S(v)$ is a sigmoid limiting transformation.

3 PSO Based Circuit Design

In this section we present the PSO algorithm in terms of the circuit encoding as a chromosome, the PSO parameters and the fitness function.

3.1 Problem Definition and Circuit Encoding

We adopt a PSO algorithm to design combinational logic circuits. A truth table specifies the circuits and the goal is to implement a functional circuit with the least possible complexity. Four sets of logic gates have been defined, as shown in Table 1, being Gset 2 the simplest one and Gset 6 the most complex gate set. Logic gate named WIRE means a logical no-operation. In the PSO scheme the circuits are encoded as a rectangular matrix [10] **A** ($row \times column = r \times c$) of logic cells. Three genes represent each cell: *input1input2gate type*, where *input1* and *input2* are one of the circuit inputs, if they are in the first column, or one of the previous outputs, if they are in other columns. The gate type is one of the elements adopted in the gate set. The chromosome is formed with as many triplets as the matrix size demands (e.g. triplets $= 3 \times r \times c$).

Table 1: Gate sets

Gate set	Logic gates
Gset 2	{AND,XOR,WIRE}
Gset 3	{AND,OR,XOR,WIRE}
Gset 4	{AND,OR,XOR,NOT,WIRE}
Gset 6	{AND,OR,XOR,NOT,NAND,NOR,WIRE}

3.2 PSO Parameters

The initial population of circuits (particles) has a random generation. The initial velocity of each particle is initialized with zero. The following velocities are calculated applying equation (3) and the new positions result from using equation (4). In this way, each potential solution, called particle, flies through the problem space. For each gene is calculated the corresponding velocity. Therefore, the new positions are as many as the number of genes in the chromosome. If the new values of the input genes result out of range, then a re-insertion function is used. If the calculated gate gene is not allowed a new valid one is generated at random. These particles then have memory and each one keeps information of its previous best position (*pbest*) and its corresponding fitness. The swarm has the *pbest* of all the particles and the particle with the greatest fitness is called the global best (*gbest*).

The basic concept of the PSO technique lies in accelerating each particle towards its *pbest* and *gbest* locations with a random weighted acceleration. However, in our case we also use a kind of mutation operator that introduces a new cell in 10% of the population. This mutation operator changes the characteristics of a given cell in the matrix. Therefore, the mutation modifies the gate type and the two inputs, meaning that a completely new cell can appear in the chromosome.

To run the PSO we have also to define the number P of individuals to create the initial population of particles. This population is always the same size across the generations, until reaching the solution.

3.3 The Fitness Functions

The calculation of the fitness function F_s in (6) has two parts, f_1 and f_2, where f_1 measures the functionality and f_2 measures the simplicity. In a first phase, we compare the output $\mathbf{Y}$ produced by the PSO-generated circuit with the required values $\mathbf{Y}_R$, according with the truth table, on a bit-per-bit basis. By other words, f_1 is incremented by one for each correct bit of the output until f_1 reaches the maximum value f_{10}, that occurs when we have a functional circuit. Once the circuit is functional, in a second phase, the algorithm tries to generate circuits with the least number of gates. This means that the resulting circuit must have as much genes *gate type* $\equiv$ *wire* as possible. Therefore, the index f_2, that measures the simplicity (the number of null operations), is increased by *one* (*zero*) for each *wire* (*gate*) of the generated circuit, yielding:

- First phase, circuit functionality:

$$f_{10} = 2^{ni} \times no \tag{9}$$

$$f_1 = f_1 + 1, \quad if\ \{bit\ i\ of\ \mathbf{Y}\} = \{bit\ i\ of\ \mathbf{Y}_R\}, \quad i = 1,...,f_{10} \tag{10}$$

- Second phase, circuit simplicity:

$$f_2 = f_2 + 1,\ if\ gate\ type = wire \tag{11}$$

$$F_s = \begin{cases} f_1, & F_s < f_{10} \\ f_1 + f_2, & F_s \geq f_{10} \end{cases} \tag{12}$$

where $i = 1,\ldots,f_{10}$, and *ni* and *no* represent the number of inputs and outputs of the circuit, respectively.

The concept of dynamic fitness function F_d results from an analogy between control systems and the GA case, where we master the population through the fitness function. The simplest control system is the proportional algorithm; nevertheless, there can be other control algorithms, such as, for example, the proportional and the differential scheme.

In this line of thought expression (6) is a static fitness function F_s and corresponds to using a simple proportional algorithm. Therefore, to implement a proportional-derivative evolution the fitness function needs a scheme of the type:

$$F_d = F_s + KD^{\mu}[F_s] \tag{13}$$

where $0 \leq \mu \leq 1$ is the differential fractional-order and $K \in \Re$ is the 'gain' of the dynamical term.

The generalization of the concept of derivative $D^{\mu}[f(x)]$ to noninteger values of μ goes back to the beginning of the theory of differential calculus. In fact, Leibniz, in his correspondence with Bernoulli, L'Hôpital and Wallis, had several notes about its calculation for $\mu = 1/2$. Based on these concepts, the adoption of the Fractional Calculus (FC) in control algorithms has been recently studied using the frequency and discrete-time domains.

The mathematical definition of a derivative of fractional order μ has been the subject of several different approaches. For example, Eq. (11) represent the Grünwald–Letnikov definition of the fractional derivative of order μ of the signal $x(t)$:

$$D^{\mu}[x(t)] = \lim_{h \to 0} \frac{1}{h^{\mu}} \sum_{k=0}^{\infty} \frac{(-1)^k \Gamma(\mu+1)}{k!\Gamma(\mu-k+1)} x(t-kh) \tag{14}$$

where Γ is the gamma function and h is the time increment. This Grünwald–Letnikov formulation inspired a discrete-time calculation algorithm, based on the approximation of the time increment h through the sampling period T and a r-term truncated series yielding the equation:

$$D^{\mu}[x(t)] \approx \frac{1}{T^{\alpha}} \sum_{k=0}^{r} \frac{(-1)^k \Gamma(\mu+1)}{k!\ \Gamma(\mu-k+1)} x(t-kT) \tag{15}$$

This technique was adopted in [7] with $r = 50$.

In this paper the fractional derivative is calculated through a Padé fraction approximation of Euler transformation:

$$\begin{aligned} D^{\mu}(z) &= \left(\tfrac{1}{T}\right)^{\mu} Padé\{(1-z^{-1})^{\mu}\}_{m,n} \\ &= \left(\tfrac{1}{T}\right)^{\mu} \tfrac{P_m(z^{-1})}{Q_n(z^{-1})} \\ &= \left(\tfrac{1}{T}\right)^{\mu} \tfrac{p_0+p_1z^{-1}+\cdots+p_mz^{-m}}{q_0+q_1z^{-1}+\cdots+q_mz^{-n}} \end{aligned} \tag{16}$$

where $m, n \in \aleph$ are the orders of the polynomials and z^{-1} represents the discrete time sampling.

4 Experiments and Results

A reliable execution and analysis of an EE algorithm usually requires a large number of simulations to provide a reasonable assurance that the stochastic effects are properly considered. Therefore, in this study are developed $n = 20$ simulations for each case under analysis.

The experiments consist on running the three algorithms {GA, MA, PSO} to generate a typical combinational logic circuit, namely a 2-to-1 multiplexer ($M2-1$), a 1-bit full adder (*FA1*), a 4-bit parity checker (*PC4*) and a 2-bit multiplier (*MUL2*), using the fitness scheme described in (6) and (10). The circuits are generated with the gate sets presented in Table 1 and $P = 3{,}000$, $w = 0.5$, $c_1 = 1.5$ and $c_2 = 2$.

4.1 Static Fitness Function

In this sub-section we present the results obtained using the static fitness function.

Figures 4 and 5 depict the standard deviation of the number of generations to achieve the solution $S(N)$ *versus* the average number of generations to achieve the solution $Av(N)$ for the algorithms {GA, MA, PSO}, the four circuits and the four gate sets. In these figures, we can see that the *MUL2* circuit is the most complex one, while the *PC4* and the $M2-1$ are the simplest circuits. It is also possible to conclude that Gset 6 is the less efficient gate set for all algorithms and circuits.

Figure 4 reveals that the plots follow a power law:

$$S(N) = a\,[Av(N)]^b \quad a,b \in \Re \tag{17}$$

Table 2 presents the numerical values of the parameters (a, b) for the three algorithms.

Table 2: The parameters (a, b) and (c, d) for the three algorithms

Algorithm	a	b	c	d
GA	0.0365	1.602	0.1526	1.1734
MA	0.0728	1.2602	0.2089	1.3587
PSO	0.2677	1.1528	0.0141	1.1233

In terms of $S(N)$ *versus* $Av(N)$, the MA algorithm presents the best results for all circuits and gate sets. In what concerns the other two algorithms, the PSO is superior (inferior) to the GA for complex (simple) circuits.

Figure 5 depict the average processing time to obtain the solution $Av(PT)$ *versus* the average number of generations to achieve the solution $Av(N)$ for the algorithms {GA, MA, PSO}, the four circuits and the four gate sets. When analyzing these charts it is clear that the PSO algorithm demonstrates to be around ten times faster than the MA and the GA algorithms.

These plots follow also a power law:

$$Av(PT) = c\,[Av(N)]^d \quad c,d \in \Re \tag{18}$$

Table 2 shows parameters (c, d) and we can see that the PSO algorithm has the best values.

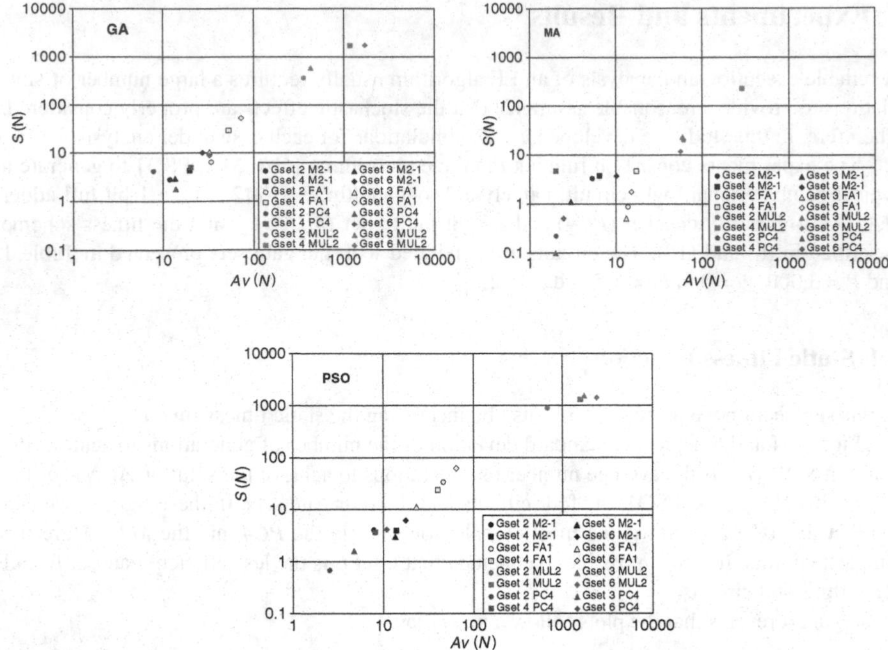

Fig. 4: *S*(*N*) *versus* *Av*(*N*) with $P = 3{,}000$ and F_s for the GA, the MA and the PSO algorithms

4.2 Dynamic fitness function

In this sub-section we analyze the performance of the algorithm when using the dynamic fitness function F_d.

Figures 6 depict *S*(*N*) and *Av*(*PT*), *versus* *Av*(*N*) for the PSO algorithm using F_d, the four circuits and the four gate sets. We conclude that F_d leads to better results in particular for the *MUL2* circuit and for the index *Av*(*PT*).

Furthermore we verify that Figure 7 presents a comparison between F_s and F_d.

5 Scalability Analysis

An issue that emerges with the increasing number of the circuit inputs and outputs is the scalability problem. Since the truth table grows exponentially, the computational burden to achieve the solution increases dramatically [20].

Figures 8 and 9 show the evolution of *S*(*N*) *versus* *Av*(*N*) and *Av*(*PT*) *versus* *Av*(*N*), respectively, for the parity checker family of circuits, for an increasing number of bits. The parity checker family is {2, 3, 4, 5, 6} bit.

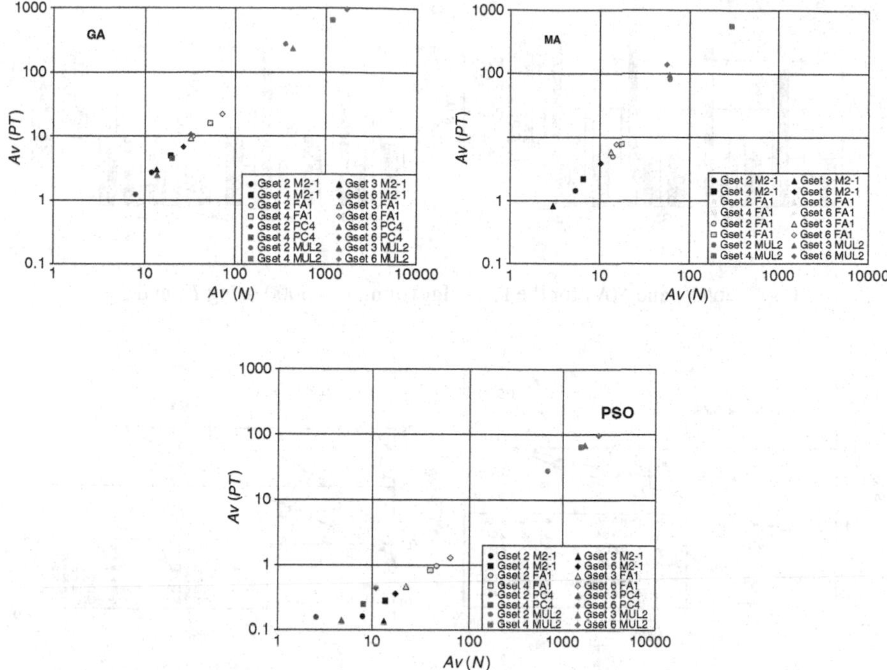

Fig. 5: *Av(PT) versus Av(N)* with $P = 3{,}000$ and F_s for the GA, the MA and the PSO algorithms

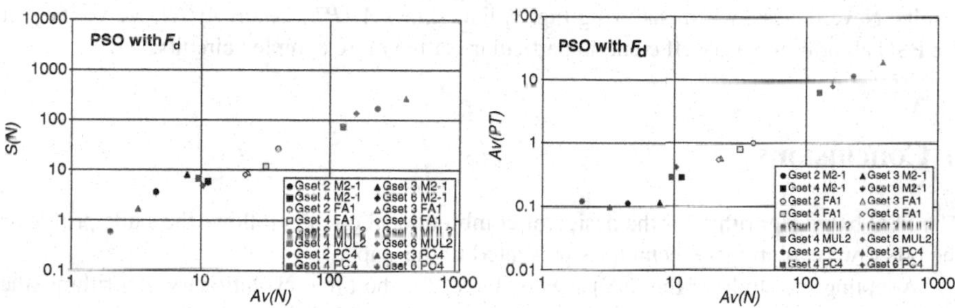

Fig. 6: *S(N) versus Av(N)* and *Av(PT) versus Av(N)* for the PSO algorithm, $P = 3{,}000$ and F_d

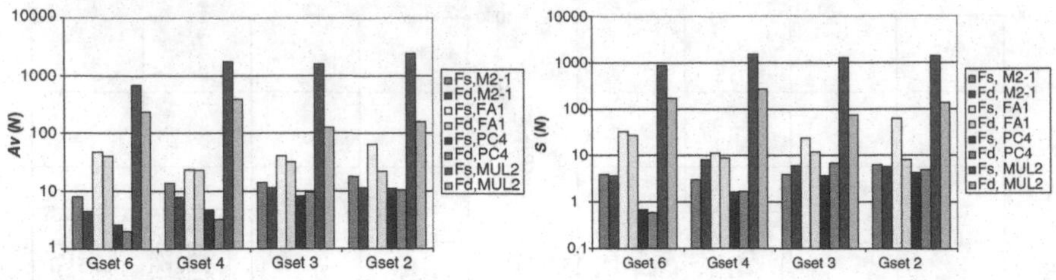

Fig. 7: *Av*(*N*) and *S*(*N*) for the PSO algorithm, $P = 3000$ using F_s and F_d

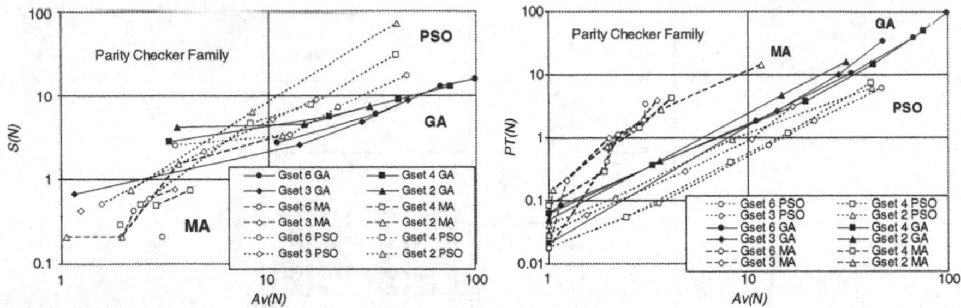

Fig. 8: *S*(*N*) *versus Av*(*N*) and *Av*(*PT*) *versus Av*(*N*) for the parity checker family, for the GA, the MA and the PSO algorithms and for the Gsets under evaluation for $P = 3{,}000$

In terms of *S*(*N*) *versus Av*(*N*) it is possible to say that the MA algorithm presents the best results. Nevertheless, when analyzing Fig. 9, that shows *Av*(*PT*) *versus Av*(*N*), we verify that the PSO algorithm is very efficient, in particular for the more complex circuits.

6 Conclusions

The PSO based algorithm for the design of combinational circuits follows the same profile as the other two evolutionary techniques presented in this paper.

Adopting the study of the *S*(*N*) *versus Av*(*N*) for the three evolutionary algorithms, the MA algorithm presents better results over the GA and the PSO algorithms. However, in what concerns the processing time to achieve the solutions the PSO outcomes clearly the GA and the MA algorithms. Moreover, applying the dynamic fitness the results obtained are improved further in all gate sets and in particular for the more complex circuits.

Acknowledgments

The authors thank GECAD – *Grupo de Investigação em Engenharia do Conhecimento e Apoio à Decisão*, for the financial support to this work.

References

1. Louis S and Rawlins G (1991) Designer Genetic Algorithms: Genetic Algorithms in Structure Design. In Proceedings of the Fourth International Conference on Genetic Algorithms, San Mateo, CA, USA
2. Goldberg D E (1989) Genetic Algorithms in Search Optimization and Machine Learning. Addison-Wesley, Reading, MA
3. Zebulum R S, Pacheco M A and Vellasco M M (2001) Evolutionary Electronics: Automatic Design of Electronic Circuits and Systems by Genetic Algorithms, CRC Press
4. Koza J R (1992) Genetic Programming. On the Programming of Computers by means of Natural Selection, MIT Press, Cambridge, MA
5. Kennedy J and Eberhart R C (1995) Particle Swarm Optimization. In Proceedings of the IEEE International Conference Neural Networks, pp 1942–1948, Perth, Western Australia.
6. Shi Y and Eberhart R C (1998) A Modified Particle Swarm Optimizer. In Proceedings of the 1998 International Conference on Evolutionary Computation, Anchorage, Alaska, pp. 69–73
7. Reis C, Machado J A T and Cunha J B (2005) Evolutionary Design of Combinational Circuits Using Fractional-Order Fitness. In Proceedings of the Fifth EUROMECH Nonlinear Dynamics Conference, Eindhoven, Netherlands, pp. 1312–1321
8. Coello C A, Christiansen A D and Aguirre A H (1996) Using Genetic Algorithms to Design Combinational Logic Circuits. Intelligent Engineering Through Artificial Neural Networks. Vol. 6, pp. 391–396
9. Clerc M and Kennedy J (2002) The Particle Swarm: Explosion, Stability, and Convergence in a multi-dimensional Complex Space. In IEEE Transactions on Evolutionary Computation Vol. 6, pp. 58–73
10. Cecília Reis J A Tenreiro Machado, and J. Boaventura Cunha (2004) Evolutionary Design of Combinational Logic Circuits, JACIII, Fuji Tec. Press, Vol. 8, No. 5, pp. 507-513, Sept
11. Reis C, Machado J A T and Cunha J B. An Evolutionary Hybrid Approach in the Design of Combinational Digital Circuits. In WSEAS Transactions on Systems, 4(12), 2338–2345

From Cybernetics to Plectics: A Practical Approach to Systems Enquiry in Engineering

Béla Pátkai[1], József K. Tar[2], and Imre J. Rudas[2]

[1] Distributed Information and Automation Lab, Institute for Manufacturing, University of Cambridge, Mill Lane, Cambridge, CB2 1RX, UK
bela@patkai.org

[2] Budapest Tech, John von Neumann Faculty of Informatics, Institute of Intelligent Engineering Systems, 1034 Budapest, Bécsi út 96/B, Hungary
tar.jozsef@nik.bmf.hu, rudas@bmf.hu

Summary. The most prominent systems theories from the 20th century are reviewed in this chapter and the arguments of complex system theorists is supported who use the term "plectics" instead of the overused and ambiguous "systems science" and "systems theory". It is claimed that the measurement of complex systems cannot be separated from their modelling as the boundaries between the specific steps of the scientific method are necessarily blurred. A critical and extended interpretation of the complex system modelling method is provided and the importance of discipline-specific paradigms and their systematic interdisciplinary transfer is proposed.

1 Introduction

The enormous industrial, economic and social development of the last three centuries has been built on the solid foundations of science. This development has not only been fruitful for the welfare of society but also coevolved with science itself, giving rise to various theories and methods, viewpoints and disciplines [1]. In the mentioned period of interest these disciplines were typically distinct fields of study until about the first part of the twentieth century when the more and more refined methods of measurement – in physical, biological, socio-economic systems – revealed again the inherent connectedness of natural phenomena and motivated the formation of multi- and interdisciplinary research teams, systematically tearing down the barriers between scientific disciplines. These teams have been working on domain problems but aimed at finding and using common methods that were correctly assumed to provide significant improvements of results. Due to the nature of this development a central element of the scientific method, i.e. measurement, has remained in the focus of domain experts and has not enjoyed the multidisciplinary attention it deserves. Especially in the development of systems theories – and the morphing discipline of "Complex Systems" or "Plectics" a fresh look at the idea of measurement and modelling seems unavoidable, despite fears of failure and difficulties.

For all these reasons, we can assume that measurement of Complex Systems is necessarily a modelling task, and the scientific language available up to now is in itself not completely

J.A.T. Machado et al. (eds.), *Intelligent Engineering Systems and Computational Cybernetics,*

adequate for describing it, hence its extension by a methodology for transferring analogies between disciplines in a systematic way is necessary.

2 The Evolution of Systemic Enquiry

In Fig. 1 a simple classification of systems is represented by Stafford Beer – an early proponent and eminent architect of management cybernetics methods [2]. This classification demonstrates clearly that the idea of approaching systems from the complexity point of view is not new, it merely wore a different label and occasionally a special emphasis on a certain discipline or formal method.

Systems	Simple	Complex	Exceedingly Complex
Deterministic	Window catch	Electronic digital computer	Empty
	Billiards	Planetary system	
	Machine-shop layout	Automation	
Probabilistic	Penny tossing	Stockholding	The economy
	Jellyfish movements	Conditional reflexes	The brain
	Statistical quality control	Industrial profitability	The company

Fig. 1: Stafford Beer's classification of systems [2]

2.1 A Question of Paradigms?

In Fig. 2 a simple comparison is made between Cybernetics, General Systems Theory and System Dynamics, the most well-known and influential system theories. The tables point out vividly that the birth of systemic thinking in different disciplines have had a different origin and different domains of application. While the motivation of all these system theories was the same, their form was naturally shaped by the circumstances of their evolution. We suggest that the often noted confusion in systems research is due to these differences and could be usually resolved very simply by careful communication. The next section refers to Complex Systems, the discipline that has taken this communication concept ahead the most, mainly due to its relatively late development when relevant experience was already present in the research community.

Cybernetics (1940s)

Math	Links	Science
Pure	General Systems Theory	Physical
Applied	Cybernetics	Biological
Computational	System Dynamics	Social

General Systems Theory (1940s)

Math	Links	Science
Pure	General Systems Theory	Physical
Applied	Cybernetics	Biological
Computational	System Dynamics	Social

System Dynamics (1950s)

Math	Links	Science
Pure	General Systems Theory	Physical
Applied	Cybernetics	Biological
Computational	System Dynamics	Social

Fig. 2: Comparison of dominant system science movements

2.2 Complex Systems or Plectics

Complex systems have been defined in various ways, e.g. "complex systems are systems that cannot be described in a unique way" [7]. However, the definition and its aim by Murray Gell-Mann bravely extends the interest of this discipline to include connections between very distant levels of description, as it is depicted in Fig. 3 that is based on [7] and is used by [8] to build a systematic method for transferring analogies between application domains.

As Gell-Mann puts it in [6]: "Complex Systems (or Plectics) illuminates the chain of connections between, on the one hand, the simple underlying laws that govern the behavior of all matter in the universe and, on the other hand, the complex fabric that we see around us, exhibiting diversity, individuality, and evolution. The interplay between simplicity and complexity is the heart of our subject". Such a far-reaching inquiry requires us to stretch the limits of the scientific method beyond straightforward and simple experiments and models, mainly because there is nothing else that can be done. The previous definition demonstrates that – in case we take it seriously – we are no longer able to live with simplifications such as "isolating the problem" or "verification by direct, always repeatable measurement".

The classical understanding of the scientific method and the well-known rule of William Occam suggest a minimalistic approach to modelling. This approach is not just useful but also elegant and improves the efficiency of interpersonal communication. Complex System methods – however – often do the opposite – if only temporarily – and intentionally use models that are not the most parsimonious ones to find a way to represent the abundance of relationships having a potential influence on the studied system [4]. This idea is taken further in the context of measurement in the next section.

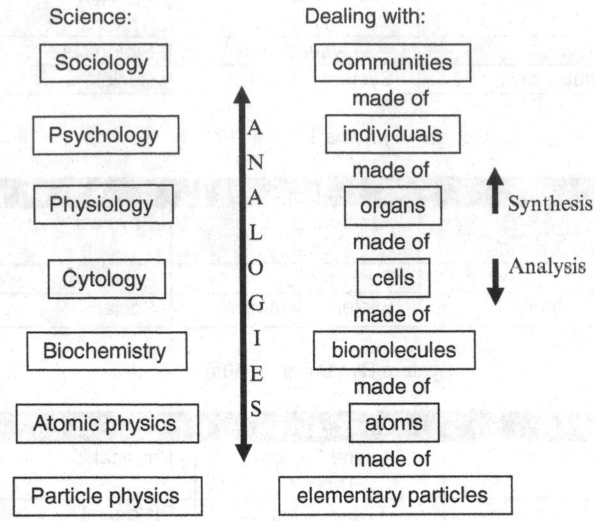

Fig. 3: Analogy transfer in a hierarchical view of scientific disciplines

3 Measurement of Complex Systems

Measurement – or experimentation in general – is an indispensable part of the classical scientific method. A measurement is an interaction between a measured system and an apparatus or equipment and it provides a quantitative description of some quantities pertaining to the system. As such, the interactions can modify the values measured after the measurement act. Experiment designers exercise great care in order to limit such effects, but in many cases they cannot be avoided. Without measurement there is no scientific investigation – this realization is generally accepted for at least half a millennium and it has laid the foundations for the development of the so-called hard sciences. However, as Galileo has already realized, measurement provides an abstraction of the real world and this measurement only represents a limited aspect of reality. A measurement device maps a measurable aspect of reality to a scientific symbol of measurement. This scheme – also called the Galileo scheme – overcomes the "ambiguity of language", as there is a one to one correspondence between aspects of physical reality and scientific symbols.

This naïve scheme was extended by lots of practical experience and philosophical thought in the centuries following Galileo. The breakthrough in Physics (and also in other scientific disciplines) in the 20th century meant that the concept of measurement was generalized and its potential and limitations were discovered in detail. Developments in Quantum Physics and philosophy (e.g. that of Emmanuel Kant)have been transferred to other scientific disciplines, and the realization that measurements change the measured quantities became commonly accepted in the scientific community.

The following list contains the practical problems, summarized to show the profoundness of the issues of measuring complex systems:

1. Scientific work requires measurements, but in the realm of Complex Systems this problem does not usually have appropriate emphasis (giving rise to speculations).
2. Classical measurements provide "static" results, i.e.:
    (a) The interdependency of measured quantities is not represented (only as a separate step on the modelling side).
    (b) Dynamic and nonlinear aspects of measurements are not investigated.
    (c) Measurement is completely separated from modeling, therefore:
        i. These aspects of measurements are not well documented, the modeler does not exactly know about important details, limitations, errors, uncertainties.
        ii. Modelers – and consequently decision makers – often do not design the measurement, they work from what is given.
        iii. Modelers often do not communicate with those doing measurements, provide no feedback and receive no more input than raw data.
3. Complex Systems models are usually based on single high level measurements, there is no vertical integration of other models.
4. Measurement of Complexity/Complex Systems is often regarded as impossible and relatively very little attention is paid in most domains of applications.
5. There is no generally accepted and well-formulated language for describing complex systems.
6. Measurement problems of complex systems is rarely discussed in a multi-disciplinary environment.

4 Modelling Complex Systems

The objective of complex system modelling is somewhat different than that of non-complex systems; instead of optimization and control the goal is understanding, optimal influence and "systemic control". This "systemic control" is realized – or more correctly "emerges", rather than is implemented – by a complex system model, a complex decision support methodology and a set of appropriate actions prescribed by the previous two.

The general requirements for such a complex systemic modelling framework are as follows:

1. Collaborative
    (a) Decision makers and other modellers should have software and simulation support for making decisions together or exert influence on others [3].
    (b) Modellers with different background should have the possibility to share their models, simulations, metaphors/analogies.
2. Integrated
    (a) Provide software, methodological and conceptual/verbal platform for running different models together [5].
    (b) Users can run simulations by incorporating others' simulation models.
3. Distributed
    (a) Models are made and reside locally (where they are used, e.g. the factory scheduling model is developed by the shop-floor manager)

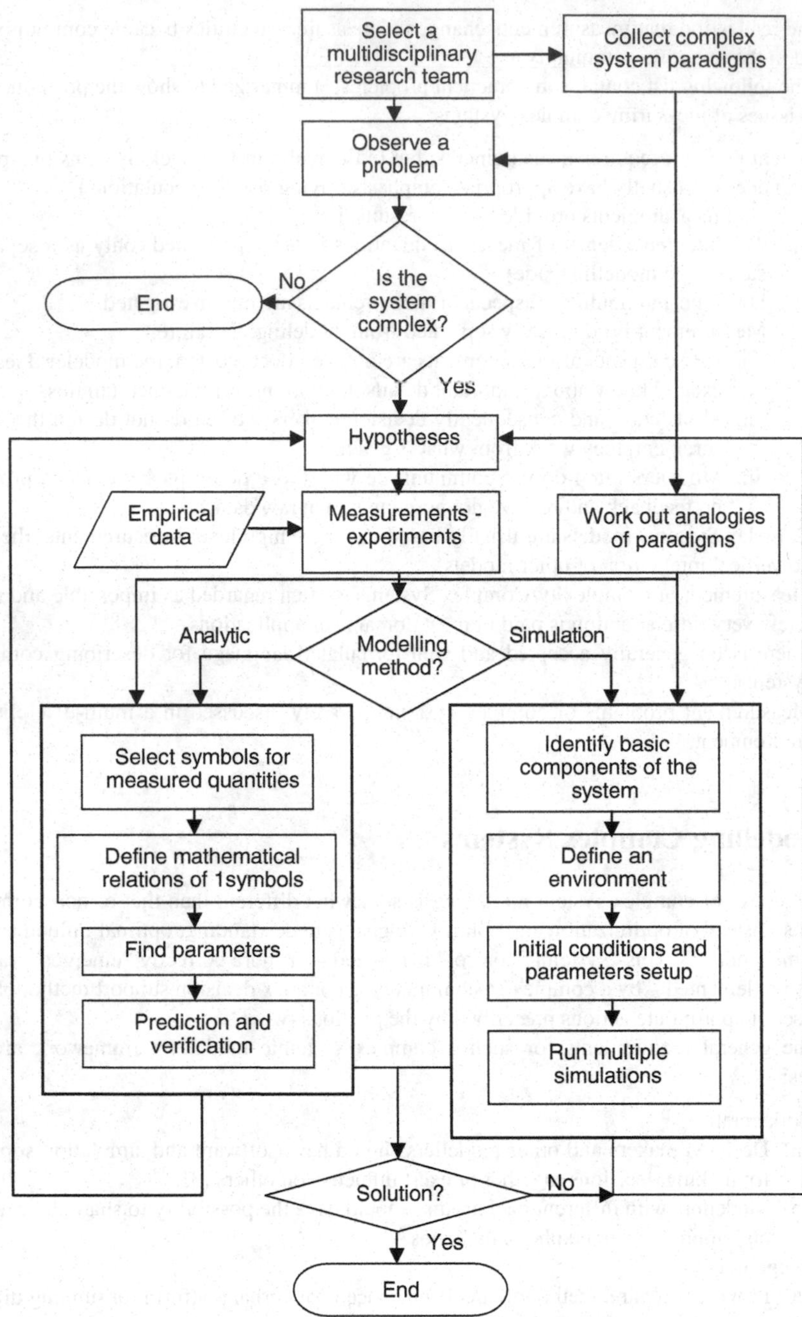

Fig. 4: The extended complex system modelling method

A methodology that addresses the challenges of multidisciplinary communication and systemic measurements is proposed in Fig. 4. Here the boundaries of the methodology are extended significantly and the otherwise unmodeled actions influencing research group formation and the systematic use of analogies are also considered. It is important to note that this methodology is based on the methods worked out in [8, 9]. Also as early as the dawn of cybernetics some preliminary ideas were presented by Stafford Beer about the importance of intentional research group design, where the participants in research were selected as part of the solution method. This proposed method was partially evaluated at the Tampere University of Technology, Finland where a collaboration project[1] on Complex Systems was systematically established on the use of a harmonized vocabulary of complexity and domain problems and general methods were discussed with an interdisciplinary audience.

5 Conclusions

The extended interpretation of Complex Systems (as "Plectics") has substantial consequences in the redefinition of the concept of measurement, experimentation and modeling of systems. The main challenges were discussed in this chapter and a methodology was proposed. The verification of this method is naturally indirect, but the measurement of its success is possible.

References

1. Kuhn TS (1970) The Structure of Scientific Revolutions. University of Chicago Press, Chicago, IL
2. Beer S (1958) Cybernetics and Management. The English Universities Press Ltd, London
3. Fowler JW, Rose O (2004) Grand Challenges in Modeling and Simulation of Complex Manufacturing Sytems. SIMULATION 80(9):469–476
4. Dooley K (1997) A Complex Adaptive Systems Model of Organization Change. Nonlinear Dynamics, Psychology, & Life Science 1(1):69–97
5. Special Issue on Computer Automated Multi-Paradigm Modeling. ACM Transactions on Modeling and Computer Simulation, Vol 12, No 4, October 2002
6. Gell-Mann M, Let's call it Plectics, Complexity Journal, Vol. 1, No. 5 (1995/96).
7. Arecchi FT (1992) Complexity in Science: Models and Metaphors. In: The Emergence of Complexity in Mathematics, Physics, Chemistry and Biology. Pontifica Academia Scientiarvm, 27–31 October, 1992. Princeton University Press Princeton, NJ
8. Pátkai B (2004) An Integrated Methodology for Modelling Complex Adaptive Production Networks. Ph. D thesis. Tampere University of Technology, Finland
9. Pátkai B (2005) Analogy-Based Methodology for Complex Adaptive Production Network Modelling. International Journal of Advanced Computational Intelligence and Intelligent Informatics (JACIII), Fuji Technological Press, Tokio, Japan, 9(4), 399–408

[1] The website of the project can be found at `www.tut.fi/complex`.

Part VI

Mathematical Methods and Models

Extending the Spatial Relational Model PLA to Represent Trees

Ágnes B. Novák[1] and Zsolt Tuza[2]

[1] Péter Pázmány Catholic University, Faculty of Information Technology, H-1088 Budapest, Práter u. 50/A, Hungary
b_novak@digitus.itk.ppke.hu

[2] Computer and Automation Institute, Hungarian Academy of Sciences, H-1111 Budapest, Kende u. 13-17, Hungary; and
Department of Computer Science, University of Pannonia, H-8200 Veszprém, Hungary
tuza@sztaki.hu

Summary. The spatial relational PLA-Model is very handy to store two-connected planar graphs, but in its original form it is not capable at all to store any other important classes of planar graphs. In this paper we enhance the model and describe a method to represent trees. The approach is based on Halin graphs, that are planar graphs of minimum degree 3 from which a tree is obtained when the boundary edges of the infinite region are removed. We show how Halin graphs can be identified from the model, and how trees can be represented as virtual Halin graphs in order to store them without major alteration of the original PLA-Model.

1 Introduction

This paper is one in the series of a systematic investigation on the PLA spatial relational model. The PLA-Model, that will be described in detail in Section 3, was introduced by the US Bureau of the Census [1] for representing topological properties of geographic data (maps). The properties of PLA were first investigated in detail in [2]. The paper [3] describes various types of further models, too. The interested reader may find a detailed introduction to spatial information systems in [4].

In the papers [5–9] several algorithms have been published in connection with the PLA-Model, moreover a temporal version and a 'weighted' version of it have been described. For example, we dealt with identifying two-connectivity, bipartite graphs, Hamiltonian cycles of radially connected planar graphs, spanning trees, imbeddings, etc.

Due to the difficulties of a regular imbedding of an unrestricted planar graph, the PLA-Model is supposed to store *two-connected planar graphs* only. This did not mean a serious restriction in the original applications because the aim of PLA was to represent 'typical' geographical maps, and these usually are two-connected.

In case if one needs to use the model to store other information, however, it could be of essential importance to represent not only two-connected, but any type of (i.e., non-two-connected

J.A.T. Machado et al. (eds.), *Intelligent Engineering Systems and Computational Cybernetics,*

or even disconnected) planar graphs. In this paper we mostly consider representations of trees. Further investigation will be needed to represent the intermediate class of planar graphs with cut-edges and those with several connected components; some ideas in this direction are described in Section 6. Our approach to trees is based on Halin graphs, that will be described in the next section. The present work is the extended version of the conference paper [17].

In spite of the fact that Halin graphs show a rather simple structure, they can be applied in several different areas. Halin introduced this type of graphs in his paper [10]. There he dealt generally with properties of minimal connectivity in graphs; to be more precise, he proved that his graphs are minimally three-edge-connected. This and other interesting properties of Halin graphs can be found in the references [11–15]. We will mention further properties of them in Section 2.

In the next section we review the necessary notions and notation for graphs in general, and in connection with Halin graphs in particular. In Section 3 the PLA-Model is described. Results concerning trees and other types of non-two-connected planar graphs are discussed in Sections 4, 5 and 6, the latter dealing with an outline of some concepts for the general solution. Various potential applications are mentioned in the concluding section.

2 Halin Graphs

A graph $G = (V,E)$ consists of a set V of vertices and a set E of edges, where each edge is an unordered pair of vertices, $E \subseteq \binom{V}{2}$. The degree of a vertex is the number of edges incident with it. When a graph is considered with its geometric representation, we shall often use the terms point and line for vertex and edge.

The union G^* of graphs G_1 and G_2 is $G^* = G_1 \cup G_2 := (V_1 \cup V_2, E_1 \cup E_2)$, where V_1, V_2 and E_1, E_2 are the sets of vertices and edges of G_1 and G_2, respectively.

A Halin graph H is a particular type of planar graph, a union of a 'special' tree T, and a cycle that we shall denote by C, that is: $H = T \cup C$. Tree T is special in the sense that it is required to have the following properties:

(a) T does not have any vertices of degree 2

(b) T has at least one vertex of degree 3 or larger.

If properties (a) and (b) are satisfied, we will call T an *H-type* tree.

Cycle C joins all the vertices of degree 1 in T. That is, the leaves of the tree are connected by C, providing them with a regular imbedding in which C would look like the boundary of an area criss-crossed by tree T.

Figure 1 exhibits an example for a Halin graph; the bold edges belong to tree T, while the dotted edges form cycle C. Vertex labels a, d, e and g are unimportant at this point; they will serve later to illustrate a preparation process described in Section 4.

It is known that Halin graphs can be recognized in polynomial time. In our approach the representation of the graph will store a more complete information, and it will allow to design a method of recognition with linear running time.

Checking whether a given graph is a Halin graph or not, is of interest, because such graphs belong to a so-called three-terminal recursive class. It is proved in [14] that many, otherwise extremely difficult graph problems, like chromatic number, dominating set, minimum maximal matching, etc., can be solved with linear-time algorithms on Halin graphs.

On the other hand, unfortunately, to decide whether a graph has a spanning Halin subgraph (having all the vertices, but just a subset of edges), or whether a subgraph is a part of a Halin graph, are NP-complete problems.

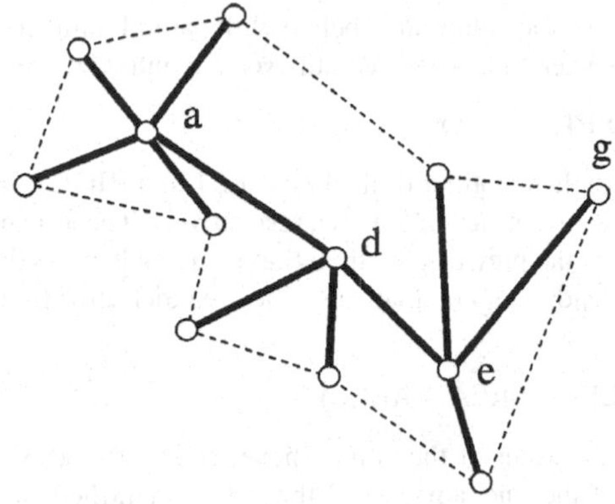

Fig. 1: Halin graph: tree plus cycle

3 The PLA-Model

In this section we briefly review the PLA-Model, citing some parts of Section II of [8]. As indicated already, this model is a spatial relational one.

Under the word 'spatial' we usually mean that the stored object has some well-defined geometrical extent, represented in the model in some way. However, it is not completely true for the PLA-Model.

The PLA-Model belongs to the spatial models in a generalized sense. The aim of the PLA-Model is the representation of the topology of a 'typical' map–or, equivalently, a planar graph together with its regions (areas)–in terms of relational databases. So the geometrical realization of the graph is not stored explicitly.

The PLA database consists of four relations, but three of them can be reconstructed exclusively from R3, as it is proved in [8]. For this reason, we begin the description with this most important relation.

(*i*) R3(AREA, POINT, LINE, LABEL)

Relation R3 stores the border of an area (face). The first attribute is the name of the area whose neighbourhood is represented. It is assumed that every area is surrounded by an alternating cycle of lines and points. So the second attribute means the identifier of the point (vertex), the third one is the identifier of the line on the border of that area, and the last attribute, LABEL is the serial number with respect to the area named in the first attribute. The labels are taken consistently with the orientation, that is, the enumeration of points and lines in the cycle corresponds to the clockwise orientation, except for the infinite area, for which the orientation is counterclockwise. In each row of the relation, the point and the line are incident and the line is the successor of the point in the corresponding orientation.

Next we list the other three PLA-relations, which still are useful tools, and will be used in our algorithms.

(*ii*) R1(LINE, FROM, TO)

LINE is the identifier of the edge whose 'starting' point identifier is FROM, and the 'end' point identifier is TO. The objects stored in the database are undirected, but in order to obtain a unique representation of the areas there is a need for some orientation. In the sequel, to avoid ambiguity, we suppose that the points are labelled with natural numbers, and the starting point of each line is always the vertex labelled with the *smaller* number.

(*iii*) R2(LINE, LEFT, RIGHT)

LINE corresponds to the identifier of the line, LEFT and RIGHT are the identifiers of the area on left side and on the right side of the line, respectively. The left and right sides are determined in accordance with the previous relation; that is, line e is traversed from point i to point j if $i < j$, and facing to endpoint j on that edge, the area identified by LEFT is geometrically on the left.

(*iv*) R4(POINT, LINE, AREA, LABEL)

POINT means the identifier of the point whose neighborhood is stored in this relation. LINE is the identifier of the line, adjacent to the vertex identified by POINT, AREA is the identifier of the area having vertex POINT and edge LINE on its border, and LABEL is a serial number, storing the alternating sequence of lines and areas in the given imbedding. Similarly to the previous paragraph, the orientation is clockwise (but now for all points, without any exception), and of course the line is on the boundary of the area.

4 Method for Storing Trees in Enhanced PLA

The problem of storing trees in the PLA-Model lies in the fact that in lack of two-connectivity, the 'boundary' of the graph is difficult to recognize and would even require to relax the conditions imposed on the model. Otherwise, in case of two-connected graphs the infinite area is recognizable easily from R3, see e.g. [8], and from that on, the already elaborated algorithms can be used for reconstructing the stored graph. In case of trees, this infinite area would be the only area we store, yielding a degenerate representation of the tree in its PLA database.

This means that we need to split the only infinite area of a tree into two types of virtual areas. One area type corresponds to the 'inside', bounded areas, the other is the outside, infinite area. If we imagine a regular imbedding of a tree, this view leads us to a natural solution using Halin graphs.

Consequently, the tree has to be completed with a 'boundary' perimeter, a cycle incident with the leaves of the tree. However, it is not trivial, which leaf should be joined to which other leaves.

Before answering this question, we also have to deal with the problem that a given tree could be not of H-type, due to the presence of degree-2 vertices.

At least, we know that if T is not a path then at least one non-peripheral vertex of H has degree greater than 2. However, the vertices having two incident edges must be eliminated. If vertex b has edges ab and bc incident with it, then these two edges are contracted to the single edge ac. The new graph, now without vertex b, is topologically isomorphic to the original one. By the repeated application of this step we can eliminate all vertices of degree 2. In this way each maximal simple chain gets contracted to a single edge.

Figure 2 shows a tree T' by bold lines. This tree is not eligible for completing it to a Halin graph, since producing the bordering cycle C' (by dotted lines) through the leaves, it would contain vertices having degree 2 (the nodes filled by black). These vertices will be melted into one as follows: edge ab and edge bc are substituted first by edge ac, then ac and cd are united as ad. In a similar way, we get from edges ef and fg the new edge eg. The resulting new graph is shown in Fig. 1.

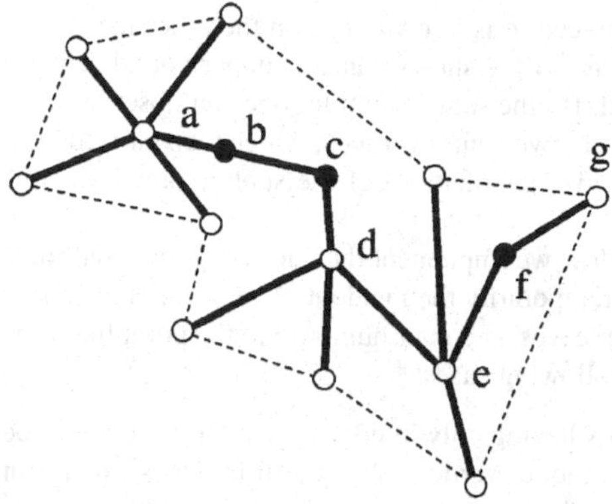

Fig. 2: An example for edge elimination of a non-Halin graph

This minor technical problem can be solved for example by applying the weighted extension of PLA, the so-called PLAX [9]. In that model we have weights for the edges in R1:

RX1(LINE, FROM, TO, WEIGHT).

Let us define the weight function $W : E \rightarrow N$ (here N stands for the set of natural numbers) in the following way:

W(LINE) := 1 if it is a normal edge having degrees different from 2 in T

W(LINE) := 0 if it is a virtual edge (of C), and finally

W(LINE) := k if it is contracted from k edges of degree 2

The weighted extension is definitely needed for reconstructing the graph, but for storing other properties, it usually suffices just to project RX1 onto the first three attributes. So, after preparing the tree to obey H-type, we may concentrate on finding the appropriate perimeter for it. Once we found it, we add those virtual edges in order to get a virtual Halin graph.

Using a rooted tree turns out to be useful. For that purpose we are going to find an appropriate vertex, which will play the role of the root. It is important from the point of view of reconstruction.

Among several possibilities, let us mention the three most convenient ones:

– Root 1: the vertex having the largest degree

– Root 2: the middle vertices/vertex of a longest path

– Root 3: the vertex in the 'centre'

Root 1 can be obtained easily from R3, just counting the occurrences of the vertex identifiers.

Root 2 can be found using well-known algorithms that find a longest path in a tree.

Finding Root 3 is quite interesting, so we show here a method for it. The *centre of a tree* is a vertex with the property that deleting that vertex from the tree, in each of the remaining components the number of vertices is at most half of the original number of vertices. If a vertex does not have this property, it means that after its removal one of the components contains more than half of the vertices.

The existence of the centre is known and, on the other hand, the 'double star' with two adjacent vertices of degree $|V|/2$ shows that this upper bound cannot be strengthened to any smaller value in general. By the same example, one can observe further that a tree has either just one centre or precisely two centre vertices, which then have to be adjacent. (The degree(s) of the centre(s) and the size distributions of the subtree branches attached to them are not at all determined.)

In the algorithm below we implement the following informal idea. For a vertex, which is not a centre, we orient temporarily the incident edge towards the component not smaller than $|V|/2$. Starting with the leaves, and continuing up to the point that each vertex is investigated, we will get one of the following cases:

– There is one vertex having only in-edges: then this vertex will be chosen as the root.

– There are two neighbour vertices, having all in-edges except for the one joining them, which is oriented in both directions. In this case it is practical to choose the one whose deletion results in smaller subtrees apart from the largest one.

This informal description can be captured by counting the branch sizes of the possible subtrees rooted at the different vertices. For instance, if we cut off leaves, then at each cutting we would have a tree with just one vertex. This is reflected in the degrees of the vertices, so in the algorithm the path that we will follow is the smallest sum of sizes of the attached branches.

The output of Algorithm 1 is the centre of the tree; this will be chosen for root.

In the description below, a subset of vertices to be processed will be stored in an auxiliary heap H. For each vertex in H a value is computed, and TopH is an element with minimum value in H. An element is inserted into H when all but one of its neighbours have been processed. We use the notation H+v and H−v for the insertion into H and the extraction from H, respectively, of vertex v, also assuming that the heap property gets updated in each such step. The variables associated with vertex v have the following meaning: count(v) is one smaller than the number of neighbours of v not processed so far, and value(v) is the total number of vertices in the 'branches' considered so far that would directly be attached to v if v were chosen to be the root at the moment.

ALGORITHM 1
for finding the centre of a tree on at least three vertices

```
INPUT: Graph tree G = (V,E)
1. /* initializing the values for the vertices */
V* = V, H = ∅
DO WHILE NOT(V = ∅)
   for each v ∈ V
      count(v) = degree(v) − 1
      value(v) = 0
      IF count(v) = 0
      THEN H = H+v, V* = V* \ {v}
END DO
2. /* sequentially eliminating vertices not to be selected as root */
DO WHILE |H| > 1
      x = TopH
      y = unique neighbour of x in H∪V*
      H = H−x
```

value(y) = value(x) + value(y) + 1
count(y) = count(y) − 1
IF count(y) = 0 THEN H = H+y, $V^* = V^* \setminus \{v\}$
END DO

3. /* specifying the root */
OUTPUT R=TopH

Having R at hand, no matter which technique has been used (Root 1, Root 2, or Root 3), now we have to set the order of leaves for getting the peripheric cycle C.

The main idea is depth-first search, driven by the smallest label. So we follow the path determined by the vertices with the smallest labels (in one path we get a monotone decreasing sequence of labels). Since depth-first search is well-known, we just specify the role of smallest label informally.

Starting at the root, go to the neighbour with smallest label. Once we choose a vertex, go to the next minimally labelled vertex, and so on. In this way we find a leaf, L1. This will be the 'first' vertex in cycle C. Then step back by one level 'up', say to vertex u, and find the neighbour of u with the smallest label, but being different from the one we already visited. Following the path and always choosing the vertex with the smallest label, we reach another leaf, say L2. Then L1 and L2 form the first edge of C. This edge will be put into XR1 by the tuple

<ID#1, L1, L2, 0>

It is important to number the tuples from one, and always to increase the ID number by one. It is also important to put the vertices into XR1 in the same order as we found them. So the auxiliary orientation must follow the order we get the vertices.

The last number in that tuple stands for the weight. Since it is not a real edge, just a virtual one, it has weight 0. XR1 can be built up by this method.

Accordingly, R3 has to be constructed. In XR1 we have the infinite area stored by the edges of weight 0. For the infinite area we just have to follow the constructed IDs in XR1; that is, their value can be copied into attribute LABEL in R3. For example, suppose that the following two tuples are in XR1:

<ID#1, L1, L2, 0>

<ID#2, L2, L3, 0>

Then the corresponding entries in R3 would be:

<OUT, L1, ID#1, 1>

<OUT, L2, ID#2, 2>

This has to be continued for all the edges having weight 0.

In the next step we have to represent the inside areas. We have as many inside areas as the number of leaves.

Technically, the easiest way to build up R3 is just to use the already constructed depth-first search tree in the following way.

First, suppose that we only have 0 and 1 weights in XR1, and furthermore, we found leaf L1, and stepped already back to vertex u. Then we must have an edge with weight 1 with endpoints L1 and u, with some ID number i. This implies to put the following tuples into R3:

<IN1, u, i, #1>

<IN1, L1, ID#1, #2>

This is in accordance with the model description for R3 (see Section 3), and also with the previous construction of XR1, see the first tuple of it above. The process has to be continued until the number of inside areas reaches the number of edges in C.

5 Retrieving a Tree from Enhanced PLA

As it will be described in Subsection 7.1 in a greater detail, testing whether a stored object is a Halin graph is a very important issue.

If the enhanced or original PLA database is given, it is easy to figure out whether the stored object is a Halin graph or not. First we need to identify the boundary of the infinite area. Then we could test whether the number of inside areas would match with the number of edges of the infinite area, and whether all vertex degrees are greater than 2. If the answer is yes, then the object is a Halin graph provided that each inside area shares exactly one edge on its boundary with the infinite area. This search in R3 can be made in linear time.

The decision, whether it is a virtual Halin graph or an ordinary one, can be made from XR1. Obviously, if all edges of the infinite area are weighted with zero, then it is a virtual Halin graph.

The imbedding of the (virtual) Halin graph can be obtained applying one of the algorithms in [8]. In case of virtual Halin graphs, edges with zero weights must be deleted after imbedding the graph.

Further, it is necessary to deal with trees for which we needed to apply the 'contracting' process. Using the reconstruction process described above, we have XR1 with the weights. We need to pick the edges with weights greater than one, and simply insert the required number of vertices on that edge. Even the standard R1 can be constructed in this way, if needed.

Finally, let us mention that also a mixed virtual-real Halin graph may serve practical purposes.

6 Representing Non-two-connected Planar Graphs

Since the original version of the PLA-Model is planned only for storing two-connected planar graphs, it is an important issue to solve the representation of unrestricted planar graphs, such as connected graphs with cut-edges or disconnected ones with several components.

Due to space limitation, in this paper we concentrate on the main aspect of the problem, namely, how to store and retrieve trees. It is because the method we described in Section 4 for representing trees will serve as a basis for the general solution.

The main idea in our present method is to complete a connected graph with virtual edges in order to get a two-connected graph.

In case of planar graphs in general, we follow the same concept, using the model introduced above. Once the components are identified, it can be tested efficiently, which of them is an isolated point, or two-connected, or connected only.

Then the components should be connected to each other by virtual edges. To achieve two-connectivity, more virtual edges should be introduced. To find the most appropriate virtual completion of the graph, however, may depend on various aspects.

Even in the simplest case, where one component is an isolated vertex and the other one is a two-connected subgraph, it is not straightforward, which points should become the neighbours of the isolated point after setting the new virtual edges.

In some other cases new virtual edges might be inserted componentwise, but for some special types of graphs it may be significantly simpler to apply virtualization for the entire graph. This will be discussed in a forthcoming paper.

7 Conclusions

In this paper a method for storing trees in the PLA spatial relational model is presented.

The original model is extended with weights for its first relation. The role of these weights is to store information about the nature of an edge: is it a real, a virtual, or a real but 'contracted' one? This extension is needed because trees can be stored by a virtual completion of the given tree to Halin graphs.

The problem of tree storage has been solved, and also it has been pointed out, how to recognize Halin graphs from the PLA-Model. The method can be improved further to an algorithm for general planar graphs. In this context it would be of interest to analyze, how one can extend to large classes the important algorithms that are known to be linear for Halin graphs; see, for example, those listed in Section 2.

7.1 Some Applications

Virtual Halin graphs (see Section 4) may be widely applied, since each problem representable by trees has a chance to be modelled by them. For example, in biology, phylogenetic trees have the properties prescribed for T in Section 2 above.

In database theory, hierarchical data structures may be considered as they form T at some prescribed circumstances. By completing them with perimetric cycle C, the underlying undirected structure is a Halin graph.

Trees serve very often as data models. There is an example for this in the TriStarp system [16]. (However, in TriStarp there occur directed trees.) A similar example is the HUB technique, in a Sequence Retrieval Systems in bioinformatics.

Acknowledgments

Research of the second author was supported in part by the Hungarian Scientific Research Fund, OTKA grant T-049613.

References

1. Corbett J P (1979) Topological Principles in Cartography, Technical paper 48, US Bureau of the Census, Washington, DC
2. Kuijpers B, Paredaens J, Van den Bussche (1995) Lossless Representation of Topological Spatial Data. Lecture Notes in Computer Science 951: 1–13
3. Paredaens J (1995) Spatial Databases. Lecture Notes in Computer Science 893: 14–35

4. Laurini R, Thomson D (1992) Fundamentals of Spatial Information Systems. Academic Press, APIC series: 37
5. Novák Á B, Tuza Zs (1999) Objects Reconstruction and Its Complexity Problems in the PLA Spatial Database Model. In: Proceedings of IEEE 10th International Conference on Intelligent Engineering Systems: 219–223, Stara Lesma, Slovakia
6. Novák Á B, Fehér T (2000) Constraints for Bipartite and Eulerian Graphs in a Spatial Relational Model. In: Proceedings of 4th IEEE Conference on Intelligent Engineering Systems: 107–110, Portoroz, Slovenia
7. Novák Á B, Tuza Zs (1999) Representation, Constraints and Complexity of Graphs in the PLA Spatial Database Model. In: Proceedings of the Jubilee International Conference of Bánki Donát Polytechnic, Budapest, Hungary: 57–60
8. Novák Á B, Tuza Zs (2002) Reconstruction Graphs and Testing Their Properties in a Relational Spatial Database. Computers and Mathematics with Applications 43: 1391–1406
9. Novák Á B, Tuza Zs (2002) Representing Directed and Weighted Graphs and Their Properties in a Relational Spatial Database Model. In: Proceedings of 6th IEEE Conference on Intelligent Engineering Systems: 357–361, Opatija, Croatia
10. Halin R (1971) Studies on minimally n-connected graphs. In: Welsh D J A (ed), Combinatorial Mathematics and Its Applications. Academic: 129–136, Academic Press, London.
11. Lovász L, Plummer M D (1975) On a Family of Planar Bicritical Graphs. In: Proceedings London Math Soc 30: 160–175
12. Bondy J A, Lovász L (1985) Length of Cycles in Halin Graphs. J. Graph Theory 8: 397–410
13. Bondy J A (1975) Pancyclic Graphs: Recent Results. In: Infinite and Finite Sets 1, Colloq. Math. Soc. János Bolyai, 10, North-Holland: 181–187
14. Borie R B, Parker R G, Tovey C A (1992) Automatic Generation of Linear-Time Algorithms from Predicate Calculus Descriptions of Problems on Recursively Constructed Graph Families. Algorithmica 7: 555–581
15. Horton S B, Parker R G, Borie R B (1992) On Some Results Pertaining to Halin Graphs. Congressus Numerantium 93: 65–87
16. King P J H, Derakhshan M, Poulovassilis S, Small C (1990) TriStarp - An Investigation into the Implementation and Exploitation of Binary Relational Storage Structures. In: Brown A, Hitchcock B (eds), BNCOD-8 Proceedings of the 8th British National Conference on Databases, Pitman: 64–84
17. Novák Á B, Tuza Zs (2006) A Method to Represent Restricted Classes of Planar Graphs in the Enhanced Spatial Relational Model PLA, Based on Halin Graphs In: Proceedings of IEEE 10th International Conference on Intelligent Engineering Systems: 173–177, London, UK

Quantity Competition in a Differentiated Duopoly

Fernanda A. Ferreira[1,2], Flávio Ferreira[1], Miguel Ferreira[2], and Alberto A. Pinto[2]

[1] Departamento de Matemática, ESEIG - Instituto Politécnico do Porto
Rua D. Sancho I, 981, 4480-876 Vila do Conde, Portugal
fernandaamelia@eseig.ipp.pt, flavioferreira@eseig.ipp.pt

[2] Departamento de Matemática Pura, Faculdade de Ciências da Universidade do Porto
Rua do Campo Alegre, 687, 4169-007 Porto, Portugal
miguel.ferreira@fc.up.pt, aapinto@fc.up.pt

Summary. In this paper, we consider a Stackelberg duopoly competition with differentiated goods, linear and symmetric demand and with unknown costs. In our model, the two firms play a non-cooperative game with two stages: in a first stage, firm F_1 chooses the quantity, q_1, that is going to produce; in the second stage, firm F_2 observes the quantity q_1 produced by firm F_1 and chooses its own quantity q_2. Firms choose their output levels in order to maximise their profits. We suppose that each firm has two different technologies, and uses one of them following a certain probability distribution. The use of either one or the other technology affects the unitary production cost. We show that there is exactly one perfect Bayesian equilibrium for this game. We analyse the variations of the expected profits with the parameters of the model, namely with the parameters of the probability distributions, and with the parameters of the demand and differentiation.

1 Introduction

One of the most widely cited models of non-cooperative oligopoly behaviour is the Stackelberg model, developed by Heinrich von Stackelberg in 1934. Researchers in industrial organization have used the Stackelberg model to analyse various important economic issues, such as entry and strategic precommitment. In the Stackelberg model, firms choose outputs sequentially, and thus, the technology may change or the prices of inputs may vary over time, giving rise to cost differentials among the firms (see, for instance [11]. Related works are done in [2, 4, 5, 6, 7, 9]).

Let F_1 and F_2 be two firms, each producing a differentiated product, and competing on quantities. Von Stackelberg [12] proposed a dynamic model of duopoly in which a dominant (leader) firm moves first and a subordinate (follower) firm moves second. In the case of complete information, it is well-known that the leading firm has advantages over the follower (see [3]). The timing of the game is as follows: (i) The leading firm (F_1) chooses a quantity level $q_1 \geq 0$; (ii) The follower (F_2) observes q_1, and then chooses a quantity level $q_2 \geq 0$. We will study this model by considering that each firm has two different technologies, and uses

J.A.T. Machado et al. (eds.), *Intelligent Engineering Systems and Computational Cybernetics,*

one of them following a certain probability distribution. The use of either one or the other technology affects the unitary production cost. We suppose that firm F_1's unitary production cost is c_A with probability ϕ and c_B with probability $1-\phi$ (where $c_A > c_B$), and firm F_2's unitary production cost is c_H with probability θ and c_L with probability $1-\theta$ (where $c_H > c_L$). Both probability distributions of unitary production costs are common knowledge. In this work, we determine the quantities in the perfect Bayesian equilibrium for the above model, and we show that, in contrast to the case with complete information, the follower firm may have a higher expected profit than the leading. Assuming that firm F_1 has to produce more than $(\alpha - c_H)/\gamma$ and less than $(\alpha - c_L)/\gamma$, we observe the existence of five regions on the parameters space such that in each one of them the equilibrium is different. We compute the perfect Bayesian equilibrium for each of the five regions. We analyse the variations of the expected profits with the parameters of the probability distributions, for different degrees of product differentiation. We also analyse the effect of other parameters of the model on the expected profits.

2 The Model and the Equilibrium

We consider an economy with a monopolistic sector with two firms, F_1 and F_2. Firm F_i produces a differentiated product i at a constant marginal cost. We present the sequential-move model, with incomplete information, in which firms choose quantities. The timing of the game is as follows: (i) Firm F_1 (leader) chooses a quantity $q_1 \geq 0$ for its good; (ii) Firm F_2 (follower) observes q_1 and then chooses a quantity $q_2 \geq 0$ for its good.

The inverse demands are

$$p_i = \alpha - q_i - \gamma q_j,$$

provided that the prices p_i are non-negative, with $i, j \in \{1,2\}$ and $i \neq j$, where $\alpha > 0$ and $0 \leq \gamma \leq 1$. The parameter γ expresses the degree of product differentiation (see [10]), and since $\gamma \leq 1$, "cross effects" are dominated by "own effects" (see [1]). Moreover, if $\gamma = 1$, then the goods are homogeneous, and if $\gamma = 0$, then the goods are independent. Firm F_i's profit, π_i, is given by

$$\pi_i(q_i, q_j) = q_i(p_i - c_i) = q_i(\alpha - q_i - \gamma q_j - c_i),$$

where $0 < c_i < \alpha$ is the unitary production cost for firm F_i. We suppose that each firm has two different technologies, and uses one of them following some probability distribution. The use of either one or the other technology affects the unitary production cost. The following probability distributions of unitary production costs are common knowledge among both firms:

$$C_1 = \begin{cases} c_A & \text{with probability } \phi \\ c_B & \text{with probability } 1-\phi \end{cases}, \qquad C_2 = \begin{cases} c_H & \text{with probability } \theta \\ c_L & \text{with probability } 1-\theta \end{cases}.$$

We suppose that $c_B < c_L < c_A < c_H < \alpha$. Firms' profits are given by

$$\pi_1(q_1(c_1), q_2(c_2)) = (\alpha - q_1(c_1) - \gamma q_2(c_2) - c_1)q_1(c_1),$$

$$\pi_2(q_1(c_1), q_2(c_2)) = (\alpha - q_2(c_2) - \gamma q_1(c_1) - c_2)q_2(c_2),$$

where c_i is firm F_i's unitary production cost, and the quantity $q_i(c_i)$ depends upon the unitary production cost c_i of firm F_i, for $i \in \{1,2\}$.

Firm F_1 should choose a quantity, $q_1^*(c_A)$ or $q_1^*(c_B)$, depending on its unitary production cost, to maximize its expected profit; and firm F_2, knowing firm F_1's decision, should choose a quantity, $q_2^*(c_H|q_1^*(c_A))$, $q_2^*(c_L|q_1^*(c_A))$, $q_2^*(c_H|q_1^*(c_B))$ or $q_2^*(c_L|q_1^*(c_B))$, depending on its unitary production cost, to maximize its expected profit. In the next theorem we compute the quantities for the perfect Bayesian equilibrium. We find five parameter regions such that in each one of them the equilibrium is different.

Theorem 1. *Let $E(C_2) = \theta c_H + (1-\theta)c_L$ be the firm F_2's expected unitary production cost. Suppose that*

$$\frac{\alpha - c_H}{\gamma} \leq q_1(c_i) \leq \frac{\alpha - c_L}{\gamma}, \quad \text{with } i \in \{A,B\}.$$

(a) If

$$(\alpha - 2c_H + c_L)(1-\theta)\gamma^2 + 2(\alpha - c_A)\gamma - 4(\alpha - c_H) \geq 0 \tag{1}$$

and

$$(\alpha - 2c_H + c_L)(1-\theta)\gamma^2 + 2(\alpha - c_B)\gamma - 4(\alpha - c_L) \leq 0, \tag{2}$$

then the quantities in the perfect Bayesian equilibrium are given by

$$\begin{aligned}
q_1^*(c_A) &= \frac{(2-\gamma(1-\theta))\alpha - 2c_A + \gamma(1-\theta)c_L}{2(2-\gamma^2(1-\theta))},\\
q_1^*(c_B) &= \frac{(2-\gamma(1-\theta))\alpha - 2c_B + \gamma(1-\theta)c_L}{2(2-\gamma^2(1-\theta))}\\
q_2^*(c_H|q_1^*(c_A)) &= 0,\\
q_2^*(c_L|q_1^*(c_A)) &= \frac{(4-2\gamma-(1-\theta)\gamma^2)\alpha + 2\gamma c_A - (4-(1-\theta)\gamma^2)c_L}{4(2-(1-\theta)\gamma^2)},\\
q_2^*(c_H|q_1^*(c_B)) &= 0,\\
q_2^*(c_L|q_1^*(c_B)) &= \frac{(4-2\gamma-(1-\theta)\gamma^2)\alpha + 2\gamma c_B - (4-(1-\theta)\gamma^2)c_L}{4(2-(1-\theta)\gamma^2)}.
\end{aligned}$$

(b) If

$$(\alpha - 2c_H + c_L)(1-\theta)\gamma^2 + 2(\alpha - c_A)\gamma - 4(\alpha - c_H) \geq 0, \tag{3}$$

$$(\alpha - 2c_H + c_L)(1-\theta)\gamma^2 + 2(\alpha - c_A)\gamma - 4(\alpha - c_L) \leq 0 \tag{4}$$

and

$$(\alpha - 2c_H + c_L)(1-\theta)\gamma^2 + 2(\alpha - c_B)\gamma - 4(\alpha - c_L) \geq 0, \tag{5}$$

then the quantities in the perfect Bayesian equilibrium are given by

$$\begin{aligned}
q_1^*(c_A) &= \frac{(2-\gamma(1-\theta))\alpha - 2c_A + \gamma(1-\theta)c_L}{2(2-\gamma^2(1-\theta))},\\
q_1^*(c_B) &= \frac{\alpha - c_L}{\gamma},\\
q_2^*(c_H|q_1^*(c_A)) &= 0,\\
q_2^*(c_L|q_1^*(c_A)) &= \frac{(4-2\gamma-(1-\theta)\gamma^2)\alpha + 2\gamma c_A - (4-(1-\theta)\gamma^2)c_L}{4(2-(1-\theta)\gamma^2)},\\
q_2^*(c_H|q_1^*(c_B)) &= 0,\\
q_2^*(c_L|q_1^*(c_B)) &= 0.
\end{aligned}$$

(c) If

$$(\alpha - 2c_H + c_L)(1-\theta)\gamma^2 + 2(\alpha - c_A)\gamma - 4(\alpha - c_H) \leq 0, \tag{6}$$

$$(\alpha - 2c_H + c_L)(1-\theta)\gamma^2 + 2(\alpha - c_B)\gamma - 4(\alpha - c_H) \geq 0, \tag{7}$$

$$(\alpha - 2c_H + c_L)(1-\theta)\gamma^2 + 2(\alpha - c_B)\gamma - 4(\alpha - c_L) \leq 0 \tag{8}$$

and

$$(c_H - c_L)(1-\theta)\gamma^2 - 2(\alpha - c_A)\gamma + 2(\alpha - c_H) \le 0, \tag{9}$$

then the quantities in the perfect Bayesian equilibrium are given by

$$q_1^*(c_A) = \frac{\alpha - c_H}{\gamma},$$
$$q_1^*(c_B) = \frac{(2-\gamma(1-\theta))\alpha - 2c_B + \gamma(1-\theta)c_L}{2(2-\gamma^2(1-\theta))},$$
$$q_2^*(c_H|q_1^*(c_A)) = 0,$$
$$q_2^*(c_L|q_1^*(c_A)) = \frac{c_H - c_L}{2},$$
$$q_2^*(c_H|q_1^*(c_B)) = 0,$$
$$q_2^*(c_L|q_1^*(c_B)) = \frac{(4-2\gamma-(1-\theta)\gamma^2)\alpha + 2\gamma c_B - (4-(1-\theta)\gamma^2)c_L}{4(2-(1-\theta)\gamma^2)}.$$

(d) If

$$(\alpha - 2c_H + c_L)(1-\theta)\gamma^2 + 2(\alpha - c_B)\gamma - 4(\alpha - c_H) \le 0 \tag{10}$$

and

$$(c_H - c_L)(1-\theta)\gamma^2 - 2(\alpha - c_A)\gamma + 2(\alpha - c_H) \le 0, \tag{11}$$

then the quantities in the perfect Bayesian equilibrium are given by

$$q_1^*(c_A) = \frac{\alpha - c_H}{\gamma}, \qquad q_1^*(c_B) = \frac{\alpha - c_H}{\gamma},$$
$$q_2^*(c_H|q_1^*(c_A)) = 0, \qquad q_2^*(c_L|q_1^*(c_A)) = \frac{c_H - c_L}{2},$$
$$q_2^*(c_H|q_1^*(c_B)) = 0, \qquad q_2^*(c_L|q_1^*(c_B)) = \frac{c_H - c_L}{2}.$$

(e) If

$$(\alpha - 2c_H + c_L)(1-\theta)\gamma^2 + 2(\alpha - c_A)\gamma - 4(\alpha - c_H) \le 0, \tag{12}$$
$$(\alpha - 2c_H + c_L)(1-\theta)\gamma^2 + 2(\alpha - c_B)\gamma - 4(\alpha - c_L) \ge 0 \tag{13}$$

and

$$(c_H - c_L)(1-\theta)\gamma^2 - 2(\alpha - c_A)\gamma + 2(\alpha - c_H) \le 0, \tag{14}$$

then the quantities in the perfect Bayesian equilibrium are given by

$$q_1^*(c_A) = \frac{\alpha - c_H}{\gamma}, \qquad q_1^*(c_B) = \frac{\alpha - c_L}{\gamma},$$
$$q_2^*(c_H|q_1^*(c_A)) = 0, \qquad q_2^*(c_L|q_1^*(c_A)) = \frac{c_H - c_L}{2},$$
$$q_2^*(c_H|q_1^*(c_B)) = 0, \qquad q_2^*(c_L|q_1^*(c_B)) = 0.$$

Proof. Firms' payoff functions are given by

$$\pi_1(q_1(c_A), q_2) = q_1(c_A)(\alpha - q_1(c_A) - \gamma q_2 - c_A),$$
$$\pi_1(q_1(c_B), q_2) = q_1(c_B)(\alpha - q_1(c_B) - \gamma q_2 - c_B),$$
$$\pi_2(q_1, q_2(c_H)) = q_2(c_H)(\alpha - q_2(c_H) - \gamma q_1 - c_H),$$
$$\pi_2(q_1, q_2(c_L)) = q_2(c_L)(\alpha - q_2(c_L) - \gamma q_1 - c_L).$$

Using backwards-induction, we will first compute firm F_2's reaction, $q_2^*(q_1)$, to an arbitrary quantity q_1 fixed by firm F_1. We are going to consider, separately, the cases where the production cost of firm F_1 is (i) c_A and (ii) c_B.

(i) Let us suppose that firm F_1 used the most expensive technology, i.e the quantity q_1 depends upon c_A, that we represent by $q_1(c_A)$. If firm F_2's unitary production cost is high, then

$$q_2^*(c_H|q_1(c_A)) = \arg\max_{q_2 \geq 0}(\alpha - q_2 - \gamma q_1(c_A) - c_H)q_2;$$

and if it is low, then

$$q_2^*(c_L|q_1(c_A)) = \arg\max_{q_2 \geq 0}(\alpha - q_2 - \gamma q_1(c_A) - c_L)q_2.$$

(ii) Let us suppose that firm F_1 used the cheapest technology, i.e the quantity q_1 depends upon c_B, that we represent by $q_1(c_B)$. If firm F_2's unitary production cost is high, then

$$q_2^*(c_H|q_1(c_B)) = \arg\max_{q_2 \geq 0}(\alpha - q_2 - \gamma q_1(c_B) - c_H)q_2;$$

and if it is low, then

$$q_2^*(c_L|q_1(c_B)) = \arg\max_{q_2 \geq 0}(\alpha - q_2 - \gamma q_1(c_B) - c_L)q_2.$$

Hence,

$$q_2^*(c_H|q_1(c_A)) = \max\left\{0, \frac{\alpha - c_H - \gamma q_1(c_A)}{2}\right\}, \quad q_2^*(c_L|q_1(c_A)) = \max\left\{0, \frac{\alpha - c_L - \gamma q_1(c_A)}{2}\right\},$$

$$q_2^*(c_H|q_1(c_B)) = \max\left\{0, \frac{\alpha - c_H - \gamma q_1(c_B)}{2}\right\}, \quad q_2^*(c_L|q_1(c_B)) = \max\left\{0, \frac{\alpha - c_L - \gamma q_1(c_B)}{2}\right\}.$$

By the hypothesis

$$\frac{a - c_H}{\gamma} \leq q_1(c_i) \leq \frac{\alpha - c_L}{\gamma}, \quad \text{with } i \in \{A, B\},$$

we obtain that

$$q_2^*(c_H|q_1(c_A)) = 0, \qquad q_2^*(c_L|q_1(c_A)) = \frac{\alpha - c_L - \gamma q_1(c_A)}{2},$$

$$q_2^*(c_H|q_1(c_B)) = 0, \qquad q_2^*(c_L|q_1(c_B)) = \frac{\alpha - c_L - \gamma q_1(c_B)}{2}.$$

So, we already have that

$$q_2^*(c_H|q_1(c_A)) = q_2^*(c_H|q_1(c_B)) = 0,$$

in all the items $(a - e)$. Firm F_1 can anticipate $q_2^*(q_1)$ and then use this value to compute q_1^*, and we get the results presented in the theorem. □

In the next corollary, that follows immediately from Theorem 1, we give the expected profits at equilibrium for each firm.

Corollary 1. *Let $E(C_1) = \phi c_A + (1 - \phi)c_B$ and $E(C_2) = \theta c_H + (1 - \theta)c_L$ be the expected unitary production cost of, respectively, firm F_1 and firm F_2. Let $V(C_i)$ be the variance of firm F_i's unitary production cost, for $i \in \{1, 2\}$. Under the hypothesis*

$$\frac{a - c_H}{\gamma} \leq q_1(c_i) \leq \frac{a - c_L}{\gamma}, \quad \text{with } i \in \{A, B\},$$

in Theorem 1, the expected profits at equilibrium for each firm are as follows:

(a) If inequalities (1) and (2) hold, then firm F_1's expected profit $E(\pi_1^*)$ and firm F_2's expected profit $E(\pi_2^*)$ are, respectively, given by

$$E(\pi_1^*) = \frac{((2-(1-\theta)\gamma)\alpha + (1-\theta)\gamma c_L - 2E(C_1))^2}{8\,(2-(1-\theta)\gamma^2)} + \frac{V(C_1)}{2\,(2-(1-\theta)\gamma^2)}$$

and

$$E(\pi_2^*) = \frac{\gamma^2 V(C_1)(1-\theta)}{4\,(2-(1-\theta)\gamma^2)^2} + \frac{((4-2\gamma-(1-\theta)\gamma^2)\alpha-(4-(1-\theta)\gamma^2)c_L+2\gamma E(C_1))^2\,(1-\theta)}{16\,(2-(1-\theta)\gamma^2)^2}.$$

(b) If inequalities (3)-(5) hold, then firm F_1's expected profit $E(\pi_1^*)$ and firm F_2's expected profit $E(\pi_2^*)$ are, respectively, given by

$$E(\pi_1^*) = \frac{((2-(1-\theta)\gamma)\alpha-2c_A+\gamma(1-\theta)c_L)^2\,\phi}{8\,(2-(1-\theta)\gamma^2)} - \frac{(\alpha-c_L)\,((1-\gamma)\alpha+\gamma c_B-c_L)\,(1-\phi)}{\gamma^2}$$

and

$$E(\pi_2^*) = \frac{((4-2\gamma-(1-\theta)\gamma^2)\alpha+2\gamma c_A-(4-(1-\theta)\gamma^2)c_L)^2\,(1-\theta)\phi}{16\,(2-(1-\theta)\gamma^2)^2}.$$

(c) If inequalities (6)-(8) hold, then firm F_1's expected profit $E(\pi_1^*)$ and firm F_2's expected profit $E(\pi_2^*)$ are, respectively, given by

$$E(\pi_1^*) = \frac{((2-(1-\theta)\gamma)\alpha-2c_B+\gamma(1-\theta)c_L)^2\,(1-\phi)}{8\,(2-(1-\theta)\gamma^2)} - \frac{(\alpha-c_H)\,(2(1-\gamma)\alpha+2\gamma c_A-(2-\gamma^2)c_H-\gamma^2 E(C_2))\,\phi}{2\gamma^2}$$

and

$$E(\pi_2^*) = \frac{(c_H-c_L)^2(1-\theta)\phi}{4} + \frac{((4-2\gamma-(1-\theta)\gamma^2)\alpha+2\gamma c_B-(4-(1-\theta)\gamma^2)c_L)^2\,(1-\theta)(1-\phi)}{16\,(2-(1-\theta)\gamma^2)^2}.$$

(d) If inequality (10) holds, then firm F_1's expected profit $E(\pi_1^*)$ and firm F_2's expected profit $E(\pi_2^*)$ are, respectively, given by

$$E(\pi_1^*) = \frac{(\alpha-c_H)\,(-2(1-\gamma)\alpha+(2-\gamma^2)c_H-2\gamma E(C_1)+\gamma^2 E(C_2))}{2\gamma^2}$$

and

$$E(\pi_2^*) = \frac{(c_H-c_L)^2(1-\theta)}{4}.$$

(e) If inequalities (12) and (13) hold, then firm F_1's expected profit $E(\pi_1^*)$ and firm F_2's expected profit $E(\pi_2^*)$ are, respectively, given by

$$E(\pi_1^*) = \frac{(\alpha - c_H)\left(-2(1-\gamma)\alpha - 2\gamma c_A + (2-\gamma^2)c_H + \gamma^2 E(C_2)\right)\phi}{2\gamma^2} - \frac{(\alpha - c_L)((1-\gamma)\alpha + \gamma c_B - c_L)(1-\phi)}{\gamma^2}$$

and

$$E(\pi_2^*) = \frac{(c_H - c_L)^2(1-\theta)\phi}{4}.$$

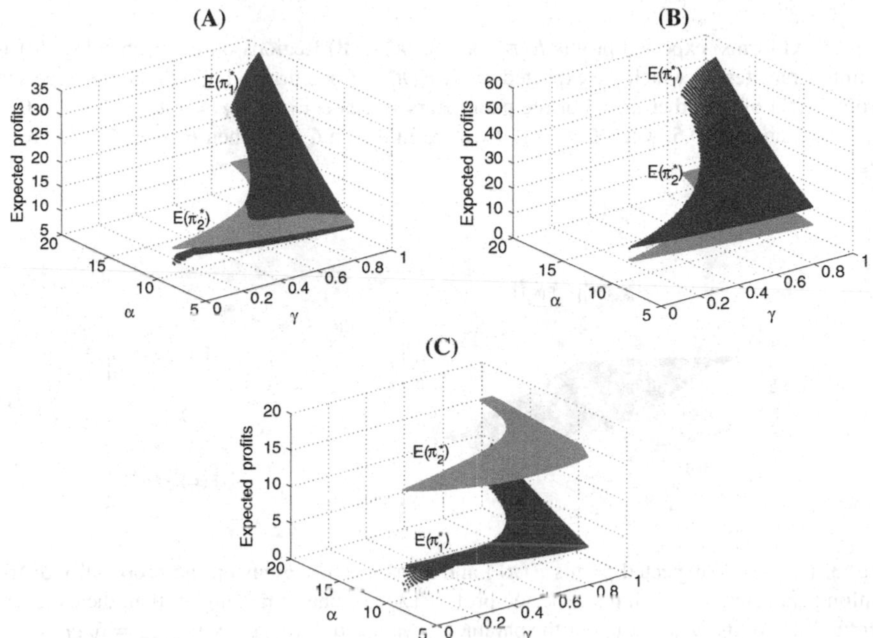

Fig. 1: Firms' expected profits $E(\pi_1^*)$ and $E(\pi_2^*)$. Parameters values $c_H = 9$, $c_A = 5$, $c_L = 0.1$, $c_B = 0.05$, and (**A**) $\theta = 0.4$ and $\phi = 0.6$; (**B**) $\theta = 0.5$ and $\phi = 0.3$; (**C**) $\theta = 0.2$ and $\phi = 0.9$

In Fig. 1, we show the variations of the expected profits with the demand parameter α, and with the degree γ of differentiation of the goods. We see that three different cases arise: a first case illustrated by Fig. 1(A), where for some pairs (γ, α) firm F_1 has a higher expected profit than firm F_2, and for the remaining pairs (γ, α) firm F_2 has a higher expected profit than firm F_1; a second case where firm F_1 always has a higher expected profit than firm F_2 (see Fig. 1(B); setting a maximum admissible value for α, a third situation occurs in which firm F_2 always has a higher expected profit than firm F_1 (see Fig. 1(C).

We will refer to the set of points satisfying conditions (1) and (2) as region A. The set of points satisfying (3), (4) and (5), will be called region B. To the set of points satisfying (6), (7), (8) and (9) we will call region C. The set of points satisfying (10) and (11) will be called region D, and to the set of points satisfying (12), (13) and (14) we will call region E (see Figs. (2–7)).

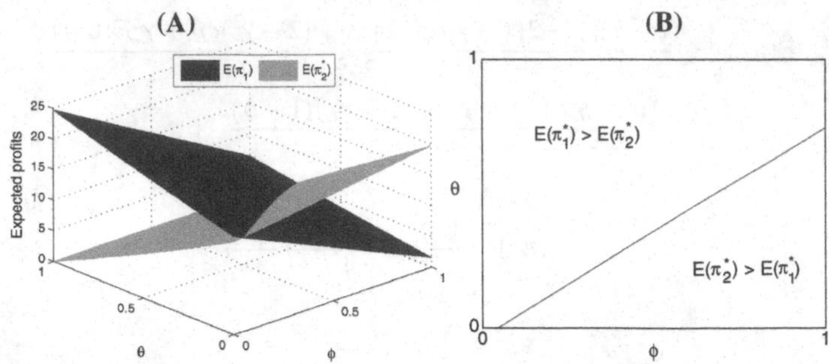

Fig. 2: (**A**) Firms' expected profits $E(\pi_1^*)$ and $E(\pi_2^*)$. (**B**) Regions on the probability distribution parameters in which the expected profit $E(\pi_i^*)$ of one firm is higher than the expected profit $E(\pi_j^*)$ of the other firm. For the parameters values $\alpha = 10$, $c_H = 9$, $c_A = 5$, $c_L = 0.1$, $c_B = 0.05$ and $\gamma = 0.5$, when $\theta \leq 39/79$ we are in region C and when $\theta \geq 39/79$ we are in region A

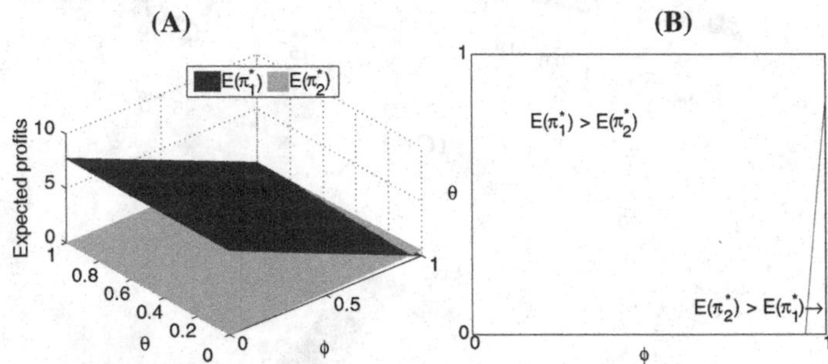

Fig. 3: (**A**) Firms' expected profits $E(\pi_1^*)$ and $E(\pi_2^*)$. (**B**) Regions on the probability distribution parameters in which the expected profit $E(\pi_i^*)$ of one firm is higher than the expected profit $E(\pi_j^*)$ of the other firm. For the parameters values $\alpha = 6.6$, $c_H = 6.5$, $c_A = 6$, $c_L = 5.1$, $c_B = 1$ and $\gamma = 0.6$, when $\theta \leq 37/117$ we are in region E, and when $\theta \geq 37/117$ we are in region B

In the region of parameters (ϕ, θ), we observe one sub-region where firm F_1 has a higher expected profit than firm F_2, $E(\pi_1^*) > E(\pi_2^*)$, and another sub-region where the opposite happens, $E(\pi_2^*) > E(\pi_1^*)$ (see Figs. (2–6)). Moreover, the region where $E(\pi_2^*) > E(\pi_1^*)$ can be empty (see Fig. 7). We note that the region where firm F_1 has a higher expected profit than firm F_2 is bigger than the region where firm F_2 has a higher expected profit than firm F_1. This differs from what is observed in the model with complete information. In the Stackelberg model with complete information, firm F_1, the leading firm, always has a higher profit than firm F_2, the follower.

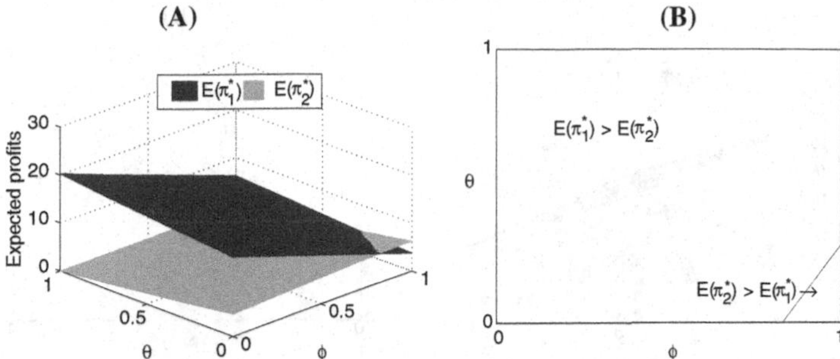

Fig. 4: (**A**) Firms' expected profits $E(\pi_1^*)$ and $E(\pi_2^*)$. (**B**) Regions on the probability distribution parameters in which the expected profit $E(\pi_i^*)$ of one firm is higher than the expected profit $E(\pi_j^*)$ of the other firm. For the parameters values $\alpha = 10$, $c_H = 9$, $c_A = 5$, $c_L = 4$, $c_B = 1$ and $\gamma = 0.4$, for all $0 \leq \theta \leq 1$, we are in region C

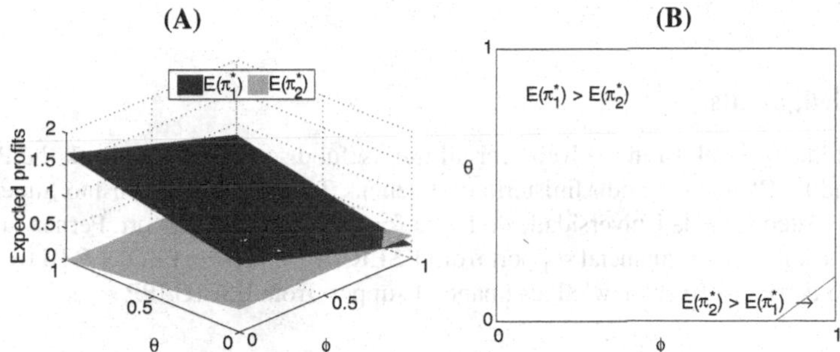

Fig. 5: (**A**) Firms' expected profits $E(\pi_1^*)$ and $E(\pi_2^*)$. (**B**) Regions on the probability distribution parameters in which the expected profit $E(\pi_i^*)$ of one firm is higher than the expected profit $E(\pi_j^*)$ of the other firm. For the parameters values $\alpha = 10$, $c_H = 9$, $c_A = 8$, $c_L = 7.8$, $c_B = 7.4$ and $\gamma = 0.7$, for all $0 \leq \theta \leq 1$, we are in region D

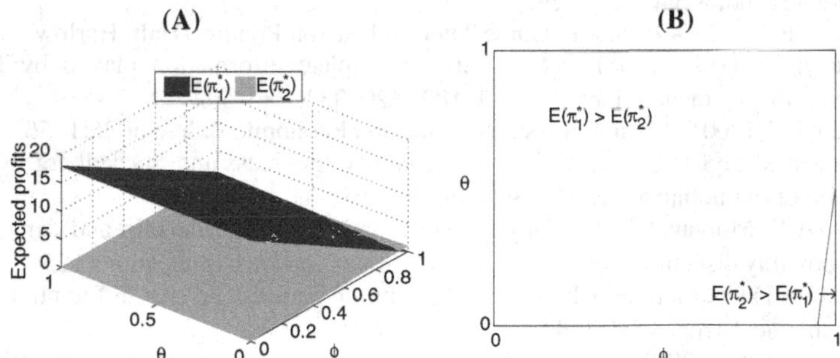

Fig. 6: (**A**) Firms' expected profits $E(\pi_1^*)$ and $E(\pi_2^*)$. (**B**) Regions on the probability distribution parameters in which the expected profit $E(\pi_i^*)$ of one firm is higher than the expected profit $E(\pi_j^*)$ of the other firm. For the parameters values $\alpha = 10$, $c_H = 9$, $c_A = 8$, $c_L = 7$, $c_B = 1$ and $\gamma = 1$, for all $0 \leq \theta \leq 1$, we are in region E

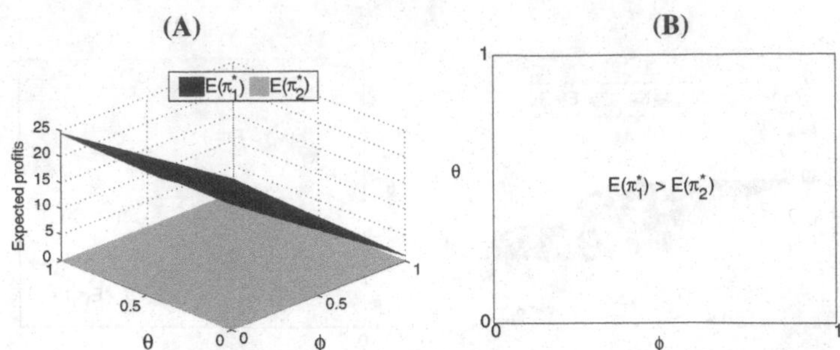

Fig. 7: (**A**) Firms' expected profits $E(\pi_1^*)$ and $E(\pi_2^*)$. (**B**) Regions on the probability distribution parameters in which the expected profit $E(\pi_i^*)$ of one firm is higher than the expected profit $E(\pi_j^*)$ of the other firm. For the parameters values $\alpha = 10$, $c_H = 5$, $c_A = 4.5$, $c_L = 4.4$, $c_B = 0.1$ and $\gamma = 1$, when $\theta \leq 9/22$ we are in region E, when $9/22 \leq \theta \leq 21/22$ we are in region C and when $\theta \geq 21/22$ we are in region D

Acknowledgments

We would like to thank Bruno Oliveira for all the useful discussions. We thank the Programs POCTI and POCI by FCT and Ministério da Ciência, Tecnologia e do Ensino Superior, and Centro de Matemática da Universidade do Porto for their financial support. Fernanda Ferreira gratefully acknowledges financial support from ESEIG/IPP and from PRODEP III by FSE and EU. Flávio Ferreira also acknowledges financial support from ESEIG/IPP.

References

1. Gal-Or E (1985) First-mover and second-mover advantages. International Economic Review 26(3):649–653
2. Gal-Or E (1987) First-mover disadvantages with private information. Review of Economic Studies 54(2):279–292
3. Gibbons R (1992) A Primer in Game Theory. Pearson Prentice-Hall, Harlow
4. Harsanyi JC (1967–1968) Games with incomplete information played by Bayesian players. Management Science 14:159–182, 320–334, 486–502
5. Harsanyi JC (2001) Memorial issue. Games and Economic Behavior 36:1–56
6. Hirokawa M (2001) Strategic choice of quantity stickiness and Stackelberg leadership. Bulletin of Economic Research 53(1):19–34
7. Mallozzi L, Morgan J (2005) Oligopolistic markets with leadership and demand functions possibly discountinuous. Journal of Optimization and Applications 125(2):393–407
8. Pal D (1991) Cournot duopoly with two prodution periods and cost differentials. Journal of Economic Theory 55:441–448
9. Pal D, Sarkar J (2001) A Stackelberg oligopoly with nonidentical firms. Bulletin of Economic Research 53(2):127–134
10. Singh N, Vives X (1984) Price and quantity competition in a differentiated duopoly. RAND Journal of Economics 15:546–554
11. Tirole J (1988) The Theory of Industrial Organization. MIT Press, Cambridge, MA
12. von Stackelberg H (1934) Marktform und Gleichgewicht. Julius Springer, Vienna

On the Fractional Order Control of Heat Systems

Isabel S. Jesus, J. A. Tenreiro Machado, and Ramiro S. Barbosa

Institute of Engineering of Porto, Department of Electrotechnical Engineering
Rua Dr. António Bernardino de Almeida, n. 431, 4200-072 Porto, Portugal
{isj, jtm, rsb}@isep.ipp.pt

Summary. The differentiation of non-integer order has its origin in the seventeenth century, but only in the last two decades appeared the first applications in the area of control theory. In this paper we consider the study of a heat diffusion system based on the application of the fractional calculus concepts. In this perspective, several control methodologies are investigated namely the fractional PID and the Smith predictor. Extensive simulations are presented assessing the performance of the proposed fractional-order algorithms.

1 Introduction

Fractional calculus (FC) is a generalization of integration and differentiation to a complex order α, being the fundamental operator ${}_aD_t^\alpha$, where a and t are the limits of the operation [1,2]. The FC concepts constitute a useful tool to describe several physical phenomena, such as heat, flow, electricity, magnetism, mechanics or fluid dynamics and presently it is applied in almost all areas of science and engineering.

In the last years, FC has been used increasingly to model the constitutive behavior of materials and physical systems exhibiting hereditary and memory properties. This is the main advantage of fractional-order derivatives in comparison with classical integer-order models, where these effects are neglected.

It is well-known that the fractional-order operator $s^{0.5}$ appears in several types of problems [3]. The transmission lines, the heat flow or the diffusion of neutrons in a nuclear reactor are examples where the half-operator is the fundamental element. On the other hand, diffusion is one of the three fundamental partial differential equations of mathematical physics [4].

In this paper we investigate the heat diffusion system in the perspective of applying the FC theory. Several control strategies based on fractional-order algorithms are presented and compared with the classical schemes. The adoption of fractional-order controllers has been justified by its superior performance, particularly when used with fractional-order dynamical systems, such as the case of the heat system under study. The fractional-order PID ($\mathrm{PI}^\alpha\mathrm{D}^\beta$ controller) involves an integrator of order $\alpha \in \Re+$ and a differentiator of order $\beta \in \Re+$. It

J.A.T. Machado et al. (eds.), *Intelligent Engineering Systems and Computational Cybernetics*,

was demonstrated the good performance of this type of controller, in comparison with the conventional PID algorithms.

Bearing these ideas in mind, the paper is organized as follows. Section 2 gives the fundamentals of fractional-order control systems. Section 3 introduces the heat diffusion system. Section 4 points out several control strategies for the heat system and discusses their results. Finally, Section 5 draws the main conclusions and addresses perspectives towards future developments.

2 Fractional-Order Control Systems

Fractional-order control systems are characterized by differential equations that have, in the dynamical system and/or in the control algorithm, an integral and/or a derivative of fractional-order. Due to the fact that these operators are defined by irrational continuous transfer functions, in the Laplace domain, or infinite dimensional discrete transfer functions, in the Z domain, we often encounter evaluation problems in the simulations. Therefore, when analyzing fractional-order systems, we usually adopt continuous or discrete integer-order approximations of fractional-order operators [5–7].

The mathematical definition of a fractional-order derivative and integral has been the subject of several different approaches [1, 2]. One commonly used definition for the fractional-order derivative is given by the Riemann–Liouville definition ($\alpha > 0$):

$$ {}_aD_t^\alpha f(t) = \frac{1}{\Gamma(n-\alpha)} \frac{d^n}{dt^n} \int_a^t \frac{f(\tau)}{(t-\tau)^{\alpha-n+1}} d\tau, \quad n-1 < \alpha < n, \tag{1} $$

where $f(t)$ is the applied function and $\Gamma(x)$ is the Gamma function of x. Another widely used definition is given by the Grünwald–Letnikov approach ($\alpha \in \Re$):

$$ {}_aD_t^\alpha f(t) = \lim_{h\to 0} \frac{1}{h^\alpha} \sum_{k=0}^{\left[\frac{t-a}{h}\right]} (-1)^k \binom{\alpha}{k} f(t-kh) \tag{2a} $$

$$ \binom{\alpha}{k} = \frac{\Gamma(\alpha+1)}{\Gamma(k+1)\Gamma(\alpha-k+1)} \tag{2b} $$

where h is the time increment and $[x]$ means the integer part of x.

The "memory" effect of these operators is demonstrated by (1) and (2), where the convolution integral in (1) and the infinite series in (2), reveal the unlimited memory of these operators, ideal for modeling hereditary and memory properties in physical systems and materials.

An alternative definition to (1) and (2), which reveals useful for the analysis of fractional-order control systems, is given by the Laplace transform method. Considering vanishing initial conditions, the fractional *differintegration* is defined in the Laplace domain, $F(s) = L\{f(t)\}$, as:

$$ L\{{}_aD_t^\alpha f(t)\} = s^\alpha F(s), \quad \alpha \in \Re \tag{3} $$

An important aspect of fractional-order algorithms can be illustrated through the elemental control system with open-loop transfer function $G(s) = Ks^{-\alpha}$ ($1 < \alpha < 2$) in the forward path and unit feedback. The open-loop Bode diagrams of amplitude and phase have a slope of -20α dB/dec and a constant phase of $-\alpha\pi/2$ rad over the entire frequency domain. Therefore, the closed-loop system has a constant phase margin of $\mathrm{PM} = \pi(1-\alpha/2)$ rad that is independent of the system gain K. Therefore, the closed-loop system will be robust against gain variations exhibiting step responses with an iso-damping property [8].

In this paper we adopt discrete integer-order approximations to the fundamental element s^{α} ($\alpha \in \Re$) of a fractional-order control (FOC) strategy. The usual approach for obtaining discrete equivalents of continuous operators of type s^{α} adopts the Euler, Tustin and Al-Alaoui generating functions [5].

It is well known that rational-type approximations frequently converge faster than polynomial-type approximations and have a wider domain of convergence in the complex domain. Thus, by using the Euler operator $w(z^{-1}) = (1 - z^{-1})/T$, and performing a power series expansion of $\left[w(z^{-1})\right]^{\alpha} = \left[(1-z^{-1})/T\right]^{\alpha}$ gives the discretization formula corresponding to the Grünwald–Letnikov definition (2):

$$D^{\alpha}\left(z^{-1}\right) = \left(\frac{1-z^{-1}}{T}\right)^{\alpha} = \sum_{k=0}^{\infty} h^{\alpha}(k) z^{-k} \tag{4a}$$

$$h^{\alpha}(k) = \left(\frac{1}{T}\right)^{\alpha} \binom{k-\alpha-1}{k} \tag{4b}$$

where the impulse response sequence $h^{\alpha}(k)$ is given by the expression (4b) ($k \geq 0$).

A rational-type approximation can be obtained by applying the Padé approximation method to the impulse response sequence (4b) $h^{\alpha}(k)$, yielding the discrete transfer function:

$$H\left(z^{-1}\right) = \frac{b_0 + b_1 z^{-1} + \ldots + b_m z^{-m}}{1 + a_1 z^{-1} + \ldots + a_n z^{-n}} = \sum_{k=0}^{\infty} h(k) z^{-k}, \tag{5}$$

where $m \leq n$ and the coefficients a_k and b_k are determined by fitting the first $m+n+1$ values of $h^{\alpha}(k)$ into the impulse response $h(k)$ of the desired approximation $H(z^{-1})$. Thus, we obtain an approximation that has a perfect match to the desired impulse response $h^{\alpha}(k)$ for the first $m+n+1$ values of k [8]. Note that the above Padé approximation is obtained by considering the Euler operator but the determination process will be exactly the same for other types of discretization schemes.

3 Heat Diffusion

The heat diffusion is governed by a linear partial differential equation (PDE) of the form:

$$\frac{\partial u}{\partial t} = k\left(\frac{\partial^2 u}{\partial x^2} + \frac{\partial^2 u}{\partial y^2} + \frac{\partial^2 u}{\partial z^2}\right), \tag{6}$$

where k is the diffusivity, t is the time, u is the temperature and (x,y,z) are the space cartesian coordinates. The system (6) involves the solution of a PDE of parabolic type for which the standard theory guarantees the existence of a unique solution [9].

For the case of a planar perfectly isolated surface we usually apply a constant temperature U_0 at $x = 0$ and we analyze the heat diffusion along the horizontal coordinate x. Under these conditions, the heat diffusion phenomenon is described by a non-integer order model, yielding:

$$U(x,s) = \frac{U_0}{s} G(s),\ G(s) = e^{-x\sqrt{\frac{s}{k}}}, \tag{7}$$

where x is the space coordinate, U_0 is the boundary condition and $G(s)$ is the system transfer function.

In our study, the simulation of the heat diffusion is performed by adopting the Crank–Nicholson implicit numerical integration based on the discrete approximation to differentiation [10–13].

4 Control Strategies for Heat Diffusion Systems

In this section we consider three strategies for the control of the heat diffusion system. In the first two subsections, we analyze the system of Fig. 1 by adopting classical PID and fractional PID$^\beta$ controllers, respectively. In the third subsection we use a Smith predictor (SP) structure with a fractional PID$^\beta$ controller (Fig. 5).

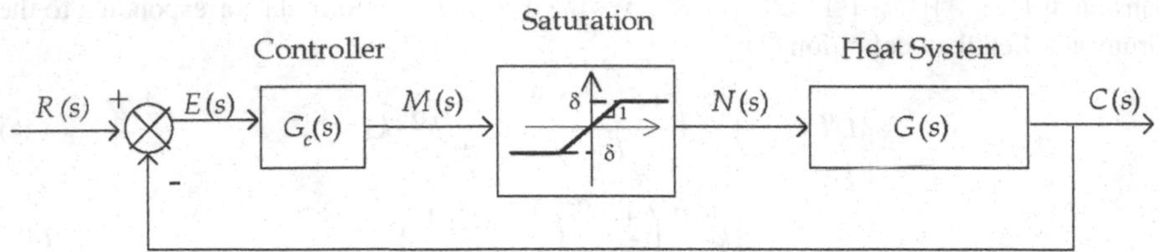

Fig. 1: Closed-loop system with PID or PID$^\beta$ controller $G_c(s)$

The generalized PID controller $G_c(s)$ has a transfer function of the form:

$$G_c(s) = K\left[1 + \frac{1}{T_i s^\alpha} + T_d s^\beta\right], \tag{8}$$

where α and β are the orders of the fractional integrator and differentiator, respectively. The constants K, T_i and T_d are correspondingly the proportional gain, the integral time constant and the derivative time constant.

Clearly, taking $(\alpha, \beta) = \{(1, 1), (1, 0), (0, 1), (0, 0)\}$ we get the classical {PID, PI, PD, P} controllers, respectively. Other PID controllers are possible, namely: PD$^\beta$ controller, PI$^\alpha$ controller, PID$^\beta$ controller, and so on. The PI$^\alpha$D$^\beta$ controller is more flexible and gives the possibility of adjusting more carefully the closed-loop system characteristics.

4.1 The PID Controller

In this subsection we analyze the closed-loop system with a conventional PID controller given by the transfer function (8) with $\alpha = \beta = 1$. Often, the PID parameters (K, T_i, T_d) are tuned by using the so-called Ziegler–Nichols open loop (ZNOL) method. The ZNOL heuristics are based on the approximate first-order plus dead-time model:

$$\hat{G}(s) = \frac{K_p}{\tau s + 1} e^{-sT} \tag{9}$$

A step input is applied at $x = 0.0$ m and the output $c(t)$ analyzed for $x = 3.0$ m, without the saturation. The resulting parameters are $\{K_p, \tau, T\} = \{0.52,\ 162, 28\}$ leading to the PID constants $\{K,\ T_i,\ T_d\} = \{18.07, 34.0, 8.5\}$. We verify that the system with a PID controller, tuned according with the ZNOL heuristics, does not produce satisfactory results giving a significant overshoot, and a large settling time. Moreover, the step response reveals a considerable time delay. The poor results obtained indicate that the method of tuning as well the structure of the system may not be the most adequate for the control of the heat system under consideration. In fact, the inherent fractional-order dynamics of the system lead us to consider other configurations. In this perspective, we propose the use of fractional-order controllers and the SP to achieve a superior control performance.

4.2 The PID$^\beta$ Controller

In this subsection we analyze the closed-loop system with a PID$^\beta$ controller given by the transfer function (8) with $\alpha = 1$. The fractional-order derivative term $T_d s^\beta$ in (8) is implemented by using a fourth-order Padé discrete rational transfer function of type (5). It is used a sampling period of $T = 0.1$ s.

The PID$^\beta$ controller is tuned by minimization of an integral performance index. For that, we adopt the integral square error (ISE) and the integral time square error (ITSE) criteria defined as:

$$\mathrm{ISE} = \int_0^\infty [r(t) - c(t)]^2 \, dt \tag{10}$$

$$\mathrm{ITSE} = \int_0^\infty t\,[r(t) - c(t)]^2 \, dt \tag{11}$$

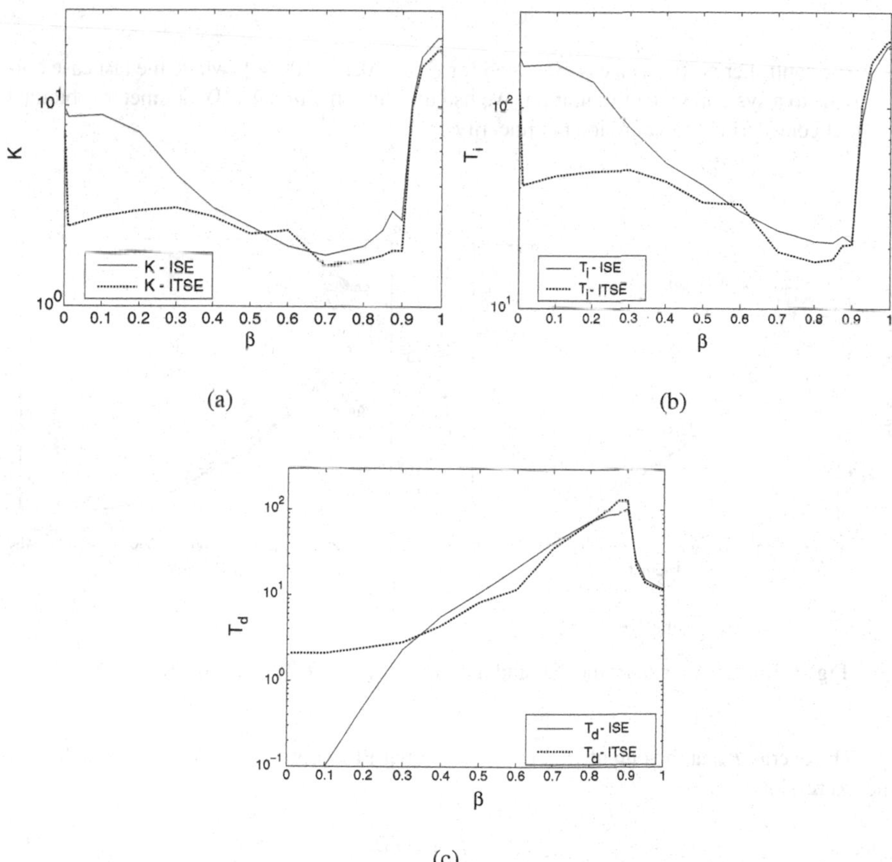

Fig. 2: The PID$^\beta$ parameters (K, T_i, T_d) versus β for the ISE and the ITSE criteria

We can use other integral performance criteria such as the integral absolute error (IAE) or the integral time absolute error (ITAE) but the ISE and the ITSE criteria have produced the best results. Furthermore, the ITSE criterion enable us to study the influence of time in the error generated by the system.

A step reference input $R(s) = 1/s$ is applied at $x = 0.0$ m and the output $c(t)$ is analyzed for $x = 3.0$ m, without considering saturation. The heat system is simulated for 3,000 s. Figure 2 illustrates the variation of the fractional PID parameters (K, T_i, T_d) as function of the order's derivative β, tuned according with the ISE and the ITSE criteria.

The curves reveal that for $\beta < 0.4$, the parameters (K, T_i, T_d) are slightly different, for the two ISE and ITSE criteria, but for $\beta \geq 0.4$ they almost lead to similar values. This fact indicates a large influence of a weak order derivative on system's dynamics.

To further illustrate the performance of the fractional-order controllers a saturation nonlinearity is included in the closed-loop system of Fig. 1 and inserted in series with the output of the PID controller $G_c(s)$. The saturation element is defined as:

$$n(m) = \begin{cases} +\delta, & m > \delta \\ m, & |m| \leq \delta \\ -\delta, & m < -\delta \end{cases}$$

The controller performance is evaluated for $\delta = \{20, \ldots, 100, \infty\}$, where the last case corresponds to a system without saturation. We use the same fractional-PID parameters obtained without considering the saturation nonlinearity.

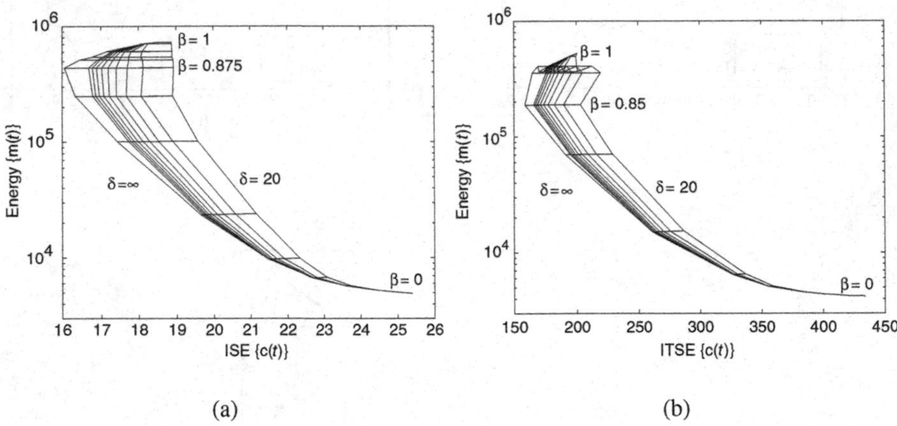

(a) (b)

Fig. 3: Energy E_m *versus* the ISE and the ITSE for $\delta = \{20, \ldots, 100, \infty\}$, $0 \leq \beta \leq 1$

The energy E_m at the output $m(t)$ of the fractional PID controller is also analyzed through the expression:

$$E_m = \int_0^{T_s} m^2(t)dt, \tag{12}$$

where T_s is the time corresponding to the 5% settling time criterion of the system output $c(t)$.

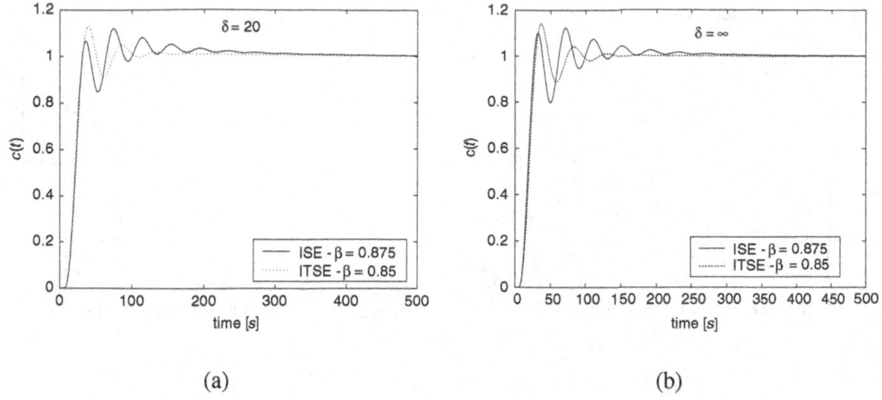

Fig. 4: Step responses of the closed-loop system for the ISE and the ITSE indices, with a PID$^{\beta}$ controller, for $\delta = \{20, \infty\}$ and $x = 3.0$ m

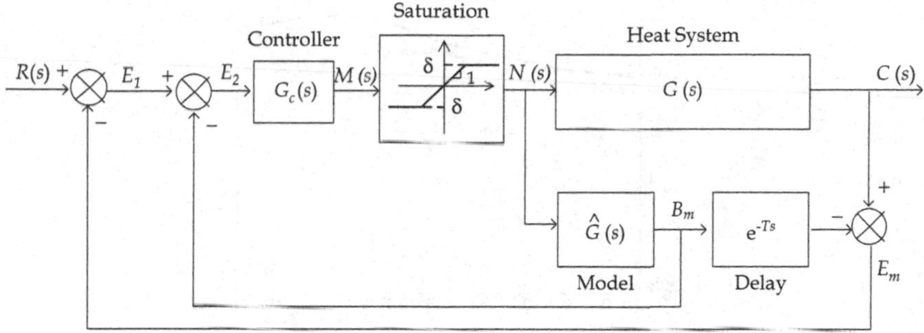

Fig. 5: Closed-loop system with a Smith predictor and a PID$^{\beta}$ controller $G_c(s)$

Figure 3 depicts the energy E_m as function of the ISE and the ITSE indices for $0 \le \beta \le 1$. As can be seen, the energy changes smoothly for different values of δ when considering a given order β. However, fixing the value δ, we verify that the energy increases significantly with β.

On the other hand, we observe that the ISE decreases with δ for $\beta \le 0.875$, while for $\beta > 0.875$ the ISE increases very quickly. The same conclusions can be outlined relatively to the ITSE criterion. The results confirm the good performance of the system for low values of the fractional-order derivative term.

When comparing the two indices, we also note that the value of the ITSE is generally larger than the corresponding for the ISE case. This occurs because of the large simulation time needed to stabilize the system, which is $T_s \sim 700$ s.

Figure 4 shows the step responses of the closed-loop system for the PID$^{\beta}$ tuned in the perspective of ISE and ITSE.

The controller parameters $\{K, T_i, T_d, \beta\}$ correspond to the minimization of those indices leading to the values ISE: $\{K, T_i, T_d, \beta\} \equiv \{3, 23, 90.6, 0.875\}$ and ITSE: $\{K, T_i, T_d, \beta\} \equiv \{1.8, 17.6, 103.6, 0.85\}$.

The step responses reveal a large diminishing of the overshoot and the rise time when compared with the integer PID, showing a good transient response and a zero steady-state error. As should be expected, the PID$^{\beta}$ leads to better results than the PID controller tuned

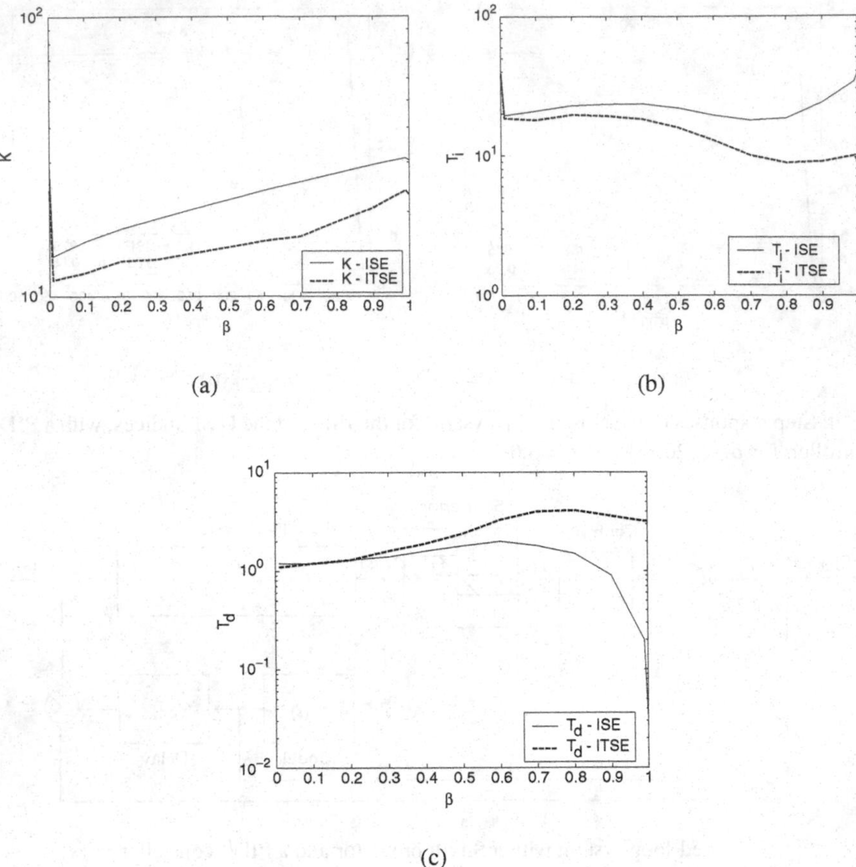

Fig. 6: The PID$^{\beta}$ parameters (K, T_i, T_d) versus β for the ISE and the ITSE criteria

through the ZNOL method. These results demonstrate the effectiveness of the fractional-order algorithms when used for the control of fractional-order systems.

4.3 The Smith Predictor with a PID$^{\beta}$ Controller

In this subsection we adopt a fractional PID$^{\beta}$ controller inserted in a SP structure, as represented in Fig. 5. This algorithm, developed by O. J. M. Smith in 1957, is a dead-time compensation technique, very effective in improving the control of processes having time delays [14, 15].

The transfer function $\hat{G}(s)$, inserted in the second branch of the SP, is described by the following first-order plus dead-time model:

$$\hat{G}(s) = \frac{0.52}{162s+1} e^{-28s} \tag{13}$$

where the parameters $(K_p, \tau, T) = (0.52,\ 162,\ 28)$ were estimated by a least-squares fit between the frequency responses of the numerical model of the heat system and the SP model $\hat{G}(s)$.

The PID$^{\beta}$ controller is tuned by applying the ISE and the ITSE integral performance criteria, as described in the previous subsection.

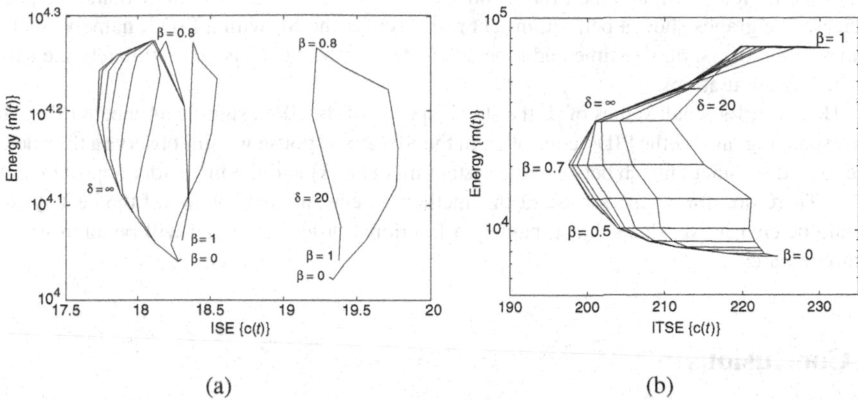

Fig. 7: Energy E_m versus the ISE and the ITSE indices for $\delta = \{20, ..., 100, \infty\}$, $0 \leq \beta \leq 1$

Figure 6 illustrates the variation of the PID$^{\beta}$ parameters $(K,\ T_i,\ T_d)$ as function of the order's derivative β, for the ISE and the ITSE indices, without the occurrence of saturation.

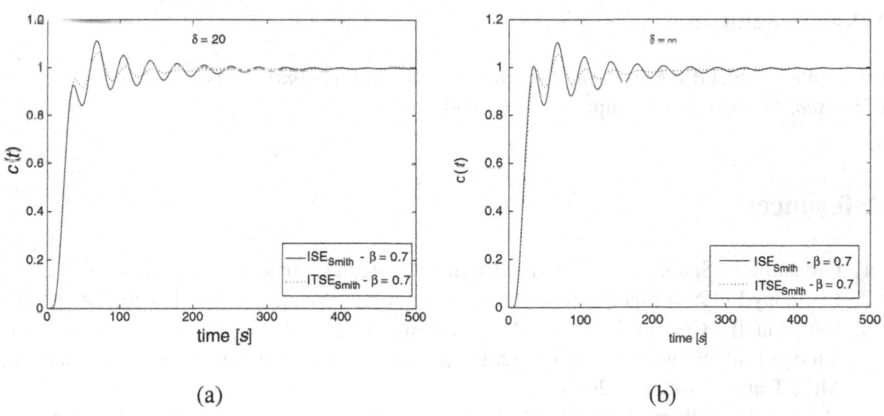

Fig. 8: Step responses of the closed-loop system for the PID$^{\beta}$ and the Smith predictor with PID$^{\beta}$, for the ISE and the ITSE indices, $\delta = \{20, \infty\}$, $x = 3.0$ m

Figure 7 shows the relation between the energy E_m and the values of the ISE and ITSE indices. We verify that the best case is achieved when $\beta = 0.7$, revealing that, with a SP structure, the effectiveness of the fractional-order controller is more evident.

Since the effect of the system time delay is diminished by the use of the SP, the PID^β controller and, more specifically, the fractional-order derivative D^β will be more effective and, consequently, produce better results.

Figure 8 illustrates the step responses for $x = 3.0$ m when applying a step input $R(s) = 1/s$ at $x = 0.0$ m, for the SP and the PID^β controller with $\beta = 0.7$, both for the ISE and the ITSE indices. The graphs show a better transient response for the SP with a PID^β, namely smaller values of the overshot, rise time and time delay. The settling time is approximately the same for both configurations.

However, for small values of β, the step response of the SP is significantly better than the corresponding one for the PID^β controller. In the SP step response we still observe a time delay due to an insufficient match between the system model $G(s)$ and the first-order approximation $\hat{G}(s)$. Therefore, for a superior use of this method a better approximation of the heat system should be envisaged. This subject, namely a fractional order SP model will be addressed in future research.

5 Conclusions

This paper presented the fundamental aspects of the FC theory. We demonstrated that FC is a tool for modelling physical phenomena having superior capabilities than those of traditional methodologies. In this perspective, we studied a heat diffusion system, which is described through the fractional-order operator $s^{0.5}$. The dynamics of the system was analyzed in the perspective of FC, and some of its implications upon control algorithms and systems with time delay were also investigated.

Acknowledgments

The authors thank GECAD – *Grupo de Investigação em Engenharia do Conhecimento e Apoio à Decisão*, for the financial support to this work.

References

1. Oldham KB, Spanier J (1974) The Fractional Calculus. Academic, London
2. Podlubny I (1999) Fractional Differential Equations. Academic, San Diego, CA
3. Battaglia JL, Cois O, Puigsegur L, Oustaloup A (2001) Solving an inverse heat conduction problem using a non-integer identified model. International Journal of Heat and Mass Transfer 44:2671–2680
4. Courant R, Hilbert D (1962) Methods of Mathematical Physics, Partial Differential Equations. Wiley Interscience II, New York
5. Barbosa RS, Machado JAT, Silva MF (2006) Time domain design of fractional differintegrators using least-squares. Signal Processing, EURASIP/Elsevier, 86(10):2567–2581
6. Petras I, Vinagre BM (2002) Practical application of digital fractional order controller to temperature control. Acta Montanistica Slovaca 7(2):131–137

7. Podlubny I, Petras I, Vinagre BM, Chen Y-Q, O'Leary P, Dorcak L (2003) Realization of fractional order controllers. Acta Montanistica Slovaca 8(3):233–235
8. Barbosa RS, Machado JAT, Ferreira IM (2004) Tuning of PID controllers based on Bode's ideal transfer function. Nonlinear Dynamics 38(1/4):305–321
9. Machado JT, Jesus I, Boaventura Cunha J, Tar JK (2006) Fractional Dynamics and Control of Distributed Parameter Systems. In: Intelligent Systems at the Service of Mankind. Vol 2, 295–305
10. Crank J (1956) The Mathematics of Diffusion. Oxford University Press, London
11. Gerald CF, Wheatley PO (1999) Applied Numerical Analysis. Addison-Wesley, Reading, MA
12. Jesus IS, Barbosa RS, Machado JAT, Cunha JB (2006) Strategies for the Control of Heat Diffusion Systems Based on Fractional Calculus. In: IEEE International Conference on Computational Cybernetics. Estonia
13. Jesus IS, Machado JAT, Cunha JB (2007) Fractional Dynamics and Control of Heat Diffusion Systems. In: MIC 2007 – The 26th IASTED International Conference on Modelling, Identification and Control. Innsbruck, Austria
14. Smith OJM (1957) Closed control of loops with dead time. Chemical Engineering Process 53:217–219
15. Majhi S, Atherton DP (1999) Modified Smith predictor and controller for processes with time delay. IEE Proceedings or Control Theory and Applications 146(5):359–366

Restricting Factors at Modification of Parameters of Associative Engineering Objects

László Horváth

Budapest Tech Polytechnical Institution
H-1034 Budapest, Bécsi út 96/b, Hungary
horvath.laszlo@nik.bmf.hu

Summary. Advancements in product development have reached full integration of engineering activities and processes in product lifecycle management (PLM) systems. PLM systems are based on high-level modeling, simulation and data management. Despite significant development of modeling in PLM systems, a strong demand was recognized for improved decision assistance in product development. Decision assistance can be improved by application of methods from the area of computer intelligence. In order for a product development company to stay competitive, it is important for its modeling system to be relied on local even personal knowledge. The authors analyzed current PLM systems for shortcomings and possibilities for extended intelligence at decision-making during product development. They propose methods in order to increase suitability of current modeling systems to accommodate knowledge based IT at definition of sets of parameters of modeled objects and in the management of frequent changes of modeled objects. In the center of the proposed methodology, constrained parameters act as restricting factors at definition and modification of parameters of associative engineering objects. Paper starts with an outlook to modeling in current engineering systems and preliminary results by the authors. Following this, groups of essential information as handled by he proposed modeling are summarized and procedures for processing of that groups of information are detailed. Next, management of chains of changes along chains of associative product objects and a new style of decision assistance in modeling systems are explained. Changes are created or verified by behavior analysis. Finally, behavior analysis, human intent combination, product data view creation, and change management are discussed as the proposed integrated and coordinated methodology for enhanced support of decision-making in product development.[1]

1 Introduction

One of the dynamically developed areas in computer applications is engineering. CAD/CAM systems with capability of creating tool paths for manufacturing using purposeful geometric description have developed into full-integrated systems for lifetime management of product

[1] The authors gratefully acknowledge the grant provided by the OTKA Fund for Research of the Hungarian Government. Project number is T 037304.

J.A.T. Machado et al. (eds.), *Intelligent Engineering Systems and Computational Cybernetics*,

information (PLM – product lifecycle management). Comprehensive functionality serves digital product definition, product data management, and communication of engineers. However, requirements against engineering systems are changing. More and more engineers work at distant workstations. The expected innovation cycles of products are short and high number of product variants is to be developed. In order to cope with new requirements, powerful assistance of decisions by using of enhanced intelligent characteristics of engineering systems is needed. Current PLM systems provide restricted support for decisions mainly because they have not means to survey complex and unstructured sets of associative connections amongst parameters of engineering objects. Moreover, intelligent decision-making must rely on company and product specific knowledge and this knowledge cannot be built into general-purpose modeling systems. The utmost objective is called as virtual manufacturing or virtual prototyping where advanced modeling and simulation are applied for improving and optimizing of products, inside a computer system, without any physical manufacturing and measurement operations. Institute of Intelligent Engineering Systems at the John von Neumann Faculty of Informatics, Budapest Tech has founded the Laboratory for Intelligent Engineering Systems to support research for enhanced intelligence of current engineering systems. The laboratory is an experimental installation of advanced and comprehensive industrial professional engineering system and is equipped with environment for experiments in application of intelligent computing. As a standard environment for experiments with intelligent engineering processes, highly integrated application of product modeling, human-computer interaction, collaborative communication, product data management, Internet portal, and intelligent information processing are considered. As a contribution to enhanced intelligence at product modeling systems, this paper concentrates on selected issues about changes of modeled engineering objects during development of products. The proposed methodology relies on definition, specification and analysis of behaviors of modeled product objects. Behavior describes design objectives for product development. Information for a change on an affected object a along change chain in a product is carried by adaptive action entity. Paper starts with an outlook to modeling in current engineering systems and preliminary results by the authors. Following this, groups of essential information as handled by he proposed modeling are summarized and procedures for processing of that groups of information are detailed. Next, management of chains of changes along chains of associative product objects and a new style of decision assistance in modeling systems are explained. Changes are created or verified by behavior analysis. Finally, behavior analysis, human intent combination, product data view creation, and change management are discussed as the proposed integrated and coordinated methodology for enhanced support of decision-making in product development.

2 Preliminaries and Background

Functionality of a typical conventional product modeling system is configured for digital definition of product objects and application of these definitions at analysis of parts and structures, and programming of production equipment control. By now, product definition covers full-integrated modeling for lifecycle of products including knowledge based advising, model driven analysis and control, and advanced human-computer interaction (Fig. 1). Enhanced lifecycle management of product information is possible by advanced management of product data, engineering processes and collaborative communication of engineers as it is outlined in Fig. 1.

The proposed modeling of product object change process was conceptualized considering several prospective focus techniques in product modeling. For this purpose, the authors

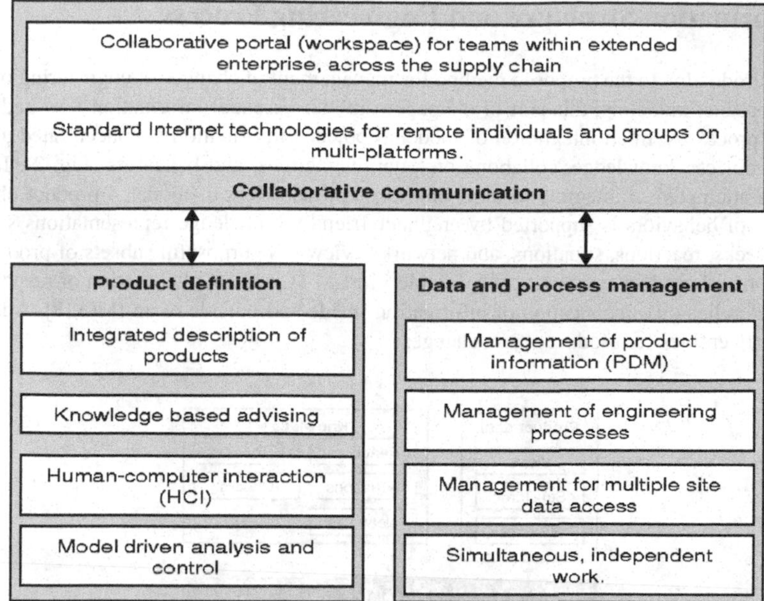

Fig. 1: Functionality of product lifecycle management system

did a systematic analysis of recent techniques in product modeling and the results were published in the book [1]. Besides essential product information about elementary, structural, and relationship model entities, product modeling increasingly relies on more advanced concepts such as reference models, modeling resources, and application protocols [2]. Majority of products to be modeled is shape centered. In a shape-centered model, form features are applied as application oriented definition of shape [3]. Any other information can be mapped to shape description. Product behavior based descriptions [4] and agents [5] represent techniques for modeling towards intelligent engineering.

Some relevant earlier results by the authors are cited in the following. An analysis of modeling in CAD/CAM systems was concentrated on application features of parts, adaptive engineering processes, and definition of associative product model entities [6]. The authors developed new modeling methods for the following purposes.

- Application of behavior and adaptive action model entities at decision assistance [7]
- Integrated description of a group of closely related engineering objects in an integrated model object (IMO) [8]
- Modeling of intent of responsible and other engineers as background information for decision-making [9]
- Integration of human intent descriptions in product models [9]
- Propagation of effect of change of product objects [10]

During the work in above developments, the authors of this paper concluded that survey of all affecting factors and affected objects is necessary at processing of product changes during product development. Very complex and unstructured sets of relationships of parameters of product and other related engineering objects must be handled.

3 Information Structure and Engineering Process

As an introduction to the proposed method for management of changes of engineering objects during product modeling, this part of the paper outlines essential information flow and engineering processes. In an integration of modeled objects, information is concentrated around product objects, knowledge, collaboration, adaptive actions, and behaviors (Fig. 2). Intelligence is attempted to integrate in modeling through behaviors of modeled product objects. Analysis of behaviors is supported by engineer friendly knowledge representations such as rules, checks, reactions, situations, and networks. Views as purposeful subsets of product information and product structures are available in PLM systems. Collaboration of responsible engineers relies upon connection of information inside and outside of an IMO [8]. Adaptive actions are entities for product object changes.

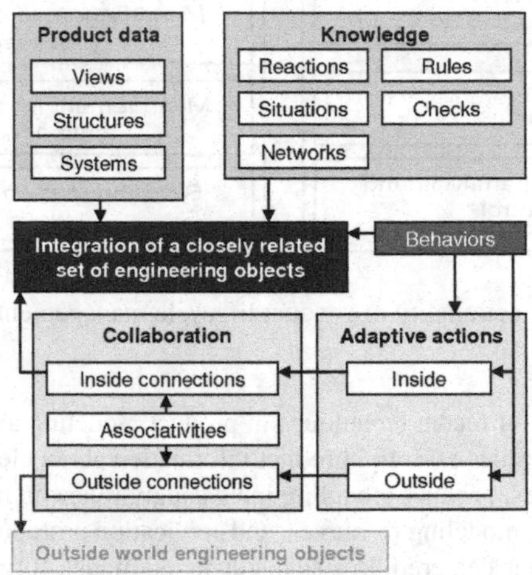

Fig. 2: Information structure

Engineering processes are organized around product description and managements (Fig. 3). Description of engineering objects using modeling procedures is supported by extensions for building associative connections, definition of knowledge, and machine learning. In close connection with descriptions, behavior management drives action generation. Action management propagates changes by using of associativity management. Product data management uses techniques for project management, product structuring, and creation of views. Authorization provides support for project management. Communication between engineers and PLM procedures are established on Internet portal. Collaboration of participants, human-computer interaction (HCI), information exchange, and other engineering processes are organized around portals. This up-to-date environment for PLM is assumed at the reported research.

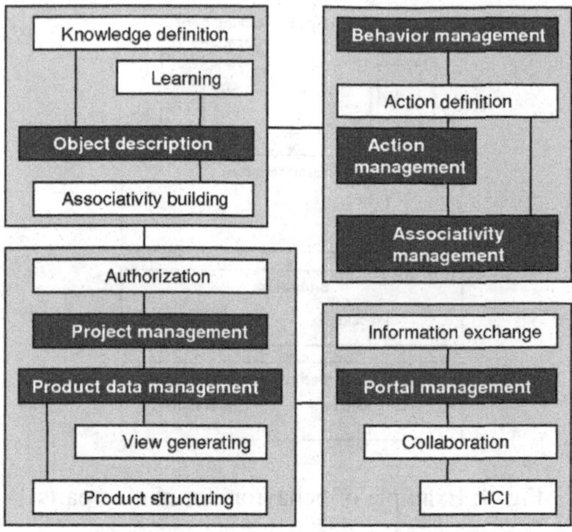

Fig. 3: Processes for engineering

4 Construction of Product Model by Chains of Changes

Engineers specify sets, attributes, and associative connections of product model entities in order to describe and relate physical and logical engineering objects in PLM. Main objective is to establish product with specified behaviors of modeled objects. Some level of built-in intelligence is needed for modeling procedures or existing models to answer the question that how will modeled objects behave. Engineers are responsible for their work so that they must have any information about the decision process at any time. Consequently, the authors consider intelligent change management that acts as advanced navigator and not as design automata. Using engineer controlled machine intelligence, engineers have much more chance to find conflict free solution than in conventional modeling.

Behaviors of a modeled product object represents a design objective. Behavior may change with changed specification of relevant design objective. It is assumed that any change of any modeled product object may affect one or more behaviors of that object and any other related product objects. Objects need repeated evaluation for changed behaviors. Behavior definition is connected to description of engineering objects by situations and circumstances. A situation is composed by parameters of modeled object for which a behavior is defined. If a change of an engineering object modifies situation defined for any behavior, that behaviors needs repeated evaluation.

Examples in Figs. 4 and 5 explain extended definition of behaviors as it is applied by the authors for design objectives. Two typical sets of behaviors are defined for a self-locking cone connection and a three-dimensional surface. Two behaviors of the self-locking joint are the connection and collision free path for assembly and subassembly. Situations for the evaluation of these behaviors are placing parts and moving volumes, accordingly. In the other example, a surface has shape and continuity behaviors. Shape behavior depends on control means of the shape while continuity behavior depends on connection with adjoining surfaces in the boundary of the modeled part of the product.

Product development, variant creation, and correction activities by engineers constitute sequences of changes of engineering objects. Consequences of a change of a modeled object

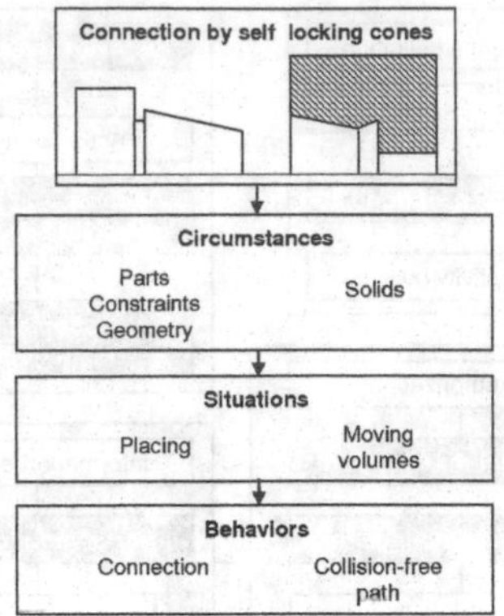

Fig. 4: Example of behaviors for mating parts

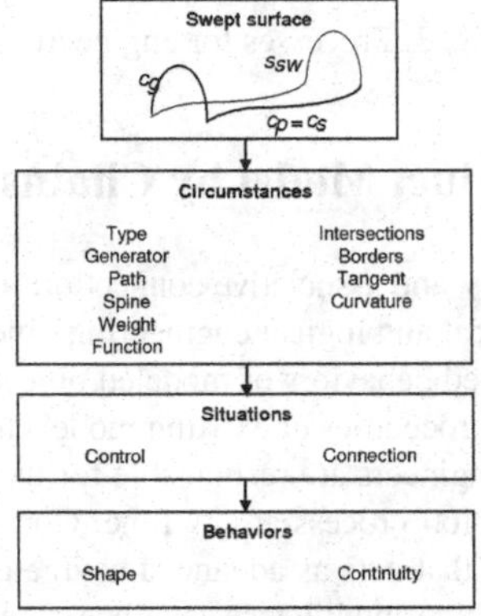

Fig. 5: Example of behaviors for a surface in boundary of a part

extend to more or less modeled objects, inside and outside of an IMO. Any modeled object that is in associative connection with a changed modeled object is considered to be in the affect zone of the changed object. Any object in the affect zone outside of the actual IMO must be accessed from the world outside of it. Analysis for changed behavior in the affect zone may suggest three consequences.

- Change of the elementary or composition modeled object is allowed because it does not break behavior specifications.
- Change request is rejected due to improper changes of specified behaviors.
- Resolution of conflicts is often required.

The proposed change management is outlined in Fig. 6. Changes are communicated with modeling procedures with less or more capabilities for intelligent processing. Responsible humans control modeling procedures. Alternatively, stored knowledge may represent humans. Change information is carried by adaptive action entity. Adaptive actions are generated by modeling procedures, received from engineers or attempted by world outside of the actual IMO. Conditional adaptive actions are accepted or rejected. These are states of adaptive actions. Effects of a change are analyzed in the affect zone for changed behaviors. Effect analysis generally proposes additional changes and extension of the affect zone by new associative definitions for modeled objects. In the world outside of he actual IMO, change attempts are accepted or rejected, and new changes are proposed.

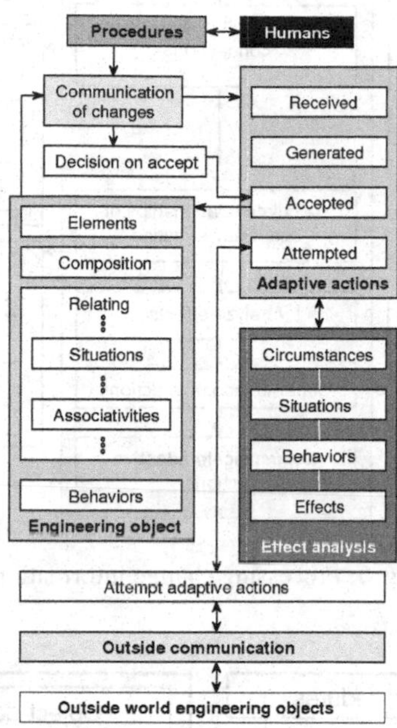

Fig. 6: Management of changes

Product modeling processes generate instances of generic definitions for product objects, situations, associativities, adaptive actions, and behaviors. Processing of change information starts with identification of affected product objects, circumstances, and situations (Fig. 7). Inside, outside and unavailable associativities are selected in order to describe affect zone of the changed engineering object. Modeling procedures for interaction by responsible engineer may identify unavailable associativities. Values of associativities are calculated then passed to behavior analysis where adaptive actions are accepted, rejected and generated. Sometimes unknown associativities are to be estimated. Development of an autonomous IMO is a consequence of inside or outside change attempts.

Activities for decision support and the related model descriptions are summarized in Fig. 8. In the proposed modeling for management of product model changes decisions are supported by four essential methods as follows.

- Analysis of changed behaviors
- Creation of views of product data
- Combination of intents of responsible engineers
- Change management

Human knowledge sources and links to outside knowledge sources are applied to complete knowledge information. Conventional knowledge representations such as rules, checks,

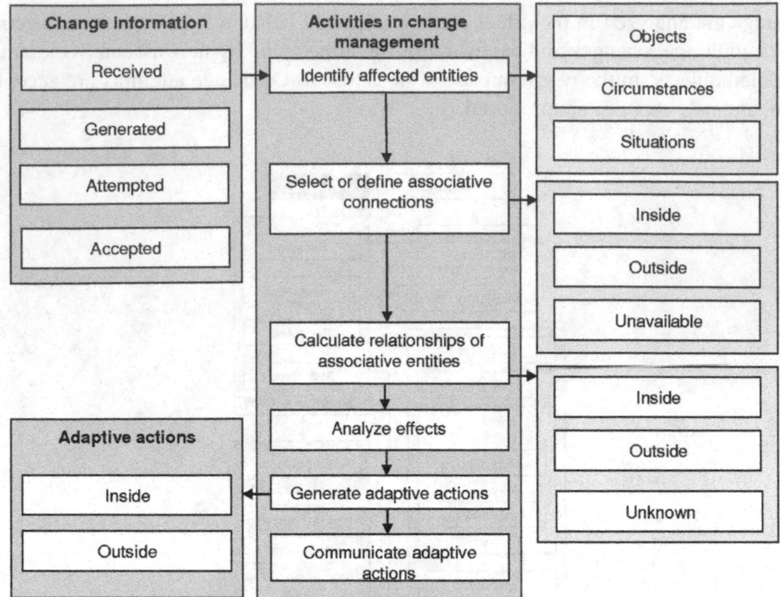

Fig. 7: Processing change information

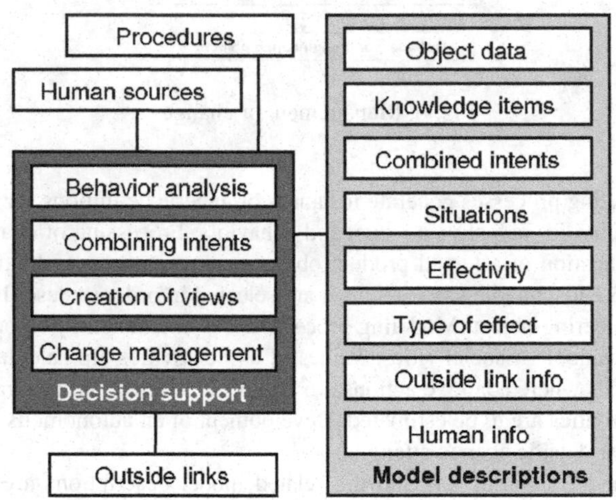

Fig. 8: Activities and information for decision support

etc. are extended by change management specific knowledge representations such as situations for behavior analysis, typical combined intents to assist combining intents, effectivities to make views, and affect definitions. In addition, information for links to human and outside knowledge is included. Views are created for inquiries regarding subsets of modeled information filtered according to specification.

5 Functionality for Decision Assistance

A coordinated application of the essential decision support methods determines functionality of the proposed change management. Decision is controlled by intent of authorized engineers. However, a simple decision, for example on a single dimension in a part, may use intents of scientists, experts, local instructors, and customers. Multiple humans control a single decision, including intents from responsible engineers, legislation, standard, and company measures. Activity for combining of partial intents often includes resolution of conflicts. Direct control by responsible engineer is completed by indirect controls. Aim of combined intent is to avoid traditional problems about decisions with complex human intent background.

Implementation of change management requires functionality and information tailored for product and development process specific demands. In order to gain their coordinated definition for the proposed modeling, functions and information are arranged in sectors. In Fig. 9, one of the possible arrangements includes sectors for object, parameter, learning, behavior, and interface. In the object sector, criteria and results of digital product definition are handled. Design objectives, limitations, and associative entity definitions constitute actual set of constraints for engineering processes. In the behavior sector, situations are generated, behaviors are analyzed, solutions are optimized, and adaptive actions are created. Patterns and rules are learned in order to collect knowledge and experience. Parameter section coordinates object parameters by handling their interactions, combinations, and influence. Associative connections of product model and the related objects are defined in the form of equations, relationships, rules, and networks. Communications, mainly about effects of changes, are organized in the interface sector.

6 Conclusions and Future Research

The authors proposed a method for decision assistance at change management in product modeling. Design objectives, limitations, and associative connections constitute sets of constraints for modeled objects. Constrained parameters are restricting factors for creation and modification of parameters of associative engineering objects. As an application of earlier research results [8], closely related engineering objects are handled as IMO.

As extensions to conventional knowledge representations, the proposed methods apply behaviors of modeled engineering objects as design objectives for the evaluation of changes in product model. Complex and unstructured sets of associative connections amongst parameters of engineering objects should be handled. In order to give a more transparent functionality of the proposed modeling, functions and information for change management are arranged in sectors. An example but representative set of object, parameter, learning, behavior, and interface sectors is also detailed in the paper (Fig. 9).

The proposed change management methodology can be implemented as extension to functionality of a PLM system. Digital product definition, product data management, and

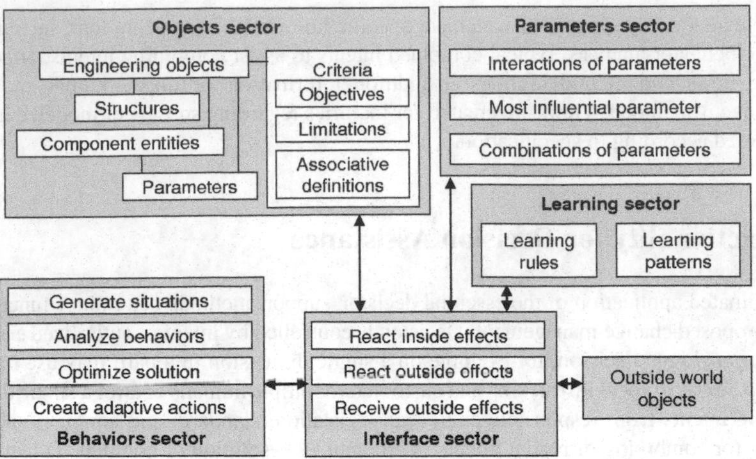

Fig. 9: Sectors for functions and information in the proposed modeling

collaborative communication functions and related product data sets can be accessed and communicated through API of open architecture PLM systems. Including in an IMO or applied as separate modeling tool accessible for several IMOs, change management programs can be developed by using of application programming environment of PLM installation. Contributions of this paper are as follows.

- An organized and purposeful arrangement of modeling functions and model information for change management
- Coordinated and interconnected activities for processing of changes
- Outline of modeled object behavior driven product development

The question that how to elaborate modeling procedures for the proposed behavior related methodology has remained open so that the answer is not an issue in this paper. An additional analysis is needed before starting of an experimental implementation. Main issue in this analysis will be application of the proposed change management in coordination with current product modeling and computational intelligence. Other primary issues for the future research in the proposed modeling are relationships of parameter sets for typical engineering objects and coordination of adaptive actions along chains of associative object definitions in product models.

References

1. Horváth L and Rudas IJ (2004) Modeling and Problem Solving Methods for Engineers. Elsevier, Academic. New York, USA
2. Mannistö T, Peltonen H, Martio A, and Sulonen R (1998) Computer-Aided Design 30(14):1111–1118
3. Shah Jami J, Mantyla M, and Shah Jamie J (1995) Parametric and Feature-Based Cad/Cam: Concepts, Techniques, and Applications. Wiley, New York

4. Hasegawa Y, and Fukuda T (1999) Motion Coordination of Behavior-based Controller for Brachiation Robot. In the Proceedings of the 1999 IEEE International Conference on Systems, Man, and Cybernetic, Human Communication and Cybernetics, IEEE, Tokyo, 6:896–901
5. Tambe M, Johnson WL, Jones R, Koss F, Laird J, Rosenbloom P, and Schwamb K (1995) AI Magazine 16(1):1079–1090
6. Horváth L, Rudas IJ (2001) Journal of Advanced Computational Intelligence 5(5):269–278
7. Horváth L, Rudas IJ, and Hancke G (2003) Feature Driven Integrated Product and Robot Assembly Modeling. In the Proceedings of the Seventh International Conference on Automation Technology, Automation 2003, Chia-yi, Taiwan, 906–911
8. Horváth L and Rudas IJ (2004) Active Description of Engineering Objects for Modeling in Extended Companies. In the Proceedings of the 2004 IEEE International Conference on Systems, Man & Cybernetics, SMC 2004, The Hague, The Netherlands, 3312–3317
9. Horváth L, Rudas IJ, and Couto C (2001) Integration of Human Intent Model Descriptions in Product Models. In: Digital Enterprise - New Challenges Life-Cycle Approach in Management and Production. Kluwer, Budapest, Hungary, 1–12
10. Horváth L, Rudas IJ (2004) Journal of Advanced Computational Intelligene and Intelligent Informatics 8(5):544–551

Flexibility in Stackelberg Leadership

Fernanda A. Ferreira[1,2], Flávio Ferreira[1], and Alberto A. Pinto[2]

[1] Departamento de Matemática, ESEIG – Instituto Politécnico do Porto
Rua D. Sancho I, 981, 4480-876 Vila do Conde, Portugal
fernandaamelia@eseig.ipp.pt, flavioferreira@eseig.ipp.pt

[2] Departamento de Matemática Pura, Faculdade de Ciências da Universidade do Porto
Rua do Campo Alegre, 687, 4169-007 Porto, Portugal
aapinto@fc.up.pt

Summary. We consider a Stackelberg model with demand uncertainty, only for the first mover. We study the advantages of leadership and flexibility with the variation of the demand uncertainty. Liu proved for demand uncertainty parameter greater than three that the follower firm can have an advantage with respect to the leading firm for some realizations of the demand intercept. Here, we prove that for demand uncertainty parameter less than three the leading firm is always in advantage.

1 Introduction

The Stackelberg model (see [17]) is one of the most widely used models in industrial organization for analyzing firms' behavior in a competitive environment. It studies the strategic situation where firms sequentially choose their output levels in a market. The question we ask is: Do first movers really have strategic advantage in practice? The belief of first-mover advantage was widely held among entrepreneurs and venture capitalists, but is now questioned by numerous practitioners. We do see some examples of successful pioneering firms as describing in [11] (see also [1, 2, 3, 4, 5, 6, 7, 8, 9, 10, 12, 13, 14, 15, 16]). Dell was the first to introduce the direct-sale business model into the PC market, and it achieved great success, growing from Mr. Dell's small-dorm business into a giant in the PC market. However, we can find many counterexamples. During the dot-com booming era, Pets.com, Webvan, Garden.com and eToys were all first movers in their respective market segments, but they all ended up burning through their investment capital before attracting enough customers to sustain a business (see [18]). Why do these pioneering firms get very different results? The probability of success of pioneering in a market clearly depends on many factors, including technology, marketing strategy and market demand. Several research papers focus their attention in giving answers to such question. Usually, the followers in markets get more market information than first movers before sinking their investments. In some industries that we consider to have fairly stable and predictable market demand, the pioneering firm tends to be the biggest player. However, if a market has a high degree of uncertainty, the followers can wait and see the customers' response to the new product introduced by the first movers.

J.A.T. Machado et al. (eds.), *Intelligent Engineering Systems and Computational Cybernetics,*

As in Liu's model (see [11]), we consider that only the first mover (leading firm) faces demand uncertainty. The demand uncertainty is given by a random variable uniformly distributed, with mean μ and standard deviation σ characterizing the *demand uncertainty parameter* $\theta = (\mu + \sqrt{3}\sigma)/(\mu - \sqrt{3}\sigma)$. By the time the second mover chooses its output level, that uncertainty is resolved. Therefore, the leading firm possesses first-mover advantage, but the second mover enjoys an informational advantage because it can adjust the production level after observing the realized demand (flexibility). Liu has proved that for demand intervals $[a,b]$ with $b/a = \theta \geq 3$, there is a perfect Bayesian equilibrium such that for some extreme values of the realized demand the inversion of the profits of the firms occurs (see [11]). We prove the uniqueness of Liu's perfect Bayesian equilibrium. We also analyze the case of demand intervals $[a,b]$ with $\theta < 3$, showing the existence and uniqueness of the perfect Bayesian equilibrium for each realized demand, and we check that in this case the leading firm is never in disadvantage.

2 Perfect Bayesian Equilibrium

We consider two firms, F_1 and F_2, each producing a homogeneous product. The demand, for simplicity, is linear, namely

$$p_i = \alpha - q_i - q_j, \tag{1}$$

with $\alpha > 0$, where p_i is the price and q_i the amount produced of product i, for $i, j \in \{1,2\}$ with $i \neq j$. Firms have the same constant marginal cost c. We consider from now on prices net of marginal costs. This is without loss of generality since if marginal cost is positive, we may replace α by $\alpha - c$. We consider that the demand intercept is a random variable uniformly distributed in the interval $[a,b]$, with $b > a > 0$. We note that the demand uncertainty parameter θ defined in the Introduction is equal to the ratio b/a. The distribution of α is of common knowledge. Profit π_i of firm F_i is given by

$$\pi_i = q_i p_i = q_i(\alpha - q_i - q_j). \tag{2}$$

The timing of the game is as follows:

(i) Firm F_1 chooses a quantity level $q_1 \geq 0$ without knowing the value of the demand realization.
(ii) Firm F_2 first observes the demand realization and observes q_1, and then chooses a quantity level $q_2 \geq 0$.

The quantity pair (q_1^*, q_2^*) is a *perfect Bayesian equilibrium*, if

$$\begin{cases} q_1^* = \arg\max_{q_1 \geq 0} E(\pi_1(q_1, q_2^*)) \\ q_2^* = \arg\max_{q_2 \geq 0} E(\pi_2(q_1^*, q_2)) \end{cases}.$$

Theorem 1. *Consider a Stackelberg duopoly facing the demand system (1), where the parameter α is uniformly distributed in the interval $[a,b]$. Then, there is a unique perfect Bayesian equilibrium given as follows.*

1. If $b \leq 3a$, then

$$q_1^* = \frac{a+b}{4} \quad \text{and} \quad q_2^* = \frac{\alpha}{2} - \frac{a+b}{8}.$$

2. *If $b \geq 3a$, then*

$$q_1^* = \frac{\Delta}{3}$$

and

$$q_2^* = \begin{cases} (3\alpha - \Delta)/6, & \text{if } \alpha \in [a, q_1^*] \\ 0, & \text{if } \alpha \in [q_1^*, b] \end{cases},$$

where

$$\Delta = 4a - 2b + \sqrt{10a^2 - 16ab + 7b^2}.$$

In Fig. 1 we show the localization of the quantity q_1^* in the interval $[0,b]$, in both cases $b \leq 3a$ and $b \geq 3a$.

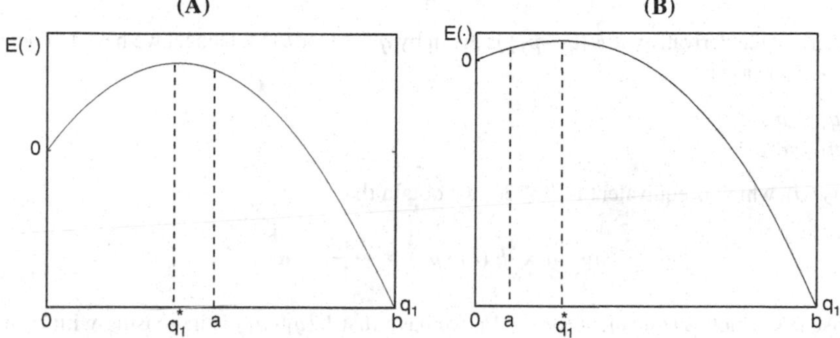

Fig. 1: Localization of the quantity q_1^* in the interval $[0,b]$, in the cases (**A**) $b \leq 3a$; and (**B**) $b \geq 3a$

Proof of Theorem 1. Using backwards-induction, we first compute firm F_2's reaction, $q_2^*(q_1,\alpha)$, to an arbitrary quantity q_1 fixed by firm F_1, and to the realized demand parameter α. The quantity $q_2^*(q_1,\alpha)$ is given by

$$q_2^*(q_1,\alpha) = \arg\max_{q_2 \geq 0} q_2(\alpha - q_1 - q_2),$$

which yields

$$q_2^*(q_1,\alpha) = \max\left\{0, \frac{\alpha - q_1}{2}\right\}. \tag{3}$$

Therefore, firm F_1's problem in the first stage of the game amounts to determine

$$q_1^* = \arg\max_{q_1 \geq 0} E\left(q_1 \cdot p_1\right),$$

where $p_1 = \alpha - q_1 - q_2^*(q_1,\alpha)$ and $E(\bullet)$ is the expectation with respect to the demand intercept α. We observe that $q_1^* \leq b$, because otherwise $\pi_1 < 0$. Now, we are going to find

(i)

$$\arg\max_{0 \leq q_1 \leq a} E\left(q_1 \cdot p_1\right)$$

and

(ii)

$$\arg\max_{a\leq q_1\leq b} E\left(q_1\cdot p_1\right).$$

Note that the density function of α's distribution is $1/(b-a)$.
Case (i): $0\leq q_1\leq a$. Since, in this case, $q_1\leq a$ and $a\leq\alpha$, by (3), we get that

$$q_2^*(q_1,\alpha)=\frac{\alpha-q_1}{2}.$$

Therefore, we have that

$$\begin{aligned}E\left(q_1\cdot p_1\right)&=\int_a^b q_1\left(\alpha-q_1-\frac{\alpha-q_1}{2}\right)\frac{1}{b-a}\,d\alpha\\&=-\frac{1}{2}q_1^2+\frac{a+b}{4}q_1.\end{aligned}$$

The zero of the derivative of $E\left(q_1\cdot p_1\right)$ is given by $\tilde{q}_1=(a+b)/4$. Hence, we have to consider two distinct cases:

(a) $\tilde{q}_1\leq a$ and
(b) $\tilde{q}_1\geq a$

In case *a*), which is equivalent to $b\leq 3a$, we obtain that

$$\arg\max_{0\leq q_1\leq a} E\left(q_1\cdot p_1\right)=\frac{a+b}{4}\equiv q_{1,A}.$$

In case *(b)*, which is equivalent to $b\geq 3a$, we have that $E\left(q_1\cdot p_1\right)$ is increasing with α in the interval $[0,a]$, and so

$$\arg\max_{0\leq q_1\leq a} E\left(q_1\cdot p_1\right)=a\equiv q_{1,B}.$$

Case (ii): $a\leq q_1\leq b$. In this case, by (3), we get that

$$q_2^*(q_1,\alpha)=\begin{cases}(\alpha-q_1)/2, & \text{if } \alpha\in[a,q_1]\\ 0, & \text{if } \alpha\in[q_1,b]\end{cases}.$$

Hence,

$$E\left(q_1\cdot p_1\right)=\int_a^{q_1} q_1(\alpha-q_1)\frac{1}{b-a}\,d\alpha+\int_{q_1}^b q_1\left(\alpha-q_1-\frac{\alpha-q_1}{2}\right)\frac{1}{b-a}\,d\alpha.$$

By [11], the zeros of the derivative of $E\left(q_1\cdot p_1\right)$ are the solutions of

$$-3q_1^2-4(b-2a)q_1+\left(b^2-2a^2\right)=0.$$

The non-negative solution is given by

$$\tilde{q}_1=\frac{4a-2b+\sqrt{10a^2-16ab+7b^2}}{3}.$$

Hence, we have to consider two distinct cases:

(c) $\tilde{q}_1\leq a$ and
(d) $\tilde{q}_1\geq a$

In case (c), which is equivalent to $b \leq 3a$, we have that $E(q_1 \cdot p_1)$ is decreasing with α in the interval $[a,b]$, and so

$$\arg \max_{a \leq q_1 \leq b} E(q_1 \cdot p_1) = a \equiv q_{1,C}.$$

In case (d), which is equivalent to $b \geq 3a$, we obtain that

$$\arg \max_{a \leq q_1 \leq b} E(q_1 \cdot p_1) = \frac{4a - 2b + \sqrt{10a^2 - 16ab + 7b^2}}{3} \equiv q_{1,D}.$$

Now, we observe that

1. if $b \leq 3a$, then

$$E(q_{1,A} \cdot p_1) \geq E(q_{1,C} \cdot p_1).$$

Thus,

$$q_1^* = q_{1,A} \equiv \frac{a+b}{4} \text{ and } q_2^* = \frac{\alpha}{2} - \frac{a+b}{8};$$

2. if $b \geq 3a$, then

$$E(q_{1,D} \cdot p_1) \geq E(q_{1,B} \cdot p_1).$$

Thus,

$$q_1^* = q_{1,D} \equiv \frac{\Delta}{3}$$

and

$$q_2^* = \begin{cases} (3\alpha - \Delta)/6, & \text{if } \alpha \in [a, q_1^*] \\ 0, & \text{if } \alpha \in [q_1^*, b] \end{cases},$$

which ends the proof. □

Corollary 1. *1. If $b \leq 3a$, then*

$$\pi_2^* \leq \pi_1^*, \text{ for all } \alpha \in [a,b].$$

2. If $b \geq 3a$, then

$$\begin{cases} \pi_2^* \leq \pi_1^*, & \text{if } \alpha \in [\Delta/3, \Delta] \\ \pi_2^* \geq \pi_1^*, & \text{if } \alpha \in [a, \Delta/3] \cup [\Delta, b] \end{cases}.$$

The two different situations described in this corollary are illustrated in Fig. 2.

Proof of Corollary 1. The statement 2 is proved in [11]. We are going to prove statement 1. If $b \leq 3a$, then the profits π_1^* and π_2^* are given by

$$\pi_1^* = \frac{a+b}{4}\left(\frac{\alpha}{2} - \frac{a+b}{8}\right) \text{ and } \pi_2^* = \left(\frac{\alpha}{2} - \frac{a+b}{8}\right)^2.$$

Thus,

$$\pi_1^* - \pi_2^* = \frac{(4\alpha - (a+b))(3(a+b) - 4\alpha)}{64},$$

which is non-negative. □

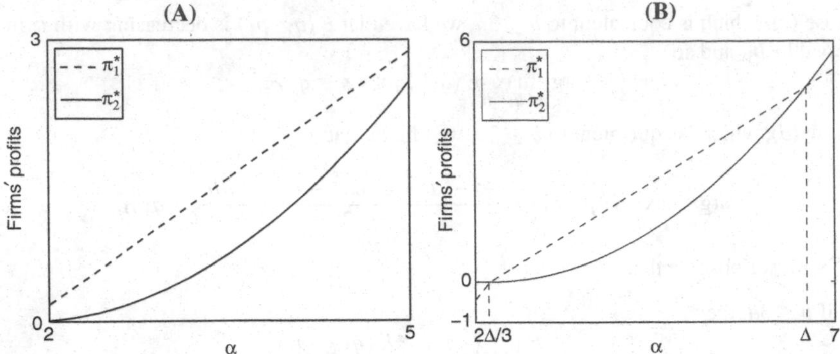

Fig. 2: The profits of both firms when the interval of the uniform distribution of the parameter α is as in each situation of Corollary 1: (**A**) $[a,b] = [2,5]$; (**B**) $[a,b] = [2,7]$

3 Conclusions

We showed that the Stackelberg model considered with demand uncertainty for the first mover has a unique perfect Bayesian equilibrium. Moreover, we determined this equilibrium, which is given by distinct expressions in the case of $\theta < 3$ and in the case of $\theta \geq 3$. We also checked that in the first case the leading firm is never in disadvantage, but in the second one it can lose its leadership advantage.

Acknowledgments

We thank the Programs POCI and POSI by FCT and Ministério da Ciência, Tecnologia e do Ensino Superior, and Centro de Matemática da Universidade do Porto for their financial support. Fernanda Ferreira gratefully acknowledges financial support from ESEIG/IPP and from PRODEP III by FSE and EU. Flávio Ferreira also acknowledges financial support from ESEIG/IPP.

References

1. Gal-Or E (1985) First-mover and second-mover advantages. International Economic Review 26(3):649–653
2. Gal-Or E (1987) First-mover disadvantages with private information. Review of Economic Studies 54(2):279–292.
3. Ghemawat P, Nalebuff B (1985) Exit. The RAND Journal of Economics 16(2):184–194.
4. Hirokawa M (2001) Strategic choice of quantity stickiness and Stackelberg leadership. Bulletin of Economic Research 53(1):19–34.
5. Hirokawa M, Sasaki D (2000) Endogeneous co-leadership when demand is uncertain. Australian Economic Papers 39(3):278–290.
6. Hirokawa M, Sasaki D (2001) Endogeneously asynchronous entries into an uncertain industry. Journal of Economics and Management Strategy 10(3):435–461.

7. Hoppe H (2002) The timing of new technology adoption: theoretical models and empirical evidence. Manchester School 7:56–76.
8. Hoppe H, Lehmann-Grube U (2001) Second-mover advantages in dynamic quality competition. Journal of Economics and Management Strategy 10:419–433.
9. Hoppe H, Lehmann-Grube U (2005) Innovation timing games: a general framework with applications. Journal of Economic Theory 121(1):30–50.
10. Huck S, Konrad K, Muller W (2001) Big fish eat small fish: on merger in Stackelberg markets. Economics Letters 73(2):213–217.
11. Liu Z (2005) Stackelberg leadership with demand uncertainty. Managerial Decision Economics 26:345–350.
12. Maggi G (1996) Endogenous leadership in a new market. RAND Journal of Economics 27:641–659.
13. Mallozzi L, Morgan J (2005) Oligopolistic markets with leadership and demand functions possibly discountinuous. Journal of Optimization and Applications 125(2):393–407.
14. Pal D (1991) Cournot duopoly with two prodution periods and cost differentials. Journal of Economic Theory 55:441–448.
15. Pal D, Sarkar J (2001) A Stackelberg oligopoly with nonidentical firms. Bulletin of Economic Research 53(2):127–134.
16. Robson A (1990) Stackelberg and marshall. American Economic Review 80:69–82.
17. von Stackelberg H (1934) Marktform und Gleichgewicht. Julius Springer, Vienna.
18. Stalter K (2002) Moving first on a winning idea doesn't ensure first-place finish: pioneers find pitfalls. Investor's Business Daily, October 7.

Investing to Survive in a Duopoly Model

Alberto A. Pinto[1], Bruno M. P. M. Oliveira[1,2], Fernanda A. Ferreira[1,3], and Miguel Ferreira[1]

[1] Departamento de Matemática Pura, Faculdade de Ciências da Universidade do Porto
Rua do Campo Alegre, 687, 4169-007 Porto, Portugal
aapinto@fc.up.pt, miguel.ferreira@fc.up.pt

[2] Faculdade de Ciências da Nutrição e Alimentação da Universidade do Porto
R. Dr. Roberto Frias, 4250-465 Porto, Portugal
bmpmo@fcna.up.pt

[3] Departamento de Matemática, ESEIG – Instituto Politécnico do Porto
Rua D. Sancho I, 981, 4480-876 Vila do Conde, Portugal
fernandaamelia@eseig.ipp.pt

Summary. We present deterministic dynamics on the production costs of Cournot competitions, based on perfect Nash equilibria of nonlinear R&D investment strategies to reduce the production costs of the firms at every period of the game. We analyse the effects that the R&D investment strategies can have in the profits of the firms along the time. We show that small changes in the initial production costs or small changes in the parameters that determine the efficiency of the R&D programs or of the firms can produce strong economic effects in the long run of the profits of the firms.

1 Introduction

We present deterministic dynamics on the production costs of Cournot competitions, based on R&D investment strategies of the firms. At every period of time, the firms involved in a Cournot competition invest in R&D projects to reduce their production costs. This competition is modeled by a two stages game for which the perfect Nash equilibria are computed. By deciding, at every period, to use the perfect Nash investment equilibria of the R&D strategies, the firms give rise to deterministic dynamics on the production costs characterizing the duopoly competition.

We observe that small changes in the initial production costs or small changes in the parameters that determine the efficiency of the firms or of the corresponding R&D programs, the firms can produce drastic economic effects in the long run of the profits of the firms. For instance, we show that a firm, F_1, with a disadvantage in the initial production costs but with a higher quality of R&D programs is able to recover along the time to a level of production costs lower than the other firm, F_2. However, for small changes in the initial production costs of the firms, or in the efficiency of the R&D investment programs, the firm F_1 is not able anymore to recover. Related works are done in [1, 2, 3, 4, 5, 6].

J.A.T. Machado et al. (eds.), *Intelligent Engineering Systems and Computational Cybernetics*,

2 R&D Investments on Costs

The Cournot competition with R&D investment programs consists of two subgames in one period of time. The first subgame is an R&D investment program, where both firms have initial production costs and choose, simultaneously, the R&D investment strategies to obtain new production costs. The second subgame, for simplicity of the model, is a Cournot competition with production costs equal to the reduced cost determined by the R&D investment program. As it is well known, the second subgame has a unique perfect Nash equilibrium. In some parameter region of our model, the game presents a unique Perfect Nash equilibrium, except for initial costs far away of the minimum attainable reduced production cost where the uniqueness of the equilibrium is broken.

We are going to find the perfect Nash equilibria of the Cournot competition with R&D investment programs by first discussing the Cournot competition and then the R&D investment strategies. Let us consider an economy with a monopolistic sector with two firms, F_1 and F_2, each one producing a differentiated good. As considered by Singh and Vives [7], we take the representative consumer preferences described by the following utility function

$$U(q_1,q_2) = \alpha_1 q_1 + \alpha_2 q_2 - \left(\beta_1 q_1^2 + 2\gamma q_1 q_2 + \beta_2 q_2^2\right)/2,$$

where q_i is the amount of good produced by the firm F_i, and $\alpha_i, \beta_i > 0$, for $i \in \{1,2\}$. The inverse demands are linear and, letting p_i be the price of the good produced by the firm F_i, they are given by

$$p_1 = \alpha_1 - \beta_1 q_1 - \gamma q_2,$$
$$p_2 = \alpha_2 - \gamma q_1 - \beta_2 q_2,$$

in the region of quantity space where prices are positive. The goods are substitutes, independent, or complements according to whether $\gamma > 0$, $\gamma = 0$, or $\gamma < 0$, respectively. Demand for good i is always downward sloping in its own price and increases (decreases) the price of the competitor, if the goods are substitutes (complements). When $\alpha_1 = \alpha_2$, the ratio $\gamma^2/\beta_1\beta_2$ expresses the degree of product differentiation ranging from zero, when the goods are independent, to one, when the goods are perfect substitutes. When $\gamma > 0$ and $\gamma^2/\beta_1\beta_2$ approaches to one, we are close to a homogeneous market.

The firm F_i invests an amount v_i in an R&D program that reduces the production costs to

$$a_i = c_i - \varepsilon_i(c_i - c_{L_i})\frac{v_i}{\lambda_i + v_i},$$

where the parameter $c_{i_L} > 0$ is the minimum attainable production cost for the firm F_i, and c_i is the firm F_i's unitary production cost at the beginning of the period satisfying $c_{i_L} \le c_i < \alpha_i$. (See Fig. 1.) The maximum reduction $\Delta_i = \varepsilon_i(c_i - c_{i_L})$ of the production cost is a percentage $0 < \varepsilon_i < 1$ of the difference between the current cost, c_i, and the lowest possible production cost, c_{i_L}. The parameter $\lambda_i > 0$ can be seen as a "measure"of the inverse of the quality of the R&D program of the firm F_i, since a smaller λ_i will result in a bigger reduction of the production costs for the same investment. In particular, $c_i - a_i(\lambda_i)$ gives half $\Delta_i/2$ of the maximum possible reduction Δ_i of the production cost for firm F_i.

The profit $\pi_i(q_i, q_j)$ of the firm F_i is given by

$$\pi_i(q_i,q_j) = q_i\left(\alpha_i - \beta_i q_i - \gamma q_j - a_i\right) - v_i, \tag{1}$$

for $i, j \in \{1,2\}$ and $i \neq j$. The optimal output level $q_i^* = q_i^*(q_j)$ of the firm F_i in response to the output level q_j of the firm F_j is given by

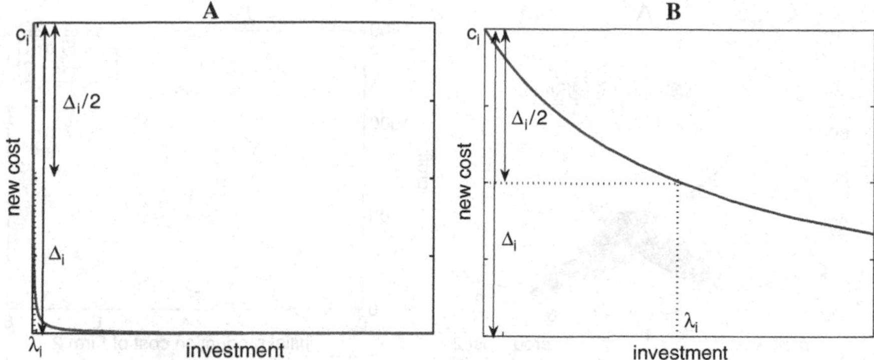

Fig. 1: New production costs as a function of the investment. (A): The maximum reduction in the production costs is Δ_i, obtained for an infinite investment $v_i = +\infty$. (B) (zoom of the left part of (A)): When the investment is $v_i = \lambda_i$ the reduction in the production costs is equal to $\Delta_i/2$

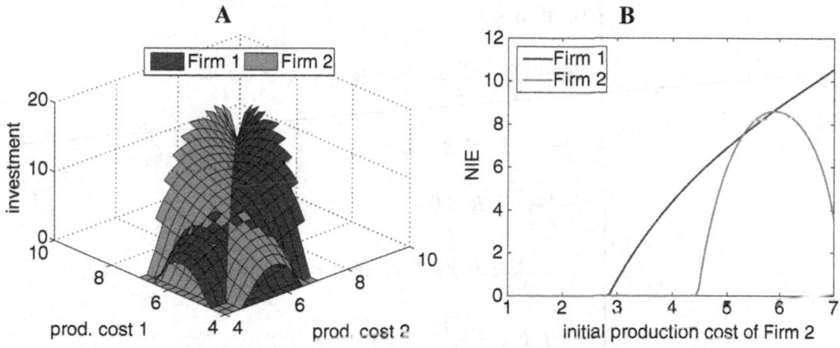

Fig. 2: The effect of the production costs on the Nash investment equilibrium. Identical firms, producing homogeneous goods, with parameters $\alpha_i = 10$, $\beta_i = 0.013$, $c_{i_L} = 4$, $\varepsilon_i = 0.1$ and $\lambda_i = 0.1$. In Fig. (**B**) we show a transversal cut of Figure (**A**), taking an initial production costs of firm F_1 equal to $c_1 = 5.3$

$$q_i^* = \arg \max_{0 \le q_i \le \alpha_i/\beta_i} q_i \left(\alpha_i - \beta_i q_i - \gamma q_j - a_i\right) - v_i.$$

Hence, the optimal output level of firm F_i is given by

$$q_i^* = \max\left\{0, \frac{\alpha_i - a_i - \gamma q_j}{2\beta_i}\right\}.$$

Let

$$R = R_1 = \frac{2\beta_2\alpha_1 - \gamma\alpha_2 - 2\beta_2 a_1 + \gamma a_2}{4\beta_1\beta_2 - \gamma^2}.$$

The Nash equilibrium output (q_1^*, q_2^*) is given by

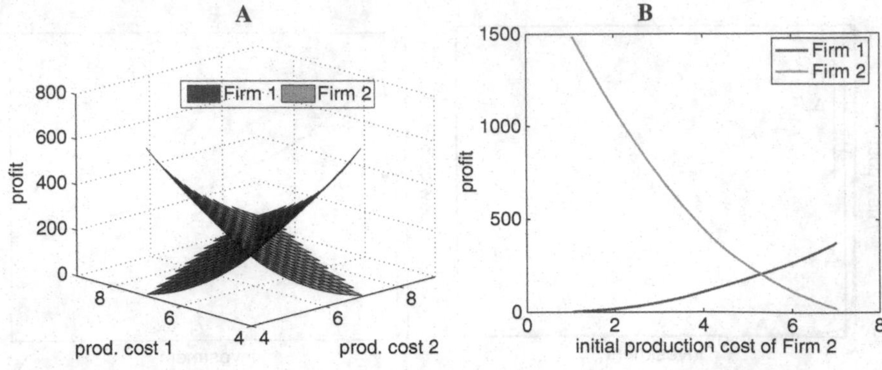

Fig. 3: The effect of the production costs in the profits. Identical firms producing homogeneous goods, with parameters $\alpha_i = 10$, $\beta_i = 0.013$, $c_{i_L} = 4$, $\varepsilon_i = 0.1$ and $\lambda_i = 0.1$. In Fig. (**B**) we show a transversal cut of Fig. (**A**), taking an initial production costs of firm F_1 equal to $c_1 = 5.3$

$$q_1^* = \begin{cases} 0, \text{ if } R < 0 \\ \frac{2\beta_2\alpha_1 - \gamma\alpha_2 - 2\beta_2 a_1 + \gamma a_2}{4\beta_1\beta_2 - \gamma^2}, \text{ if } 0 \leq R < \frac{\alpha_2 - a_2}{\gamma} \\ \frac{\alpha_1 - a_1}{2\beta_1}, \text{ if } R \geq \frac{\alpha_2 - a_2}{\gamma} \end{cases}$$

$$q_2^* = \begin{cases} \frac{\alpha_2 - a_2}{2\beta_2}, \text{ if } R < 0 \\ \frac{2\beta_1\alpha_2 - \gamma\alpha_1 - 2\beta_1 a_2 + \gamma a_1}{4\beta_1\beta_2 - \gamma^2}, \text{ if } 0 \leq R < \frac{\alpha_2 - a_2}{\gamma} \\ 0, \text{ if } R \geq \frac{\alpha_2 - a_2}{\gamma} \end{cases}.$$

Thus, firm F_1 has profit

$$\pi_1^*(q_1^*, q_2^*) = \begin{cases} -v_1, \text{ if } R < 0 \\ \frac{\beta_1(2\beta_2(\alpha_1 - a_1) - \gamma(\alpha_2 - a_2))^2}{(4\beta_1\beta_2 - \gamma^2)^2} - v_1, \text{ if } 0 \leq R < \frac{\alpha_2 - a_2}{\gamma} \\ \frac{(\alpha_1 - a_1)^2}{4\beta_1} - v_1, \text{ if } R \geq \frac{\alpha_2 - a_2}{\gamma} \end{cases},$$

and F_2 has profit

$$\pi_2^*(q_1^*, q_2^*) = \begin{cases} \frac{(\alpha_2 - a_2)^2}{4\beta_2} - v_2, \text{ if } R < 0 \\ \frac{\beta_2(2\beta_1(\alpha_2 - a_2) - \gamma(\alpha_1 - a_1))^2}{(4\beta_1\beta_2 - \gamma^2)^2} - v_2, \text{ if } 0 \leq R < \frac{\alpha_2 - a_2}{\gamma} \\ -v_2, \text{ if } R \geq \frac{\alpha_2 - a_2}{\gamma} \end{cases}.$$

The perfect Nash investment equilibrium (v_1^*, v_2^*) of the first subgame is given by

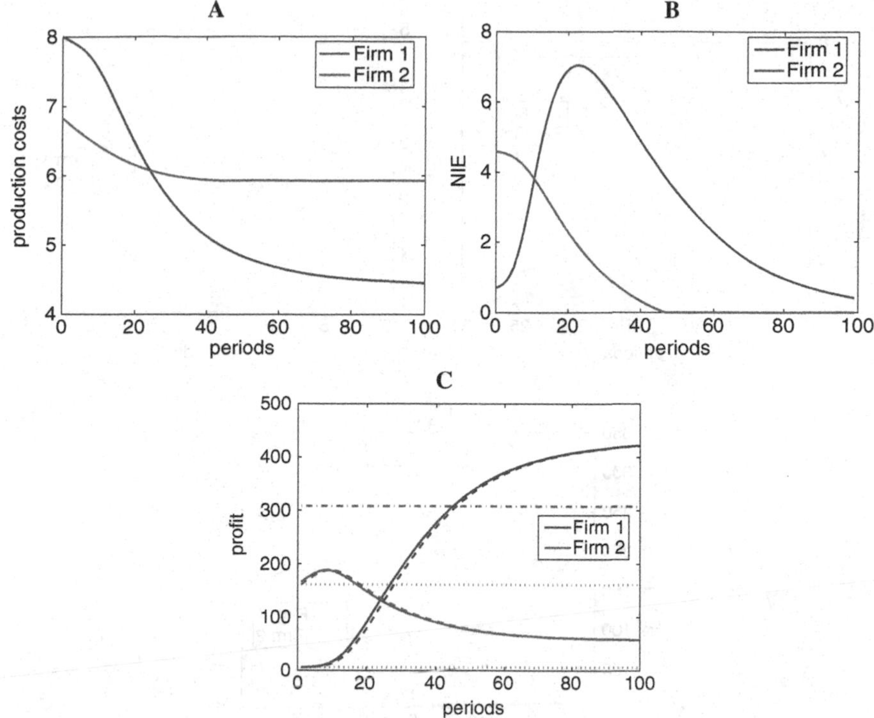

Fig. 4: Firms with different R&D investment programs. (**A**) Production costs. (**B**) Nash Investment Equilibrium. (**C**) Profit: (i) The "lines"correspond to the evolution of the profits determined by the stochastic dynamics; (ii) The "dashes"determine the profits of the firms in the case where both firms do not invest in that period; (iii) The "dots"determine the profits of the firms in the case where both firms never invest; (iv) The "dash-dot"is the profit in the case where both firms play the Nash equilibrium of the Cournot game with the minimum attainable production cost c_{L_i}. For initial production costs $c_{1_1} = 8$, $c_{2_1} = 6.83$ and parameters $\alpha_i = 10$, $\beta_i = 0.013$, $c_{i_l} = 4$, $\varepsilon_1 = 0.2$, $\varepsilon_2 = 0.1$ and $\lambda_i = 10$

$$v_1^* = \arg\max_{v_1 \geq 0} \pi_1$$

and

$$v_2^* = \arg\max_{v_2 \geq 0} \pi_2.$$

In Fig. 2(**A**), we present the perfect Nash equilibrium investment (v_1^*, v_2^*) for non-identical firms. In Fig. 2(**B**), we fix the production cost of the firm F_1 equal to 5.3 and we observe that the firm F_1 starts investing when the production cost of firm F_2 are close to 3, and increases its investment with the production costs of firm F_2. Firm F_2 starts investing when its production costs are close to the production cost of firm F_1, and it is not able to invest when its production costs are too large comparing with the production cost of the firm F_1.

In Fig. 3(**A**) we present the profits (π_1^*, π_2^*) for identical firms when they choose the perfect Nash investment equilibrium as a function of the production costs of both firms. Figure 3**B** is a cross section of Fig. 3(**A**) at theproduction costs c_1 of firm F_1 equal to 5.3. In Fig. 3(**B**)

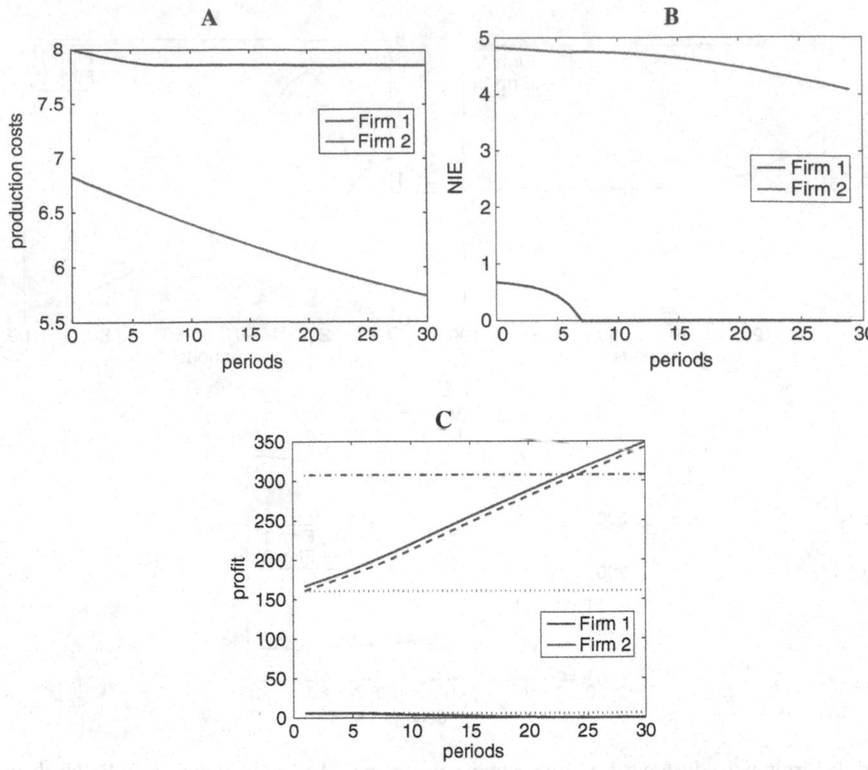

Fig. 5: Firms with different R&D investment programs. (**A**) Production costs. (**B**) Nash Investment Equilibrium. (**C**) Profit: (i) The "lines"correspond to the evolution of the profits determined by the stochastic dynamics; (ii) The "dashes"determine the profits of the firms in the case where both firms do not invest in that period; (iii) The "dots"determine the profits of the firms in the case where both firms never invest; (iv) The "dash-dot"is the profit in the case where both firms play the Nash equilibrium of the Cournot game with the minimum attainable production cost c_{L_i}. For initial production costs $c_{1_1} = 8$, $c_{2_1} = 6.83$ and parameters $\alpha_i = 10$, $\beta_i = 0.013$, $c_{i_L} = 4$, $\varepsilon_1 = 0.2$, $\varepsilon_2 = 0.103$ and $\lambda_i = 10$

we observe that with the increase of the production costs c_2 of firm F_2, the profits π_2^* of firm F_2 decrease and the profits π_1^* of firm F_1 increase.

3 Deterministic Dynamics

The deterministic dynamics on the production costs of the duopoly competition appear from the firms deciding to play the perfect Nash equilibrium in the Cournot competition with R&D investment programs, period after period.

We consider that firm F_1 has a better R&D quality than firm F_2 ($\varepsilon_1 > \varepsilon_2$) but that firm F_1 has a higher initial production cost than firm F_2. We see that the advantage of the higher quality of the R&D program of firm F_1 allows along the time firm F_1 to decrease its production

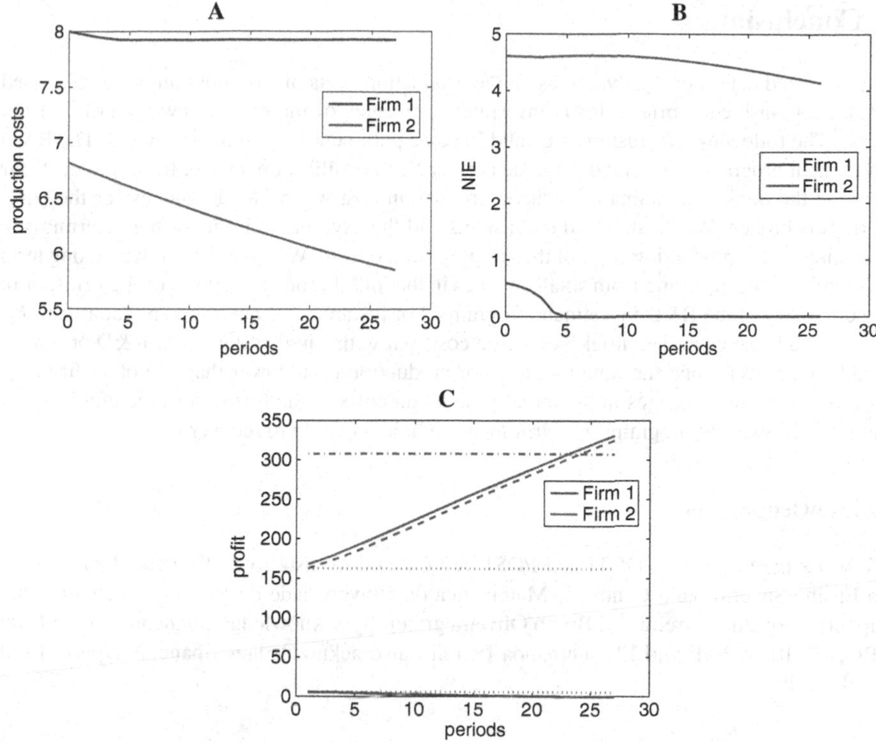

Fig. 6: Firms with different R&D investment programs. (**A**) Production costs. (**B**) Nash Investment Equilibrium. (**C**) Profit. (i) The "lines" correspond to the evolution of the profits determined by the stochastic dynamics; (ii) The "dashes" determine the profits of the firms in the case where both firms do not invest in that period; (iii) The "dots" determine the profits of the firms in the case where both firms never invest; (iv) The "dash-dot" is the profit in the case where both firms play the Nash equilibrium of the Cournot game with the minimum attainable production cost c_{L_i}. For initial production costs $c_{1_1} = 8$, $c_{2_1} = 6.82$ and parameters $\alpha_i = 10$, $\beta_l = 0.013$, $c_{i_L} = 4$, $\varepsilon_1 = 0.2$, $\varepsilon_2 = 0.1$ and $\lambda_i = 10$

costs to a level lower than firm F_2 and like that recovering from the initial advantage of firm F_2 (see Fig. 4).

A small increase in the R&D quality of firm F_2 provokes that firm F_1 is not able anymore to recover along the time the production costs (see Fig. 5). Moreover, the gap between the production costs of the two firms increases along the time.

A small decrease in the initial production costs of firm F_2 provokes that firm F_1 is not able anymore to recover along the time the production costs (see Fig. 6). Moreover, the gap between the production costs of the two firms increases along the time. Hence, in this case, a small decrease in the initial production costs of firm F_2 has similar effects in the long run of the profits of the firms to a small increase in the R&D quality of firm F_2.

4 Conclusions

We presented deterministic dynamics on the production costs of Cournot competitions, based on perfect Nash equilibria of R&D investment strategies of the firms at every period of the game. The following conclusions are valid in some parameter region of our model. The R&D investment programm presented a unique perfect Nash equilibrium, except for initial costs far away of the minimum attainable reduced production cost where the uniqueness of the equilibrium is broken. We illustrated the transients and the asymptotic limits of the deterministic dynamics on the production costs of the duopoly competition. We have shown strong long term economic effects resulting from small changes in the initial production costs of the firms, or in the efficiency of the R&D investment programs. For instance, we have shown that a firm, F_1, with a disadvantage in the initial production costs but with a higher quality of R&D programs is able to recover along the time to a level of production costs lower than the other firm, F_2. However, for small changes in the initial production costs of the firms, or in the efficiency of the R&D investment programs, the firm F_1 is not able anymore to recover.

Acknowledgments

We thank the Programs POCTI and POSI by FCT and Ministério da Ciência, Tecnologia e do Ensino Superior, and Centro de Matemática da Universidade do Porto for their financial support. Fernanda Ferreira and Bruno Oliveira gratefully acknowledge financial support from PRODEP III by FSE and EU. Fernanda Ferreira also acknowledges financial support from ESEIG/IPP.

References

1. Amir R, Evstigneev I, Wooders J (2001) Noncooperative versus cooperative R&D with endogenous spillover rates. Core Discussion Paper 2001/50, Louvain-la-Neuve, Belgium
2. Bischi GI, Gallegati M, Naimzada A (1999) Symmetry-breaking bifurcations and representative firm in dynamic duopoly games. Annals of Operations Research 89:253–272
3. Brander JA, Spencer BJ (1983) Strategic commitment with R&D: the symmetric case. The Bell Journal of Economics 14:225–235
4. Cournot A (1838) Recherches sur les Principes Mathématiques de la Théorie des Richesses, Paris. English edition: (1897) Researches into the Mathematical Principles of the Theory of Wealth. Edited by N. Bacon, New York, Macmillan
5. Pinto AA, Oliveira B, Ferreira FA, Ferreira F (2007) Stochasticity favoring the effects of the R&D strategies of the firms. In: Machado JT, Patkai B, Rudas IJ (eds) Intelligent Engineering Systems and Computational Cybernetics. Springer, New York
6. Qiu DL (1997) On the dynamic efficiency of Bertrand and Cournot equilibria. Journal of Economic Theory 75:213–229
7. Singh N, Vives X (1984) Price and quantity competition in a differentiated duopoly. RAND Journal of Economics 15:546–554

Stochasticity Favoring the Effects of the R&D Strategies of the Firms

Alberto A. Pinto[1], Bruno M. P. M. Oliveira[1,2], Fernanda A. Ferreira[1,3], and Flávio Ferreira[3]

[1] Departamento de Matemática Pura, Faculdade de Ciências da Universidade do Porto
Rua do Campo Alegre, 687, 4169-007 Porto, Portugal
aapinto@fc.up.pt, miguel.ferreira@fc.up.pt

[2] Faculdade de Ciências da Nutrição e Alimentação da Universidade do Porto
R. Dr. Roberto Frias, 4250-465 Porto, Portugal
bmpmo@fcna.up.pt

[3] Departamento de Matemática, ESEIG - Instituto Politécnico do Porto
Rua D. Sancho I, 981, 4480-876 Vila do Conde, Portugal
{fernandaamelia, flavioferreira}@eseig.ipp.pt

Summary. We present stochastic dynamics on the production costs of Cournot competitions, based on perfect Nash equilibria of nonlinear R&D investment strategies to reduce the production costs of the firms at every period of the game. We analyse the effects that the R&D investment strategies can have in the profits of the firms along the time. We observe that, in certain cases, the uncertainty can improve the effects of the R&D strategies in the profits of the firms due to the non-linearity of the profit functions and also of the R&D parameters.

1 Introduction

The R&D program presented in this paper follows closely the properties proposed for R&D programs in Amir's paper [1]. However, we introduce a novelty in the R&D program comparing with Qiu's R&D program (see [5]), because the function determining the investments accordingly to the decrease on the production costs has not an infinity derivative for zero investments. At every period of time, the firms involved in a Cournot competition invest in R&D projects to reduce their production costs. This competition is modeled by a two stages game for which the perfect Nash equilibria are computed. By deciding, at every period, to use the perfect Nash investment equilibria of the R&D strategies, with uncertainty, the firms give rise to stochastic dynamics on the production costs characterizing the duopoly competition. We focus our attention in the effects of the uncertainty in the profits of the firms. We observe that, in certain cases, the uncertainty can improve the effects of the R&D strategies in the profits of the firms due to the non-linearity of the profit functions and also to the non-linearity of the R&D program. Related works are done in [2, 3, 4].

J.A.T. Machado et al. (eds.), *Intelligent Engineering Systems and Computational Cybernetics,*

2 R&D Investments on Costs

The Cournot competition with R&D investment programs consists of two subgames in one period of time. The first subgame is an R&D investment program, where both firms have initial production costs and choose, simultaneously, the R&D investment strategies to obtain new production costs. The second subgame, for simplicity of the model, is a Cournot competition with production costs equal to the reduced cost determined by the R&D investment program. As it is well known, the second subgame has a unique perfect Nash equilibrium. In some parameter region of our model, the game presents a unique Perfect Nash equilibrium, except for initial costs far away of the minimum attainable reduced production cost where the uniqueness of the equilibrium is broken.

We are going to find the perfect Nash equilibria of the Cournot competition with R&D investment programs by first discussing the Cournot competition and then the R&D investment strategies. Let us consider an economy with a monopolistic sector with two firms, F_1 and F_2, each one producing a differentiated good. As considered by Singh and Vives [6], we take the representative consumer preferences described by the following utility function

$$U(q_1,q_2) = \alpha_1 q_1 + \alpha_2 q_2 - \left(\beta_1 q_1^2 + 2\gamma q_1 q_2 + \beta_2 q_2^2\right)/2,$$

where q_i is the amount of good produced by the firm F_i, and $\alpha_i, \beta_i > 0$, for $i \in \{1,2\}$. The inverse demands are linear and, letting p_i be the price of the good produced by the firm F_i, they are given by

$$p_1 = \alpha_1 - \beta_1 q_1 - \gamma q_2,$$
$$p_2 = \alpha_2 - \gamma q_1 - \beta_2 q_2,$$

in the region of quantity space where prices are positive. The goods are substitutes, independent, or complements according to whether $\gamma > 0$, $\gamma = 0$, or $\gamma < 0$, respectively. Demand for good i is always downward sloping in its own price and increases (decreases) the price of the competitor, if the goods are substitutes (complements). When $\alpha_1 = \alpha_2$, the ratio $\gamma^2/\beta_1\beta_2$ expresses the degree of product differentiation ranging from zero, when the goods are independent, to one, when the goods are perfect substitutes. When $\gamma > 0$ and $\gamma^2/\beta_1\beta_2$ approaches to one, we are close to a homogeneous market.

The firm F_i invests an amount v_i in an R&D program that reduces the production costs to

$$a_i = c_i - \varepsilon_i (c_i - c_{L_i}) \frac{v_i}{\lambda_i + v_i},$$

where the parameter $c_{i_L} > 0$ is the minimum attainable production cost for the firm F_i, and c_i is the firm F_i's unitary production cost at the beginning of the period satisfying $c_{i_L} \le c_i < \alpha_i$ (see Fig. 1). The maximum reduction $\Delta_i = \varepsilon_i (c_i - c_{i_L})$ of the production cost is a percentage $0 < \varepsilon_i < 1$ of the difference between the current cost, c_i, and the lowest possible production cost, c_{i_L}. The parameter $\lambda_i > 0$ can be seen as a "measure" of the inverse of the quality of the R&D program of the firm F_i, since a smaller λ_i will result in a bigger reduction of the production costs for the same investment. In particular, $c_i - a_i(\lambda_i)$ gives half $\Delta_i/2$ of the maximum possible reduction Δ_i of the production cost for firm F_i.

The profit $\pi_i(q_i, q_j)$ of the firm F_i is given by

$$\pi_i(q_i, q_j) = q_i \left(\alpha_i - \beta_i q_i - \gamma q_j - a_i\right) - v_i, \tag{1}$$

for $i, j \in \{1,2\}$ and $i \neq j$. The optimal output level $q_i^* = q_i^*(q_j)$ of the firm F_i in response to the output level q_j of the firm F_j is given by

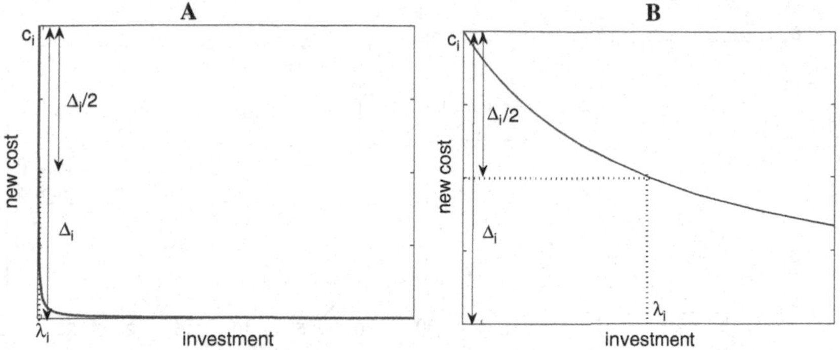

Fig. 1: New production costs as a function of the investment. (A): The maximum reduction in the production costs is Δ_i, obtained for an infinite investment $v_i = +\infty$. (B) (zoom of the left part of (A)): When the investment is $v_i = \lambda_i$ the reduction in the production costs is equal to $\Delta_i/2$

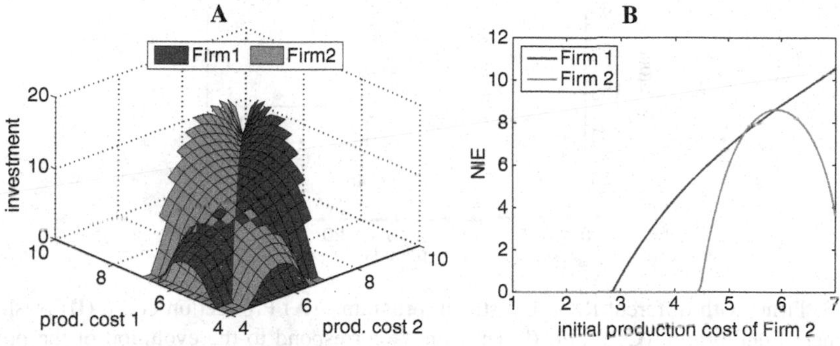

Fig. 2: The effect of the production costs on the Nash investment equilibrium. Identical firms, producing homogeneous goods, with parameters $\alpha_i = 10$, $\beta_i = 0.013$, $c_{i_L} = 4$, $\varepsilon_i = 0.1$ and $\lambda_i = 0.1$. In Fig. (**B**) we show a transversal cut of Fig. (**A**), taking an initial production costs of firm F_1 equal to $c_1 = 5.3$

$$q_i^* = \arg \max_{0 \le q_i \le \alpha_i/\beta_i} q_i \left(\alpha_i - \beta_i q_i - \gamma q_j - a_i\right) - v_i.$$

Hence, the optimal output level of firm F_i is given by

$$q_i^* = \max \left\{ 0, \frac{\alpha_i - a_i - \gamma q_j}{2\beta_i} \right\}.$$

Let

$$R = R_1 = \frac{2\beta_2 \alpha_1 - \gamma \alpha_2 - 2\beta_2 a_1 + \gamma a_2}{4\beta_1 \beta_2 - \gamma^2}.$$

The Nash equilibrium output (q_1^*, q_2^*) is given by

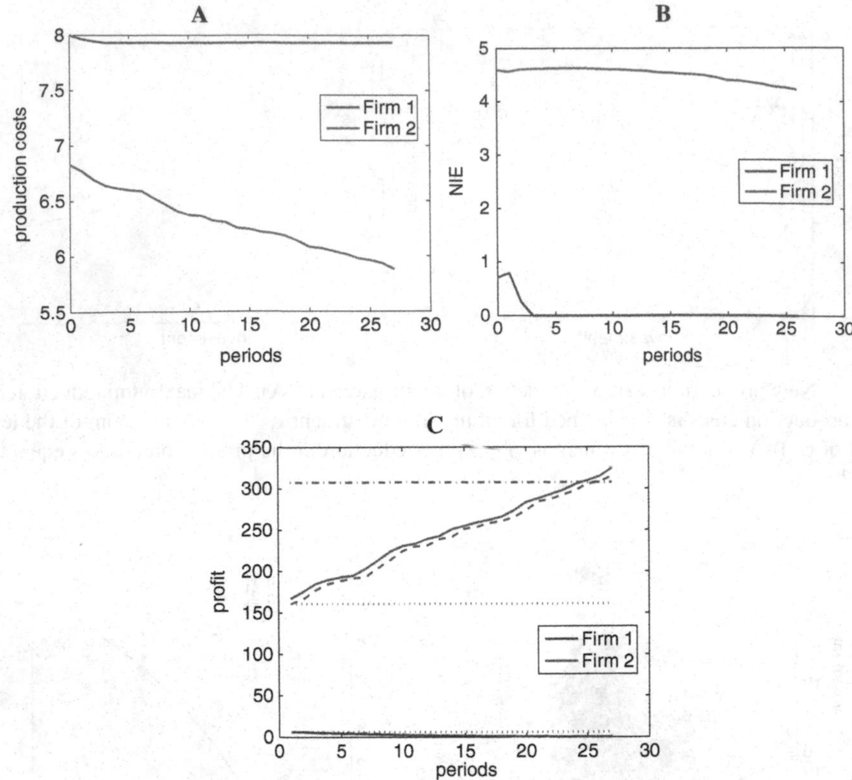

Fig. 3: Firms with different R&D investment programs. (**A**) Production costs. (**B**) Nash Investment Equilibrium. (**C**) Profit: (i) The "lines" correspond to the evolution of the profits determined by the stochastic dynamics; (ii) The "dashes" determine the profits of the firms in the case where both firms do not invest in that period; (iii) The "dots" determine the profits of the firms in the case where both firms never invest; (iv) The "dash-dot" is the profit in the case where both firms play the Nash equilibrium of the Cournot game with the minimum attainable production cost c_{L_i}. For initial production costs $c_{1_1} = 8$, $c_{2_1} = 6.83$ and parameters $\alpha_i = 10$, $\beta_i = 0.013$, $c_{i_L} = 4$, $\varepsilon_1 = 0.2$, $\varepsilon_2 = 0.1$ and $\lambda_i = 10$

$$q_1^* = \begin{cases} 0, \text{ if } R < 0 \\ \frac{2\beta_2\alpha_1 - \gamma\alpha_2 - 2\beta_2 a_1 + \gamma a_2}{4\beta_1\beta_2 - \gamma^2}, \text{ if } 0 \le R < \frac{\alpha_2 - a_2}{\gamma} \\ \frac{\alpha_1 - a_1}{2\beta_1}, \text{ if } R \ge \frac{\alpha_2 - a_2}{\gamma} \end{cases}$$

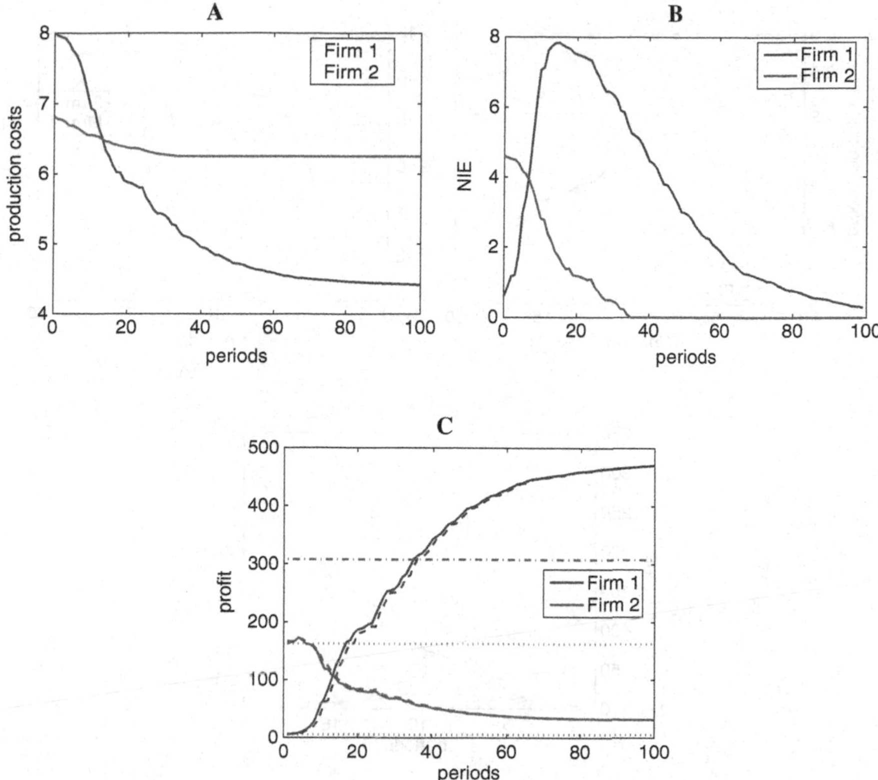

Fig. 4: Firms with different R&D investment programs. (**A**) Production costs. (**B**) Nash Investment Equilibrium. (**C**) Profit: (i) The "lines" correspond to the evolution of the profits determined by the stochastic dynamics; (ii) The "dashes" determine the profits of the firms in the case where both firms do not invest in that period; (iii) The "dots" determine the profits of the firms in the case where both firms never invest; (iv) The "dash-dot" is the profit in the case where both firms play the Nash equilibrium of the Cournot game with the minimum attainable production cost c_{L_i}. For initial production costs $c_{1_1} = 8$, $c_{2_1} = 6.82$ and parameters $\alpha_i = 10$, $\beta_i = 0.013$, $c_{i_L} = 4$, $\varepsilon_1 = 0.2$, $\varepsilon_2 = 0.1$ and $\lambda_i = 10$

$$q_2^* = \begin{cases} \frac{\alpha_2 - a_2}{2\beta_2}, \text{ if } R < 0 \\ \frac{2\beta_1\alpha_2 - \gamma\alpha_1 - 2\beta_1 a_2 + \gamma a_1}{4\beta_1\beta_2 - \gamma^2}, \text{ if } 0 \leq R < \frac{\alpha_2 - a_2}{\gamma} \\ 0, \text{ if } R \geq \frac{\alpha_2 - a_2}{\gamma} \end{cases}.$$

Thus, firm F_1 has profit

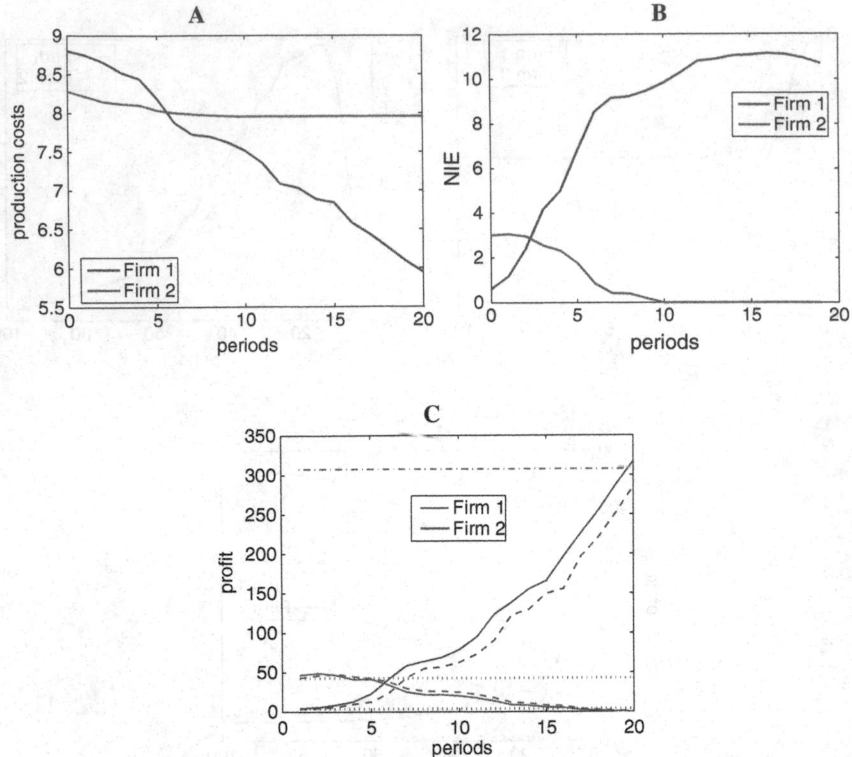

Fig. 5: Firms with different R&D investment programs. (**A**) Production costs. (**B**) Nash Investment Equilibrium. (**C**) Profit: (i) The "lines" correspond to the evolution of the profits determined by the stochastic dynamics; (ii) The "dashes" determine the profits of the firms in the case where both firms do not invest in that period; (iii) The "dots" determine the profits of the firms in the case where both firms never invest; (iv) The "dash-dot" is the profit in the case where both firms play the Nash equilibrium of the Cournot game with the minimum attainable production cost c_{L_i}. For initial production costs $c_{1_1} = 8.8$, $c_{2_1} = 8.28$ and parameters $\alpha_i = 10$, $\beta_i = 0.013$, $c_{i_L} = 4$, $\varepsilon_1 = 0.2$, $\varepsilon_2 = 0.1$ and $\lambda_i = 10$

$$\pi_1^*(q_1^*, q_2^*) = \begin{cases} -v_1, \text{ if } R < 0 \\ \frac{\beta_1(2\beta_2(\alpha_1 - a_1) - \gamma(\alpha_2 - a_2))^2}{(4\beta_1\beta_2 - \gamma^2)^2} - v_1, \text{ if } 0 \le R < \frac{\alpha_2 - a_2}{\gamma} \\ \frac{(\alpha_1 - a_1)^2}{4\beta_1} - v_1, \text{ if } R \ge \frac{\alpha_2 - a_2}{\gamma} \end{cases},$$

and F_2 has profit

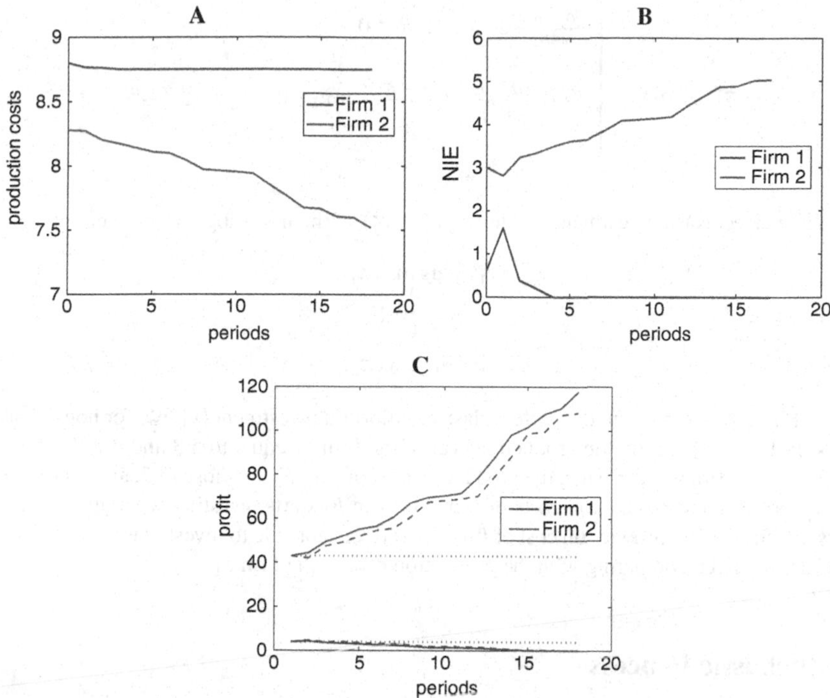

Fig. 6: Firms with different R&D investment programs. (**A**) Production costs. (**B**) Nash Investment Equilibrium. (**C**) Profit: (i) The "lines" correspond to the evolution of the profits determined by the stochastic dynamics; (ii) The "dashes" determine the profits of the firms in the case where both firms do not invest in that period; (iii) The "dots" determine the profits of the firms in the case where both firms never invest; (iv) The "dash-dot" is the profit in the case where both firms play the Nash equilibrium of the Cournot game with the minimum attainable production cost c_{L_i}. For initial production costs $c_{1_1} = 8.8$, $c_{2_1} = 8.28$ and parameters $\alpha_i = 10$, $\beta_i = 0.013$, $c_{i_L} = 4$, $\varepsilon_1 = 0.2$, $\varepsilon_2 = 0.1$ and $\lambda_i = 10$

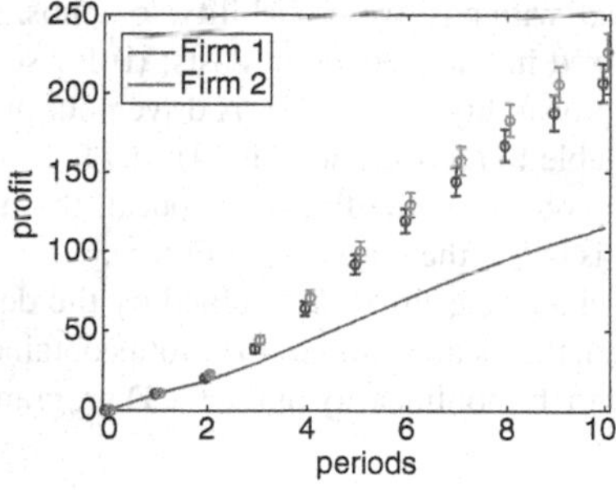

Fig. 7: Profit from the deterministic dynamics and from 1,000 stochastic simulations. For initial production costs $c_{i_1} = 9$ and parameters $\alpha_i = 10$, $\beta_i = 0.018$, $c_{L_i} = 4$, $\varepsilon_i = 0.1$ and $\lambda_i = 1$. Line: deterministic profit; Balls: mean profit of the simulations; Bars: 95% confidence intervals of the simulations

$$\pi_2^*(q_1^*, q_2^*) = \begin{cases} \frac{(\alpha_2 - a_2)^2}{4\beta_2} - v_2, & \text{if } R < 0 \\ \frac{\beta_2(2\beta_1(\alpha_2 - a_2) - \gamma(\alpha_1 - a_1))^2}{(4\beta_1\beta_2 - \gamma^2)^2} - v_2, & \text{if } 0 \leq R < \frac{\alpha_2 - a_2}{\gamma} \\ -v_2, & \text{if } R \geq \frac{\alpha_2 - a_2}{\gamma} \end{cases}.$$

The perfect Nash investment equilibrium (v_1^*, v_2^*) of the first subgame is given by

$$v_1^* = \arg\max_{v_1 \geq 0} \pi_1$$

and

$$v_2^* = \arg\max_{v_2 \geq 0} \pi_2.$$

In Fig. 2(**A**), we present the perfect Nash equilibrium investment (v_1^*, v_2^*) for non-identical firms. In Fig. 2(**B**), we fix the production cost of the firm F_1 equal to 5.3 and we observe that the firm F_1 starts investing when the production cost of firm F_2 are close to 3, and increases its investment with the production costs of firm F_2. Firm F_2 starts investing when its production costs are close to the production cost of firm F_1, and it is not able to invest when its production costs are too large comparing with the production cost of the firm F_1.

3 Stochastic Process

The deterministic dynamics on the production costs of the duopoly competition appear from the firms deciding to play the perfect Nash equilibrium in the Cournot competition with R&D investment programs, period after period. The deterministic dynamics give rise to a stochastic process on the production costs of the duopoly competition by the firms deciding to play the Cournot competition with uncertainty in the results of their R&D investment programs, period after period, as we pass to describe: Given production costs c_1 and c_2, at period n, the stochastic process determines new production costs, at period $n+1$, that are the realizations of the random variables A_1 and A_2 uniformly distributed, with mean a_1 and a_2, and standard deviation $(c_1 - a_1)/2\sqrt{3}$ and $(c_2 - a_2)/2\sqrt{3})$, respectively.

The stochasticity can lead, with positive probability, to opposite evolutions of the production costs of the firms for fixed initial production costs: (i) for some fixed parameter values of the model, with positive probability, the firm F_1 is driven out of the market (see Fig. 3) or, with positive probability, is able to recover (see Fig. 4); (ii) for some fixed parameter values of the model, with positive probability, the firm F_1 is out of the market (see Fig. 5) or, with positive probability, firm F_2 is out of the market (see Fig. 6).

We observe that the profits of the firms determined by the deterministic dynamics have values significantly lower than the mean values of the profits obtained in the stochastic simulations (see Fig. 7). This is due to the nonlinearity of the R&D program and of the profit function.

4 Conclusions

We presented stochastic dynamics on the production costs of Cournot competitions, based on perfect Nash equilibria of R&D investment strategies ofthe firms at every period of the game. The following conclusions are valid in some parameter region of our model. The R&D investment program presented a unique perfect Nash equilibrium, except for initial costs far away of the minimum attainable reduced production cost where the uniqueness of the equilibrium is broken. We illustrated the stochastic dynamics on the production costs of the duopoly competition. We observed that in the presence of the uncertainty in the model, there are initial production costs such that the firms have opposite outcomes with positive probabilities: (i) The firm F_1 is driven out of the market, with positive probability, or is able to recover, with positive probability; (ii) The firm F_1 is out of the market, with positive probability, or the firm F_2 is out of the market, with positive probability. We observe that the profits of the firms determined by the deterministic dynamics have values significantly lower than the mean values of the profits obtained in the stochastic simulations due to the non-linear effects.

Acknowledgments

We thank the Programs POCTI and POSI by FCT and Ministério da Ciência, Tecnologia e do Ensino Superior, and Centro de Matemática da Universidade do Porto for their financial support. F.A. Ferreira and B. Oliveira gratefully acknowledge financial support from PRODEP III by FSE and EU. F.A. Ferreira and F. Ferreira also acknowledge financial support from ESEIG/IPP.

References

1. Amir R, Evstigneev I, Wooders J (2001) Noncooperative versus cooperative R&D with endogenous spillover rates. Core Discussion Paper 2001/50, Louvain-la-Neuve, Belgium
2. Bischi GI, Gallegati M, Naimzada A (1999) Symmetry-breaking bifurcations and representative firm in dynamic duopoly games. Annals of Operations Research 89:253–272
3. Brander JA, Spencer BJ (1983) Strategic commitment with R&D: the symmetric case. The Bell Journal of Economics 14:225–235
4. Pinto AA, Oliveira B, Ferreira FA, Ferreira M (2007) Investing to survive in a duopoly model. In: Machado JT, Patkai B, Rudas IJ (eds) Intelligent Engineering Systems and Computational Cybernetics. Springer, New York
5. Qiu DL (1997) On the dynamic efficiency of Bertrand and Cournot equilibria. Journal of Economic Theory 75:213–229
6. Singh N, Vives X (1984) Price and quantity competition in a differentiated duopoly. RAND Journal of Economics 15:546–554

Part VII

New Methods and Approaches

Defining Fuzzy Measures: A Comparative Study with Genetic and Gradient Descent Algorithms

Sajid H. Alavi[1], Javad Jassbi[2], Paulo J. A. Serra[3], and Rita A. Ribeiro[4]

[1] M.C.G.-Department of Artificial Intelligence, Pasdaran, Tehran, Iran
alavi@mcg-co.com
[2] Azad University, Science and Research Campus, Hesarak, Tehran, Iran
jassbi@iaucss.org
[3] UNINOVA-CA3, Campus Universidade Nova Lisboa, Caparica, 2829-516, Portugal
pja@uninova.pt
[4] UNINOVA-CA3, Campus Universidade Nova Lisboa, Caparica, 2829-516, Portugal
rar@uninova.pt

Summary. Due to limitations of classical weighted average aggregation operators, there is an increase usage of fuzzy integrals, like the Sugeno and Choquet integrals, as alternative aggregation operators. However, their applicability has been threatened by the crux of determining the fuzzy measures in real problems. One way to determine these measures is by using learning data and optimizing the parameters. In this paper we made a comparative study of two well known optimization algorithms, Genetic Algorithm and Gradient Descent to determine fuzzy measures. Two illustrative cases are used to compare the algorithms and assess their performance.

1 Introduction

Application of Fuzzy Integrals as aggregation operators in fuzzy inference systems and multicriteria decision making systems is increasingly recognized in recent years, due to many detected problems in using a classic aggregation method like weighted average [1–5]. The main weakness of classical additive aggregations, e.g. weighted average is that they do not consider redundancy and synergy in their model [4, 6]. In this paper we focus on Choquet integral [6,7] because it has the main advantage of modeling directly the interaction (synergy) between variables or information sources [7, 8]. Choquet aggregation operator has been applied in many areas such as Classification, Multi Criteria Decision Making, Data Modeling and recently in Fuzzy Inference Systems (FIS) [4, 6, 8, 9].

Although the motivation to use fuzzy integrals for aggregation is quite high, there is a problem these integrals are suffering from, i.e. how to determine the fuzzy measure that provides the weights for representing the synergies.

Determining a fuzzy measure is a complex task and a constraint for the usage of these aggregation operators. The main difficulty lies in determining the 2^n variable weights, which are exponentially increasing with the number of input parameters. In 1974 Sugeno [2,3] proposed a fuzzy measure and fuzzy integral. Since then researchers have been investigating ways of

J.A.T. Machado et al. (eds.), *Intelligent Engineering Systems and Computational Cybernetics,*

determining fuzzy measures, from heuristic algorithms such as Genetic Algorithms [10–12] and Neural Networks [13, 15], to classic algorithms like the Gradient Descent [4].

In this work we compare two algorithms, which are commonly used for identifying fuzzy measures, Genetic Algorithm (GA) and Gradient descent (GD)-as proposed by Grabisch [4, 13–15]. The reason we selected these algorithms is two folded: (a) their successful application in real cases [4, 8, 9]; (b) they represent two different approaches, one (GA) is a heuristic algorithm and the other (GD) belongs to classic optimization algorithms. We didn't compare with Neural Networks, because they use a similar optimization algorithm like the GD.

This paper is organized as follows. In Section 2 we provide some background on fuzzy integrals and fuzzy measures. In Section 3 both algorithms are introduced and finally in section IV we compare the results obtained with GA and GD in two illustrative cases.

2 Background on Fuzzy Integrals and Measures

As mentioned before, Sugeno proposed the first notion of fuzzy measure and fuzzy integral in 1974 [2]. Crisp measures and integrals is a very important field in mathematics, with many applications related with decision making [3, 7, 17]. The additive property of a crisp measure, whilst being its fundamental propriety, is also a source of inflexibility. The concept of fuzzy measure [2, 3, 16, 17] is an extension of crisp measures, where the additive propriety has been replaced by a monotonicity condition – a weaker requirement. An important aspect of determining fuzzy measures is that they can provide weights for representing synergies between: (a) criteria in multicriteria decision making [7]; (b) rules in fuzzy inference systems [9, 18] both for partial and complete sets, as well as to express positive or negative synergies between these criteria (rules).

2.1 Definition of Fuzzy Measures

Let N be a set of interacting criteria N={1,2,...,n} , and P(N) the power set of N. A function $\eta : \mathrm{P}(N) \rightarrow [0,1]$ is called a fuzzy measure if [3]:

$$\eta(\phi) = 0, \eta(N) = 1 \tag{1}$$

$$T \subseteq S \subseteq N \Rightarrow \eta(T) \leq \eta(S) \tag{2}$$

The difference between this measure and the classical, crisp measure is the fact that in a fuzzy measure the additivity propriety -$\eta(T \cup S) = \eta(T) + \eta(S)$, for every T and S such that $T \cap S = \emptyset$- of a crisp measure has been relaxed, originating a weaker requirement [6].

2.2 Definition of Choquet Integral

Fuzzy integrals are generally defined as operators on $[0,1]^n$; the definition presented here will be restricted to the case where n = 1 [6]. The so-called Choquet integral was first presented by Murofushi and Sugeno [2, 16] using a function defined by Choquet in capacity theory [18] and is defined by [6, 7]:

$$C_\eta(x_1, \ldots, x_n) \equiv \sum \left(\eta\left(A_{(i)}\right) - \eta\left(A_{(i+1)}\right)\right) \cdot x_{(i)} \tag{3}$$

where (.) indicates a reordering for the elements x_i such that $0 \leq x_{(1)} \leq \cdots \leq x_{(n)} \leq 1$, $A_{(i)} = \left\{x_{(i)}, \ldots, x_{(n)}\right\}$ and $A_{(n+1)} = \emptyset$, with η being a fuzzy measure.

2.3 Role of Fuzzy Measure in Choquet Integral

Let η be a measure defined in N, a set of criteria. In the context of fuzzy integrals, η is supposed to represent the importance that different subsets of N should take in the final result [7]. If N has n elements, then the values of the measure η must be determined for the 2^n possible subsets of N, a task that might prove to be difficult for high values of n . Moreover, Grabisch [7] states two important considerations to have in mind when choosing a fuzzy measure to use with a fuzzy integral: first, the global importance of an element $x_{(i)} \in N$ is not only determined by $\eta\left(\left\{x_{(i)}\right\}\right)$ but also by all $\eta(A)$ such that $x_{(i)} \in A$; the same question arises for larger sets of criteria.

2.4 Identification of Fuzzy Measures

A very important point in this context is how to determine the measure to be used with a fuzzy integral, i.e., how to determine the 2^n coefficients of the fuzzy measure. Grabisch presents three approaches [7]: identification based on semantics, identification based on learning data and a combination of both.

In this work we follow the second approach, identification of fuzzy measures based on learning data (in this work simulated data) because it allows the use of optimization algorithms to determine the measures. Assuming learning data is available, the coefficients of the fuzzy measure can be considered parameters, and their values can be obtained via the minimization of a certain error criterion. Formally, supposing that learning data $(\underline{z}_k, y_k), k = 1, \ldots, \ell$ exists, where $\underline{z}_k = (z_{k1}, \ldots, z_{kn})$ represents a vector of inputs, and y_k is the expected evaluation for $\underline{z}_k$, hence the parameters can be obtained by minimizing [7]:

$$E^2 = \sum_{k=1}^{\ell} \left(C_\eta\left(\underline{z}_k\right) - y_k\right)^2 \tag{4}$$

As mentioned before, when using the Choquet Integral instead of linear aggregation like weighted average methods, in information fusion, the most important thing is to determine the values of the fuzzy measure η. In this study we compare Genetic Algorithm (GA) and Gradient Descent (GD) algorithm (a modified version of Grabisch's [4]) to find how to determine a specific fuzzy measure.

First we consider $X = \{x_1, \ldots, x_n\}$ as information sources. Then we consider the Choquet Integral as a system with multi input and single output, as shown in Fig. 1 [5]. Notice that the scheme is equivalent to a FIS [5] where inputs are the antecedents of rules and the output is the aggregation of the rules outputs using the Choquet integral. To determine the fuzzy measures

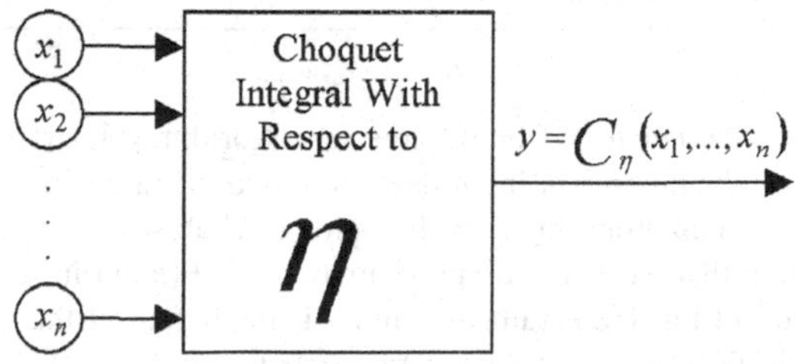

Fig. 1: Choquet integral as a multi input, one output system [10]

η , a suitable set of training data is required. We use the following $\ell \times (n+1)$ data matrix:

$$\begin{bmatrix} z_{11} & \cdots & z_{1n} & y_1 \\ \vdots & \ddots & \vdots & \vdots \\ z_{\ell 1} & \cdots & z_{\ell n} & y_\ell \end{bmatrix} \tag{5}$$

where each 'y_k' indicates the output of the system presented in Fig. 1 for each vector of n inputs.

To find the values of η we have to solve the following optimization problem, which is similar to (4) [12]:

$$\min e = \sqrt{\frac{1}{\ell}\sum_{k=1}^{\ell}(y_k - \hat{y}_k)^2} \tag{6}$$

Where $\hat{y}_k = C_\eta(\underline{z}_k), k = 1, \ldots, \ell$ is the output value of the system for a specific fuzzy measure η and y_k is the expected output of the system.

3 Optimization Algorithms: GA and GD

The Genetic Algorithm (GA) used in this work is an adaptive GA completely described in [12]. The parameters and fitness function used are described in Table 1. When we started testing

Table 1: Genetic Algorithm parameters

Error function: e=MSE
Fitness function: 2^{-5e}
Each gene resolution: m=10
Max(e),Min(e),Average(e) and Δe=Max(e)-Min(e)
Stopping criteria: $e < 2^{-2m}$ or $\Delta e < 2^{-3m}$
Adaptive mutation probability:0.3(1+b)
Adaptive crossover probability:0.4-0.3b
Realignment probability: 0.3
In above parameters: $a = \min(\frac{-\log_2(\max(e))}{m}, 1)$ $b = \min(\frac{-\log_2(\Delta e)}{m}, 1)$
Population size: S=100
Crossover type: two point
Mutation type: k-point (k is a random number)
Realignment type: three point

the Gradient Descent (GD) equation from Grabisch's algorithm [4], strange results were obtained, especially when the number of input data increased. So, after investigating the reason for these results we ended up proposing a modification to Grabisch equation. Let's clarify this problem by considering that we have an approximator $y = P(\mathbf{x}; \mathbf{w})$ in which $\mathbf{x}$ is the vector of inputs, $\mathbf{w}$ is a vector of tunable parameters and y is the output of the approximator. A set of input/output data is also taken into consideration as training data, having a matrix of the form: $[\mathbf{X}|\mathbf{Y}]_{m\times(n+1)}$. In this matrix, "X" is the set of all input vectors, each with n components; "**Y**" is the output values corresponding to each input vector of "**X**" and "m" is the number of training data. For training such an approximator (which consists in determining the parameters from the training data) we can use the gradient descent (GD) algorithm. For this algorithm the following formulae is used for tuning the parameters:

$$E = P(\mathbf{X}, \bar{\mathbf{w}}) - \mathbf{Y} \tag{7}$$

$$\bar{\mathbf{w}}^{k+1} = \bar{\mathbf{w}}^{k} - \alpha.E.\frac{\partial P(\mathbf{X}, \mathbf{w})}{\partial \mathbf{w}}\Big|_{\mathbf{w}=\bar{\mathbf{w}}^{k}} \tag{8}$$

Here, $\bar{\mathbf{w}}$ is the vector of parameters for the system that we want to attain, E is the error feature (for example mean square error on deviation from system output and desired output), α is the learning rate and k is the number of iteration. Equation (8) represents an iterative scheme in which step-by-step modification of the vector of parameters ($\mathbf{w}$), results in the reduction of the gradient magnitude of error surface. It means that by selecting a large enough value for k, we can make the error arbitrary close to its minimum value. As an example, considering a system with three inputs (n = 3) and assuming, without any loss of generality, that the values of the input vector of training data are in ascending order, we have:

$$\begin{aligned} C_{\eta}(x) &= x_1(\eta_{123} - \eta_{23}) + x_2(\eta_{23} - \eta_3) + x_3\eta_3 \\ &= \eta_{123}x_1 + \eta_{23}(x_2 - x_1) + \eta_3(x_3 - x_2) \end{aligned} \tag{9}$$

If we apply the GD modified formulae (7),(8) we have:

$$\begin{aligned} \eta_3^{k+1} &= \eta_3^{k} - \alpha.e.\frac{\partial C_{\eta}(x)}{\partial \eta_3} = \eta_3^{k} - \alpha.e.(x_3 - x_2) \\ \eta_{23}^{k+1} &= \eta_{23}^{k} - \alpha.e.\frac{\partial C_{\eta}(x)}{\partial \eta_{23}} = \eta_{23}^{k} - \alpha.e.(x_2 - x_1) \end{aligned} \tag{10}$$

and generally:

$$u^{k+1}(i) = u^{k}(i) - \alpha.e.\left(x_{(n-i+1)} - x_{(n-i)}\right) \tag{11}$$

While, Grabisch's formula [4] is:

$$u^{k+1}(i) = u^{k}(i) - \alpha.e.\left(x_{(n-i)} - x_{(n-i+1)}\right) \tag{12}$$

where $u(i)$ in both equations represents η.

Henceforth we will refer to formulas (11) and (12) as the unmodified Grabisch gradient descent algorithm (UGD) and the modified Grabish gradient descent algorithm (MGD) respectively.

It should be pointed that the difference in both formulations causes some difficulties in computing the fuzzy measures and comparing the two versions of the GD. Hence, in the next section we start by comparing the efficiency of the two versions of GD algorithm (UGD and MGD) with the Genetic Algorithm (GA), for the first example. For the second example we only compare GA and our MGD because it clearly seems to behave more appropriately than the UGD.

4 Algorithms' Performance Comparison

As mentioned before, in this work we tested two illustrative examples: the first was borrowed from Grabisch [4] and includes four inputs and the training data set is determined as described in [4] i.e. the input values for the system in Fig. 1 are the elements of the four-tuple set $\{0,0.5,1\}^4$. The second example has six inputs and we randomly generated the training data set. The initial measure values are set a priori and are shown in Table 2; the 3^6(=729) training data used for this case were generated in the same way as the training data for the case with four inputs.

4.1 Illustrative Example I

The initial values for the measure are shown in Table 2. By applying the three algorithms: the

Table 2: Initial measure values for Example I [4]

$\eta_1 = 0.1$	$\eta_{13} = 0.3235$	$\eta_{123} = 0.5$
$\eta_2 = 0.2105$	$\eta_{14} = 0.3235$	$\eta_{124} = 0.8667$
$\eta_3 = 0.2353$	$\eta_{23} = 0.7333$	$\eta_{134} = 0.8824$
$\eta_4 = 0.6667$	$\eta_{24} = 0.8070$	$\eta_{234} = 0.9474$
$\eta_{12} = 0.3$	$\eta_{34} = 0.8235$	

unmodified version of Grabisch's Gradient Descent (UGD); the modified version of Grabisch's Gradient Descent (MGD); and Genetic Algorithm (GA) – the following results were obtained:

Figs. 2–4 present the convergence diagram for determining fuzzy measures, using algorithms GA, UGD and MGD.

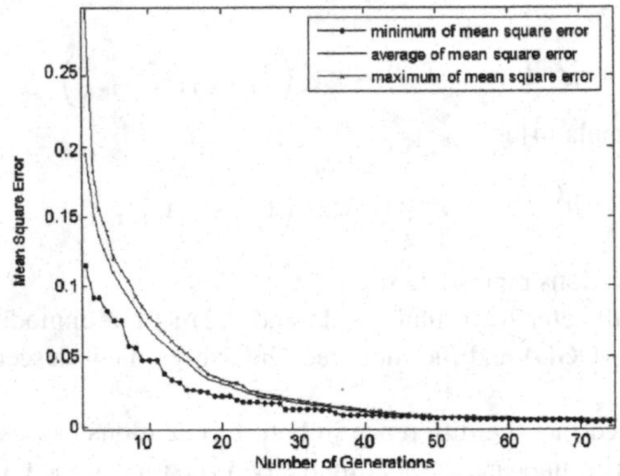

Fig. 2: Convergence diagram for example I with GA

Table 3 presents both the performance in terms of time and mean square error. The run time of each algorithm is based on a Pentium IV^{TM}-2.4 GHz with 1 MB of RAM. The error criterion calculated based on deviation between tuned system output and desired output. Table 3 shows that the MGD not only provides better results, when we consider the MSE, but also performs faster than the other two algorithms. GA and UGD have almost the same accuracy in obtaining fuzzy measures while GA is noticeably slower than than UGD.It can be concluded that MGD absolutely works better than GA and UGD both in run time and accuracy perspective.

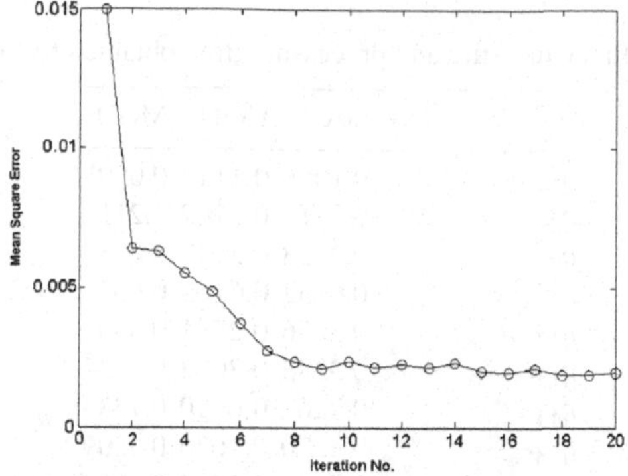

Fig. 3: Convergence diagram for example I with UGD

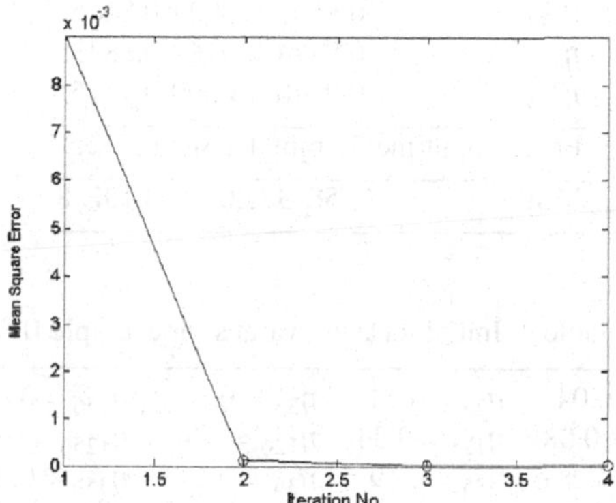

Fig. 4: Convergence diagram for example I with MGD

4.2 Illustrative Example II

This example includes 6 inputs. Table 4 depicts the initial fuzzy measures that were used as ideal values. These measures were chosen randomly and fulfils the properties of boundary conditions (1) and monotonicity (2).

We generated the training data to determine the ability of each algorithm in reproducing the measures of Table 4 by considering them as unknown variables. In this part we compared the results for GA and MGD.

Figure 5 shows the convergence diagram of GA while determining the fuzzy measures and Table 5 presents the values of them obtained by the GA. The run time for this algorithm was 8 h 43 min and the value of the MSE for the deviation between tuned system output and desired output is 0.169, which is quite acceptable.

Figure 6 and Table 6 present the same results for the modified Grabisch gradient descent Algorithm (MGD).

The values of measures obtained by MGD exactly match the values of the actual measures. The run time for the algorithm was 12.8 sec and the final valueof MSE deviation between

Table 3: Results for fuzzy measure and processing time obtained for the three algorithms

	GA	UGD	MGD
η_1	0.1018	0.1114	0.0998
η_2	0.2118	0.2767	0.2114
η_3	0.2353	0.2743	0.2353
η_4	0.6652	0.6169	0.6674
η_{12}	0.3056	0.2731	0.3004
η_{13}	0.3283	0.3083	0.3237
η_{14}	0.7266	0.6910	0.7333
η_{23}	0.4220	0.2636	0.4209
η_{24}	0.7974	0.7444	0.8067
η_{34}	0.8133	0.6169	0.8232
η_{123}	0.4999	0.5752	0.5000
η_{124}	0.8663	0.8924	0.8665
η_{134}	0.8824	0.9169	0.8823
η_{234}	0.9540	1.0000	0.9475
Processing time	27 min	1.7 sec	0.7 sec
MSE	3.5E-3	2.6E-3	3.13E-8

Table 4: Initial measure values for example II

$\eta_2 = 0.04$	$\eta_{25} = 0.14$	$\eta_{26} = 0.33$	$\eta_{256} = 0.48$
$\eta_{12} = 0.08$	$\eta_{125} = 0.34$	$\eta_{126} = 0.46$	$\eta_{1256} = 0.80$
$\eta_3 = 0.08$	$\eta_{35} = 0.19$	$\eta_{36} = 0.22$	$\eta_{356} = 0.43$
$\eta_{13} = 0.14$	$\eta_{135} = 0.26$	$\eta_{136} = 0.49$	$\eta_{1356} = 0.56$
$\eta_{23} = 0.22$	$\eta_{235} = 0.41$	$\eta_{236} = 0.58$	$\eta_{2356} = 0.71$
$\eta_{123} = 0.30$	$\eta_{1235} = 0.50$	$\eta_{1236} = 0.69$	$\eta_{12356} = 0.85$
$\eta_4 = 0.03$	$\eta_{45} = 0.26$	$\eta_{46} = 0.17$	$\eta_{456} = 0.61$
$\eta_{14} = 0.15$	$\eta_{145} = 0.34$	$\eta_{146} = 0.67$	$\eta_{1456} = 0.83$
$\eta_{24} = 0.26$	$\eta_{245} = 0.41$	$\eta_{246} = 0.37$	$\eta_{2456} = 0.76$
$\eta_{124} = 0.32$	$\eta_{1245} = 0.57$	$\eta_{1246} = 0.88$	$\eta_{12456} = 0.97$
$\eta_{34} = 0.14$	$\eta_{345} = 0.28$	$\eta_{346} = 0.67$	$\eta_{3456} = 0.73$
$\eta_{134} = 0.32$	$\eta_{1345} = 0.51$	$\eta_{1346} = 0.75$	$\eta_{13456} = 0.84$
$\eta_{234} = 0.72$	$\eta_{2345} = 0.90$	$\eta_{2346} = 0.81$	$\eta_{23456} = 0.91$
$\eta_{1234} = 0.73$	$\eta_{12345} = 0.99$	$\eta_{12346} = 0.89$	
$\eta_5 = 0.12$	$\eta_6 = 0.03$	$\eta_{56} = 0.41$	

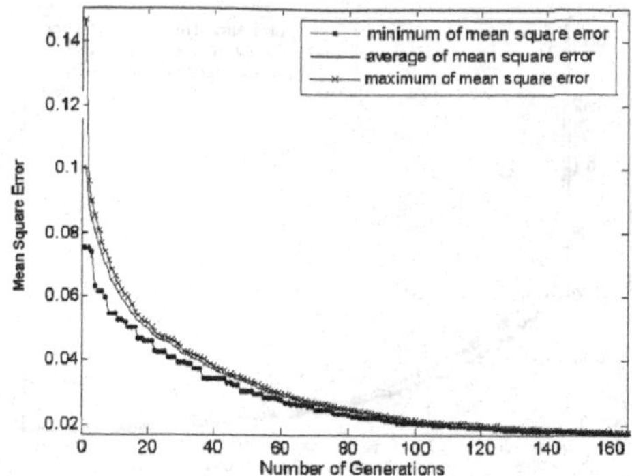

Fig. 5: Convergence diagram for example II with GA

Table 5: Obtained GA fuzzy measure for example II

$\eta_1 = 0.0684$	$\eta_{15} = 0.1397$	$\eta_{16} = 0.4414$	$\eta_{156} = 0.4883$
$\eta_2 = 0.0371$	$\eta_{25} = 0.1221$	$\eta_{26} = 0.2949$	$\eta_{256} = 0.5117$
$\eta_{12} = 0.0850$	$\eta_{125} = 0.3418$	$\eta_{126} = 0.4492$	$\eta_{1256} = 0.8174$
$\eta_3 = 0.0566$	$\eta_{35} = 0.1797$	$\eta_{36} = 0.2227$	$\eta_{356} = 0.5117$
$\eta_{13} = 0.1279$	$\eta_{135} = 0.2617$	$\eta_{136} = 0.5176$	$\eta_{1356} = 0.5684$
$\eta_{23} = 0.2568$	$\eta_{235} = 0.3897$	$\eta_{236} = 0.5801$	$\eta_{2356} = 0.6738$
$\eta_{123} = 0.3154$	$\eta_{1235} = 0.5371$	$\eta_{1236} = 0.7012$	$\eta_{12356} = 0.8447$
$\eta_4 = 0.0303$	$\eta_{45} = 0.2471$	$\eta_{46} = 0.2510$	$\eta_{456} = 0.6367$
$\eta_{14} = 0.1504$	$\eta_{145} = 0.2481$	$\eta_{146} = 0.6631$	$\eta_{1456} = 0.8066$
$\eta_{24} = 0.2363$	$\eta_{245} = 0.4111$	$\eta_{246} = 0.3691$	$\eta_{2456} = 0.7539$
$\eta_{124} = 0.3076$	$\eta_{1245} = 0.5762$	$\eta_{1246} = 0.8580$	$\eta_{12456} = 0.9981$
$\eta_{34} = 0.1377$	$\eta_{345} = 0.2676$	$\eta_{346} = 0.6611$	$\eta_{3456} = 0.7627$
$\eta_{134} = 0.3154$	$\eta_{1345} = 0.7119$	$\eta_{1346} = 0.7490$	$\eta_{13456} = 0.8106$
$\eta_{234} = 0.7012$	$\eta_{2345} = 0.8711$	$\eta_{2346} = 0.8115$	$\eta_{23456} = 0.8818$
$\eta_{1234} = 0.7393$	$\eta_{12345} = 0.9922$	$\eta_{12346} = 0.8880$	
$\eta_5 = 0.1162$	$\eta_6 = 0.0224$	$\eta_{56} = 0.4190$	

tuned system output and desired output is 8.01E-12. The results show again that MGD method is more precise and radically faster than the GA. Furthermore, for cases with many inputs, GA runtime increases exponentially and so its utilization becomes almost unviable.

5 Conclusions

In this paper we compared algorithms for calculating, automatically, fuzzy measures. Our aim was to select the most efficient algorithm to enable a successful usage of Choquet integral aggregation in fuzzy inference systems.

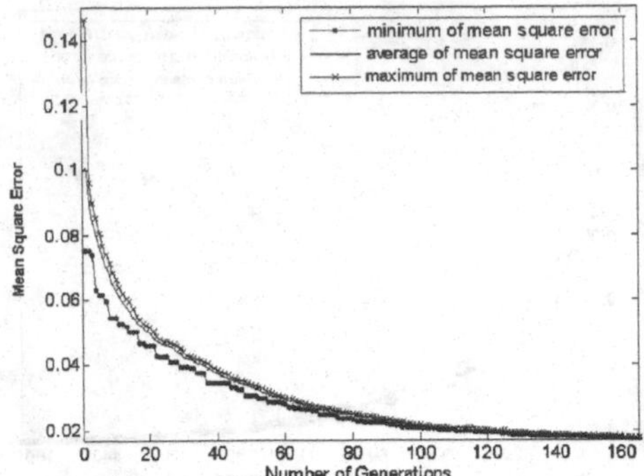

Fig. 6: Convergence diagram for example II with MGD

Table 6: Initial measure values for example II

$\eta_1 = 0.0699$	$\eta_{15} = 0.15$	$\eta_{16} = 0.42$	$\eta_{156} = 0.49$
$\eta_2 = 0.0399$	$\eta_{25} = 0.14$	$\eta_{26} = 0.33$	$\eta_{256} = 0.48$
$\eta_{12} = 0.0799$	$\eta_{125} = 0.34$	$\eta_{126} = 0.46$	$\eta_{1256} = 0.80$
$\eta_3 = 0.0799$	$\eta_{35} = 0.19$	$\eta_{36} = 0.22$	$\eta_{356} = 0.43$
$\eta_{13} = 0.14$	$\eta_{135} = 0.26$	$\eta_{136} = 0.49$	$\eta_{1356} = 0.56$
$\eta_{23} = 0.22$	$\eta_{235} = 0.41$	$\eta_{236} = 0.58$	$\eta_{2356} = 0.71$
$\eta_{123} = 0.30$	$\eta_{1235} = 0.50$	$\eta_{1236} = 0.69$	$\eta_{12356} = 0.85$
$\eta_4 = 0.0299$	$\eta_{45} = 0.26$	$\eta_{46} = 0.17$	$\eta_{456} = 0.61$
$\eta_{14} = 0.15$	$\eta_{145} = 0.34$	$\eta_{146} = 0.67$	$\eta_{1456} = 0.83$
$\eta_{24} = 0.26$	$\eta_{245} = 0.41$	$\eta_{246} = 0.37$	$\eta_{2456} = 0.76$
$\eta_{124} = 0.32$	$\eta_{1245} = 0.57$	$\eta_{1246} = 0.88$	$\eta_{12456} = 0.97$
$\eta_{34} = 0.14$	$\eta_{345} = 0.28$	$\eta_{346} = 0.67$	$\eta_{3456} = 0.73$
$\eta_{134} = 0.32$	$\eta_{1345} = 0.71$	$\eta_{1346} = 0.75$	$\eta_{13456} = 0.84$
$\eta_{234} = 0.72$	$\eta_{2345} = 0.87$	$\eta_{2346} = 0.81$	$\eta_{23456} = 0.91$
$\eta_{1234} = 0.73$	$\eta_{12345} = 0.99$	$\eta_{12346} = 0.89$	
$\eta_5 = 0.12$	$\eta_6 = 0.03$	$\eta_{56} = 0.41$	

Unmodified (original) Grabische Gradient Descent (UGD) algorithm shows some instability in converging to minimum MSE. Hence we proposed a modification on it and showed that the modified version (MGD) is noticeably accurate and a little faster than the unmodified one.

The comparison between GA and modified Grabisch Gradient Descent Algorithm (MGD) showed that the latter not only performs better in terms of processing time but also performs better regarding the MSE. The comparison also showed that when the number of inputs as well as the set of training data increases, MGD was able to find better results in less time.

In the near future we intend to compare the application of the MGD and GA to a real space monitoring problem to assess the performance on the system with both algorithms.

References

1. Detyniecki M (2001) Fundamentals on aggregation operators. In: Proceedings AGOP', Asturias, Spain, July
2. Murofushi T, Sugeno M (1989) An interpretation of fuzzy measure and the Choquet integral as an integral with respect to a fuzzy measure. In: Fuzzy Sets and Systems 29:201–227
3. Murofushi T, Sugeno M (2000) Fuzzy measures and fuzzy integrals. In: Fuzzy Measures and Integrals: Theory and Applications. Physica-Verlag:3–41
4. Grabisch M (1995) A new algorithm for identifying fuzzy measure and its application to pattern recognition. In: Proceedings of the 1995 4th IEEE International Conference on Fuzzy Systems
5. Mendel JM (2001) Uncertain Rule-Based Fuzzy Inference Systems: Introduction and New Directions. Prentice-Hall, Inc.
6. Grabisch M (1995) Fuzzy integral in multicriteria decision making. In: Fuzzy Sets and Systems 69:279–298
7. Grabisch M (1996) The application of fuzzy integrals in multicriteria decision making. In: European Journal of Operational Research 89:445–456
8. Grabisch M (2003) Modeling data by the Choquet integral. In: Information Fusion in Data Mining. Springer, Berlin/Heidelberg/New York
9. Serra P, Ribeiro RA, Marques-Pereira R, Steel R, Niezette M, Donati A (2008) Fuzzy Thermal Alarm System for venus Express. In: Encyclopedia of Decision Making and Decision Support Technology, Editors: Frederic Adam and Patrick Humphreys. Publisher: Information Science Reference, Vol I: 391–401
10. Wang Z, Leung K, Wang J (1999) A genetic algorithm for determining nonadditive set functions in information fusion. Fuzzy Sets and Systems 102–103:436–469
11. Chen T, Wang J (2001) Identification of λ-fuzzy measures using sampling design and genetic algorithms. Fuzzy Sets and Systems 123-3:321–341
12. Wang Z, Leung Wong K, Wang J (2000) A new type of nonlinear and the computational algorithm. Fuzzy Sets and Systems 112-2:223–231
13. Wang J, Wang Z (1997) Using neural networks to determine Sugeno measures by statistics. Neural Networks 10-1:183–195
14. Wang J, Wang Z (1996) Detecting constructions of nonlinear integral systems from input-output data: An application of neural network. In: Proceedings of the 1996 Biennial Conference of the North America (Fuzzy Information Processing Society), NAF1PS'96, Berkeley, California, USA:559–563
15. Wang Z, Klir GJ (1991) Fuzzy Measure Theory. Plenum, New York
16. Murofushi T, Sugeno M (1991) A theory of fuzzy measures. Representation, the Choquet integral and null sets. Journal Mathematical Analysis Applications 159:532–549
17. Marques Pereira R, Ribeiro RA, Serra P (2006) Rule correlation and Choquet integration in fuzzy inference systems. In: International Journal of Uncertainty, Fuzziness and Knowledge-Based Systems (submitted)
18. Choquet G (1953) Theory of capacities. Annals Institute Fourier 5:131–295

A Quantum Theory Based Medium Access Control for Wireless Networks

Márton Bérces[1] and Sándor Imre[2]

[1] Department of Telecommunications, Budapest University of Technology and Economics, Magyar Tudósok krt. 2, H-1117 Budapest
berces@hit.bme.hu

[2] Department of Telecommunications, Budapest University of Technology and Economics, Magyar Tudósok krt. 2, H-1117 Budapest
imre@hit.bme.hu

Summary. Medium Access Control (MAC) is an important part of wireless telecommunication systems. The main goal of a MAC protocol is to provide the best usage of the common resources for the users. One of these resources is typically the communication channel. By quantum informatics and computation – that gain more and more attention – some calculations and algorithms may become more efficient. The possible implementation of a quantum based system would lead us to great benefits, by applying it to an already existing problem. In this paper we give a model for medium access control via quantum methods.

1 Introduction

Since telecommunications has become part of everyday life, it has generated a lot of interesting questions and thus it is a great topic for researchers. On the other hand the implementation of a quantum based computing system is not yet to come. However the theoretical results of quantum computing are impressive, only a few quantum based system are manufactured and available on the market.

In this paper a possible solution is proposed for accessing the communication's medium in wireless network via quantum methods. After a short introduction the basics of nowadays used MAC layers are described. Questions like: "Why it is necessary to have a MAC layer at all? What are the tasks of it and what benefits does a user have from it?" are answered. We also present the changes that are required to have quantum based channel access control. In the third section the rules of the "quantum world" are presented. Basic principles are described that are used for every calculation. Some already existing result is mentioned and some of the benefits of a possible implementation are listed. After that the quantum based MAC model is introduced and analyzed. The paper is closed by conclusions.

2 Overview of the MAC Protocols

Medium Access Control required in a multi-user environment where the users share the same medium. That medium today is mostly the radio channel. The wireless connections have

J.A.T. Machado et al. (eds.), *Intelligent Engineering Systems and Computational Cybernetics,*

become really easy to handle for the user, but the technology behind them is constantly developing. In a conference room or a cafe bar where a public "hotspot" is available, the MAC protocol handles the usage of the access point (AP). The most commonly used MAC protocol is defined in the IEEE 802.11 standard. There are several papers available in the topic. They from a wide range of style. Surveys, introductions, reviews like [1, 2] and specialized ones too. Routing [3], QoS, etc. are also interesting topics considering ad-hoc networks.

MAC protocols can be categorized according to their centralized or distributed nature. When a distributed MAC is used, all the nodes act according to certain rules, they observe the channel and individually decide when to transmit. Centralized MAC is used when no circumstance is against a pre-defined entity (usually an Access Point (AP)) that assigns transmission rights to the users according to their previous requests. In this paper we consider the centralized scheme when two nodes and an access point are present at the same time, and they form the cell that medium access should be controlled. There are some well known problems considering this issue like the "hidden terminal problem". These can be solved usually at the cost of resources. (e.g. Medium is used also for control, and during that the users can not send or receive.) The main goal is to maximize the accessible time of the radio channel by minimizing the time used for control and yet providing an efficient and fair access to all nodes.

The 802.11 standard defines the tasks of the MAC sublayer: medium access, fragmentation, security. Beside the MAC sublayer a MAC Layer Management entity can handle synchronization, power management, roaming, etc. The position of the MAC protocol in the layer structure is shown in Fig. 1.

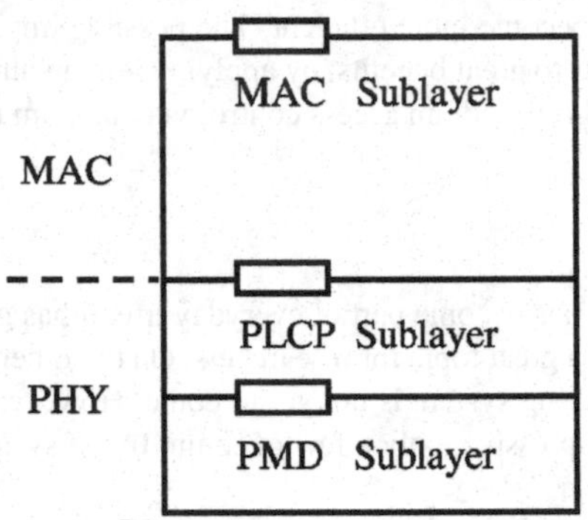

Fig. 1: Position of the MAC layer

The Physical Medium Dependent Sublayer (PMD) and the Physical Layer Convergence Protocol (PLCP) form the physical layer. PLCP converts the MAC frames into physical packets and those are modulated and coded by the PMD. Above the MAC layer are the upper layers (network, application, etc.) that provide more service to the user but here are not dealt with.

We consider a time slot system, where nodes transmit in certain time intervals. In this case synchronization is required between the terminals and the AP. Since the synchronization is solved we use that and other features of the already existing protocol. The channel access is changed, for that we will use quantum methods. By this the layer structure of the protocol stack is modified as shown in Fig. 2. The new layer will take care of the transmission rights. The new layer is called the Quantum Channel Access Sublayer. Although this structure looks similar to the one shown in Fig. 1, it differs from that in many ways. Hence a possible

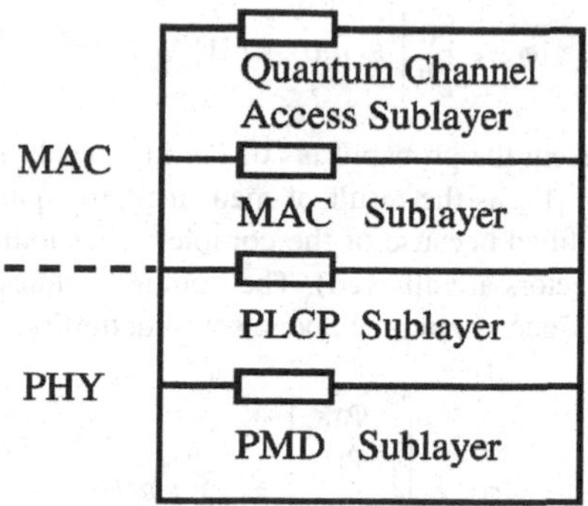

Fig. 2: Modified layer structure

quantum based communication is considered, the layers under the quantum channel access sublayer should all be quantum based. In the next section the basic rules of the quantum world are presented.

3 Basics of Quantum Computing and Communications

With quantum computing and communications gaining more attention many papers [4, 5] and books [6, 7] come out. It may be a possible solution for solving problems that are expectable from reaching the borders of the current technology in producing computing units according to Moore's law. Thus, it is important to understand the basics of quantum computing.

The basic unit of information in quantum computing is the qubit. It is like a classical bit in a way but it can "carry" more information than a regular bit. It has two basis states and the overall state of the qubit is some kind of superposition of the base states. The qubit is used to be compared to a coin. The bases are the sides and the superposition is when the coin is thrown up. When in air, it is in the superposition of the two sides. We do not know which side is up and which one is down. When the coin reaches the ground and stops spinning that can be matched to the measurement in the quantum world. With a certain probability we will see one side face up.

The already existing results of quantum informatics are impressive: Grover's algorithm provides a search in an unsorted database with $O(\sqrt{n})$ number of iterations. (The number of elements in the set is n.) We have already some results in our hands on secure key distribution [8, 9, 10] and signal processing [11, 12, 13] proving that quantum computing based solutions may overcome the classical ones.

From our "computational point of view" the principles of the "quantum world" can be easily summarized as follows. (These principles are also called the "postulates of the quantum mechanics".)

(1) Each state of a closed system can be described by means of a vector in a Hilbert space. In quantum computing the state vectors are denoted as $|\varphi\rangle$ (say 'ket phi' according to Dirac). The coordinates of $|\varphi\rangle$ are complex numbers and each of them refers to the probability amplitude of the associated basis (classical) state. The simplest quantum system is the so called qubit which replaces in the quantum world the classical information bearing unit, the bit. A qubit can be prepared in a two dimensional superposition of both classical bit values (0 and 1) as

$$|\varphi\rangle = \begin{bmatrix} a \\ b \end{bmatrix} = a|0\rangle + b|1\rangle \;\; a,b \in C \tag{1}$$

where $|a|^2$ and $|b|^2$ represent the probabilities of getting the classical bit values (orthonormal basis states) $|0\rangle$ and $|1\rangle$ as the result of measuring the qubit respectively. Of course $|a|^2 + |b|^2 = 1$ shall be fulfilled because of the complete probability law of probability theory (i.e. only unit length vectors are allowed). The conjugate transpose of $|\varphi\rangle$ is denoted by $\langle\varphi| = (|\varphi\rangle^*)^T = \left[a^* \; b^*\right]$. Generalizing the above set of definitions a quregister consisting of n qubits is described as

$$|\varphi\rangle = \begin{bmatrix} \varphi_0 \\ \varphi_1 \\ \vdots \\ \varphi_{2^n-1} \end{bmatrix} = \sum_{i=0}^{2^n-1} \varphi_i |i\rangle. \tag{2}$$

Based on (2) we emphasize that the above quregister contains all the $2^n - 1$ basis states (i.e. classical integer number) at the same time, which can be regarded as the source of quantum parallelism.

The inner (scalar) product of two states is denoted and defined as

$$\langle\varphi|\psi\rangle = \sum_{i=0}^{2^n-1} \varphi_i^* \cdot \psi_i. \tag{3}$$

Two vectors are orthogonal if and only if $\langle\varphi|\psi\rangle = 0$ and are identical if and only if $\langle\varphi|\psi\rangle = 1$. The outer (matrix) product of two states is denoted and defined as

$$|\varphi\rangle\langle\psi| = A = \begin{bmatrix} A_{00} & \cdots & A_{0(2^n-1)} \\ \vdots & \ddots & \vdots \\ A_{(2^n-1)0} & \cdots & A_{(2^n-1)(2^n-1)} \end{bmatrix}, \qquad A_{ij} = \varphi_i \cdot \psi_j^*. \tag{4}$$

(2) Having explained how to describe the system, now we present how to calculate its evolution. In quantum computing only a special type of linear operators can be used, namely the unitary operator i.e. the inverse of its matrix its adjoint (conjugate transpose) $U^{-1} = U^\dagger$. So each quantum gate can be handled as an $n \times n$ matrix and the state of the system at the output of the gate $|\varphi_{out}\rangle = U|\varphi_{in}\rangle$.

(3) The only exception under the unitarity principle of quantum gates is the measurement device which does not suffer from this restriction. Each measurement can be defined by means of a set of measurement operators $\{M_m\}$, where m refers to the different measurement outcomes. These operators can be imagined as they were in the same "gate or box" and all of them act upon measurement. Then the measured system evolves into its next state according to the measurement probabilities. Hence the mathematical representation of a measurement operator is not always unitary the measurement is not always reversible. The probability of getting a measurement result m assuming $|\varphi_{in}\rangle$ as an input state equals to

$$P(m) = \langle\varphi_{in} M_m^\dagger | M_m \varphi_{in}\rangle \tag{5}$$

and the state of the system after the measurement is

$$|\varphi_{out}\rangle = \frac{M_m|\varphi_{in}\rangle}{\sqrt{\langle\varphi_{in} M_m^\dagger | M_m \varphi_{in}\rangle}}. \tag{6}$$

The measurement operator set needs to fulfill the following condition too:

$$\sum_m M_m^\dagger M_m = I. \tag{7}$$

So the measured system will not disappear we will know its state after measurement with certain probability.

(4) The last rule provides how to merge quantum systems (e.g. qubits into a quregister) together. We shall use the tensor product ($\otimes$) between the state vectors of the individual systems. E.g. if we have two qubits $|\varphi_1\rangle = a_1|0\rangle + b_1|1\rangle$ and $|\varphi_2\rangle = a_2|0\rangle + b_2|1\rangle$ then the state of the quregister consisting of these two qubits is

$$|\varphi_1\varphi_2\rangle = |\varphi_1\rangle \otimes |\varphi_2\rangle = a_1a_2|00\rangle + b_1a_2|10\rangle + a_1b_2|01\rangle + b_1b_2|11\rangle. \tag{8}$$

The tensor product preserves the unit length for the qregister in accordance with Postulate 1. The notation $\otimes$ is often omitted or replaced by '$\cdot$'. These were the principles our calculations will be based on. In the next section the quantum based MAC protocol is described.

4 A New Model for Medium Access Control (MAC) by Quantum Methods

In this section a new model for Medium Access Control is presented. We consider a simple wireless network that consists of an Access Point (AP) and two nodes. The network is centralized, the AP controls the medium. The system uses time slots to operate, meaning that each participant is allowed to transmit only during a certain time interval. In the first subsection, calculations behind accessing one time slot are presented, after that a possible application of that idea is described in order to create an efficient MAC protocol.

4.1 Use of Entangled Pairs to Provide Access for One Slot

The idea behind a more efficient medium access control is the property of the entangled pairs. When an entangled pair is separated (split into half) then the two parts of it behave as they were one. So when one part is measured then the other part (no matter how far away it is) evolves into the same state as the measured one. Knowing this we try to create a MAC protocol for the following setup. There is a given access point (AP) that has controlling capabilities and there are two nodes that would like to use the same resource, the radio channel. Let us assume that the setup is a time slotted system, so a node or the AP can use the channel only for one slot. There are several well known protocols that are based on this. A qubit based MAC protocol could be the following.

(1) The AP sends entangled pairs to the nodes. The entangled pairs can be in the superposition of two following states

$$|0\rangle = \begin{bmatrix} 1 \\ 0 \end{bmatrix} \quad , \quad |1\rangle = \begin{bmatrix} 0 \\ 1 \end{bmatrix}. \tag{9}$$

So a general entangled pair looks like

$$|\varphi\rangle = a|01\rangle + b|10\rangle. \tag{10}$$

(2) Both nodes receive one part the entangled pair. Coefficients a and b will be determined later.

(3) Before a node uses a time slot to send data to the AP it needs to perform a measurement on its entangled pair. If the measurement gives an appropriate result the node can start sending data. How should be the measurement created so it can provide right properties for a task described above? After measuring a qubit, its state will change immediately. So when two measurements are done on the same system, the first one is actually done on the original system but the other one measures the system's state after the first measurement. Our goal is to create a system that can provide us a certain method of giving one of the nodes the access to the channel. Thus we need to construct measurement operators that work with probability 1. This will be achieved through the entanglement. When the first measurement occurs, then the state of the qubit changes immediately. After this the other node measures it. Both nodes needs to know exactly too, who will use the channel in the next time slot. Let us see a simple example for that. Let two nodes' (node A and B) measurement operator be the following matrix set

$$M_{01} = |01\rangle\langle 01| = \begin{bmatrix} 0&0&0&0 \\ 0&1&0&0 \\ 0&0&0&0 \\ 0&0&0&0 \end{bmatrix} , \qquad M_{10} = |10\rangle\langle 10| = \begin{bmatrix} 0&0&0&0 \\ 0&0&0&0 \\ 0&0&1&0 \\ 0&0&0&0 \end{bmatrix} . \tag{11}$$

These are the basic projective measurement operators. These are created as the outer product of each base state. The probability of getting a right result after measuring $|\varphi\rangle$

$$P(m=01|\varphi) = \left\langle \varphi^\dagger M_{01}^\dagger \,\middle|\, M_{01}\varphi \right\rangle = |a|^2 \tag{12}$$

$$P(m=10|\varphi) = \left\langle \varphi^\dagger M_{10}^\dagger \,\middle|\, M_{10}\varphi \right\rangle = |b|^2. \tag{13}$$

Since we want to be sure about the result of the measurement we need to choose a and b properly. Of course we can choose any complex number pair that satisfies the following restriction: $|a|^2 + |b|^2 = 1$. Let us consider the case when we have found a and b so the probability of a right measurement is one. The measurement box "contains" all the measurement operators and the measured qubit evolves according to the probabilities. Let $|\varphi_0\rangle$ denote the case when operator M_{01} gives us a certain measurement result. Thus we will know exactly the state after measurement. After a right measurement $|\varphi_0\rangle$ evolves into the following state:

$$|\varphi_0\rangle = \frac{M_{01}|\varphi_0\rangle}{\sqrt{1}} = |\varphi_0\rangle . \tag{14}$$

So $|\varphi_0\rangle$ remains the same after measurement. As a checking let us see the probability of $|\varphi_0\rangle$ being in state $|\varphi_1\rangle$:

$$P(m=10||\varphi\rangle) = \left\langle \varphi_0^\dagger M_{10}^\dagger \,\middle|\, M_{10}\varphi_0 \right\rangle = 0. \tag{15}$$

Until now we considered $|\varphi_0\rangle$ only. What changes does a measurement do to $|\varphi_1\rangle$? It can be derived that measuring $|\varphi_1\rangle$ gives us the result opposite to the results in case of $|\varphi_0\rangle$. We find that $|\varphi_1\rangle$ is in state $|0\rangle$ with 0 probability. This also means that $|\varphi_1\rangle$ is certainly in state $|1\rangle$ .Thus $|\varphi_1\rangle$ also remains the same after measurement. Let us summarize the expressions above! We created a measurement that leaves the qubit in its original state. We achieved this by the projective measurement operators for the orthogonal base states. When both nodes has a part of an entangled pair with base states $|0\rangle$, $|1\rangle$ each can measure it at the same time getting the same result. This way both nodes know when to use the channel after a measurement. For example when the result is $|0\rangle$ then node A can send or receive. When the measured state is $|1\rangle$ then node B has the opportunity to use the next time slot. In this section the method to access one slot was described. Using this in the next section the MAC protocol is introduced.

4.2 Description of the MAC Protocol

Using the results above, algorithms for medium access control can be developed. We consider only the access of the radio channel. Other functions remain in the "original" MAC sublayer. As said earlier we have an Access Point (AP) and two nodes. Let us assume that both nodes can communicate with the access point. Let the timeslot be for example 20 ms long, so in a second there is 50 time slots. In each first slot the AP sends to the nodes the entangled pairs. The nodes measure it, they get the result and will know, which one of them can use the next time slot. After the sending slot, again a reserved slot from the AP is to come.

Of course the terminals may require an extra uplink in order to demand slots. This can happen for example at every second. In the first two slots (because we have two nodes) both send their request to the AP. In this packet each states their requirements for the next second. When node A have a downlink stream, then the AP can decide to give every second slot or two third of the slots to node A. When the total requests exceed the maximum number of usable slots for the next second the AP can decide which node's request is to be partially rejected? If the nodes have a rule that states not to request more then half of the slots, then this problem doesn't exist but, may be unused slots. During a second, the number of packets a node can use is constant. In this scheme the access to the slots are shown in Figure 3. In every second

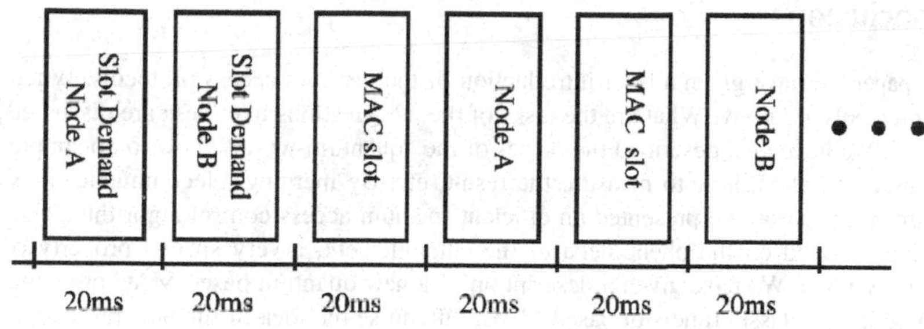

Fig. 3: Slot allocation

there are 50 slots as described before. Twenty-six of this is reserved for the MAC protocol, the other 24 can be distributed among the two nodes. So this case the 48% of the time slots are dedicated to the nodes, in other words 48% of the total bandwidth can be utilized. This result applies to the case when we have two nodes. Increasing the time slot demand period (that is 1 s now), the effectiveness is convergent to 50% upon the period goes to infinity. The AP can send whenever it intends to by not sending any entangled pair to the nodes. So the nodes have nothing to measure and they will not have access to the next slot.

5 Analysis and Comparison to Classical MAC

In this chapter the efficiency of the introduced new method is analyzed. We will investigate the number of usable time slots. First we consider the "slotted ALOHA" with two nodes. The nodes are to send immediately when they have data. If collusion occurs then both wait for a randomized number of time slots and then send again. We consider the case when they want to send in each slot with

$$p = 0,24. \tag{16}$$

We use this probability because the quantum based algorithm provides 48% effectiveness, and when we examine the system for a "long time" then this probability is measured for each node. The number of nodes is:

$$N = 2. \tag{17}$$

In each slot the probability of a successful transmission is:

$$S = p \cdot (1 - p) = 0,1824. \tag{18}$$

This applies to both nodes. Based on this the expected value of the slots that have a successful transmission:

$$E = 50 \cdot 0,1824 = 9,12. \tag{19}$$

This applies for a one second time interval. Because with quantum channel access each node gets 12 slots a second, this means that we have almost three slots less in each second that can be available for transmission. Under less load the slotted aloha is more effective, because if the transmission probability is low then the node can immediately transmit the probability of collusion is very low. In our new algorithm each node needs to wait until its next slot.

6 Conclusions

In this paper we have given a brief introduction of the medium access protocols. Where the MAC protocols are used? What are the tasks of them? Questions like these are discussed and answered. We have also described the basics of the "quantum-world", how to do simple calculations with qubits, how to measure the result, etc. By merging telecommunications and quantum computation we presented an efficient medium access control algorithm. The new algorithm is based on the phenomena of the entanglement, a very special property of the quantum systems. We have given a description of a new quantum based MAC protocol, and compared it to a classic time-slot based MAC. Of course the idea of the new protocol arises some questions. For example: How can a non-time slot MAC be created, based on the entanglement? We have shown that a quantum method can be more efficient in handling the transmission opportunities. Future work will cover the following areas, where a quantum based algorithm may be more effective than the classical ones: connection to a cell, routing, more dynamic slot allocation, etc.

References

1. Kumar S, Vinett S, Raghavan JD (2006) Medium Access Control Protocols for Ad Hoc Wireless Networks: A Survey. Ad Hoc Networks 4:326–358
2. Ye W, Heidemann J (2003) Medium Access Control in Wireless Sensor Networks. USC/ISI Technical Report, October 2003
3. Royer EM, Toh CK (1999) A Review of Current Routing Protocols for Ad Hoc Mobile Wireless Networks. IEEE Personal Communication 6(2):46–55
4. Imre S, Quantum Computing Based Feedback Channel Coding for Medium Access Control (2001) International Symposium on 3rd Generation Infrastructure and Services. Athens, Greece
5. Raussendorf R, Browne DE, Briegel HJ (2002) The One-way Quantum Computer – A Non-network Model of Quantum Computation. Journal of Modern Optics 49:1299

6. Imre S, Balázs F (2005): Quantum Computing and Communication: An Engineering Approach. John Wiley & Sons, Inc., Hoboken, NJ, ISBN: 9780470869031
7. Nielsen MA, Chuang IL (2003) Quantum Computation and Quantum Information. Cambridge University Press, Cambridge
8. Mullins J (2002) Making Unbreakable Codes. IEEE Spectrum 05:40–45
9. Bourenane M, Ljunggren D, Karlsson A, Jonsson P, Hening A, Ciscar JP (2000) Experimental Long Wavelength Quantum Cryptography: From Single-Photon Transmission to Key Extraction Protocols. Journal of Modern Optics (2/3):563–579
10. Naik DS, Peterson CG, White AG, Berglund AJ, Kwiat PG (2000) Entangled state quantum cryptography: Eavesdropping on the Ekert protocol. Physics Review Letters 84:4733
11. Eldar YC, Oppemheim AV (2002) Quantum Signal Processing. IEEE Signal Processing Magazine 19(6):12–32
12. Imre S, Balázs F (2002) Quantum Multi-User Detection. Applications and Services in the Wireless Networks. HERMES Penton Ltd, London, ISBN1-9039-9630-9:126–133
13. Imre S, Balázs F (2002) Non-coherent Multi-User Detection Based on Quantum Search. In: IEEE International Conference on Communications, April 28–May 2, 2002, New York, ISBN 0-7803-7400-2

A Concept for Optimizing Behavioural Effectiveness & Efficiency

Jan Carlo Barca, Grace Rumantir, and Raymond Li

Department of Information Technology, Monash University, Melbourne, Australia
Jcbar1@student.monash.edu.au,
Grace.Rumantir@infotech.monash.edu.au

Summary. Both humans and machines exhibit strengths and weaknesses that can be enhanced by merging the two entities. This research aims to provide a broader understanding of how closer interactions between these two entities can facilitate more optimal goal-directed performance through the use of artificial extensions of the human body. Such extensions may assist us in adapting to and manipulating our environments in a more effective way than any system known today. To demonstrate this concept, we have developed a simulation where a semi interactive virtual spider can be navigated through an environment consisting of several obstacles and a virtual predator capable of killing the spider. The virtual spider can be navigated through the use of three different control systems that can be used to assist in optimising overall goal directed performance. The first two control systems use, an onscreen button interface and a touch sensor, respectively to facilitate human navigation of the spider. The third control system is an autonomous navigation system through the use of machine intelligence embedded in the spider. This system enables the spider to navigate and react to changes in its local environment. The results of this study indicate that machines should be allowed to override human control in order to maximise the benefits of collaboration between man and machine. This research further indicates that the development of strong machine intelligence, sensor systems that engage all human senses, extra sensory input systems, physical remote manipulators, multiple intelligent extensions of the human body, as well as a tighter symbiosis between man and machine, can support an upgrade of the human form.

1 Introduction

Machines are used to perform dangerous and repetitious work in the air, under water and on the ground. These different machines are used in a broad spectrum of commercial industries and academic fields such as the oil sector, the military, car productions, the mining industry, nuclear reactors and in relation to exploration of space. Machines are used in all these fields because they can be designed to meet the specific needs of a task, perform a non-stop work routine and can be easily replaced. As such machines can perform a variety of task-specific work more effectively than humans. In contrast, individual humans exhibit features such as creativity, self-reflection and are able to perform a broader variety of tasks than one specific machine. The different individual qualities associated with man and machine can form a synergy that increases the overall effectiveness and efficiency of the goal directed performance

J.A.T. Machado et al. (eds.), *Intelligent Engineering Systems and Computational Cybernetics,*

the two separate entities can achieve. However, this raises new questions such as: to what extent will humans continue to stay in control of machines in the future and could a fusion of the two entities be an appropriate methodological approach to optimise the two systems. This research aims to explore how Artificial Life (AL), Artificial Intelligence (AI), Cybernetics, Telepresence (TP), Intelligence Augmentation (IA) and synthetic entities might ensure that humans stay in the control loop of machines and facilitate improvements of both man and machine. The basis of this concept is demonstrated through a virtual simulation incorporating a semi-interactive virtual spider, a virtual predator and a virtual world.

One of the central strengths of the demonstrated concept is the navigational features, which enable the virtual spider to move through and respond to the virtual environment both through human control (through the use of either an on screen button based interface or a touch sensor) and/or autonomously through the use of machine intelligence (in the form of an intelligent system embedded in it). The benefits of this approach to control is three-fold. First, human can be kept in the control loop at all times. Second, it enables the system to act on either human or machine intelligence depending on how well the directives support the system in achieving an overall goal. Third, it might be used as a model for real world systems where remote interactive semi autonomous extensions of the human body support individual humans in adapting to and manipulating their environments by expanding the humans spatial awareness and abilities to manipulate their environments.

2 Background Theory

Machines have been found to outperform humans in a range of features associated with goal directed performance [1]. Activities that contribute to effective goal directed performance and have proved to be performed more effectively by machines than by humans are: calculation, memory storage and task specific decision-making [1]. An-other quality that favours machines over humans is technology's ability to interact with a broader range of sensors and manipulators at any time step. The possibility to continually redesign and upgrade machines is also a factor that favours machines as it allows for improving machine performance at a faster rate than humans. This in turn, indicate that it is quite possible that machines develop in a way that makes them better than humans in more and more factors contributing to optimal goal directed performance. With the factors that favour machines in mind and when we consider that intelligence can be measured by a systems ability to solve problems [2–4], it raises the concern that machines may dominate humans if they develop more optimal goal directed performance than humans. This is because intelligence is the quality that makes humans dominate the world to-day [5]. It is therefore suggested to "develop as quickly as possible technologies that make possible a direct connection between human brain and computer, so that artificial brains contribute to human intelligence rather than oppose it" [6].

In current cases where machines are controlled directly from the human brain, control can be obtained through the use of one single sensor. This has been proven through a wide range of experiments, including: navigation of a wheelchair [7], effective rolling of a plane in a flight simulator and as a substitute for "mouse input", for driving a cursor across a computer screen [8]. The above-mentioned approach to control has also been proven effective for controlling: a "neuromuscular stimulator that controls knee extensions" [8] and a robotic hand over the Internet [9]. No matter which approach that is chosen for controlling the interactions between the two entities, the aim should be to improve the performance of the two systems and enable humans to stay in the control loop of machines.

In order to explore how the above-mentioned amalgamation can be constructed in the best possible way, the researcher has found it beneficial to develop a simulator that incorporate the basic components of this type of system. Development of such a simulator has been found important because virtual components that are constructed in a way which reflect the future real world counterparts, will produce processes that will be as genuine as the real world counterparts they simulate [3]. Simulators (such as the one presented in this text) are, for this reason, powerful tools for unravelling and identifying novel, strong, and efficient solutions for use in real world systems [10].

3 System Architecture

The virtual components of the simulator include: a virtual spider, a virtual predatory snake and a virtual environment with sticky surfaces for the spider to navigate through. The spider can move and interact with this virtual environment in three possible ways. Two of these are based on human control, while the third operates the spider through the use of machine intelligence.

The first option for human based control utilizes an on screen button-based interface as a means for interacting with the spider. Each button has a shape that corresponds to the action the spider will perform if the button is pressed (e.g. shaped as arrows pointing in different directions). A second option for human navigation of the spider is in the use of a single touch sensor (Fig. 1 illustrates how the touch sensor is stimulated). Stimulation of this sensor triggers a signal that controls how the spider moves. The triggered action differs depending on the state of an inbuilt sequential state machine. The different actions the sequential state machine can trigger are visualized through textual metaphors in the user interface. These actions switch between being accessible for activation, and not, based on a pre-programmed sequence pattern implemented in the sequential state machine. As such, if the system receives stimuli from the sensor when the sequential state machine has a specific state, the spider will perform the action associated with that specific state. The user knows what action that is available for activation by looking at small virtual "lamps" situated to the right for each textual metaphor (if a "lamp" is illuminated then the associated action is "open" for activation).

The control option that utilizes machine intelligence for controlling the spider allows the spider to be controlled without human interference and provides the overall system with the benefits associated with synthetic systems. This option for control can be switched on and off by the user. If it is switched on, it will only be activated if it is found to provide a more optimal solution to a particular problem encountered in the local environment, than the solution that the human is able to provide. In this case the solution is evaluated in relation to the overall goals of survival or reaching a specific destination.

A feature that allows the user to switch between camera views is also incorporated in the system. This is done so that the human navigator can obtain a more complete awareness of the environment. The available camera views are first, second and third person view in relation to the spider.

4 Demonstration

The automatic and manual modes of navigation operate in their own distinctive way such that overall goal related performance is optimised. Figure 2 shows the manual human directed button-based approach to control. Here the buttons in the "direction control" section of the

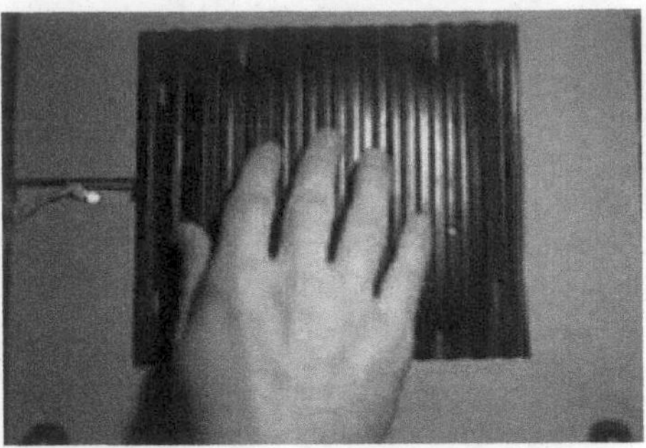

Fig. 1: Use of touch sensor

user interface can be used for enabling human interaction with the spider. The shapes of the buttons reflect the direction, in which the virtual spider will move if that specific button is pressed (e.g. arrow pointing up = forward, arrow pointing left = left, arrow pointing right = right). If the system does not detect any input, the spider will stop. Further, the sensor control section in Fig. 2 shows the manual human directed button-based approach to control. Here the buttons in the "direction control" section of the user interface can be used for enabling human interaction with the spider. The shapes of the buttons reflect the direction, in which the virtual spider will move if that specific button is pressed (e.g. arrow pointing up = forward, arrow pointing left = left, arrow pointing right = right). If the system does not detect any input, the spider will stop. Further, the sensor control section in Fig. 2 shows the interface that makes touch sensor based navigation possible. Here, each movement can be activated only when the small virtual "lamp" to the right of the textual representation of a selected movement is highlighted. Each of the different movement alternatives are "open" for activation only within a predefined time interval and the movement alternative "open" for activation is continuously cycling. If the system detects stimuli when a specific action is "open" for activation, the spider will perform the associated movement.

The machine intelligence based control system can be activated automatically if the spider encounters a situation that can change how optimal goal related performance can be obtained. The first step is to trigger a warning signal that indicates to the human navigator that it might be beneficial to evaluate the current performance to achieve an overall goal. This gives the human navigator an opportunity to suggest how to deal with the possible change in optimal goal directed performance. If the spider's performance does not change after a warning signal has been triggered, an emergency procedure can be initialised. The warning signal is displayed in the lower right section of the video screen in the interface displayed in Fig. 2. There are two situations in this simulation that can affect how optimal goal directed performance is achieved. The first of these situations occurs if the spider comes close to the above-mentioned randomly appearing predatory snake (in this case the warning signal displayed is: Predator warning!!). Figure 3 illustrates how the virtual spider automatically can change/"block" human controlled activities and perform a temporal evacuation in order to protect itself from being eaten. Here the path numbered as 1 is the spider's original direction of movement, while path number 2 is the direction in which the spider chooses to move if machine intelligence is switched on. This second path will take the spider to safety by allowing it to hide in the bush in the left section of the Fig. 3. If the spider is at a safe distance from the snake when it is within the bush, then the emergency procedure is terminated so that the human navigator once again can navigate

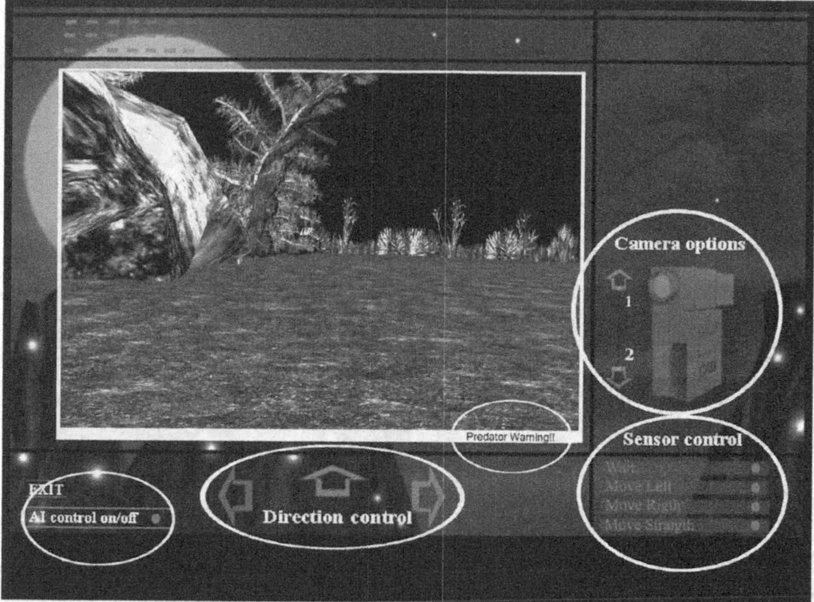

Fig. 2: Interface to manual human directed button-based approach to control

the spider. If, on the other hand, the snake still is too close to the spider when it is trying to hide in the bush, then the spider will continue to be automatically controlled by machine intelligence and proceed with its evacuation procedure by moving through the bush in order to escape the snake. The benefit of this approach to control is that the emergency procedure can be initialised even if the snake appears outside the field of view of the human navigator as the spider can "sense" its local environment in 360 degrees.

If the spider navigates towards a sticky surface, a second type of warning signal is activated (this warning signal displays: Dangerous terrain warning!!). If machine intelligence is switched on, and the human does not change the direction in which the spider is navigated after this warning signal has been triggered, then machine intelligence can navigate the spider around the sticky surface so that it avoids getting stuck. This is done automatically by allowing the spider to rotate +120 degrees around its egocentric y-axis when it reaches the boundary of a sticky surface when machine intelligence is switched on. After this the spider walks in a straight line away from the surface until it reaches a predefined distance from it. If the human at this point indicates that he/her still wish to navigate the spider in the original direction, the spider will rotate -120 degrees around its y-axis and walk towards the sticky surface again. This makes the spider move in a zigzag pattern along the boundary of the surface until it finds a safe path around it. Figure 4 illustrates this scenario. Here the original path of the spider is illustrated as line one, while line two is the path the spider will choose if machine intelligence is switched on. If the human does not indicate that he/she still wish to move in the original direction after the spider has rotated +120 degrees and is moving away from the sticky surface, then the spider will continue to move away from the this surface.

If the automatic artificial/machine intelligence (AI) functions are switched off, the warning signals will be the only indication that there is a change in optimal goal directed

Fig. 3: Virtual spider takes over control from human for survival

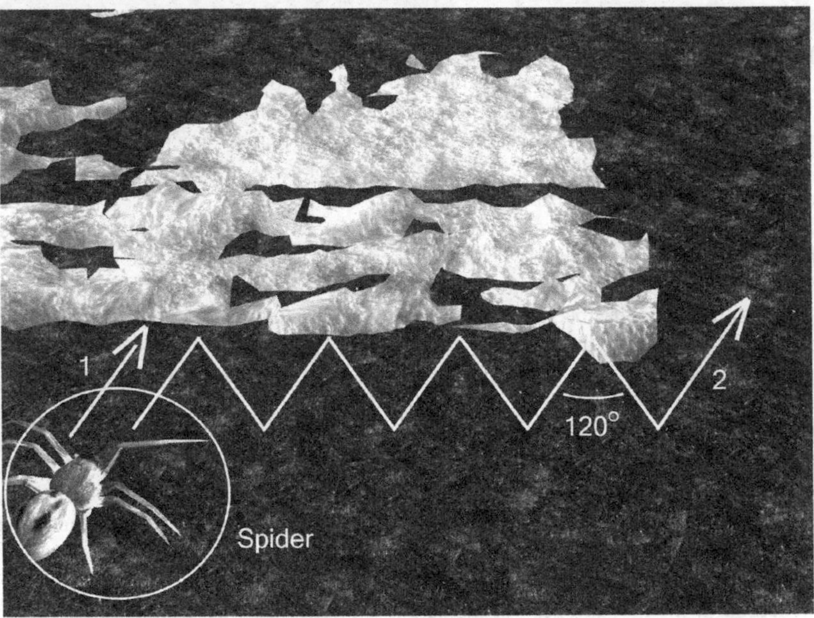

Fig. 4: Spider move in a zigzag pattern along the boundary of the surface until it finds a safe path around it towards the original direction intended by the human controller

performance. The lower left section of Fig. 2 portrays the AI control On/Off section that allows the user to control if automatic machine intelligence functions are activated or not. This enables the human to always stay in control of the system if the situation requires it, but also makes it possible for the spider to deviate from optimal goal directed performance if machine intelligence based control is switched off. A danger of switching off machine intelligence based control is that the spider might end up getting stuck in the sticky surface or being killed by the snake. Figure 5 illustrates how the spider can get stuck in a sticky surface if the automatic AI control is switched off.

Fig. 5: If the automatic AI control is switched off, the spider can get stuck in the surface

The multiple alternatives to control support a more flexible goal directed and robust navigation through the environment because the system can act faster on predefined situations and "sense" dangers in the environment that can not be directly seen than what would be possible if only a human operator was used. The human on the other hand, can support the system with more flexible solutions to encountered problems than what would be possible if only machine intelligence was used to navigate the spider. In order to provide the human navigator with as broad overview over the virtual environment as possible, a feature that supports manipulation of camera views is incorporated into the system. This feature is controlled in the Camera options section at the right side of the interface illustrated in Fig. 2. To allow for a first person view (in relation to the spider) the vertical switch in the camera options section is moved up, while in order to get a second person view the switch should be moved in a downwards position.

5 Future Developments

Further research will include obstacle avoidance where the obstacles have more complex shape than the sticky surface presented in this paper. More intricate path searching algorithms should also be investigated. It has also been found beneficial to explore the possibility of connecting the human to not only one, but multiple intelligent extensions of the human body. In order to achieve these goals, several key fields that it might prove important to explore have been identified. One of these fields is Artificial Life (AL). The field of AL is important because it aims at developing guidelines for constructing the autonomous behaviour required for effectively "driving" the synthetic extensions of the body. Concepts of interest in the field of AL include genotype-phenotype-environment interactions, mutations, supple adoption, natural selection and autonomy. These concepts are important because all of them are essential for supporting a system with continuously optimising its performance in changing environments [11]. What level of autonomy that ideally should be allowed for the artificial extensions is also an important issue. On one side one has to consider that the more autonomously the extensions of the body can operate, the more extensions the human body can be connected to because each individual extension will require less human involvement. On the contrary it is important to keep the level of machine autonomy on a level, which ensures that humans always can stay in the control loop so that humans continue to stay in control of machine performance.

The field of Artificial Intelligence (AI) should also be continuously studied to ensure that the most favourable type of machine intelligence contributes to the combined man/machine system and in turn the augmentation of overall performance. Background research into the field of AI indicate that Neural Networks (NN) supported by Genetic Algorithms (GA's) may enable future systems to provide more flexible solutions to changing problems in the environment than the currently used classical approach The reason for this is that some NN models in combination with GA's are ideal for generalizing information and change behaviour based on learning.

It has also been found that a human fused with artificial extensions in a proper way, may augment the overall level of intelligence not only by adding the artificial intelligence to the humans, but also by expanding the human brain by having multiple remote entities out there that are almost a part of the human. These entities can broaden the humans spatial awareness by semi autonomously looking out for the humans interests and helping him/her with dealing with multiple tasks [12]. Concepts from the area of TP have been identified as important to facilitate this type of IA. TP "is the experience or impression of being present at a location remote from one's own immediate physical environment" [13]. The benefit of facilitating a complete sense of TP in as man/machine symbiosis is that it can assist in broadening the humans spatial awareness by projecting the operator's senses to environments remote from its local biological body. This will add to the combined intelligence of man and machine, because "complexity in the mind is associated with complexity in sensory experiences" [14] and TP enables the fused man and machine to adapt based on a broader and more intricate spatial understanding of its environments. Therefore, if one manage to let the human operators senses "float" through communication devices to remote extensions of the body with sufficient accuracy to support "an appropriate illusion of direct interaction with the remote environment" [15], the overall system might be able to adapt to its environments much more effectively than any other entity known to day. To achieve full TP all the Aristotelian senses should be projected to the remote extensions of the body. These senses are "sight, hearing, smell, taste and touch. The last involves both kinaesthetic and tactil [12]. Concepts from the area of TP have been identified as important to facilitate this type of IA. TP "is the experience or impression of being present at a location remote from one's own immediate physical environment" [13]. Thee feedback. In some instances it may also be required to stimulate a person's vestibular system to simulate acceleration, gravity and orientation" [13]. Senses, that are not naturally included in the human body, should also be encouraged in the synthetic extensions as this will provide a considerable edge over those systems who do not have these extra senses [16]. These additional senses can be picked from a wide range of available sensor technology such as: light, infrared, proximity, acceleration, electromagnetic field and optical sensors or ultrasonic systems.

Research into alternative approaches to control than the single sensor/single signal approach and sequential state machines is also advised so that its mechanisms for data transfer and convenience of use can be improved. One way to do this may be to develop alternatives that operate more sensors in parallel, as this will allow for faster and more flexible interactions between man and machine. It seems especially interesting to continue the exploration of how such parallel approaches can be controlled by the human nervous system using either internal or external sensors attached to the human body. It is indicated the most optimal way to tap into the nervous system, may be to connect directly onto the CNS (Central Nervous System), as the brain have the ability to develop specifically devoted paths for controlling a interface between a man and a machine. This statement finds support in studies of the brain that indicates that the brain "continually adjusts its conventional neuromuscular outputs based on their outcomes; it continually tries to improve performance" [8]. This approach to control might, as explained above, support a wide range of new approaches to control, including thought control. This in turn can encourage more effective communication between man and machine, which in turn strengthens the fusion between the two. Further research into control of the remote semi intelligent entities will also include investigations in the use of body motions for controlling the artificial extensions of the human body. This will through the use of optical motion capture and the markers presented in [17].

To bring artificial extensions such as the virtual spider demonstrated in this paper out from cyberspace and into the real world, research in the development of semiautonomous physical systems, possibly manifested through robotics have been found essential.

6 Conclusions

Substantial developments in artificial systems have been observed throughout the last decades. These developments have on one hand proved to make the systems able to outperform humans in several features associated with goal directed performance, while the individual synthetic systems on the other hand still lack the flexibility and overall level of intelligence that humans exhibit. This indicates that it may be beneficial to develop guidelines, and ultimately construct a system that optimises both man and machine by merging the two entities. The researcher suggests that this fusion should include multiple remote synthetic intelligent extensions of the human body assembled in a way that facilitates Multi-presence. A strong benefit of this fusion is that it keeps humans in the control loop of the machines and therefore can ensure that synthetic systems continue to contribute to human performance, rather than surpass human abilities and in turn possibly rule us. This concern is real, as superior organisms tend to rule the inferior and machines have proved to continually improve their performance in a way that makes them exhibit several aspects of goal directed performance in a more optimal way than humans. This research has pointed out several fields that each might provide cues on how to construct the abovementioned amalgamation between human and machine in the most optimal way. These identified fields are AI, AL, IA, TP, Cybernetics and Robotics.

References

1. Warwick K (1997) March of the Machines. Century Books Limited, London
2. Bedau A, Boden M (1996) The Nature of Life. In: Boden M (eds) The Philosophy of Artificial Life. Oxford University Press, Oxford
3. Boden M (1996) The Philosophy of Artificial Life. Oxford University Press, Oxford
4. Sober E (1996) Learning from Functionalism—Prospects for Strong Artificial Life. In: Boden M (eds) The Philosophy of Artificial Life. Oxford University Press, Oxford
5. Dettmer R (1997) IEEE Review 43:88–89
6. Hawking S (2001) Hawking's plan to offset computer threat to humans [Homepage of Ananova], [Online] Available: www.ananova/news [2001, May 1]
7. Warwick K (2002) I, Cyborg. Century Books Limited, London
8. Vaughan T, Wolpaw J, Conchin E (1996) EEG-Based Communication: Prospects and Problems, IEEE Transactions on Rehabilitation Engineering 4:425–430
9. Warwick K, Gasson M, Hutt B, Goodhew I, Kyberd P, Schulzrinne H, Wu X (2004) Thought communication and control: a first step using radiotelegraphy, IEEE Proceedings on Communication 151:185–189
10. Krink T, Vollrath F (1999) Virtual Spiders guide robotic control design, IEEE Intelligent Systems 14:77–83
11. Pattee H (1996) Simulations, Realizations and Theories of Life. In: Boden M (eds) The Philosophy of Artificial Life. Oxford University Press, Oxford
12. Pattie M (2001) Intelligence Augmentation [Homepage of KurzweilAI.net], [Online] Available: www.Kurzweilai.net/articles/art0264.html [2006, Aug 6]
13. Mair M, Fryer R, Heng J, Chamberlin G (1995) An Experiment in Three Dimensional Telepresence, The Institution of Electrical Engineers IEE 1–4
14. Sharkey P, Murray D (1997) Feasibility of Using Eye Tracking to Increase Resolution for Visual Telepresence, IEEE International Conference on Systems, Man and Cybernetics, Orlando, FL, Oct 12–15
15. Warwick K, Gasson M (2004) Extending the human nervous system through internet implants-experimentation and Impact, IEEE International Conference on Systems, Man and Cybernetics 2:2046–2052
16. Baica JC, Li R (2006) Augmenting the Human Entity through Man/Machine collabration, IEEE International conference on computational cybernetics, Talliun:69–73.
17. Barca J C, Rumantir G, Li R (2006) A New Illuminated Contour-Based Marker System for Optical Motion Capture, IEEE International Conference in Innovations in Information Technology, Dubai

Index